CMP

Computational Methods in Subsurface Hydrology

Proceedings of the Eighth International Conference on Computational Methods in Water Resources, held in Venice, Italy, June 11-15 1990.

Editors: G. Gambolati
          A. Rinaldo
          C.A. Brebbia
          W.G. Gray
          G.F. Pinder

Computational Mechanics Publications, Southampton Boston

Co-published with

Springer-Verlag, Berlin Heidelberg New York London Paris Tokyo

G. Gambolati
Dept. of Mathematical Models
For Applied Science
University of Padova
Via Belzoni, 7
I - 05101
Padova
Italy

A. Rinaldo
Dept. of Civil and Environmental
Engineering
University of Trento
Mesiano di Povo
I - 38050 Trento
Italy

C.A. Brebbia
The Wessex Institute and
Computational Mechanics Institute
Ashurst Lodge
Ashurst
Southampton
SO4 2AA
UK

W.G. Gray
Department of Civil Engineering
University of Notre Dame
Notre Dame
IN 46556 - 0767
USA

G.F. Pinder
Dept. of Civil Engineering
Princeton University
Princeton
NJ 08540
USA

A CIP catalogue for this book is available from the British Library

Library of Congress Catalog Card Number 90 - 081858

ISBN 1-85312-066-9 Computational Mechanics Publications, Southampton
ISBN 0-945824-48-3 Computational Mechanics Publications, Boston, USA
ISBN 3-540-52701-X Springer-Verlag Berlin Heidelberg New York London Paris Tokyo
ISBN 0-387-52701-X Springer-Verlag New York Heidelberg Berlin London Paris Tokyo

ISBN 1 85312 073 1
ISBN 0 945824 56 4 two volume set
ISBN 3 540 52700 1
ISBN 0 387 52700 1

This work is subject to copyright. All rights are reserved, whether the whole or part of the material is concerned, specifically the rights of translation, reprinting, re-use of illustrations, recitation, broadcasting, reproduction on microfilms or in other ways, and storage in data banks. Duplication of this publication or parts thereof is only permitted under the provisions of the German Copyright Law of September 9, 1965, in its version of June 24, 1985, and a copyright fee must always be paid. Violations fall under the prosecution act of the German Copyright Law.

©Computational Mechanics Publications 1990
©Springer-Verlag Berlin Heidelberg 1990

Printed and bound by Bookcraft (Bath) Ltd, Avon

The use of registered names, trademarks etc. in this publication does not imply, even in the absence of a specific statement, that such names are exempt from the relevant protective laws and regulations and therefore free for general use.

PREFACE

The choice of Venice as the site of the VIII International Conference on Computational Methods in Water Resources is both timely and meaningful. Its timeliness and significance stem from the visibility and the complexity of the water resources problem affecting the environment of the city: the lagoon and its contributing mainland. It is not therefore by chance that among the various sponsorships, gratefully acknowledged, we particularly appreciate the major contribution from the Consorzio Venezia Nuova, i.e. the institution in charge of the rescue of Venice, its lagoon and its mainland. It is the organizers' hope that this book will also contribute to the professional background and the scientific advances needed to address the complex issues related to the use and preservation of the Venetian water resources system.

This book results from the edited proceedings of the VIII International Conference on Computational Methods in Water Resources (originally Finite Elements in Water Resources) held at the Giorgio Cini Foundation, Venice, Italy, June 1990. The Conference series was started in 1976 to serve as an internationally acknowledged forum for researchers in the - at that time - novel and emerging field of finite element methods applied to water resources. The name and the peculiar role of the ongoing Conference series were later (1986) modified to host contributions based on the increasingly diverse computational techniques being applied in water resources research. The previous meetings were held at: Princeton University, USA, 1976; the Imperial College, UK, 1978; the University of Mississippi, USA, 1984; the Laboratorio Nacional de Engenharia Civil, Portugal, 1986; and the Massachussets Institute of Technology, USA, 1988.

The 1990 Proceedings cover a wide spectrum of computational methods encompassing both theory and applications. Contaminant transport in surface and subsurface hydrology has attracted most of the researchers' interest, in a way reflecting the trends observed in the referred literature in this field and plays an important role in this book. It is significant that several papers edited in this book concern the increasingly studied field of computational stochastic hydrology .

The organizing committee of the Venice Conference wishes to express deep appreciation to the key-note invited lecturers J. Glimm, S.P. Neumann,

J.C.J. Nihoul, Y. Pomeau and I. Rodriguez-Iturbe. We are also indebted to the invited lecturers J. Carrera, V. Casulli, M.A. Celia, G. Dagan, R.E. Ewing, I. Herrera, M. Kawahara, U. Meissner, A. Quarteroni, W.M. Scheestakow, A.J. Valocchi, S.S.Y. Wang, and M.F. Wheeler. A significant contribution to the scientific fallout of the meeting came from the organizers of the Wave Propagation in Shallow Waters Forum (A. Adami and A. Noli) and the Supercomputing in Water Resources Forum (A. Peters). It is also a pleasure to acknowledge the continuing efforts and support offered by C.A. Brebbia, W.G. Gray and G.F. Pinder, of the permanent organizing committee.

The committee gratefully acknowledges the sponsorship of: AGIP, Alitalia, Aquater, Banca Popolare Veneta, Bonifica, Camera di Commercio I.A.A. di Venezia, Cassa di Risparmio di Padova e Rovigo, CISE, Comune di Venezia, Consiglio Nazionale delle Ricerche, Consorzio Venezia Nuova, Centro Sperimentale per l'Idrologia e la Meteorologia della Regione del Veneto, Dagh Watson, Digital Equipment Italia, ENEL, FIATIMPRESIT, Gruppo Acqua, Hydrodata, IBM Italia, Idroser, INC - il nuovo castoro, ISMES, Istituto di Credito Fondiario delle Venezie, Istituto Federale delle Casse di Risparmio delle Venezie, Lotti & Associati, Provincia di Venezia, Rodio, SIP, Studio Geotecnico Italiano, Technital, Tecnomare, ZF-MPM, Zollet Ingegneria. The endorsement of the following organizations is also acknowledged: AGU, AIMETA, ASCE, IAHR, Istituto Veneto di Scienze Lettere ed Arti, National Science Foundation, National Society of Computational Methods in Engineering, University of Padua - School of Engineering, University of Trent - School of Engineering, (ISME) the International Society for Computational Methods in Engineering, (ISBE) the International Society for Boundary Elements

May we finally add that the final version of the accepted papers appearing in this volume is reproduced directly from the material submitted by the authors who are therefore responsible for their content.

The Editors
June 1990

CONTENTS

SECTION 2 - UNSATURATED GROUNDWATER FLOW

SECTION 5 - GROUNDWATER TRANSPORT CONTAMINATION PROBLEMS

SECTION 6 - CHEMICAL REACTION PROBLEMS IN POROUS MEDIA

SECTION 7 - STOCHASTIC PROBLEMS IN GROUNDWATER FLOWS AND TRANSPORT

PART A

SECTION 1 - SATURATED GROUNDWATER FLOW

Use of Three-Dimensional Modelling in Groundwater Management and Protection

A. Refsgaard, G. Jørgensen

Danish Hydraulic Institute, Agern Allé 5, DK-2970, Horsholm, Denmark

ABSTRACT

During the past century the abstraction of groundwater for drinking water has increased gradually in Denmark. This has in many areas caused a need for discovering and utilizing new groundwater resources.

In the second largest county of Denmark this problem has become serious and a study of the total water capacity of the groundwater reservoirs in the county has been initiated. The study comprises all the components in the hydrological cycle including surface and subsurface processes. Meterological data in terms of precipitation and temperature as well as abstraction data are available for a 100 year period enabling simulation of long range trends and variability in the groundwater level.

This paper describes the applications of a numerical model on regional scale for groundwater management and on local scale for simulation of the leakage and spreading of contaminants from waste disposal sites. The results from the three-dimensional groundwater model on regional scale are automatically extracted and used as boundary conditions for simulations on local scale. The framework modelling is the SHE - Systemè Hydrologique Européen.

INTRODUCTION

An increasing depression of the potential head, a deteriorating water quality and decreasing streamflows during the past decades have caused a need for investigating new strategies for the groundwater abstraction as well as for discovering and utilizing new aquifers in many parts of Denmark.

The practice most often adopted in this kind of investigation is to apply a two-dimensional groundwater model with average transmissivities, storage and leakage coefficients.

In the present study, however, the abstraction takes place from different aquifers and the interaction between the surface and the subsurface flow as well as the connection between the amount and **quality** of both groundwater and surface water is of major importance. Hence a comprehensive groundwater modelling also has to include a three-dimensional groundwater flow model, a model for the unsaturated zone as well as integrated groundwater/surface water and water quality models.

The extremely fast development within computer technology in these years provides the possibility of turning the most advanced research techniques into practical engineering tools. This trend will continue for the next decade and thus give the practising engineers/geologists easy available but advanced tools in their daily work. Our purpose here is to present the results of an investigation of the groundwater resources in the county of Aarhus, Denmark using SHE - Systemè Hydrologique Européen. SHE is a modelling system that comply with all of the above mentioned requests - including a three-dimensional groundwater flow and water quality model.

SHORT DESCRIPTION OF SHE

SHE is a deterministic distributed, physically based modelling system of the entire land phase of the hydrological cycle. The initial development was done jointly by the Institute of Hydrology (UK), SOGREAH (France) and Danish Hydraulic Institute. Physically based models of the individual components in SHE have been known for years. The uniqueness of SHE is, however, that it is one of the few models integrating all submodels into one system of the entire land based part of the hydrological cycle.

The basic version of SHE was limited to application on catchments with single layered unconfined aquifers. In connection with ongoing research programmes in Denmark (modelling of leachate contaminated groundwater from landfills and environmental impact of fertilizer and manure application in agriculture) DHI has, however, developed a new, fully three-dimensional groundwater component in SHE enabling description of flow and solute transport in different layers of the aquifer. In the upper layer the aquifer is unconfined and the underlying layers are treated as confined or unconfined depending on the potential head. Exchange of water between each of the layers takes place as a function of the vertical head gradients. The overall structure of the SHE is shown in Fig. 1.

The independent three-dimensional solute transport component of SHE solves the advection-dispersion equation in an high-accuracy explicit finite-difference scheme automatically extracting the simulated velocities from the SHE result file. The three-dimensional description of the transport processes enables a much more correct simulation of the spreading of solutes as the usual vertical integration of solute concentrations is avoided. A chemical add-on module for handling chemical reactions as ion-exchange, complexsation and precipitation/dissolution of species are prepared during another research project.

The improved version of SHE, which also includes pre- and post-processor modules, enables its use for planning and management of groundwater resources, including distributed abstraction of groundwater for water supply and/or risk of groundwater contamination and remedial measures for its rehabilitation.

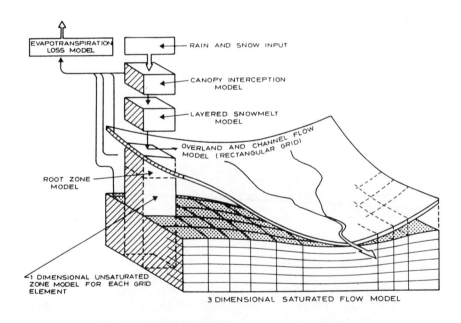

Figure 1.   Structure of the SHE.

SHORT DESCRIPTION OF THE MODELLING SETUP

The setup covers an area of approximately 800 km^2 and includes catchments for three small rivers. The catchments are characterized by glacier deposits with moderate slopes and a rather complex geological structure. The predominant vegetations are cereals and other agricultural crops. The catchments are represented by a horizontal grid, discretized optionally in 250 x 250 m^2, 500 x 500 m^2 or 1000 x 1000 m^2 squares. The setup includes 25 to 200 metres of quaternary sediments simplified in the model to 5 layers; an unsaturated zone, two aquifers and two aquitards. Fig. 2 shows the 5 layers in a typical cross-section.

Geological information about topography, horizontal and vertical extensions as well as hydrogeological information about the confined and unconfined storage coefficients and the horizontal and vertical hydraulic conductivities of each layer are necessary for setting up the subsurface part of the modelling system. Further data requirements are soil properties, space- and time-varying vegetation parameters and meteorological data. The pre-processor of SHE contains programmes for digitizing most informa-tion - including geological profiles which automatically are interpolated in space so that the data-setup consists of distrib-uted vertical extension of each layer.

The distribution of the hydraulic conductivities and storage coefficients in each layer are made by analyzing partly test-pumpings partly specific capacities of several hundreds bore holes.

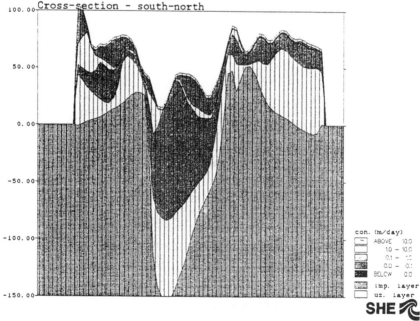

Figure 2.   South-north cross-section showing the vertical extension of the 5 layers.

PRELIMINARY AND EXPECTED RESULTS

The data basis includes time-series of streamflow discharge and hydraulic heads in aquifers from several stations with more than 50 years of records. Combined with time-series of meteorological data and information on groundwater abstraction available for a 100 year period this gives a unique possibility to calibrate the modelling system as well as to simulate and investigate the long range trends in the groundwater levels and the surface runoff in ungauged areas. As the geological setup of the sub-surface part of the model is based on physically measured geological profiles it is convenient to explore the groundwater potential of new reservoirs by model simulations as well as simulate different strategies for the ongoing groundwater abstraction.

The interaction between groundwater and surface water is another important element of the investigation as the water works without evidence may be held responsible for a reduction in stream flows which environmentally may become a serious problem.

In principle it is not necessary to calibrate a physically based hydrological model as SHE. However, as the hydrological data basis is not complete some calibration has to be done. The calibration of the unsaturated zone and the overland flow components has almost finished whereas the calibration of the subsurface part just has started.

The calibration of the surface water part has shown that it is possible with the present data basis to get rather close to the observed hydrographs, cf. Fig. 3. Furthermore, it is possible to transfer the few calibrated model parameter values unchanged from one catchment to the other catchments.

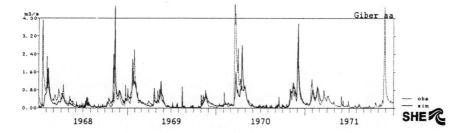

Figure 3.   Measured and simulated stream flow discharge.

On the contrary the calibration of the subsurface part of the setup has caused some difficulties due to the complexity of the geology:

* the increased groundwater abstraction results in a change from confined to unconfined aquifers

* the leakage to lower-lying reservoirs takes place in very small but high conductive areas whose location are unknown

However, intensive geological and hydrogeological investigations has given a very accurate geological setup of the aquifers, and the calibration process only includes the spatial distribution of hydraulic conductivities and storage coefficients.

When the model is properly calibrated research investigations of the effect of changes in climate, land use and other physical factors on the available groundwater resource will be continued.

SUMMARY AND DISCUSSION

Decreasing groundwater levels and groundwater quality problems as well as decreasing low flows in streams in Aarhus have initiated the presented study. The need for recognizing new strategies for groundwater abstraction or discovering new resources combined with the interaction between surface water and groundwater has caused that the problem has to be solved by applying a hydrological modelling system like SHE which enables a simulation of the entire hydrological cycle.

The calibration of the model has shown that the subsurface part of the setup is highly dependent of the geological interpretation of the composition of the glacier deposits which forms the subsurface. The complexity of the geology has caused some calibration problems but at the same time emphasized the need of applying three-dimensional solution techniques for solving the groundwater equations. The calibration of the surface water part of the setup showed encouraging results.

REFERENCES

/1/   Abbott, M.B., Bathurst, J.C., Cunge, J.A., O'Connell, P.E. and Rasmussen J.: An Introduction to the European Hydrological System - Système Hydrologique Européen - SHE - 1: History and philosophy of a physically-based distributed modelling system, and 2: Structure of a physically-based distributed modelling system. Journal of Hydrology, 87, pp. 45-77, 1986.

/2/   Aarhus Amtskommune: Hydrological modelling of groundwater management for Aarhus County, Interim report, Danish Hydraulic Institute, November 1989 (in Danish).

On the Modelling of Three-Dimensional Free Surface Seepage Problems

P. Angeloni(*), R. di Bacco(*), G. Fanelli(**), P. Molinaro(***), R. Rangogni(***)
(*) ISMES, SpA, Bergamo, Italy
(**) ENEL-SPT, Firenze, Italy
(***) ENEL-DSR-CRIS, Milano, Italy

ABSTRACT

After a few years of application of a 2D version of a code for non-steady flow analysis with free surface, the 3D version has been successfully implemented. The new code uses a fixed mesh approach where the free surface is approximated by the faceted surface obtained by grouping triangular and quadrilateral facets cut through each tethraedron element.

The development and the application of the code has been made feasible by the new advancements in mesh generators which allow complex models to be automatically prepared and scrutinized. The new generation of computers has also contributed to make available the power required to solve large non linear analyses with thousands of degrees of freedom.

The applicability of the procedure developed is presented and discussed using examples of practical applications that have been recently carried out.

INTRODUCTION

A few years ago a numerical procedure pointed to solve the non-steady flow with free surface was implemented at the Structural and Hydraulic Research Center of ENEL [1,2].

The procedure, mainly based on the work of Neuman and Witherspoon [3], considers the 2D domain under investigation as discretized in a fixed mesh of 3-noded triangular finite elements covering both the wet and dry regions.

With the fixed mesh approach, the free surface is then approximated by a piecewise linear curve consisting within each element of a straight segment. The solution procedure is based on an implicit integration algorithm, coupled to a standard

Newton-Rapson iterative scheme that allows to easily handle the obvious non-linearity resulting from the fact that at each new increment of time the location of the free surface is not 'a-priori' known.

The code developed was successfully applied for the solution of both plane and axisymmetric seepage problems, with stratified heterogeneous porous media.
At that time it clearly appeared that an extension of the code aimed to solve complete 3D problems would have been of great interest.

Unfortunately a lot of difficulties suggested to wait before to start a new development phase. In particular, although the implementation of the code was considered as easily achieved, there were two main items of concern.

The first consisted of the fact that the most straighforward extension of the code form 2D to 3D, required the substitution of 3-noded triangular elements with four noded tetrahedron elements. These elements allow the free surface to be approximated by a faceted surface made of small triangular or quadrilateral facets, but force the user to handle gigantic model hardly prepared and almost impossible to scrutinize by hand, or with the visual aid of the batch graphics available at that time.

The second item of concern consisted instead of the CPU time expected to carry out a complete transient analysis on a big model that was really unacceptable with the hardware resources of the common scalar computers of the previous generation.

IMPLEMENTATION OF THE 3D VERSION OF CODE

More recently the scenario has significantly changed and improved. The new capabilities which are now available in the most advanced pre-processors makes it possibile to easily prepare the solid model of the domain of interest, including whatever data are available to represent the actual configuration and geometry of the various layers of material. Each subvolume can be then automatically filled of thetrahedron elements with very efficient and powerful algorithms which provide the required compatibility at the common boundaries and also include severe geometry checks able to mark off every possible remaining inconsistency.

Also the power of the currently available computers is tremendously increased. The new generation of super and minisuper computers is now faster than the previous generation of orders of magnitude and what was previously a dream is now the daily routine almost on the desk of every engineer.

It is then not surprising that the reasons for postponing the new upgraded version of the originally developed 2D code, for non-steady 3D flow with free surface, have been cancelled.

Nowadays the new version of the code is completed [4] and it has been applied to solve large problems of practical interest.

The solution strategy is almost the same of the 2D version, specifically for what concerns the use in the iterative scheme of a modified Jacobian matrix, which is simplified by neglecting the inherent non-linearity of the residual vector with respect to the unknown nodal potentials [1].

The 3D version of the code has been successfully installed on the new generation computers, like CRAY and Alliant [5], and also properly interfaced for a more user's oriented application [6], as presented in the following examples.

AN EXAMPLE OF APPLICATION

The first practical application of the previously discussed procedure has been the analysis of two earth dams and a natural barrage generated from a big landslide that occurred in 1987.

Among these application, in the following the study performed on the first investigated earth dam is presented.

This dam, has an homogeneus body with impervious upstream face and a cutoff in foundation. The dam is 66 m high and has a very complicated geometry in foundation. Eleven zones of different material have been used to properly model the geometry. The finite element model, generated using an advanced mesh generator, consist of 15.000 elements and 3.200 nodes (see in fig. 1 an upstream view of the model).

Several analyses have been performed to fit the measured behaviour of the dam (piezometric heads and flow quantities).

In fig. 2 it is presented the elevation of the free surface in a planimetric view under steady-state conditions. Fig. 3 presents the piezometric head estimated from F.E.M. analysis and the piezometric head measured in the main trasversal section.

In fig. 4 it is presented a planimetric view of inflow and outflow rates through the dam foundation.

As shown in fig. 2 and fig. 4, the 3D analysis is necessary to understand the behaviour of dam. In fact, the 2D analysis previously performed for the main transversal section was found not capable to explain satisfactorily the observed behaviour of the dam, due to the evident lack of symmetry.

NEW DEVELOPMENTS

Although the goal of allowing an efficient flow analysis over a model consisting of 10.000 to 20.000 elements and requiring over 100 time steps has been met, some additional work is still expected to produce a further improvement of the methodology.

An area of interest is first of all the one of a more detailed optimization of the code to better exploit the architecture of the new computers. On this item a tremendous improvement can be still expected, provided that the code is re-structured in such a way to take more advantage of the large current availability of core memory. A great benefit can be also got from a better a more suited choice of specific algorithms which during the solution of large jobs have usually to handle arrays of 2 several thousands of entries.

The output interface can be also improved to get a better link to post-processors able to retrieve and present output data on any required cross sections cut through the model.

Finally some additional modelling capability can be also of great interest to cover the needs of handling in a more efficient way the effect of drain systems.

CONCLUSIONS

The new advancements in pre-processing, with solid modelling and volume filling generation of tetrahedron elements, as well as the increased power of the new computers have made feasible the global 3D non-steady flow analysis with free surface. A numerical procedure jointly developed by ENEL and ISMES has been briefly discussed, and results of practical applications carried out with it presented.

The procedure allows the modelling of the flow with very large models and is expected to be of special interest for the application to the problem of earth dams.

BIBLIOGRAPHY

1.   A. Di Monaco. Il codice di calcolo EFFIGE 2D - Parte I - Teoria, procedimenti numerici e struttura. ENEL-DSR-CRIS Report 3316 - 1985.

2.   R. Rangogni. Il codice di calcolo EFFIGE 2D - Parte II - Problemi di prova per la verifica e la qualificazione in garanzia di qualita. ENEL-DSR-CRIS Report 3333 - 1985.

3.   S.P. Neuman. P.A. Witherspoon, Analysis of non-steady flow
      with free surface using the finite element method. Water
      Resources Research, Vol. 7, No. 3, 1971.

4.   P.   Angeloni,   P.   Bonaldi.   Simulazione   numerica   di
      filtrazione 3D a superficie liebera in mezzi comunque
      stratificati. Sviluppo della versione 3D del codice
      EFFIGE. ISMES Report REL-DMM-3033 - 1987.

5.   L.   Mazza,   P.   Angeloni,   P.   Bonaldi.   Adeguamento
      potenzialita' EFFIGE-3D e sua installazione su Alliant FX-
      80. ISMES Report RAT-DMM-5705, 1989.

6.   P. Angeloni, P. Bonaldi, M. Borsetto. Miglioramento delle
      potenzialita'   del   codice   EFFIGE-3D   in   vista   di
      applicazioni di tipo ingegneristico. ISMES Report REL-DMM-
      3394, 1987.

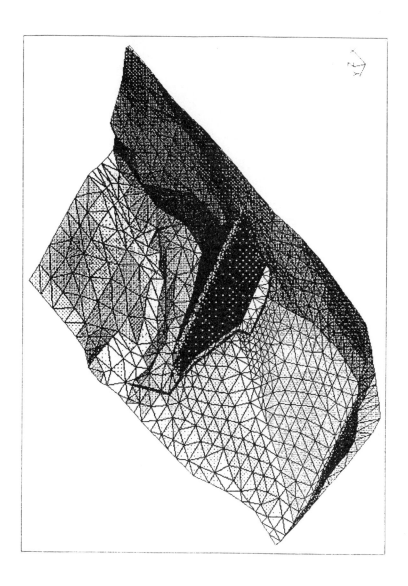

Fig. 1 - Upstream view of finite element model

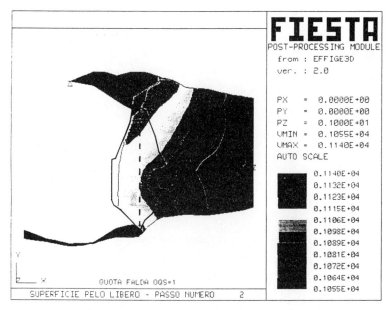

Fig. 2 - Planimetric view of free surface

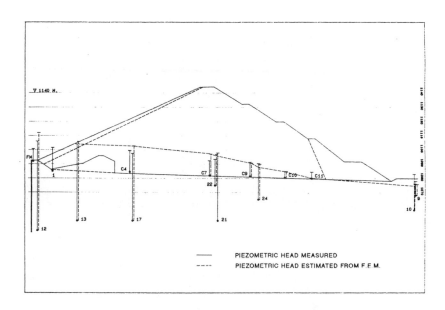

Fig. 3 - Measured and calculated values
in the main transversal section

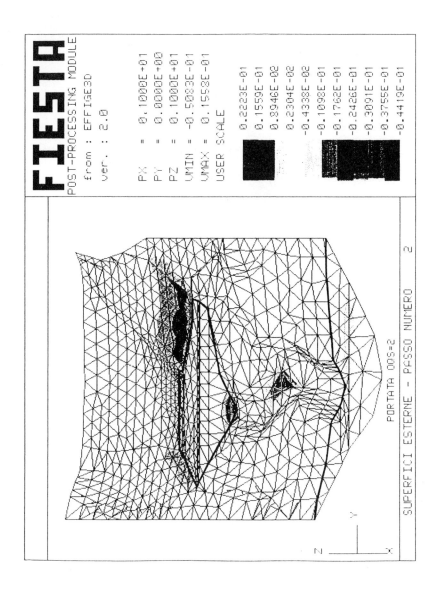

Fig. 4 - Planimetric view of inflow and outflow rates

Mixed-Hybrid Finite Elements for Saturated Groundwater Flow

E.F. Kaasschieter

TNO Institute of Applied Geoscience, PO Box 285, 2600 AG Delft, The Netherlands

1. INTRODUCTION

An accurate approximation of the specific discharge (Darcy velocity) is crucial in the numerical solution of a variety of groundwater flow problems. In approximating the specific discharge q by standard finite difference or finite element techniques, first an approximation of the piezometric head (potential) ϕ is determined as a set of cell averages, nodal values or piecewise smooth functions. This approximation of ϕ is then numerically differentiated and multiplied by the often rough tensor of hydraulic conductivity (permeability) K to obtain an approximation of q. In many cases an inaccurate specific discharge results from this approach, i.e. the approximation thus obtained does not fulfil the continuity equation sufficiently well.

In a physical context it is desirable to obtain an approximation of q, that fulfils the continuity equation as well as possible with respect to the finite difference grid or finite element mesh. Such an approximation can be determined by the mixed finite element method (see Raviart and Thomas [16], Thomas [18], Roberts and Thomas [17]). This paper will be limited to the lowest order mixed method, firstly, because higher order methods result in some conceptual complications and, secondly, because the lowest order method is comparatively easy and straightforward to use for practical problems (see Ewing *et al.* [8], Chavent and Jaffré [5]: Ch. 5).

The mixed finite element method results in a large system of linear equations. The choice of a numerical method to solve this system is restricted by the fact that its coefficient matrix is indefinite. This drawback can be circumvented by an implementation technique called hybridization, which leads to a symmetric positive definite system of linear equations (see Arnold and Brezzi [1]). Since this system is sparse, it can be solved efficiently by the preconditioned conjugate gradient method (see, e.g., Golub and Van Loan [9]: Ch. 10, Axelsson and Barker [3]: Sec. 10.4, Kaasschieter [12]).

This paper is excerpted from Kaasschieter [13]: Ch. IV in which more details can be found.

2. NOTATIONS AND PRELIMINARIES

Throughout this paper, Ω shall denote a domain in $\mathbb{R}^d$ ($d = 2$ or $d = 3$) with a piecewise smooth boundary $\partial\Omega$. Let $\partial\Omega_D$ and $\partial\Omega_N$ be mutually disjoint portions of $\partial\Omega$ (in the next section Dirichlet and Neumann boundary conditions will be defined on $\partial\Omega_D$ and $\partial\Omega_N$, respectively).

The Lebesgue spaces $L^2(\Omega)$ and $L^2(\Omega)$ contain respectively the square-integrable scalar and vectorial functions on Ω, i.e.

$$L^2(\Omega) = \{\psi: \Omega \to \mathbb{R} \mid \int_\Omega \psi^2 \, dx < \infty\}, \tag{1a}$$

$$L^2(\Omega) = \{p: \Omega \to \mathbb{R}^d \mid \int_\Omega p \cdot p \, dx < \infty\}. \tag{1b}$$

The Sobolev space $H(\text{div};\Omega)$ contains the square-integrable vectorial functions whose divergences are also square-integrable, i.e.

$$H(\text{div};\Omega) = \{p \in L^2(\Omega) \mid \nabla \cdot p \in L^2(\Omega)\}. \tag{2}$$

The linear subspace $H_N(\text{div};\Omega)$ is defined as

$$H_N(\text{div};\Omega) = \{p \in H(\text{div};\Omega) \mid n \cdot p = 0 \text{ on } \partial\Omega_N\}, \tag{3}$$

where n is the outward normal to $\partial\Omega$.

3. A MIXED VARIATIONAL FORMULATION

Consider the following boundary value problem

$$\begin{array}{|l|}
\hline
q = -K\nabla\phi, \ \nabla \cdot q = f \ \text{ in } \Omega, \\
\phi = g_D \text{ on } \partial\Omega_D, \ n \cdot q = g_N \text{ on } \partial\Omega_N, \\
\hline
\end{array} \tag{4}$$

where K is symmetric and uniformly positive definite (see Ciarlet [6]: (1.2.25)). The differential equations in problem (4) can be looked upon as Darcy's law and the continuity equation, the boundary conditions as prescribed potential and flux and in that case the function f is used to represent sources and sinks (see Bear [4]: Ch. 5).

Using Green's formula, the mixed variational formulation of problem (4) can be stated (see Thomas [18]: IX-(1.4)):
Find $(q,\phi) \in H(\text{div};\Omega) \times L^2(\Omega)$, such that $n \cdot q = g_N$ on $\partial\Omega_N$ and

$$\int_\Omega (Cq) \cdot p \, dx - \int_\Omega \phi \, \nabla \cdot p \, dx = - \int_{\partial\Omega_D} g_D \, n \cdot p \, ds \ \text{ for all } p \in H_N(\text{div};\Omega), \tag{5a}$$

$$- \int_\Omega \nabla \cdot q \, \psi \, dx = - \int_\Omega f \, \psi \, dx \ \text{ for all } \psi \in L^2(\Omega), \tag{5b}$$

where $C = K^{-1}$ is the compliance tensor. Problem (5) has a unique solution (see Thomas [18]: Th. IX-1.1).

4. A MIXED FINITE ELEMENT METHOD

Now the lowest order Raviart-Thomas discretization of problem (5) is introduced. Assume henceforth that the domain Ω is a polygon (d = 2) or a polyhedron (d = 3). A triangulation of Ω (see Ciarlet [6]: Ch. 2) is constructed by subdividing Ω in a collection S_h of mutually disjoint subdomains S, called finite elements, such that every subdomain is a triangle or a parallelogram (d = 2), or a tetrahedron, a prism or a parallelepiped (d = 3).

Let E_h be the collection of edges (d = 2) or faces (d = 3) e of all subdomains S. Assume that $\partial\Omega_D$, and thus also $\partial\Omega_N$, is the union of some edges or faces e.

For each finite element S, let $RT(S)$ be the space of linear vectorial functions p on S, such that $n_S \cdot p$ is constant on each edge or face of S, where n_S is the outward normal to the boundary ∂S of S. Define the Raviart-Thomas spaces

$$RT_{-1}(S_h) \;=\; \{p \in L^2(\Omega) \,|\, p\big|_S \in RT(S) \ \text{for all} \ S \in S_h\}, \tag{6}$$

$$RT_0(S_h) \quad= \{p \in RT_{-1}(S_h) \,|\, \text{the normal component of } p \text{ is continuous} \atop \text{across the interelement boundaries}\}, \tag{7}$$

$$RT_{0,N}(S_h) = \{p \in RT_0(S_h) \,|\, n \cdot p = 0 \ \text{on} \ \partial\Omega_N\}. \tag{8}$$

Note that $RT_0(S_h)$ and $RT_{0,N}(S_h)$ are subspaces of $H(\text{div};\Omega)$. Further, the multiplier space $M_{-1}(S_h)$ is defined as

$$M_{-1}(S_h) = \{\phi \in L^2(\Omega) \,|\, \phi\big|_S \text{ is constant for all } S \in S_h\}. \tag{9}$$

Let $g_{N,h}$ be a piecewise constant approximation of g_N, then the lowest order Raviart-Thomas mixed method for problem (5) reads as follows: Find $(q_h, \phi_h) \in RT_0(S_h) \times M_{-1}(S_h)$, such that $n \cdot q_h = g_{N,h}$ on $\partial\Omega_N$ and

$$\int_\Omega (Cq_h) \cdot p_h \, dx - \int_\Omega \phi_h \, \nabla \cdot p_h \, dx = - \int_{\partial\Omega_D} g_D \, n \cdot p_h \, dx$$
$$\text{for all } p_h \in RT_{0,N}(S_h), \tag{10a}$$

$$- \int_\Omega \nabla \cdot q_h \, \psi_h \, dx = - \int_\Omega f \, \psi_h \, dx \qquad \text{for all } \psi_h \in M_{-1}(S_h). \tag{10b}$$

Problem (10) has a unique solution (see Thomas [18]: Th. IX-2.1).

By introducing appropriate basis functions for the finite-dimensional spaces $RT_{0,N}(S_h)$ and $M_{-1}(S_h)$ (see Kaasschieter [13]: Sec. IV-6) problem (10) results in the system of linear equations

$$\begin{pmatrix} \tilde{A} & \tilde{B} \\ \tilde{B}^T & 0 \end{pmatrix} \begin{pmatrix} \tilde{Q} \\ \Phi \end{pmatrix} = \begin{pmatrix} \tilde{F}_1 \\ F_2 \end{pmatrix}. \tag{11}$$

Since $\tilde{A}$ is a symmetric positive definite matrix, system (11) yields

$$\tilde{Q} = \tilde{A}^{-1}(\tilde{F}_1 - B\Phi), \tag{12}$$

$$\tilde{B}^T \tilde{A}^{-1} \tilde{B} \Phi = \tilde{B}^T \tilde{A}^{-1} \tilde{F}_1 - F_2, \tag{13}$$

where $\tilde{B}^T\tilde{A}\tilde{B}$ is a symmetric and uniformly positive definite matrix (see Kaasschieter [13]: Th. IV-6.1).

Modelling saturated groundwater flow using the mixed finite element method may result in very large and sparse matrices $\tilde{A}$ and $\tilde{B}$, especially if the domain Ω is three-dimensional. Therefore a fast and efficient iterative method is required to solve the resulting system of linear equations. However, the coefficient matrix of equations (11) is not positive definite and $\tilde{B}^T\tilde{A}^{-1}\tilde{B}$ is not sparse. Fast and efficient methods are not yet known for these situations.

5. HYBRIDIZATION OF THE MIXED METHOD

The solution of the system of linear equations resulting from problem (10) can be simplified by enlarging the Raviart-Thomas space in which q_h is sought and introducing a Lagrange-multiplier to enforce the continuity of the normal component of q_h across the interelement boundaries.

Recall that E_h is the collection of edges ($d = 2$) or faces ($d = 3$) of all subdomains S. Let $M_{-1}(E_h)$ be the space of functions μ defined on the union of all edges or faces e, such that $\mu\big|_e$ is constant for each e. The multiplier space $M_{-1,D}(E_h)$ is defined as

$$M_{-1,D}(E_h) = \{\mu \in M_{-1}(E_h) \,|\, \mu = 0 \text{ on } \partial\Omega_D\}. \tag{14}$$

Now, it follows immediately that if $p \in RT_{-1}(S_h)$, then $p \in RT_0(S_h)$ if, and only if,

$$\sum_{S\in S_h} \int_{\partial S} n_S \cdot p\, \mu\, ds = 0 \quad \text{for all } \mu \in M_{-1,D}(E_h). \tag{15}$$

Let $g_{D,h}$ be a piecewise constant approximation of g_D, then the hybrid version of the lowest order Raviart-Thomas mixed method for problem (5) reads as follows:

Find $(q_h, \phi_h, \lambda_h) \in RT_{-1}(S_h) \times M_{-1}(S_h) \times M_{-1}(E_h)$, such that $\lambda_h = g_{D,h}$ on $\partial\Omega_D$ and

$$\int_\Omega (Cq_h)\cdot p_h\, dx - \sum_{S\in S_h} \left\{ \int_S \phi_h \nabla\cdot p_h\, dx - \int_{\partial S} \lambda_h\, n_S\cdot p_h\, ds \right\} = 0$$
$$\text{for all } p_h \in RT_{-1}(S_h), \tag{16a}$$

$$-\sum_{S\in S_h} \int_S \nabla\cdot q_h\, \psi_h\, dx = -\int_\Omega f\, \psi_h\, dx \quad \text{for all } \psi_h \in M_{-1}(S_h), \tag{16b}$$

$$\sum_{S\in S_h} \int_{\partial S} n_S\cdot q_h\, \mu_h\, ds = \int_{\partial S} g_N\, \mu_h\, ds \quad \text{for all } \mu_h \in M_{-1,D}(E_h). \tag{16c}$$

Problem (16) has a unique solution (see Arnold and Brezzi [1]: Lemma 1.3). Moreover $q_h = \tilde{q}_h$ and $\phi_h = \tilde{\phi}_h$, where $(\tilde{q}_h, \tilde{\phi}_h) \in RT_0(S_h) \times M_{-1}(S_h)$ is the unique solution of problem (10).

By introducing appropriate basis functions for the finite-dimensional spaces $RT_{-1}(S_h)$, $M_{-1}(S_h)$ and $M_{-1}(E_h)$ (see Kaasschieter [13]: Sec. IV-7) problem (16) results in the system of linear equations

$$\begin{pmatrix} A & B & C \\ B^T & 0 & 0 \\ C^T & 0 & 0 \end{pmatrix} \begin{pmatrix} Q \\ \Phi \\ \Lambda \end{pmatrix} = \begin{pmatrix} F_1 \\ F_2 \\ F_3 \end{pmatrix}. \tag{17}$$

The advantage of system (17) compared with system (11) is the block-diagonality of the symmetric positive definite matrix A. Hence A can be inverted at the finite element level. Thus, system (17) yields

$$Q = A^{-1}(F_1 - B\Phi - C\Lambda), \tag{18}$$

$$\begin{pmatrix} B^T A^{-1} B & B^T A^{-1} C \\ C^T A^{-1} B & C^T A^{-1} C \end{pmatrix} \begin{pmatrix} \Phi \\ \Lambda \end{pmatrix} = \begin{pmatrix} B^T A^{-1} F_1 - F_2 \\ B^T A^{-1} F_1 - F_3 \end{pmatrix}, \tag{19}$$

where the coefficient matrix of equations (19) is symmetric positive definite (see Kaasschieter [13]: Th. IV-7.1). Now, $B^T A^{-1} B$ is a diagonal matrix. Thus, system (19) yields

$$\Phi = (B^T A^{-1} B)^{-1} (B^T A^{-1}(F_1 - C\Lambda) - F_2), \tag{20}$$
$$D\Lambda = F, \tag{21}$$

where

$$D = C^T(A^{-1} - A^{-1} B (B^T A^{-1} B)^{-1} B^T A^{-1}) C, \tag{22a}$$
$$F = C^T A^{-1} (F_1 - B (B^T A^{-1} B)^{-1} (B^T A^{-1} F_1 - F_2)) - F_3. \tag{22b}$$

The symmetric positive definite matrix D is usually large and sparse, but not particularly well conditioned. This motivates the use of the preconditioned conjugate gradient methods (see, e.g., Golub and Van Loan [9]: Ch. 10, Axelsson and Barker [3]: Sec. 1.4, Kaasschieter [12]) to solve system (21). A variety of choices has been discussed in the literature (see, e.g., Axelsson [2], Concus et al. [7]). Popular methods for computing the preconditioning matrix are the incomplete Cholesky decomposition (see Meijerink and Van der Vorst [14,15]) and the modified incomplete Cholesky decomposition (see Gustafsson [10,11], Axelsson and Barker [3]: Sec. 1.4).

After solving system (21) the vectors Φ and Q can be computed by equations (20) and (18), respectively. This computation can be performed at the finite element level.

6. CONCLUSIONS

From numerical experiments (see Kaasschieter [13]: Sec. IV-10) it is clear that an accurate approximation of the specific discharge can be determined by the mixed finite element method. The benefits of the mixed method are apparent for problems with rough tensors of hydraulic conductivity and especially if the domain is subdivided into very flat finite elements.

Using the hybridization technique, the mixed finite element method results in a system of linear equations with a sparse and symmetric positive definite coefficient matrix. This system can be solved efficiently by the pre-conditioned conjugate gradient method.

7. REFERENCES

[1]   Arnold, D.N. and Brezzi, F. Mixed and Nonconforming Finite Element Methods: Implementation, Postprocessing and Error Estimates, Mathematical Modelling and Numerical Analysis, Vol. 19, pp. 7-32, 1985.

[2]   Axelsson, O. A Survey of Preconditioned Iterative Methods for Linear Systems of Algebraic Equations, BIT, Vol. 25, pp. 166-187, 1985.

[3]   Axelsson, O. and Barker, V.A. Finite Element Solution of Boundary Value Problems, Academic Press, New York, 1984.

[4]   Bear, J. Hydraulics of Groundwater, McGraw-Hill, New York, 1979.

[5]   Chavent, G. and Jaffré, J. Mathematical Models and Finite Elements for Reservoir Simulation, North-Holland, Amsterdam, 1986.

[6]   Ciarlet, P.G. The Finite Element Method for Elliptic Problems, North-Holland, Amsterdam, 1978.

[7]   Concus, P., Golub, G.H. and Meurant, G. Block Preconditioning for the Conjugate Gradient Method, SIAM Journal of Scientific and Statistical Computation, Vol. 6, pp. 220-252, 1985.

[8]   Ewing, R.E., Koebbe, J.V., Gonzalez, R. and Wheeler, M.F. Mixed Finite Element Methods for Accurate Fluid Velocities. Finite Elements in Fluids, (Ed. Gallagher, R.H., Carey, G.F., Oden, J.T. and Zienkiewicz, O.C.), Vol. 6, pp. 233-249, Wiley, New York, 1985.

[9]   Golub, G.H. and Van Loan, C.F. Matrix Computations, North Oxford Academic, Oxford, 1983.

[10] Gustafsson, I. A Class of First Order Factorization Methods, BIT, Vol. 18, pp. 142-156, 1978.

[11] Gustafsson, I. Modified Incomplete Cholesky (MIC) Methods. Preconditioning Methods, Theory and Applications, (Ed. Evans, D.J.), pp. 265-293, Gordon and Breach, New York, 1983.

[12] Kaasschieter, E.F. Guidelines for the Use of Preconditioned Conjugate Gradients in Solving Discretized Potential Flow Problems. Computational Methods in Water Resources. Vol. 2, Numerical Methods for Transport and Hydrological Processes, (Ed. Celia, M.A., Ferrand, L.A., Brebbia, C.A., Gray, W.G. and Pinder, G.F.), pp. 147-152, Elsevier, Amsterdam, 1988.

[13] Kaasschieter, E.F. Preconditioned Conjugate Gradients and Mixed-Hybrid Finite Elements for the Solution of Potential Flow Problems, Ph.D. Thesis, Delft University of Technology, 1990.

[14] Meijerink, J.A. and Van der Vorst, H.A. An Iterative Solution Method for Linear Systems of Which the Coefficient Matrix is a Symmetric M-Matrix, Mathematics of Computation, Vol. 31, pp. 148-162, 1977.

[15] Meijerink, J.A. and Van der Vorst, H.A. Guidelines for the Usage of Incomplete Decompositions in Solving Sets of Linear Systems as They Occur in Practical Problems, Journal of Computational Physics, Vol. 44, pp. 131-155, 1981.

[16] Raviart, P.-A. and Thomas, J.-M. A Mixed Finite Element Method for 2-nd Order Elliptic Problems. Mathematical Aspects of Finite Element Methods, (Ed. Galligani, I. and Magenes, E.), Lecture Notes in Mathematics, No. 606, pp. 292-315, Springer, Berlin, 1977.

[17] Roberts, J.E. and Thomas, J.-M. Mixed and Hybrid Finite Element Methods, to appear in Handbook of Numerical Analysis. Vol. II, Finite Element Methods, (Ed. Ciarlet, P.G. and Lions, J.-L.), North-Holland, Amsterdam.

[18] Thomas, J.-M. Sur l'Analyse Numérique des Méthodes d'Eléments Finis Hybrides et Mixtes, Ph.D. Thesis, University Pierre et Marie Curie, Paris, 1977.

Groundwater Modelling in Relation to the System's Response Time using Kalman Filtering

F.C. van Geer(*), C.B.M. te Stroet(**), M.F.P. Bierkens(***)

(*) TNO Institute of Applied Geoscience, Delft, The Netherlands

(**) Faculty of Mathematics and Informatics, Delft University of Technology, The Netherlands

(***) Geographical Institute, University of Utrecht, The Netherlands

ABSTRACT

To simulate the effects of human interventions in a geohydrological system, often stationary groundwater models are used. An important question arises: Knowing that a groundwater system is in principle dynamic, can a stationary model be applied?

In this paper this question is discussed. First, a one dimensional example is used to consider the influence of the characteristic response time of the geohydrological system on the (reliability of) model results.

Next, the use of a Kalman filter algorithm for assessing the reliability of a deterministic groundwater model is described. To integrate measurements with stationary model predictions, a stationary algorithm is used. When it is applied to fast reacting groundwater systems the same results as the (transient) Kalman filter algorithm can be obtained at only a fraction of the calculation time.

Finally, the results will be further evaluated using a "real world" case of network analysis and design.

INTRODUCTION

In recent years, at the Institute of Applied Geoscience much attention has been paid to the quantification of the reliability of groundwater flow models. The concept of this quantification is based on a combination of the non-stationary groundwater flow model MODFLOW (eg. McDonald and Harbough [1]) and Kalman filtering. A computer code was developed to compute optimal estimates of the groundwater head in space and time and the corresponding error covariance matrix, which can be seen as a measure for the model reliability.

The estimate of the groundwater head is dependent on its history and on the present inputs. The influence of the history related to the present inputs, can be seen as the memory of the groundwater system. In some cases, however, the memory is very short. Then, the estimates of the groundwater heads are almost only dependent on the present input and they can be approximated by a stationary version of the groundwaterflow model.

Analogous to the groundwater head the time dependence of the error covariance matrix is determined by the memory of the groundwater system. The time dependence of the error covariance matrix is responsable for the major part of the computation time (> 90%) in the Kalman filter scheme. For groundwater systems with a short memory computation time can be reduced considerably by calculating a stationary approximation of the error covariance matrix.

As a criterion for the memory of the groundwater system we use the characteristic response time. For example, in Van Geer and Te Stroet [2] for a single aquifer with a continuous infiltration/drainage flux caused by a dense network of watercourses (figure 1) this response time was defined as:

$$\theta :: S(L^2+\lambda^2)/kD \qquad (1)$$

where: θ characteristic response time;

S storage coefficient;

L length of aquifer;

λ leakage factor = $\sqrt{kDc}$;

kD transmissivity;

c resistance of aquitard.

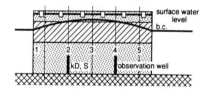

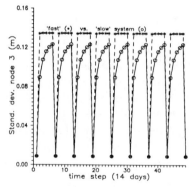

Figure 1: Schematization of example

Figure 2. Behaviour of a slow and fast reacting system

Also a dimensionless variable (response time ratio) could be defined to relate the characteristic response time to the model time step.

$$\omega = \theta/\Delta t \qquad (2)$$

where: ω indication of time behaviour of the system modelled;

Δt time step of model.

With the response time ratio, distinction can be made between a "fast reacting" system (ω < 1) and a "slow reacting" system (ω > 1). In the first case the stationary value of the model reliability is reached within one time step, wherein the second case an accumulating effect over several time steps occurs (figure 2).

This concept can be generalized to other groundwater systems. The characteristic response time is always a function of the geohydrological parameters and the geometry of the system.

$$\theta = f(S,k,D,c,L) \qquad\qquad (3)$$

The response time ratio is defined in general as given in equation 2.

In this paper, first, the algorithms are given for both the stationary and the non-stationary case. Then the response time will be derived for a synthetic situation and for this particular situation the applicability of the response time ratio as a criterion is discussed. Next, the application of the stationary approximation to a real world case is described. Finally some conclusions are drawn.

CALCULATION ALGORITHMS

The (non-stationary) Kalman filter algorithm is applicable to any system whose behaviour can be described by a state equation of the form:

$$h_k = A_k h_{k-1} + B_k u_k + w_k \qquad\qquad (4)$$

and a measurement process, which can be described by a measurement equation of the form:

$$y_k = C h_k + v_k \qquad\qquad (5)$$

where: A_k and B_k are model matrices at time step k;
  w_k is the system noise at time step k;
  y_k is a vector containing all groundwater head measurements at time step k;
  C is the measurement matrix at time step k;
  v_k is the measurement error vector at time step k.

In Van Geer, Te Stroet [3] it is decribed how the deterministic groundwaterflow model MODFLOW is incorporated in the state equation of the Kalman filter. As can be seen from equation (4) the time dependence is expressed by the term $A h_k$. For the fast reacting case this term appraoches to zero. Therefore, in the stationary algorithm this term is omitted. The two algorithms used in this study are given without proof in table 1. The symbols are defined as follows:

- $\hat{h}_k$ is the optimal linear estimate of the state vector at time k using measurements up to time step k;
- $\bar{h}_k$ is the prediction, which is the optimal linear estimate at time k using measurements up to the previous measurement time;
- P_k is the covariance matrix of the optimal estimation error: $cov\{h_k - \hat{h}_k\}$;
- M_k is the covariance matrix of the prediction error; $cov\{h_k - \bar{h}_k\}$;
- Q is the covariance matrix of system noise;
- R is the measurement error covariance matrix;

- K_k is the Kalman gain;
- I   is the unity matrix.

Table 1: Algorithms

(non-stationary) Kalman filter algorithm		stationary algorithm	
Initial conditions h_0 and P_0			
$\bar{h}_k = A_k \, \hat{h}_{k-1} + B_k \, u_k$	(6)	$\bar{h}_k = B \, u_k$	(6a)
$M_k = A_k \, P_{k-1} \, A_k^T + Q$	(7)	$M_k = Q$	(7a)
$K_k = M_k \, C^T \, \{ C \, M_k \, C^T + R \}^{-1}$	(8)	idem	
$\hat{h}_k = \bar{h}_k + K_k \, \{ y_k - C \, \bar{h}_k \}$	(9)	idem	
$P_k = \{ I - K_k \, C \} \, M_k$	(10)	idem	

The state equation is given in an explicit form. The MODFLOW model, however, is based on an implicit discretization scheme. To obtain the error covariance matrix M_k of the predictions at each time step a matrix inversion has to be performed. Although, an iterative method is used, equation (7) consumes the major part of the computation time. As can be seen from table 1 in the stationary algorithm the determination of the error covariance matrix of the prediction is reduced to a simple operation (equation (7a)).

The reliability of the calculated groundwater heads is quantified by the error covariance matrix P_k of the optimal estimates. Therefore special attention is given to the difference between the two algorithms with respect to this matrix.

APPLICABILITY

It was derived in the previous sections that, if a groundwater system is fast reacting, the behaviour of such a system can be described by a stationary algorithm. Also, optimal estimates of the piezometric levels can be obtained using the stationary algorithm. This stationary algorithm can be applied if the response time ratio

$$\omega < 1 \tag{11}$$

If the characteristic response time θ is defined properly for the groundwater system under consideration, it can serve as a sensitive and safe criterium. This can be illustrated with the following example obtained from Bierkens [4]: Water is abstracted from a semi-confined aquifer (Hantush-Jacob situation). The characteristic response time can be derived analytically:

$$\theta :: S \, (r/\lambda)^2 \, c \tag{12}$$

where: r is distance to the well
       λ is leakage facor = $\sqrt{kDc}$

In changing the parameters of equation 12 the applicability of the stationary filter algorithm for different values of ω can be investigated. This was done by varying the values of the storage coefficient S, keeping all the other geohydrological parameters constant, as well as the time step Δt. For each of the values of S, synthetic series of measurements were generated. These series were used to calibrate both the Kalman filter - as well as the stationary algorithm. In figure 3 the calibrated values of the transmissivity obtained by the original (non stationary) Kalman filter (dashed line) are given next to the transmissivity obtained with the stationary algorithm (solid line) as a function of the parameter.

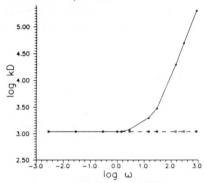

Figure 3. Calibrated transmissivities
as a function of ω.

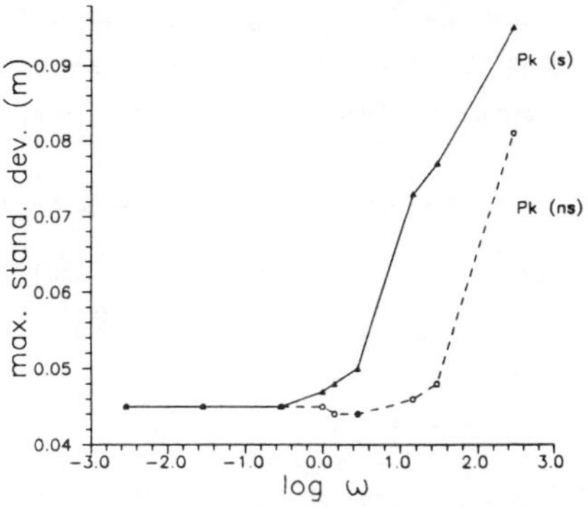

Figure 4. The standard deviation of the
estimation error

It is clear that if ω is just a little larger than 1, or in other words: Δt is just a little too small, totally wrong values of the geohydrological parameters are obtained. It is also clear that in that case both the model predictions, as well as the optimal estimates obtained by the stationary algorithm will be wrong.

In figure 4 the values of the estimation error standard deviations for one particular grid point, as calculated by both algorithms, are given as a function of the parameter ω. Again it is clear that when the stationary algorithm is used, right values are obtained only in case of fast reacting grounwater systems (i.e. ω < 1).

REAL WORLD CASE

In the north-eastern part of the Netherlands the Spannenburg groundwater pumping station is situated. A measurement network around this pumping station to monitor the changes of the groundwater head in space and time has been evaluated using the Kalman filter algorithm (eg. Van Geer and te Stroet, [2]).

The area is a part of a drainage basin, characterized by a large number of small polders with densely distributed ditches, alternating with small lakes. In the area there are two aquifers seperated by an aquitard. The groundwater is abstracted from the lower aquifer. The recharge of the aquifers is controlled by the open water levels in the ditches and lakes. The geohydrological system could be schematized in the same way as given in figure 1.

An estimation for the characteristic response time for the one dimensional case (equation (1) using Spannenburg's geohydrological characteristics can be made:

$$\theta = \frac{10^{-4} \cdot \frac{1}{2}(2500^2 + 2500.500)}{2500} = 0.15 \text{ days}$$

The measurement intervall of the monitoring network is 14 days. Because the model reliability depends solely on the system noise related to the fixed measurement intervall, also a model time step is chosen to be 14 days. In this case ω is much less than 1, so the stationary algorithm may be used and compared with the results of the Kalman filter.

In table 1 the mean and standard deviation of the series of differences of non-stationary and stationary standard deviations of optimal estimates of 12 points regularly distributed around the Spannenburg pumping station are given. As expected there is hardly any difference in the resulting reliability of the two cases.

Table 1. Differences in reliability of the two algorithms

point →	1	2	3	4	5	6
P_k $m(10^{-3})$	-0.200	-0.294	-0.060	-0.330	-0.117	0.045
$\sigma(10^{-3})$	0.435	0.592	0.144	0.815	0.434	0.213

point →	7	8	9	10	11	12
P_k $m(10^{-3})$	-0.111	-0.018	0.154	0.149	0.069	0.092
$\sigma(10^{-3})$	0.289	0.286	0.378	0.437	0.282	0.399

The computing time of the Kalman filter algorithm was about 25 minutes, whereas the stationary algorithm used only 30 seconds. So the stationary algorithm is saving computing time of about 98 %.

CONCLUSIONS

In general, one could say that the stationary algorithm and/or steady state groundwater models can be used to describe the behaviour of groundwater if the characteristic response time is smaller than the time step taken.
The stationary algorithm consumes only a few percents of the computing time of the Kalman filter.

During the investigations it is found out that it should be noted that when the system is very slow (large storage coefficient; freatric aquifer), the time series of several measurement locations almost look stationary. Now, one could assume that the system is in steady state while in fact a very long trend is abundant which can be difficult to detect in the relatively short and noisy time series. If a stationary groundwater model is used to describe such system, serious calibration errors are made according to figure 3. This results in errors of the calculated predictions of the heads, optimal estimates and associated error (co)variances (figure 4).

REFERENCES

[1]   Mc.Donald M.G., A.W. Harbaugh. A modular three-dimensional finite-difference ground-water flow model, U.S.G.S. Reston, Virginia (1984).

[2]   Geer van F.C., C.B.M. te Stroet. Program code CISKA, applied on testcace Spannenburg (in Dutch), TNO Institute of Applied Geoscience, report nr. OS 90-15-A, Delft (1990).

[3]   Geer van F.C., C.B.M. te Stroet. Kalman filtering packages for the program CISKA, a Computing model for the Interpolation of Heads with the Kalman filter Algorithm, TNO Institute of Applied Geoscience, report nr. PN 90-03-B, Delft (1990).

[4]   Bierkens F.P. An Alternative filter algorithm for fast reacting groundwater system. TNO Institute of Applied Geoscience, report nr. OS 90-07-A, Delft (1990).

Two-Dimensional Mathematical Model for Groundwater Flow Analysis

V. Cirrincione(*), L. Zoppis(*), E.R. Calubaquib(**)
(*) C. Lotti & Associati, Soc. di Ingegneria SpA, Rome, Italy
(**) Local Water Utilities Administration, Quezon City, Manila, Philippines

ABSTRACT

The structure and occurrence of an alluvial aquifer is simulated with a mathematical model in order to anticipate the effect of intensive and concentrated groundwater exploitation to meet the "Water Supply System" of the Bulacan Province (Philippines).

INTRODUCTION

Personal Computers (PCs) can provide a new approach for groundwater study and management. Indeed there are now several mathematical models for simulating aquifer structures, characteristics and hydrological system, as well as for providing a sound evaluation of aquifer potential and information for the proper planning and management thereof.

With a PC and low-cost, user-friendly software, the hydrogeologist can directly reconstruct the hydrogeological conditions of a confined or unconfined water-bearing formation, verify the validity of available hydrogeological information, check the hydrogeological balance, simulate steady-state and transient-state groundwater flow conditions, and evaluate the effects of various groundwater development hypotheses on the piezometric surface.

This paper presents a practical application of a two-dimensional computer programme for groundwater flow analysis utilized on the "Bulacan Central Water Supply Project" in the Philippines, financed by the Asian Development Bank (ADB) and executed by

C. Lotti & Associati, Rome on behalf of the Local Water Utilities Administration (LWUA), Manila, Philippines.

GENERAL

The results of the "Bulacan Central Water Supply Feasibility Study" performed by C. Lotti & Associati for the LWUA indicated the advisability of selecting the aquifer which occurs along the Angat river as the most economic source to be tapped for ten Urban Water Supply schemes in Bulacan Province.

Intensive hydrogeological investigations, including 240 Vertical Electrical Soundings, twelve test/production wells and eight observation boreholes, were thus conducted in the Angat valley to confirm the groundwater potential indicated by the Feasibility Study and to provide the basic data for implementing the mathematical model.

The Study Area is mostly devoted to rice-growing. It has a closely-knit irrigation network supplied by the Angat river.

The aquifer consists of unconsolidated deposits, mostly sand and gravel with some silt and silty clay lenses, overlying the argillaceous - tuffaceous Guadalupe Formation. It is unconfined, the water-table being at an elevation of between 2 and 15 m a.m.s.l.

The area is mostly flat. It is intersected from NE to SW by the Angat river, which is dammed for energy production and to provide water for Metro Manila (20 m³/s) and irrigation (30 m³/s).

Two Main Irrigation Canals, acting as a water divide, form the physical boundaries of the aquifer.

MATHEMATICAL MODEL

The structure and occurrence of the aquifer in the alluvial formation of the Angat Valley, has been simulated with a mathematical model in order to appraise the effect of intensive and concentrated groundwater abstraction through wells drilled or to be drilled to meet project water requirements (800-1,110 l/s).

The simulation programme adopted is a two-dimensional flow model (MHYDRO 2D, GROUNDWATER SIMULATION PACKAGE, FRANLAB, 1987). In the model, adopting Dupuit's assumption, the distribution of hydraulic head is governed by the equation of continuity as a function of transmissivity, storage coefficient and interchange of flow between the aquifer and the parts beyond the boundary.

The aquifer is schematized as a square-mesh grid with all relevant data referring to the centre of the each mesh.

Two sets of aquifer conditions can be simulated, namely steady-state flow and transient flow. In the case of the former, the data required are: inactive cells, permeability and saturated thickness of the aquifer, infiltration or recharge rate, boundary conditions and flow rate, hydraulic links and piezometric heads. When transient flow has to be simulated, the storage coefficient or specific yield of the aquifer and the time for which the flow analysis is to be made, have to be included.

The model provides for analysis of the flow equation, using the most significant variables, such as the hydraulic parameters, aquifer recharge, lateral subsurface inflow and the presence also of a surface water body (the Angat river in this case), that provides the possibility of interchanges of surface and groundwater flows (Head-Link).

It should be remarked that, as with most programs, the groundwater simulation model merely provides one method of approach to the solution of problems with several variables. Its reliability depends on the validity of the pertinent parameters used. If properly calibrated, however, it may be considered as an appropriate starting point for sound planning and groundwater management.

The mathematical model is, moreover, a valid and reliable tool for simulating actual aquifer conditions and for evaluating trends or effects due to different groundwater development assumptions.

For this purpose, however, actual aquifer recharge and abstraction data, aquifer parameters, piezometric levels, etc., should be periodically collected and updated in the model in order to check on conditions, thus permitting better evaluation of the results of the computer simulation.

INPUT DATA AND MODEL LAYOUT

The hydrogeological data provided are as follows:

a)   Aquifer parameters

the groundwater-bearing formation is considered to be unconfined and heterogeneous. The hydraulic parameters were obtained from drilling and testing of test/production wells, while the geometry of the aquifer (lateral extension and thickness) was ascertained by a Schlumberger Resistivity Survey performed along the banks of the Angat. These data were used to implement the aquifer model.

b)   Hydrological balance

To evaluate aquifer recharge reference was made to the groundwater balance, the main parameters of which are:

	IN (l/s)	OUT (l/s)
- Local recharge (from rainfall, paddy fields and secondary irrigation canals)	760	
- Recharge from main irrigation canals	355	
- Lateral subsurface inflow S. Rafael Dam)	100	
- Groundwater abstraction (private wells)		70
- Groundwater flow into the Angat river		1,160
Total	1,215	1,230

Particular attention was paid to the presence of the Angat river because, in the case of the hydrological system in question, this appears to be the natural groundwater discharge point of the aquifer. However, the river is also considered to be an important source of possible additional recharge through induced-infiltration wells. Its presence in the mathematical model, for the interchange of surface and the ground-waters, is simulated as a Head-Link Boundary.

The model layout consists of a grid system involving 180 square meshes with 500 m sides, representing the elementary hydrogeological units, with all relevant input and output data assumed to refer to the centre of each mesh.

As already mentioned, the irrigation canals, which provide the water for the paddy fields, are considered to form the water divide of the aquifer; their seepage or water losses go to form lateral recharge of the aquifer. The relevant recharge rate in the corresponding meshes or cells has been evaluated on the basis of resistivity data, bearing in mind the physical characteristics of the canals.

STEADY STATE GROUNDWATER FLOW CALIBRATION

The first runs of the simulation revealed some discrepancies in the reconstructed computer picture of water-table conditions, compared with the June 1987 piezometric map. Hence, especially for areas not covered by the experimental data, several modifications had to be made in aquifer parameters, until satisfactory agreement was attained on more than 50% of the selected control points.

Comparison of the piezometric map obtained with the computer simulation and the original plotted for June 1987, reveals satisfactory agreement.

Consequently, it can be assumed that within the project area, the simulated mathematical model of steady-state flow conditions adequately represents the real situation as regards the aquifer and its structure.

The water balance of the system computed by the simulation programme can be summarized as follows:

	IN (l/s)	OUT (l/s)
- Recharge (rainfall, paddy fields secondary irrigation canals)	730	
- Recharge from main irrigation canals and subsurface inflow	455	
- Private abstraction (wells)		72
- Groundwater flow into the Angat river		1,090
- Discharge to the western area (sub-surface outflow)		22
Total	1,185	1,184

The values show that the aquifer is in equilibrium, as defined in terms of the groundwater balance computed by the usual methods.

TRANSIENT STATE GROUNDWATER FLOW CALIBRATION

In order to detect temporal fluctuations in piezometric head, the value of the storage coefficient and the magnitude of the hydrological events and the times of the occurrence are defined in the transient state flow analysis. This is done by comparing the historical data (piezometric fluctuation) with the results of the simulation, and then by modifying the new variables, as required, until a satisfactory result is obtained.

According to the historical data and recharge computation it appears that aquifer recharge is minimal between December and May (0.87 mm/d). It averages about 1.91 mm/d in June, July, October and November and peaks at 2.36 mm/d in August and September.

According to the results of the preliminary simulations, considering that for some of the control points there is satisfactory agreement on variations in the piezometric head and the data computed by the model, no further modifications were made to the hydrological data.

GROUNDWATER DEVELOPMENT ASSUMPTIONS

The groundwater model, as calibrated, has been used to detect the effect of groundwater development in two cases:

- The firts one, Case A, concerns the development of a wellfield, consisting of 27 productions wells for a total assumed discharge, including private abstractions, of about 1.2 m³/s.

- The second one, Case B, considers a smaller wellfield of 16 production wells for a total discharge of 845 l/s.

Two other cases (A' and B') have also been considered. These are supplementary to the assumptions, as they involve a Head-Link Boundary variant. The permeability of the boundary has been decreased to a very low value to simulate the effect of no interchange between surface and ground-waters, over an unbroken period of twelve months along the Angat river.

For Cases A and B the effect of continuous
pumping for about four years, under the assumed
hydrological conditions, is not very marked.
However, if there are no interchanges with the Angat
river (Cases A' and B') the printout of the
simulation indicates very decided combined piezometric decline to a maximum of about 6 m below sea level
(Case A') in the central part of the wellfield.

Analysis of the simulation results indicates
that the piezometric head can reach a
quasi-stationary condition if the Angat river acts as
an active Head-Link Boundary for the hydrological
system. However, in the unlikely event of no surface
flow, only for short duration (one or two months),
the development programme is still acceptable in both
cases.

With reference to the hydrological balance it
ensues that in Case A a minimum of 250 l/s of water
is always required as induced recharge from the Angat
river, while for Case B the Angat river can still be
considered as a drainage outlet for the aquifer.

CONCLUSIONS

Experience acquired in the formulation and
implementation of the mathematical model for the
alluvial aquifer on the banks of the Angat river, as
well as for aquifers in other countries, shows that
although the program employed is sophisticated, it is
easy to operate even by inexperienced computer users.

A thorough knowledge of the relevant
hydrological and hydrogeological phenomena is
fundamental, however, since most of the difficulties
in modelling are conceptual rather than operational.

The results of the computer simulations reveal
the reliability of the input data and the logical
nature of the assumptions, providing water-supply
planners with the flexibility needed for operating a
new wellfield for domestic water requirements.

The model for steady - state and transient -
state groundwater flow analysis can be considered a
valid tool, ensuring a simple-to-use, relative-
ly-low-cost means of simulating aquifer behaviour.
However, where the geology and hydrogeology are
complex, as in the Angat area, it is essential to
update the model, as new additional hydrological data
and aquifer parameters become available, in order to
allow correct adjustments to the picture of aquifer
geometry and the relevant hydrological system.

A Method to Estimate Hydrogeological Parameters of Unconfined Aquifer

S. Ye, Q.R. Dong

Department of Hydrology, Hohai University, Nanjing 210024, China

ABSTRACT

The method is based on water balance equation and analysis of infiltration. Using rainfall and groundwater regime data, and assumed total field capacity(Wm), infiltration potential (Ip) and specific yield (Sy), a rise of water table ($\triangle$Hc) can be calculated. Then calculate the relative error (Re) of the rise of water table. For several rainfall events, there is an average relative error Ar for each set of parameters assumed. The set of parameters corresponding to minimum Ar are to be optimized. Results of field applications show the method is stable and reliable.

INTRODUCTION

Both the specific yield (Sy) and index of infiltration rate are important parameters in groundwater resources evaluation. They are factors effecting groundwater quantity computation. Specific yield is defined as a ratio of drained volume of water from a saturated soil to the total volume of the soil. It varies with time and space. However, specific yield is treated as a constant for a specific aquifer in practice. Pumping test suggested by Boulton [1] and Neuman [2] and numerical technique are normal ways to obtain the Sy. However, the pumping test is generally expensive and restricted in places with special boundary conditions. The numerical technique always involves bulk of vertical recharge or discharge which can not be determined accurately, as the consequence, the result of Sy may not be reliable. There are some other methods to approach Sy outlined by Todd [3] and suggested by Ye [4] but they all have limitations in use.

The present method requires data of rainfall intensities, pan evaporation and hydrograph of water table, which are normally available in many countries. The original idea about the present method was suggested by Ye [5] in 1978, which consisted only two variables i.e. Wm and Ip, the objective function is minimum deviation of Sy. Several years later, Ye et al [6] improved the method in three variables. With more field applications, the paper is a summary of the former researches.

PRINCIPLE OF THE METHOD

A rainfall event of sufficient intensity and duration will cause a rise in water table $\triangle$ H. The rainfall amount (P) of this rainfall event can be divided into four components: one satisfies the soil moisture deficit (Rd) in aeration zone, another percolates down to recharge the groundwater (Rg), a third runs off over the land surface (Rs), and a fourth loses through direct evaporation and interception(Ri). In case of bare land or crop field, the Ri is relatively small and can be neglected. For a unit area, the balance of these components is expressed as :

$$P = Rd + Rg + Rs \tag{1}$$

The Rd is defined as

$$Rd = Wm - Wo \tag{2}$$

where Wm--the total field capacity[L], which equals product of field capacity times average thickness of soil--water zone; Wo--the soil moisture content antecedent the rainfall event [L].

The proposed method includes four steps. Firstly, calculate Rd by using estimated value of Wo and assumed value of Wm. Secondly, separate Rg from the remainder (P--Rd), through comparing with the measured rainfall intensities I and the assumed infiltration potential Ip. Thirdly, estimate a rise of water table $\triangle$Hc, which equals Rg divided by an assumed Sy. Using the observed rise of water table $\triangle$H and the estimated $\triangle$Hc, a relative error Re is obtained. For several rainfall events, the average relative error Ar is computed. Finally, search the objective function, i.e. the minimum Ar and its corresponding parameters Wmo, Ipo and Syo.

1. Estimation of Rd.

The Rd is calculated from Equation (2). Wm is one of the parameters to be optimized, and is known by assumption. Wo is either obtained by measurement or estimation. Since the measured data are normally inavailable, the estimaion of Wo is neccessory. During an interval $\triangle$T (let $\triangle$T = 1 day), a water balance equation can be written for the aeration zone :

$$Wo_i - Wo_{i-1} = Eg_{i-1} - Es_{i-1} + P_{i-1} \tag{3}$$

Where. i,i-1 -- sequences of days; Eg-- daily groundwater discharge amount to the aeration zone [L]; Es-- daily land--surface evaporation amount [L]; p-- daily rainfall amount [L]. The methods to estimate Es and Eg are topics to be studied in disciplines of Hydrology and Hydrogeology. In the present study, Zhao's empirical formula of Es and Averianov's empirical formula of Eg are used:

$$Es_{i-1} = Em_{i-1} \frac{Wo_{i-1}}{Wm} \tag{4}$$

$$Eg_{i-1} = Em_{i-1}(1 - \frac{\triangle_{i-1}}{\triangle m}) \tag{5}$$

Where Em-- daily pan evaporation[L]; $\triangle$-- depth to water table [L]; $\triangle$m-- depth below which groundwater evaporation ceases[L], $\triangle$m can be obtained by analysis of the groundwater regime data. Substitution of Equation (4) and (5) into (3), yields:

$$Wo_i = K \cdot Wo_{i-1} + Em_{i-1}(1 - \frac{\triangle_{i-1}}{\triangle m}) + p_{i-1} \tag{6}$$

where $K = 1 - Em_{i-1} / Wm$

Equation (6) states that the soil moisture content on the (i)th day is equal to that of (i−1)th day decreased by factor K, plus the daily groundwater vertical discharge and rainfall. Value of Wm should be assumed. By Equation (6), the values of Wo can be calculated day−by−day until the last day prior to the rainfall event. Wo is not allowed to exceed Wm, let Wo = Wm when Wo > Wm. Once the antecedent Wo is estimated, the Rd is thus obtained by Equation (2).

2. Calculation of Rg and Rs.

When Rd is subtracted from P, the remainder is (Rg+Rs). This remainder can be divided into Rg and Rs by comparing rainfall intensities (I) and an assumed infiltration potential (Ip). In practice, the rainfall intensity (I) is defined as $\triangle P / \triangle t$ and measured in mm / hr, while infiltration potential (Ip) is defined as rg / $\triangle t$ in mm / hr. Where $\triangle P$ and rg are rainfall amount and infiltration recharge amounts respectively during interval $\triangle t$. Ip, as used here, is an index of recharge from rainfall rather than the infiltration capacity, and is a parameter that depends on both the soil characteristics and topography. Ip can be relatively large for a soil with a low infiltration capacity if the topography is insufficient to drain water from the surface. In this case, one hour rainfall may take several hours to infiltrate and thus the value of Ip is several times the infiltration capacity. On the contrary, the value of Ip can be smaller than that of the infiltration capacity when the land slope is very large.

Rd is the function of Wm. Different values of assumed Wm will cause the vertical broken line in Figure 1 slightly shifts left or right. Figure 1 also shows that the parts of rainfall intensities below the line of Ip contribute to groundwater, i.e. Rg = $\sum$ rg; while the sum of the upper parts is Rs. Obviously, values of Rg depend on assumed values of Ip. When a rainfall intensity is divided into two parts by the vertical broken line, the left part of this rainfall intensity contributes to Rd, while the right part $\triangle S$ produces either groundwater rg or both groundwater rg and surface runoff rs. There is a time interval $\triangle t'$ (only a portion of $\triangle t$) corresponding to $\triangle S$. The

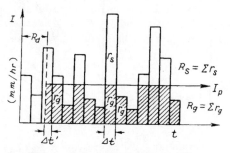

Fig. 1   Calculation of Rg and Rs

$rg = \triangle s$ and $rs = 0$ when $Ip > I$. If $Ip < I$, $rg = Ip \cdot \triangle t'$. Because $I = \triangle s / \triangle t'$, $rg = Ip \cdot (\triangle s / I)$, and $rs = \triangle s - rg$. In short, the total recharge amount of a rainfall event Rg is equal to the sum of rg.

3. Computation of Ar and optimization technique.

The computed Rg is based on the two assumed parameters Wm and Ip. By assuming a third parameter , the specific yield Sy, a calculated rise of water table $\triangle Hc$ is obtained, i.e. $\triangle Hc = Rg / Sy$. There is a relative error (Re) between the calculated $\triangle$ Hc and the measured $\triangle H$ from observation well: $Re = (\triangle Hc - \triangle H) / \triangle H$. Taking n rainfall events into consideration, each set of assumed values of Wm, Ip and Sy, will have an average relative error (Ar) which is defined as :

$$Ar = \frac{1}{n} \sum_{1}^{n} \left| \frac{\triangle Hc_i - \triangle H_i}{\triangle H_i} \right| \tag{7}$$

Of all sets of parameters assumed, the set of Wmo, Ipo and Syo corresponding to the minimum value of Ar are optimized. Using the optimized parameters, the rainfall recharge amount Rg and surface runoff amount Rs can be calculated.

Since the method only involves three parameters to be optimized, scanning method or other searching method can be used. When the reasonable ranges of these parameters are assigned, the storage of a micro—computer is large enough to solve the problem.

RESULTS OF FLELD APPLICATIONS

The problem area is yellow river deposit—plain, located in Juancheng, Dongming and Ciping Counties of Shandong Province. The unconfined aquifer in this area is composed of layers of silt, clayed silt and clay. (See Fig. 2).

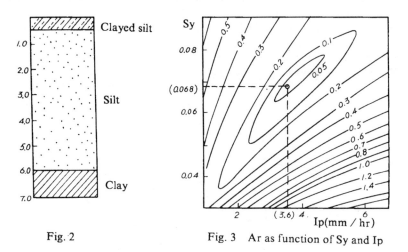

Fig. 2   Fig. 3   Ar as function of Sy and Ip

The depths to water table fluctuate from 0.5 m to about 3.0 m. Four field applications of the method in this area present excellent results. Table 1 and **Figure 3** is result of NO.67 observation well in Juancheng County.

Table 1   result of No.67 Observation well

Date of rainfall event	P (mm)	Pd (mm)	Pg (mm)	Rs (mm)	Re
1984.7.6	68.1	9.3	36.6	22,.2	−0.003
1984.7.10	43.9	4.8	20.4	18.7	0.000
1984.7.24	58.5	2.0	26.9	29.6	−0.011
1984.8.9	174.6	0.8	36.4	137.4	−0.163
1984.9.8	39.9	7.6	19.9	12.4	0.009
1984.9.23	40.6	1.5	39.1	0	−0.008
optimized parameters	Wmo = 120mm, Ipo = 3.6mm / hr Syo = 0.068 Ar = 0.033				

The other three examples give very close results which are summarised in Table

Table 2   Results of other three field applications

Location of Obs. well		Dongming NO.69	Ciping Jiabai	Ciping Lizhuang
Year of rainfall occuring		1984 1985	1976 1977	1976 1977
Number of rainfall events		6	7	6
Optimized results	Ar	0.0703	0.098	0.073
	Wmo	120	200	200
	Ipo	15.5	7	48
	Syo	0.071	0.067	0.066
Range of $|Re|$		0.000~ 0.180	0.001~0.160	0.010~0.180

The four field applications show that the values of Sy range from 0.066 to 0.071 for a silt unconfined aquifer. These values are stable and reasonable. The average relative error Ar is less then 10%, and the largest relative error Re for a rainfall event is about 18%, and in most cases, the Re have only a few percent. All of the examples indicate high sensitivity of Ar to changes in values of the optimized parameter Ip and Sy (See Fig. 3), but low sensitivity of Ar to changes in Wm. In example of No. 67 observation well, when Ip = 3.6 mm / hr and Sy = 0.068, the Ar are 0.0358, 0.0326, 0.0340 and 0.0348 corresponding to the Wm values of 80, 120, 160, and 200 mm. Since Wm is an insensitive factor, its value can be given empirically without optimization. The optimized Ip values depend on drainage conditions, the Jiabei well is sited by the side of a drainage canal, Ip is only 7mm / hr, while Lizhuang located in a depression area without sufficient drain trenches, Ip reaches 49 mm / hr.

The four examples also show that the contour lines of Ar form a single depression and has only one minimum Ar.

CONCLUSION

The present method has several apparent advantages: Its principles and mathematics are simple and easy to understand; the data required are normally available in regions with observation—well network; the cost to calculate the parameters is cheap; the results are reasonable and stable, and have relatively high accuracy. However, the method will not work in areas with large depth to water table because the rises of water tables caused by rainfall percolation are difficult to observe accurately. Besides, the heterogeneous layers may result in large value of Ar.

REFERENCES

1. Boulton, N.S.; Analysis of Data from Nonequilibrium Pumping Test Allowing for Delayed Yield From Storage. Proc. Inst. Civil Engs., London, Vol. 26, NO. 6699, 1963.
2. Neuman, S.P.; Theory of Flow in Unconfined Aquifers Considering Delayed Respones of Water Table. Water Resources Res., Vol 8(4). 1031—1045, 1972.
3. Todd, D.K., Ground Water Hydrology, pp. 23—24,1959.
4. Ye, Shuiting; Model to approach hydrogeological parmeters by Multiple Regression Analysis, Groundwater, Vol.2, pp. 28—45, 1978. (in Chinese).
5. Ye, Shuiting; Methods of Specific Yield Estimation. Symposium of Theories and Methods on Groundwater Resources Evaluation. Geological Publishing House, pp. 251—254. 1982. (in Chinese).
6. Ye, Shuiting, Corbet, T. and To'th, J., A Model to Estimate Specific Yield By Optimization, Journal Of Hydraulic Engineering, No.1, PP.1—9, 1985. (in Chinese).

Regional Versus Local Computations of Groundwater Flow

M. Nawalany

Warsaw University of Technology, Nowowiejska 20, 00-653 Warsaw, Poland and *TNO Institute of Applied Geoscience, PO Box 285, 2600 AG Delft, The Netherlands*

INTRODUCTION

A dilemma of how to calculate groundwater flow on a regional scale and estimate accurately local flows at the same time was and still is manifesting itself in most of the papers on computer methods applied in hydrogeology and environmental protection. The grid refinement and the high polynomial approximations used within a framework of the finite element method, though straightforward, require in many instances prohibitively large computer memories and/or long computing times.

There is, however, an alternative based on a decomposition of the groundwater system being simulated into its regional and local components. In such a hierarchical approach one can first compute a regional flow (in terms of piezometric head or velocity components) and then descend into any of the existing local systems (e.g. to a vicinity of wells) assuming that the boundary conditions necessary for calculations in the local system are specified already with the use of the numerical solution for the regional flow. The numerical method used in the local system can obviously be changed to satisfy the requirements of accuracy.

The approach has been applied to the realistic case in which a three-dimensional groundwater flow has been calculated for some hydrogeological regional system driven by the natural recharge and the abstraction from a number of wells. This was done with the use of *Classical Finite Element Solver* with the piezometric head as a primary variable. In the vicinity of wells several local systems have been distinguished. For their boundaries piezometric heads calculated were assumed to be the related boundary conditions. Finally the *Velocity Oriented Approach* (FLOSA-3FE model) was used to calculate the velocity field accurately in the vicinity of wells.

By combining the regional and the local solutions one could, for instance, answer questions on the residence time distributions, calculate accurate inverse trajectories etc. The case calculated was supported by a three-dimensional graphical postprocessor which allowed to trace trajectories of particular water particles or groups of particles.

REGIONAL VS. LOCAL GROUNDWATER SYSTEMS

There are two kinds of systems considered and analysed by FLOSA-FE (Fig. 1):
- *the regional system* and
- *the local system.*

Also there are different leading state variables chosen for the regional and for the local systems:

(i)   piezometric head is assumed to be a primary state variable in regional flow systems while components of specific discharge (obtainable through the Darcy Law) are merely the secondary state variables;

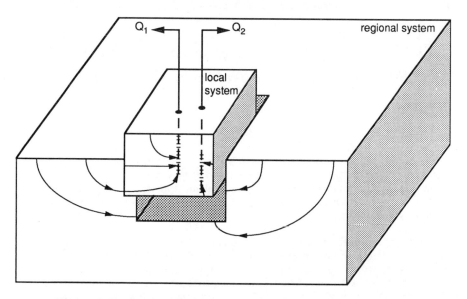

Figure 1. Regional and local systems. The local system is extracted from the regional one.

(ii) contrary to the regional approach there are components of specific discharge used as the primary state variables in the local flow system while the piezometric head is the secondary variable.

FLOW EQUATIONS

Classical formulation
The classical approach to the groundwater steady-state flow problem uses *the mass continuity equation* and Darcy's Law (cf. J. Bear [1]) which, when combined, lead to the following flow equation for homogeneous fluid:

$$\nabla \cdot (K\nabla\Phi) + Q = 0 \quad \text{in } \Omega \tag{1}$$

where Ω : flow domain

$K = \rho g k/\mu$: hydraulic conductivity, (m/s) (2)

$\Phi = (p/\rho g - z)$: piezometric head, (m) (3)

ρ : water density (assumed constant), (kg/m^3)

Q   : intensity of the source, $(m^3/s/m^3) = (1/s)$.

If a screen of the well can be considered a point located at $(x_w, y_w, z_w) \in \Omega$ then

$$Q(x,y,z) = Q_w \cdot \delta(x - x_w, y - y_w, z - z_w) \qquad (4)$$

where Q_w : production of the well, (m^3/s)
       $\delta(\cdot,\cdot,\cdot)$: delta-Dirac function, (m^{-3}).

If on the other hand a screen of the well must be described more realistically (i.e., by taking into account its finite length) then

$$Q(x,y,z) = Q_w \cdot \delta(x - x_w, y - y_w) \cdot \eta(z; z_t, z_b)/L_w \qquad (4')$$

where L_w : length of the screen, (m)
      $\delta(\cdot,\cdot)$: delta-Dirac function, (m^{-2})
      $\eta(z; z_t, z_b)$: step function equal to zero everywhere except interval
      $[z_t, z_b]$ in which $\eta = 1$, (-).

The resultant field of the volumetric flux can be obtained from the appropriate form of Darcy's Law:

$$\underline{q}_v = -K\nabla\Phi \quad \text{in } \Omega \qquad (5)$$

provided that the boundary conditions $\alpha_1\Phi + \alpha_2 q_{vn} = 1$ has been specified on $\partial\Omega$.

Velocity Oriented Approach (VOA)
W. Zijl (W. Zijl [4]) proposed so-called Velocity Oriented Approach to groundwater flow by considering q_v as the primary variable. The resultant Laplacian-like equation for q_v can be solved numerically with the standard finite-element method without any numerical differentiation of and thus without the degradation in accuracy (W. Zijl *et al.* [5]). In this approach the piezometric head Φ becomes the secondary variable and can be calculated afterwards.

If a density of water flowing in Ω can be considered constant (ρ = constant) the VOA equations and the VOA boundary conditions can be written as follows:

$$\left|\begin{array}{l}
\nabla \cdot (K\nabla S_{vx}) = KS_{vx} \dfrac{\partial^2\alpha}{\partial x^2} + KS_{vy} \dfrac{\partial^2\alpha}{\partial x\partial y} + q_{vz} \dfrac{\partial^2\alpha}{\partial x\partial z} \\[3mm]
\nabla \cdot (K\nabla S_{vy}) = KS_{vx} \dfrac{\partial^2\alpha}{\partial x\partial y} + KS_{vy} \dfrac{\partial^2\alpha}{\partial x^2} + q_{vz} \dfrac{\partial^2\alpha}{\partial y\partial z} \\[3mm]
\nabla \cdot (\dfrac{1}{K} \nabla q_{vz}) = S_{vx} \dfrac{\partial^2\alpha}{\partial x\partial z} + S_{vy} \dfrac{\partial^2\alpha}{\partial y\partial z} - \dfrac{1}{K} q_{vz} \left(\dfrac{\partial^2\alpha}{\partial x^2} + \dfrac{\partial^2\alpha}{\partial y^2}\right)
\end{array}\right. \qquad (6)$$

where α = ln (g/K)
      $\underline{S}_v = -\nabla\Phi$. $\qquad (7)$

Equations (6) are the ones which *are actually used in FLOSA-FE* to calculate groundwater flow in the local systems. Hydraulic conductivity K (the only parameter involved) is describing water-soil interactions. It is assumed to be *isotropic* but *heterogeneous*.

The VOA boundary conditions are formulated as follows:

Top/bottom boundary (assumed horizontal):

If Φ is specified on the boundary then:

$$\left|\begin{array}{l} S_{vx} = -\dfrac{\partial \Phi}{\partial x} \\[2mm] S_{vy} = -\dfrac{\partial \Phi}{\partial y} \\[2mm] \dfrac{\partial q_{vz}}{\partial n} = n_z \left\{ \dfrac{\partial}{\partial x} \left(K \dfrac{\partial \Phi}{\partial x} \right) + \dfrac{\partial}{\partial y} \left(K \dfrac{\partial \Phi}{\partial y} \right) \right\}. \end{array}\right. \tag{8}$$

If q_{vn} is specified on the boundary then:

$$\left|\begin{array}{l} \dfrac{\partial S_{vx}}{\partial n} = \dfrac{\partial}{\partial x} \left(\dfrac{1}{K} q_{vn} \right) \\[2mm] \dfrac{\partial S_{vy}}{\partial n} = \dfrac{\partial}{\partial y} \left(\dfrac{1}{K} q_{vn} \right) \\[2mm] q_{vz} = n_z \cdot q_{vn}. \end{array}\right. \tag{9}$$

Side boundary (assumed vertical):

If Φ is specified on the boundary then:

$$\left|\begin{array}{l} q_{vz} = -K \dfrac{\partial \Phi}{\partial z} \\[2mm] S_{vy} = -n_x \dfrac{\partial \Phi}{\partial \xi_1} \\[2mm] \left(\dfrac{\partial K}{\partial n} \right) S_{vx} + K \dfrac{\partial S_{vx}}{\partial n} = n_x \left\{ \dfrac{\partial}{\partial \xi_1} \left(K \dfrac{\partial \Phi}{\partial \xi_1} \right) + \dfrac{\partial}{\partial \xi_2} \left(K \dfrac{\partial \Phi}{\partial \xi_2} \right) \right\}. \end{array}\right. \tag{10}$$

If q_{vn} is specified on the boundary then:

$$\left|\begin{array}{l} S_{vx} = n_x \dfrac{1}{K} q_{vn} \\[2mm] \dfrac{\partial S_{vy}}{\partial n} = n_x \dfrac{\partial \left(\dfrac{1}{K} q_{vn} \right)}{\partial \xi_1} \\[2mm] \dfrac{1}{K} \dfrac{\partial q_{vz}}{\partial n} + \dfrac{\partial \left(\dfrac{1}{K} \right)}{\partial n} = \dfrac{\partial}{\partial \xi_2} \left(\dfrac{1}{K} q_{vn} \right). \end{array}\right. \tag{11}$$

where (ξ_1, ξ_2, ξ_3): local system of coordinates on the side boundary.

FINITE ELEMENT GALERKIN APPROXIMATION TO THE SOLUTIONS OF FLOW EQUATIONS

Φ-equations

When approximating the unknown solution Φ to the flow equation (1) by a simple element-wise linear function $\hat{\Phi}$:

$$\hat{\Phi}(x,y,z) = \sum_{i=1}^{N} \Phi_i \phi_i(x,y,z)$$

where Φ_i: (approximate) value of piezometric head at the i-th finite element node

$\quad\ \phi_i$: i-th basic function

$\quad\ N$: number of finite element nodes

and applying the Galerkin Finite Element Method one reduces a problem of solving a partial differential equation to a simpler problem of solving a set of linear algebraic equations, i.e.

$$\boxed{\begin{array}{c} \nabla \cdot (\underline{\underline{K}} \nabla \Phi) + Q = 0 \\ + \text{ boundary conditions} \end{array}} \tag{12a}$$

$$\downarrow$$

$$\boxed{\underline{\underline{P}}\,\underline{\Phi} = \underline{b}} \tag{12b}$$

where $\underline{\Phi} = (\Phi_1,...,\Phi_N)$: vector of unknown (approximate) values of piezometric head at the nodes

$\quad\ \underline{\underline{P}}$: symmetric and positive definite global matrix summarizing an information on geometry and hydraulic parameters of the flow domain Ω

$\quad\ \underline{b}$: RHS-vector representing input variables (source terms and boundary conditions).

VOA-equations

When calculating numerical solutions to the Velocity Oriented Groundwater Flow equations (6) with the Galerkin Finite Element Method it is assumed in FLOSA-FE that the following approximations are defined over finite elements (tetrahedrons):

$$\hat{S}_x(x,y,z) = \sum_{i=1}^{N} s_{xi} \cdot \phi_i(x,y,z)$$

$$\hat{S}_y(x,y,z) = \sum_{i=1}^{N} s_{yi} \cdot \phi_i(x,y,z) \tag{13}$$

$$\hat{q}_z(x,y,z) = \sum_{i=1}^{N} q_{zi} \cdot \phi_i(x,y,z).$$

(subscripts v are omitted from now on).

The basic functions are assumed to be element-wise linear while $\{s_{xi}, s_{yi}, q_{zi}: i = 1,...,N\}$ represent unknown values of transformed fluxes in N nodes of the finite element grid. By applying the Galerkin technique and using approximations (13) the flow equations (6) can be converted into a set of coupled algebraic equations

$$\nabla \cdot (K \nabla S_x) = K S_x \frac{\partial^2 \alpha}{\partial x^2} + K S_y \frac{\partial^2 \alpha}{\partial x \partial y} + q_z \frac{\partial^2 \alpha}{\partial x \partial z}$$

$$\nabla \cdot (K \nabla S_y) = K S_x \frac{\partial^2 \alpha}{\partial x \partial y} + K S_y \frac{\partial^2 \alpha}{\partial y^2} + q_z \frac{\partial^2 \alpha}{\partial y \partial z} \tag{14a}$$

$$\nabla \cdot (\frac{1}{K} \nabla q_z) = S_x \frac{\partial^2 \alpha}{\partial x \partial z} + S_y \frac{\partial^2 \alpha}{\partial y \partial z} - \frac{1}{K} q_z \left(\frac{\partial^2 \alpha}{\partial x^2} + \frac{\partial^2 \alpha}{\partial y^2} \right)$$

+ boundary conditions

$\downarrow$

$$\underline{\underline{S}}_x \underline{s}_x = \underline{b}_x$$
$$\underline{\underline{S}}_y \underline{s}_y = \underline{b}_y \tag{14b}$$
$$\underline{\underline{Q}}_x \underline{q}_z = \underline{b}_z$$

where $\left. \begin{array}{l} \underline{s}_x = (s_{x1},...,s_{xN}) \\ \underline{s}_y = (s_{y1},...,s_{yN}) \\ \underline{q}_z = (q_{z1},...,q_{zN}) \end{array} \right\}$ vectors of unknowns

$\left. \begin{array}{l} \underline{b}_x = (b_{x1},...,b_{xN}) \\ \underline{b}_y = (b_{y1},...,b_{yN}) \\ \underline{b}_z = (b_{z1},...,b_{zN}) \end{array} \right\}$ RHS-vectors

$\left. \begin{array}{l} \underline{\underline{S}}_x = \{S_{ij}\} \\ \underline{\underline{S}}_y = \{S_{ij}\} \\ \underline{\underline{Q}}_z = \{Q_{ij}\} \end{array} \right\}$ matrices of geometric and hydraulic properties of the flow domain Ω.

Assembling procedures for matrices $\underline{\underline{S}}_x$, $\underline{\underline{S}}_y$ and $\underline{\underline{Q}}_z$ are exactly the same as for classical approach, i.e. for matrix $\underline{\underline{P}}$. Also contributions to the RHS-vectors due to boundary conditions are calculated in the similar way as for piezometric head flow equations.

Both Φ- and VOA-algebraic equations are being solved using the ICCG universal solver developed by M. Nawalany (M. Nawalany [3]). Firstly, the equations for Φ are solved for a regional system, then boundary conditions are calculated for the local system and finally the VOA-simulator calculates velocity components within the local system.

Heterogeneity problem
When hydraulic conductivity K changes in space the simple analytical formulae resulting from the Galerkin finite element formulae cannot be applied and, in general case, numerical integrations must be performed in order to calculate elements of the global matrices $\underline{\underline{P}}$, $\underline{\underline{S}}_x$, $\underline{\underline{S}}_y$ or $\underline{\underline{Q}}_z$ as well as corresponding RHS-vectors. Still, if linear model of hydraulic conductivity is assumed some other analytical formulae can be found and the numerical integration avoided.

Model of hydraulic conductivity
It is assumed in FLOSA-FE that $K = K(x,y,z)$ is element-wise linear, i.e.

$$K^e(x,y,z) = \sum_{\ell=1}^{4} K_\ell^e \cdot \phi_\ell(x,y,z), \quad (e = 1,...,N_e) \tag{15}$$

where K_ℓ^e, ($\ell = 1,2,3,4$): values of hydraulic conductivity at four nodes of the
e-th element
N_e : number of elements.

With the above definition one can calculate integrals for Φ- and VOA-equations.

Φ-equations
By assuming isotropy and using (15) of the *local matrix* $\underline{\underline{P}}^e$ one obtains the following expression for elements p_{ij}^e:

$$p_{ij}^e = [\beta_i^e \beta_j^e + \gamma_i^e \gamma_j^e + \delta_i^e \delta_j^e] \cdot \hat{K}^e, \quad (i,j = 1,2,3,4) \tag{16}$$

where β_i^e, γ_i^e, δ_i^e : coefficients related to the i-th basic function
restricted to the e-th element

$$\hat{K}^e = \frac{1}{4} \sum_{\ell=1}^{4} K_\ell^e \ : \ \text{(arithmetic) mean hydraulic conductivity} \tag{17}$$
within the e-th element (tetrahedron).

The elements of the global matrix $\underline{\underline{P}}$ can be calculated element-wise out of the elements p_{ij}^e given by formula (16).

VOA-equations
Structures of the global matrices $\underline{\underline{S}}_x$ and $\underline{\underline{S}}_y$ are identical to the matrix $\underline{\underline{P}}$ if (6) equations are applied. Therefore only the assembling of the matrix $\underline{\underline{Q}}_z$ is discussed here.

Since $1/K$ is the parameter for the third equation of (6) and model (15) is assumed for hydraulic conductivity K the global matrix $\underline{\underline{Q}}_z$ can be assembled out of the following local elements:

$$Q_{ij} = \sum_{e}{}' q_{ij}^e \tag{18}$$

where

$$q_{ij}^e = [\beta_i^e \beta_j^e + \gamma_i^e \gamma_j^e + \delta_i^e \delta_j^e] \cdot J^e \tag{19}$$

$$J^e = \int_{\Omega^e} \frac{d\Omega}{4 \sum_{\ell=1}^{4} K_\ell^e \cdot \phi_\ell(x,y,z)} . \tag{20}$$

Closed formula has been found for calculating the integral (20) analytically — it reads:

$$J^e = 6V^e \cdot \frac{1}{2} \sum_{\ell=1}^{4} K_\ell^e \ln K_\ell^e \frac{1}{\prod_{m \neq \ell} (K_\ell - K_m)} \tag{21}$$

where V^e: volume of the e-th element.

It can be shown that formula (21) has the following good features:

(i)   it is positive.
(ii) integral J^e is independent of cyclic permutation of the indices of nodes in the element
(iii) it changes reprocically if hydraulic conductivity K is scaled by factor m, i.e.

$$J^e (mK_1, mK_2, mK_3, mK_4) = \frac{1}{m} J^e (K_1, K_2, K_3, K_4) \tag{22}$$

(iv) in limiting process it gives proper formulae for J^e when for instance $K_1 \rightarrow K_2$ or $K_1 \rightarrow K_2 \neq K_3 \rightarrow K_4$ etc. In particular case, when $K_1 = K_2 = K_3 = K_4$ formula has its limit $K_1 \rightarrow K_2 = K_3 = K_4 = K_0$ equal to

$$J^e (K_0, K_0, K_0, K_0) = V^e / K_0. \tag{23}$$

In FLOSA-FE simpler solution has been taken: unless $K_1 = K_2 = K_3 = K_4$ or all K_1, K_2, K_3, K_4 are different *per se* the ones which are equal to each other are changed (disturbed) by small percentage of their values so the equality does not hold. Then formula (21) applies without the danger of real overflow. Computer experiments have shown that by changing values of K by 1% (which is acceptable for practitioners) one does not change the value of J^e more than by 1%.

RESULTS

A simple regional system has been chosen in order to demonstrate 'regional-to-local' methodology used in FLOSA-FE.

Regional system

origin of the system	:	$(x_{r0}, y_{r0}, z_{r0}) = (0,0,0)$
size of the system	:	(L_x, L_y, L_z) = (100 m, 100 m, 10 m)
hydraulic conductivity	:	K = 10 m/d
porosity	:	m = 0.2
number of wells	:	N_w = 3
positions of the wells	:	$(x_{w1}, y_{w1}, z_{w1}^t, z_{w1}^b) = (20$ m, 20 m, - , -)
		$(x_{w2}, y_{w2}, z_{w2}^t, z_{w2}^b) = (40$ m, 60 m, - , -)
		$(x_{w3}, y_{w3}, z_{w3}^t, z_{w3}^b) = (70$ m, 80 m, - , -)
boundary conditions	:	for W, N, E, S, B: $q_n = 0$ m/d
		for T : specified pressure in a form of a hill in the middle of the top boundary.

Local system

origin of the system	:	$(x_{\ell 0}, y_{\ell 0}, z_{\ell 0})$ = (10 m, 10 m, -1 m)
size of the system	:	$(L_{x\ell}, L_{y\ell}, L_{z\ell})$ = (55 m, 55 m, 8 m)
hydraulic conductivity	:	K = 10 m/d
porosity	:	m = 0.2
number of wells	:	N_w = 2
positions of the wells	:	as before
boundary conditions	:	for W, N, E, S, B and T: pressure specified from the regional solution.

Regional and local systems are shown in Fig. 2.

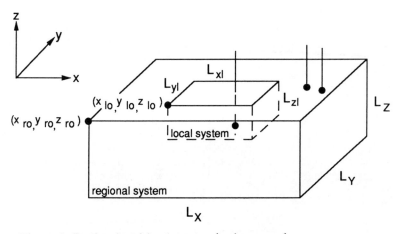

Figure 2. Regional and local systems in the example.

Figures 3a and 3b show the distributions of the piezometric head along the horizontal planes at the 5 m depths in the *regional* and *local systems*.
Figures 4a and 4b present a set of pathlines which start on the top boundary and either are attracted by the wells in the local system or they leave the system.

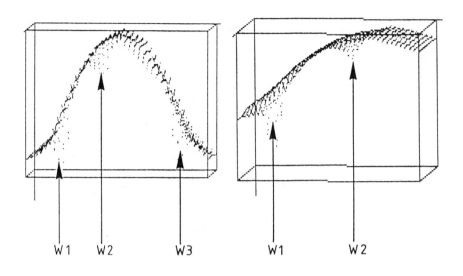

Figure 3a.

Figure 3b.

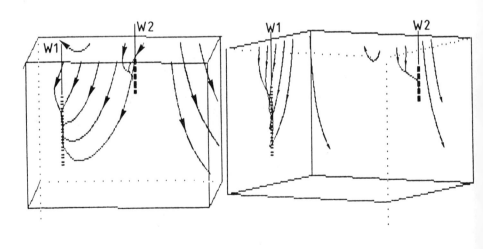

Figure 4a.

Figure 4b.

FINAL REMARKS

1. By introducing global and local systems it become possible to overcome unfavourable geometric aspect ratio between the dimensions of the flow domain and distances between the wells. Calculating accurate pathlines within a local system is now feasible.

2. Also by distributing wells, i.e., by representing their screens as the series of point sinks/sources one can obtain a more realistic picture of the piezometric head and pathlines in the flow domain.

3. Introduction of the linear model for permeability made it possible to deal with heterogeneity and to have an analytical (and quick) method of assembling the VOA equations. Accuracy of this model still need some further investigations.

REFERENCES

[1] Bear, J. Hydraulics of Groundwater, McGraw-Hill International Book Company, New York, 1979.
[2] Nawalany, M. Numerical Model for the Transport Velocity Representation of Groundwater Flow, VIth International Conference on Finite Elements in Water Resource, Lisbon, June 1986.
[3] Nawalany, M. FLOSA-FE, Users' Manual, OS 89-12, Delft, 1989.
[4] Zijl, W. Finite Element Methods Based on a Transport Velocity Representation for Groundwater Motion, Water Resources Research, Vol. 20, No. 1, 1984.
[5] Zijl, W. and M. Nawalany Vector Field Approach for Finite Differences and First-Order Finite Elements, Expert Meeting on New Developments in Groundwater Modelling, Delft, 14-16 September, 1988.

Flow Systems Analysis in Stratified Porous Media

W. Zijl

TNO Institute of Applied Geoscience, PO Box 285, 2600 AG Delft, The Netherlands

1. INTRODUCTION

Flow systems analysis was for the first time applied by Tóth [1]. Further developments deal with the application to migration of hydrocarbons in sedimentary basins and genesis of stratabound ore deposits. Stuurman *et al.* [2] apply flow systems analysis in the context of environmental protection.

In numerical regional flow systems analysis based on finite difference or finite element computer codes, the horizontal spatial scale of the grid squares is large (say the kilometre-scale). Nonuniformities in the phreatic groundwater level cannot be represented on scales smaller than the large-scale grid square. Based upon this large-scale phreatic level, the large-scale groundwater flow velocity field, flow paths and residence times can be determined; see Fig. 1.

The important question is whether these calculated flow paths are representative of the real flow paths which are affected by the small-scale velocity pattern superimposed on the large-scale velocity field. It will be shown that in the deeper layers of the basin the calculated large-scale velocities and resulting flow paths are close to the real flow paths. The large-scale relief in the phreatic groundwater level is related to the steady-state topography of the land surface, whereas the small-scale nonuniformities in the relief are dependent on seasonally fluctuating rates of effective precipitation and fluctuating levels of surface waters. This means that the deeper flow paths, which are driven by the large-scale Fourier modes of the phreatic level, exhibit a relatively steady-state character. Therefore, it makes sense to produce flow systems maps showing large-scale net recharge and discharge regions with their connecting flow paths.

2. BASIC VECTOR FIELD EQUATIONS

The continuity equation and Darcy's Law are simplified by neglecting storage, compaction and density-driven flow; only stratified groundwater basins will be considered.

The lower boundary $z = d$ of the modelling domain $0 < z < d$ is chosen as a plane where the vertical flux component $q_z(x,y,d)$ is sufficiently small with respect to the horizontal flux components $q_x(x,y,z)$ and $q_y(x,y,z)$, and is also sufficiently small with respect to the vertical flux component $q_z(x,y,z)$ far above the lower boundary. In well conducting layers (aquifers) the horizontal flux components are large compared to the horizontal flux components in poorly conducting layers (aquitards). Therefore, the lower boundary of the modelling domain must be chosen as an interface which is the bottom plane of an aquifer and the top plane of an aquitard. Since the vertical flux component on $z = d$ is assumed to be sufficiently small, and since the solutions to the equations change only slightly under slight perturbations in the boundary conditions, the solutions do not change significantly when this small vertical flux component is replaced by the boundary condition $q_z(x,y,d) = 0$. The aquitard on the top of which this boundary condition is applied, is generally called the "impervious base". In reality, however, this boundary condition is always an approximation. The choice of the depth of the impervious base is strongly dependent on the order of magnitude of the fluxes in which we are interested. For instance, when interested in the shallow, local flow systems with relatively high magnitudes of the fluxes, a relatively shallow impervious base may be chosen at a depth where the vertical flux component on that base is small enough to be neglected with respect to the higher fluxes in the shallower parts of the basin. On the other hand, when interested in the deep flow systems with relatively small fluxes, the impervious base must be chosen at a much greater depth.

The phreatic level $z = f(x,y)$ is defined as the plane where the groundwater pressure is atmospheric. When the basin is sufficiently thick, the boundary condition $\phi(x,y,0) = f(x,y)$ for the potential $\phi(x,y,z)$ may be applied on $z = 0$.

The above-discussed equations and boundary conditions lead to the following three decoupled sets of Laplace-type equations for the vector field (e,j):

$$
\left.
\begin{aligned}
\partial(k_h\,\partial e/\partial h)/\partial h + \partial(k_z\,\partial e/\partial z)/\partial z = 0 & , \quad 0 < z < d \\
e = -f\,'(h) & , \quad z = 0 \\
\partial e/\partial z = 0 & , \quad z = d
\end{aligned}
\right\} \quad (1a)
$$

$$
\left.
\begin{aligned}
\partial(k_z^{-1}\,\partial j/\partial h)/\partial h + \partial(k_h^{-1}\,\partial j/\partial z)/\partial z = 0, & \quad 0 < z < d \\
\partial j/\partial z = k_h(0)\,f\,''(h) & , \quad z = 0 \\
j = 0 & , \quad z = d
\end{aligned}
\right\} . \quad (1b)
$$

In equations (1) h is short-hand notation for the pair of horizontal Cartesian coordinates (x,y); e is short-hand notation for the pair of Cartesian components $(\partial\phi/\partial x, \partial\phi/\partial y)$; $j = q_z$, and $k_h(z)$ and $k_z(z)$ are the horizontal and vertical hydraulic conductivities. Equations (1) state that $e(h,z)$ and $j(h,z)$ are continuous functions of h and z, even on planes where $k_h(z)$ and $k_z(z)$ are discontinuous. Equations (1) also state that the vectors $(k_h\,\partial e/\partial x, k_h\,\partial e/\partial y, k_z\,\partial e/\partial z)$ and $(k_z^{-1}\,\partial j/\partial x, k_z^{-1}\,\partial j/\partial y, k_h^{-1}\,\partial j/\partial z)$ are divergence-free. This means that on a plane $z = \delta$ of discontinuous hydraulic conductivities, both $k_z\,\partial e/\partial z$ and $k_h^{-1}\,\partial j/\partial z$ are continuous.

A Fourier component of the phreatic level is given by $f(h) = a\,\exp(-ix/\ell)$, $f\,'(h) = \{(-ia/\ell)\exp(-ix/\ell), 0\}$, $f\,''(h) = -(a/\ell^2)\exp(-ix/\ell)$, $i = \sqrt{-1}$. The function $\exp(-ix/\ell)$ has magnitude one, but a and ℓ may have any magnitude. Now define the dimensionless quantities $H = h/\ell$, $F(H) = f(h)/a$, $F\,'(H) = f\,'(h)\,\ell/a$ and $F\,''(H) = f\,''(h)\,\ell^2/a$; these dimensionless quantities have order of magnitude one. Furthermore, define the dimensionless coordinate $Z = z/d$ and the dimensionless hydraulic conductivities $K_H(Z) = k_h(z)/k_{hc}$ and $K_Z(Z) = k_z(z)/k_{zc}$, where k_{hc} and k_{zc} are the characteristic hydraulic conductivities (see section 4). Finally, define $E(H,Z) = e(h,z)\, \ell/a$ and $J(H,Z) = j(h,z)\, \ell^2/(d\,k_{hc}a)$. Then the set of vector field equations (1) written in dimensionless variables is:

$$\left.\begin{array}{ll} \sigma\partial(K_H\partial E/\partial H)/\partial H + \partial(K_Z\partial E/\partial Z)/\partial Z = 0 , & 0 < Z < 1 \\[4pt] E = -F\,'(H) & , \quad Z = 0 \\[4pt] \partial E/\partial Z = 0 & , \quad Z = 1 \end{array}\right\} \qquad (2a)$$

$$\left.\begin{array}{ll} \sigma\partial(K_Z^{-1}\partial J/\partial H)/\partial H + \partial(K_H^{-1}\partial J/\partial Z)/\partial Z = 0, & 0 < Z < 1 \\[4pt] \partial J/\partial Z = K_H(0)\, F\,''(H) & , \quad Z = 0 \\[4pt] J = 0 & , \quad Z = 1 \end{array}\right\} . \qquad (2b)$$

In equations (2) all terms have order of magnitude one, except for the dimensionless number $\sigma = (d/\ell)^2\,(k_{hc}/k_{zc})$ which may have any order of magnitude.

3. SOLUTIONS FOR STRATIFIED AND HOMOGENEOUS BASINS

To obtain ordinary differential equations, the vector field components E and J in equations (2) are expanded in an infinite power series of the σ-number:

$$E(H,Z) = E_0(H,Z) + \sigma E_1(H,Z) + \sigma^2 E_2(H,Z) + \dots \qquad (3a)$$

$$J(H,Z) = J_0(H,Z) + \sigma J_1(H,Z) + \sigma^2 J_2(H,Z) + \dots . \qquad (3b)$$

Substitution of the series (3) in equations (2), and equating terms with the same power of σ, yields the following infinite hierarchy of equations:

$$\left.\begin{array}{ll} \partial(K_H\partial E_{n-1}/\partial H)/\partial H + \partial(K_Z\partial E_n/\partial Z)/\partial Z = 0 , & 0 < Z < 1 \\[4pt] E_n = -F\,'(H) \quad \text{if } n = 0, \text{ else } 0 & , \quad Z = 0 \\[4pt] \partial E_n/\partial Z = 0 & , \quad Z = 1 \end{array}\right\} \qquad (4a)$$

$$\left.\begin{array}{ll} \partial(K_Z^{-1}\partial J_{n-1}/\partial H)/\partial H + \partial(K_H^{-1}\partial J_n/\partial Z)/\partial Z = 0, & 0 < Z < 1 \\[4pt] \partial J_n/\partial Z = K_H(0)\, F\,''(H) \quad \text{if } n = 0, \text{ else } 0 & , \quad Z = 0 \\[4pt] J_n = 0 & , \quad Z = 1 \end{array}\right\} . \qquad (4b)$$

In the above hierarchy $n = 0,1,2,\dots$, and $E_{-1} = J_{-1} = 0$.
The solutions to the hierarchy of equations (4) are given by:

$$E_n(H,Z) = -R_n(Z)\, F^{(2n+1)}(H) \qquad (5a)$$

$$J_n(H,Z) = -T_n(Z)\, F^{(2n+2)}(H) \qquad (5b)$$

with:

$$R_0(Z) = 1 \qquad (6a)$$

$$T_0(Z) = \int_Z^1 K_H(Z')\, dZ' \qquad (6b)$$

$$R_n(Z) = \int_0^Z T_{n-1}(Z')/K_Z(Z')\,dZ', \quad n > 0 \tag{6c}$$

$$T_n(Z) = \int_Z^1 R_n(Z')\,K_H(Z')\,dZ', \quad n > 0. \tag{6d}$$

The characteristic hydraulic conductivities k_{hc} and k_{zc} are chosen in such a way that $T_0(0) = 1$ and $T_1(0) = 1/3$. The zeroth-order solutions ($n = 0$) are the well-known Dupuit approximations, and the solutions for $n > 0$ are the n-th order corrections to the Dupuit approximations.

Let us consider the most general Fourier mode $F(H) = \exp(-i\Omega_X X) \cdot \exp(-i\Omega_Y Y)$ with $\Omega^2 = \Omega_X^2 + \Omega_Y^2 = 1$. Then it follows from equations (3) and (5) that:

$$E(H,Z) = - \sum_{n=0}^{\infty} \{(-\sigma)^n R_n(Z)\}\, F'(H) \tag{7a}$$

$$J(H,Z) = - \sum_{n=0}^{\infty} \{(-\sigma)^n T_n(Z)\}\, F''(H). \tag{7b}$$

With the aid of expressions (6) equations (7) lead to the solutions for any stratified basin. However, only a simple example will be shown here. Consider a homogeneous porous medium; in that case it follows from expressions (6) that the series expansions in equations (7) are equal to the series expansions of the following functions:

$$E(H,Z) = - \{\cosh(Z\sqrt{\sigma}) - \tanh(\sqrt{\sigma})\sinh(Z\sqrt{\sigma})\}\, F'(H) \tag{8a}$$

$$J(H,Z) = \{\sinh(Z\sqrt{\sigma}) - \tanh(\sqrt{\sigma})\cosh(Z\sqrt{\sigma})\}\, F''(H)/\sqrt{\sigma}. \tag{8b}$$

For $\sqrt{\sigma} \ll 1$, $E(H,Z) = -F'(H)$ and $J(H,Z) = -(1 - Z)\,F''(H)$. For $\sqrt{\sigma} \gg 1$, $E(H,Z) = -\exp(-Z\sqrt{\sigma})\,F'(H)$ and $J(H,Z) = -\exp(-Z\sqrt{\sigma})\,F''(H)/\sqrt{\sigma}$. This means that low σ-number modes have a non-damped, Dupuit-like behaviour, whereas high σ-number modes are exponentially damped with increasing depth. This damping makes it possible to neglect high σ-number modes at a sufficient depth resulting in nested flow systems; see Fig. 2. For stratified porous media the rate of damping changes discontinuously from layer to layer, which makes it possible to introduce the concept of "impervious base"; the general character of this will be discussed in section 5.

4. NESTED FLOW SYSTEMS

Let us consider the Fourier mode $f(h) = \exp(-i\omega h)$ of the phreatic groundwater level. Provided that the porous medium is piecewise homogeneous, the following expressions approximately characterize the solutions to equations (1):

$$e(h,z) \simeq e^0(\omega_e)\,\exp\{-(i\omega h + \omega_e z)\} \tag{9a}$$

$$j(h,z) \simeq j^0(\omega_j)\,\exp\{-(i\omega h + \omega_j z)\}. \tag{9b}$$

For the purpose of illustrating the general character of the solutions, only the terms which exponentially damp with increasing depth z are given; the terms which exponentially grow with increasing depth are assumed to be small, similar to the solution (8) for $\sqrt{\sigma} \gg 1$.

The bottom $z = \delta$ of the homogeneous top layer represents an interface where the hydraulic conductivities discontinuously jump from one value to another. From equations (1) it follows that e, j, $k_z(\delta) \, \partial e/\partial z$ and $k_h^{-1}(\delta) \, \partial j/\partial z$ are continuous at the interface $z = \delta$. Consequently, it follows from expressions (9) that $\omega_e(\delta+) \, k_z(\delta+) = \omega_e(\delta-) \, k_z(\delta-)$ and $\omega_j(\delta+)/k_h(\delta+) = \omega_j(\delta-)/k_h(\delta-)$, where $\delta+$ ($\delta-$) is the z-coordinate just below (above) the interface. This means that when a layer with lower hydraulic conductivities (an aquitard) is encountered, a higher damping factor ω_e is found for e, and a lower damping factor ω_j is found for j. Inversely, when a layer with higher hydraulic conductivities (an aquifer) is encountered, a lower damping factor ω_e is found for e, and a higher damping factor ω_j is found for j. Of course, this is repeated for the deeper layers. This discontinuous rate of damping makes it possible to introduce "impervious bases" dependent on the wave number ω, and, as a consequence, it also explains the occurrence of nested flow systems.

5. CONCLUSIONS

In a basin the vector components e and j are exponentially damped with increasing depth. In a homogeneous medium this damping is gradual, but in a stratified medium the rate of damping suddenly changes at the interfaces between aquifers and aquitards. The damping is such that the short wave modes (high σ-number) are damped stronger than the long wave modes (low σ-numbers). This also means that the impervious base is at a shallower position for the short wave modes than it is for the long wave modes.

Consequently, instead of a series of σ-numbers, a series of "impervious bases" $z = d_1$, $z = d_2$, $z = d_3$, etc. can be distinguished, where $d_1 < d_2 < d_3$, etc. are interfaces between aquifers and aquitards, in such a way that an aquitard is below an aquifer. The regions above these impervious bases have their own σ-numbers $\sigma_1 = (d_1/\ell_1)^2 \, (k_{hc1}/k_{zc1})$, $\sigma_2 = (d_2/\ell_2)^2 \, (k_{hc2}/k_{zc2})$, etc. The impervious base $z = d_1$ is the base for short wave modes for which $\sigma_1 \approx 1$; the impervious base $z = d_2$ is the base for medium wave modes for which $\sigma_2 \approx 1$; the impervious base $z = d_3$ is the base for long wave modes for which $\sigma_3 \approx 1$, etc.; $\ell_1 < \ell_2 < \ell_3$, etc. For any wave mode the flow above the impervious base may be calculated with the Dupuit approximations, eventually supplemented with one or more higher-order corrections, and the flow below the impervious base is equal to zero.

6. REFERENCES

1. Tóth, J., A theoretical analysis of groundwater flow in small drainage basins, Journal of Geophysical Research, Vol. 68, No. 16, pp. 4795-4812, 1963.
2. Stuurman, R.J., Biesheuvel, A. and Van der Meij, J.L., The application of regional hydrological systems analysis in water management, In: Regional Characterization of Water Quality (Ed. Ragone, S.), IAHS Publication No. 183, pp. 45-57, 1989.

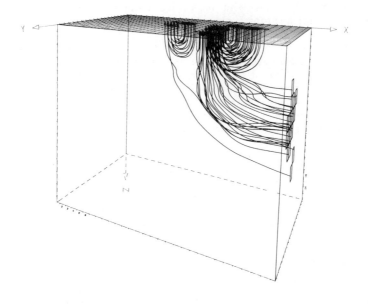

Figure 1.   Subsurface flow paths for environmental protection study; see Stuurman *et al.* [2].

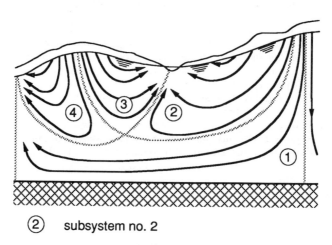

②   subsystem no. 2

⋯⋯⋯   subsystem's boundary

⌣   flow path

Figure 2.   Nested flow systems in homogeneous, strongly anisotropic basin; horizontal kilometre-scale, vertical metre-scale.

Groundwater Response to the Po River Canalization Study Analyzed by a Finite Element Model

M. Cargnelutti, M. Gonella

Unità EDP-2, Applicazioni di modellistica numerica, HYDRODATA, Via Pomba 23, 10123 Turin, Italy

ABSTRACT

A 2D finite element model has been developed and applied to simulate the groundwater behavior around a barrage planned in the canalization study of the Po River. Three situations have been studied: the situation before and after the construction of the barrage, and the situation with different hydraulic structure interventions to control the rise of piezometric levels. The results show how these structures can reduce the areas of the impact due to the raised piezometric levels and which are the best solutions.

INTRODUCTION

In the Po River canalization study, some baragges are planned in order to maintain relatively high water levels for improving inland navigation and for generating hydroelectric power. These modified water levels alter the natural conditions of flow in the nearmost groundwater system, in particular the rise of piezometric levels can cause some problems for the agriculture. To avoid these problems some structures like cutoff walls and drainage ditches can be planned. These kinds of structures can only be effective if a realistic prediction of the acquifer behavior and the actual groundwater levels are provided by appropriate model simulations.

A finite element model has been developed and applied to simulate the groundwater flow surrounding the Camatta barrage (Mantova district), northeastern Italy. The simulations are done under three conditions: the undisturbed situation before the construction of the barrage, the situation after the construction, and the modified situation due to the execution of adequate interventions to reduce the rise of piezometric levels. Then the corresponding results are compared to each other and analysed.

FINITE ELEMENT MODEL

The model, which allows for the treatment of confined and unconfined flow in both steady and unsteady conditions, has been applied in its steady state form

$$\frac{\partial}{\partial x}(K_x \frac{\partial h}{\partial x}) + \frac{\partial}{\partial y}(K_y \frac{\partial h}{\partial y}) = f(x,y) \qquad (1)$$

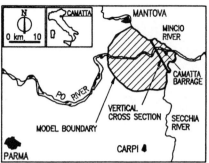

Fig. 1. Modelling area and location of the Camatta barrage.

where h is the hydraulic potential, k_x and k_y are the principal components of the hydraulic conductivity, and f(x, y) is a forcing function accounting for distributed sources and sinks (Zienkiewicz [1]).

The numerical solution is achieved by a finite element technique using triangular elements.   The model equations are solved by symbolic factorization, numerical factorization,   and backward-forward substitution after an optimal preliminary reordering (Gambolati et al. [2]). The model has been applied in two different ways, to simulate the flow in a horizontal plane and in a vertical cross section, using two different meshes. In the vertical cross section an iterative technique has been applied with a deformable mesh to find the position of the free surface.

FLOW IN THE HORIZONTAL PLANE

The Camatta barrage is located in the Po River before the confluence of Mincio and Secchia Rivers as is shown by the map of Figure 1. This figure also shows the location of the flow domain over which the model has been applied and the location of the considered vertical cross section. Figure 2 shows the triangular finite element mesh of the acquifer underlying the Camatta barrage. The grid is made up of 1570 elements for a total of 855 nodes; it is quite dense, especially close to the barrage and to the Po River banks.

A schematic cross section of the aquifer is shown in Figure 3 and has been derived on the basis of the lithostratigraphic series evidenced in the observation wells drilled in that zone (CNR-IRSA [3]). The subsurface system is composed by alternating layers of sands and silts. The upper sandy formation has a thickness of 25-30 m and is confined on the upper side by a clayey-silty layer whose thickness is between 8-14 m. The Po River is hydraulically connected with the aquifer because the river bed remains under the lower level of the first silty layer. So the piezometric level is largely influenced by the water levels in the Po River (Pellegrini [4]). In the model the acquifer has been schematized with a confined layer of 27 m thickness and an average value of hydraulic conductivity $k_x = k_y = 5 \times 10^{-5}$ m/s, obtained from laboratory tests.

<u>Undisturbed situation before the construction of the barrage</u>
As a reference state usefull to compare the groundwater flow with or without the presence of the barrage, an average condition over a period of 20 years is considered. Dirichlet boundary conditions have been assigned to the model considering the mean values of groundwater levels, and water levels in Po River, recorded in the area.

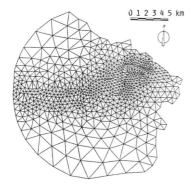
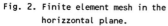

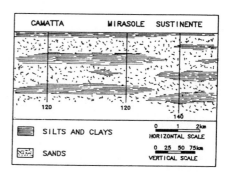

Fig. 2. Finite element mesh in the
horizzontal plane.

Fig. 3. Schematic cross section of the
aquifer.

Figure 4 shows the main results obtained in this way: piezometric contour lines and
velocity vectors field.

Situation after the construction of the barrage
The barrage produces a fixed water level in the Po River that is about 1.90 m higher
than the yearly mean value. The hydraulic profile upstream and downstream the
barrage, is a direct function of the river discharge. So two extreme conditions have
been considered, minimum and maximum regulated water levels, with a jump respectively
4.9 m and 2.0 m in the hydraulic profile. Boundary conditions have been assumed as
constant water levels along the Po, Mincio and Secchia Rivers, and as constant flux
calculated in the undisturbed situation, assigned along the remaining boundary.
Figure 5 shows the piezometric contour lines in the case of minimum regulated water
levels in the Po river and the contour lines calculated as difference between the
situation after the construction of the barrage and the undisturbed situation. With
this rappresentation it is possible to locate the areas with permanent rise of the
piezometric levels.

Modified situation due to the cutoff walls
To reduce the areas with permanent rise, cutoff walls of different lengths have been
simulated in the model as shown in Table 1. The walls, like the barrage, have been
schematized by impermeable line segments with the nodes falling on the internal walls
characterized by two numbers. The same boundary conditions of the previous situation
have been considered.

CASE N°	1	2	3	4	5
L_{left}	2.25	1.70	1.25	0.80	2.80
L_{right}	2.80	2.20	1.35	0.60	3.50

Table 1 - Different cutoff wall lengths L (km)

Figure 6 shows the main results with the case 1. It may be noted that the walls
deflect significantly the flow and reduce both the hydraulic gradients and the areas
with permanent piezometric rise. Figure 7 illustrates a comparison between the results
obtained with the different lengths of wall considering the areas with permanent rise
according to different intervals of rise values. Figure 8 shows as the determination

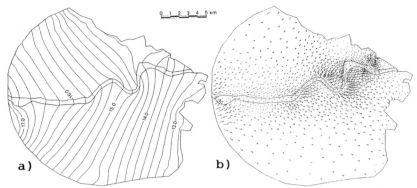

Fig. 4. Undisturbed situation: a) piezometric contour lines and b) velocity vectors field.

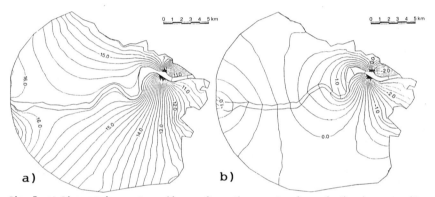

Fig. 5. a) Piezometric contour lines after the construcion of the barrage; b) piezometric head differences compared to the undisturbed situation (Fig. 4 a).

of the optimal length of the cutoff walls may be related to a parameter P.R. that is a function of the weighted average value of rise H_{ave} calculated using areas as weights.

FLOW IN THE VERTICAL CROSS SECTION

The location of the vertical section chosen for the simulations is just at the upstream side of the barrage, where the water level remains essentially constant. The Mincio and the Secchia Rivers are the boundaries of the section (Figure 1). The section has a length of about 10 km, and a depth of 35 m above an impervious boundary of clay layer. The finite element mesh can describe different kinds of river sections and different depths of the superficial layer. The mesh has 2372 elements and 1277 nodes; it is made up three parts representing the acquifer under and at both side parts of the river respectively (Figure 9). The side parts have a flexible shape to calculate the free surface position using an iterative technique (Bear and Verruijt [5]). The hydraulic conductivities of the different layers are 5×10^{-7} m/s for the superficial silt and 5×10^{-5} m/s for the acquifer. The mesh can describe different positions of the cutoff walls with respect to the river (near it or under the embankments) and different depths.

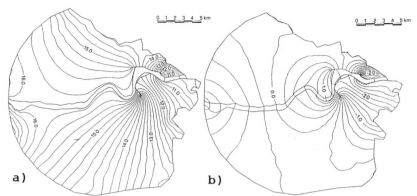

Fig. 6. Modified situation due to the cutoff walls, case 1: a) piezometric contour
lines and b) piezometric head differences compared to the undisturbed
situation (Fig. 4 a).

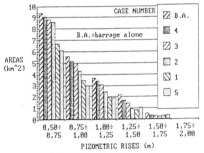

Fig. 7. Comparison between areas with
permanent rise of piezometric
level in different cases.

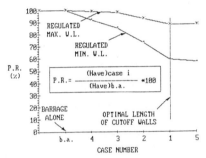

Fig. 8. Optimal length of the cutoff
walls.

The first serie of tests simulates the natural groundwater flow and the situation
with the Po River water levels regulated by the barrage. The boundary conditions are
the water level in the Po, in the Mincio and in the Secchia Rivers: these conditions
are regarded as constant in every simulation. The results agree with that of the horizontal plane simulations. Other tests are performed to design the interventions for
reducing the piezometric levels. Cutoff walls with different depths and drainage ditches are simulated. The comparison of the results is made with respect to the average
rise of the piezometric level.

Cutoff walls
Cutoff walls with two different depths are tested: 35 m, enough to reach the clay
layer, and 32 m just above it. The average rises are respectively 0.90 m and 1.40 m
(56% and 90% of the rise produced by the barrage alone). Figure 10 shows the free
surface of the two cases. The positive effects are remarkable only if the cutoff
walls obstruct completely the acquifer, because of the low hydraulic gradient; in this
situation the head loss due to the cutoff walls is relatively small.

Drainage ditches
In addition, drainage ditches togheter with cutoff walls have been simulated. The

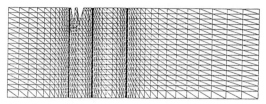

Fig. 9. Finite element mesh in the vertical cross section.

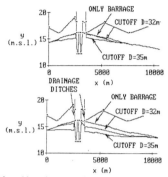

Fig. 10. Piezometric head profiles in
different cases.

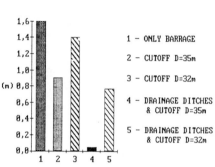

Fig. 11. Comparison between rises of
the piezometric level in the
vertical cross section.

average rise of the piezometric level is 0.05 m with drainage ditches and cutoff
walls 35 m deep, and 0.77 m if the cutoff walls depht is only 20 m (respectively 3%
and 48% of the rise produced by the barrage alone). Figure 11 shows a comparison
between the results.

CONCLUSION

The effects of the Camatta barrage on the groundwater behavior have been analysed with
a finite element model . To reduce the impact due to the raised piezometric levels,
different hydraulic structures have been simulated. It is found that the cutoff walls
together with drainage ditches provide the best solution for its effectiveness. The
modeling approach is proved to be an effective tool in helping to design such
hydraulic stuctures.

REFERENCES

1.Zienkiewicz, O.C. The Finite Element Method, Third edition, Mc Graw-Hill, 1977.
2.Gambolati, G., Toffolo, F. and Uliana, F. Groundwater Response Under an Electronuclear
Plant to a River Flood Wave Analyzed by a Nonlinear Finite Element Model, Water Resources Research, Vol. 20, pp. 903-913, 1984.
3.CNR - IRSA. Indagine sulle falde acquifere profonde della pianura padana, Quaderni
dell'Istituto di Ricerca sulle Acque, Vol. 28 - P/335, 1976.
4.Pellegrini, M.- Interazioni tra corsi d'acqua e falde idriche: responso idrodinamico e
idrochimico dell'acquifero, Seminario di aggiornamento Ecologia dell'ambiente fluviale,
Reggio Emilia, pp 23-55, 1985.
5.Bear, I. and Verruijt, A. Modeling Groundwater Flow and Pollution, D. Reidel Publishing Company, Dordrecht, Holland, 1987.

Modelling Interactions between Groundwater and Surface Water: a Case Study

Ph. Ackerer(*), M. Esteves(**), R. Kohane(***)

(*) Institut de Mécanique des Fluides, Université Louis Pasteur, URA CNRS 854, 2, rue Boussingault, F - 67083 Strasbourg, Cédex, France
(**) CEREG, Université Louis Pasteur, URA CNRS 95, 3, rue de l'Argonne, F - 67083 Strasbourg, Cédex, France
(***) Institut für Wasserbau, Universität Stuttgart, Pfaffenwaldring 61, D - 7000, Stuttgart 80, France

ABSTRACT

Detailed measurements are made in the "Ried de Colmar" (Alsace, France) in order to study water exchanges between groundwater and the river Ill. Water balances underline the importance of the different part of the hydrosystem : at low water, the groundwater provides a great part of the flow in the river ; at high water, the water fluxes reverse. In order to quantify and to locate the exchanges, a groundwater flow coupled with an open-channel flow model is developed. The classical approach is succesful in the simulation of water tables (river and aquifer) but not in the simulation of the exchanged fluxes. An original physically based approach is described in order to simulate the groundwater-surface water interactions and built in the model. Validation of the model is performed by the simulation of measured heads in the river and in the groundwater and the flow rate in the river.

INTRODUCTION

Surface water-groundwater interactions at a local scale has to be studied in detail in order to estimate water fluxes between these two hydrosystems and to evaluate the possible contamination of one of this system by the other. The purpose of this paper is to present the modelling of the water exchanges between the river and the aquifer and a validation of the proposed model.

CANARI, A COUPLED GROUNDWATER-SURFACE WATER MODEL.

CANARI is a physically based modular model which simulates the flow in the river (1D) and in the aquifer (2D) and the water exchanges between the two. The river modul is based on the St-Venant continuity and momentum equations :

$$\partial Q/\partial x + \partial A/\partial t = q_g$$
$$\partial Q/\partial t + \partial(Q^2/A)/\partial x + g A \, \partial z/\partial x = g A \, (Io - If) - (Vq - V) \, q_g \qquad (1)$$

where A is the area of flow normal to the channel invert, g is the gravitational acceleration, q_g is the groundwater inflow per unit length of channel, Q is the channel discharge, Io is the slope of the channel invert, If is the friction slope, t is time, V=Q/A is the average velocity, Vq is the groundwater inflow velocity which is neglected compare to V, x is the abscisse of the channel invert and z the depth of flow relative to the invert. The friction slope is estimated by using empirical Manning's equation. The differential equations are solved by using a classical finite differences scheme (e.g. Ligget and Cunge [1], Kohane [2]). The description of the river geometry is given by measured transverse profiles. Interpolations between two measured profiles can be performed if measured profiles delimit more or less homogeneous river sections. Local energy losses, due to damps for example, can be taken into account by the model.

The groundwater flow modul is based on the diffusivity equation :

$$S\, \partial h/\partial t = \partial/\partial x\,(\,Txx\,\partial h/\partial x + Txy\,\partial h/\partial y) + \partial/\partial y\,(\,Tyx\,\partial h/\partial x + Tyy\,\partial h/\partial y) + q_s \qquad (2)$$

where h is the piezometric head, q_s a source/sink term per unit area, S the storage coefficient, Tij the transmissivity tensor, x, y horizontal coordinates. The differential equation is solved using a finite element scheme based on linear interpolation over triangular elements (e.g. Pinder and Gray [3]) This technique has two main advantages over finite differences : great flexibility of space discretization which allows a better description of the river geometry and the piezometric head can be calculated everywhere. The mesh is build up so that groundwater nodes and transverse profiles of the river have the same location.

Based on the Darcy law, the usual equation used to calculate water fluxes between the groundwater and the surface water is the following (Herbert [4], Pinder and Sauer [5]) :

$$q = k\,(hr - h)\ \text{if}\ h \geq hf$$
$$q = k\,(hr - hf)\ \text{if}\ h < hf \qquad\qquad\qquad\qquad\qquad (3)$$

where hr is the water level in the river, h the groundwater piezometric head, hf the channel invert, k the leakage factor, q the exhange rate per unit area. The only well defined parameter is hr. The definition of hf for natural rivers will not be discussed here.

The definition of the leakage factor k is quite vague. It is often considered as the rate of the hydraulic conductivity over the thickness of the sediment of the river bed (Pinder and Sauer [5], Cunningham and Sinclair [6]). It may also be defined (Moreyl-Seytoux et al. [7]) as an empirical coefficient depending on the shape of the river bed and on the hydrodynamical properties of porous material (river bed and aquifer).

In many groundwater models, the head in the aquifer is calculated at nodes corresponding to river nodes which is a quite rough assumption. In the neighborhood of the river, the flow becomes 3D and the Dupuits-Forchheimer assumption is no longer valid. Defining h at a certain distance D from the river is a better approximation (Morel-Seytoux et al. [7], Bouwer [8]).

In our surface water-subsurface water interactions modelling, the calculation of q is performed by equation (3) but the subsurface piezometric head is calculated at a distance equal to 10 times the width of the river free surface. The factor 10 has been determined by 2D vertical simulation. In fact, the distance D varies in time depending on the water level in the river. We choose a constant value, the error due to this assumption is low because of the slope of the groundwater table at this distance from the river. The exchange area is equal to the wetted perimeter P time the river section length L.

The coupling of surface and subsurface flow proceeds as described in Pinder and Sauer [5]. The solution is supposed to be known at time t. At t+Δt, the St-Venant equations are solved iteratively with q_g set at its value at time t. The surface water heads at time t+Δt are used to calculate the groundwater heads. Then, an improved estimation of q_g can be calculated, followed by a solution to the St Venant equations again using the new q_g. This iterative process continues until successive values of q_g differ by an acceptably small amount. The initial values are obtained by steady state solutions of equation (1) and (2). It has been shown that the model is very sensitive to the initial conditions due to the small time periods in which floods occur.

INTERACTIONS BETWEEN THE RIVER ILL AND THE GROUNDWATER.

The area studied is situated in the Middle Alsatian Plain in the North-East of France. Its surface is about 40 km^2 and the river Ill flows in the middle of the area from South to North. The aquifer geometry and hydrodynamical properties are well known, the hydraulic conductivity is about 10^{-3} m/s to 10^{-2} m/s and its thickness 150 m. The groundwater table is close to the soil surface (less then 1 m in average). The river Ill supplies the aquifer in the upper part of the studied area and drains the aquifer in the lower part. Three dams more or less regulate the flow in the river. The groundwater level always stays above the Ill invert. Many "phreatic" rivers are supplied by the groundwater. They take their source within the studied area (fig 1). The measurement network is made up by 8 limnigraphs, 4 piezographs and some piezometers where measurements are performed every week. Several river gaugings were also performed. A detailed presentation of the experimental site with whole measurements over three years is given in Esteves [9]. Interactions between the water level fluctuations in the river and the head variations in the aquifer are strong. Actually, the importance of the groundwater head fluctuations decreases with distance to the river. The response time is short. In fact, when the groundwater level reaches the soil (mainly loess of about 80 cm thickness), it can be considered as a partially captive aquifer.

The CANARI code was used to simulate surface and subsurface flow over a period of 50 days with one flood. Two approaches have been used to calculate the exchange rate between both hydrosystems :
• q = k (hf - h) where hf and h are calculated at the same node,
• q = k (hf - h) where h is determined at a distance equal to 10 times the width of the free surface of the river and hf is considered as fixed head for the groundwater flow simulation.

The first approach was successfull to simulate water levels in the river and in the aquifer. But, it failed in the calculation of the river discharge. Increasing the leakage coefficient k did not improve the results. It is possible to simulate correctly the heads in the river and in the aquifer without simulating the discharge because of the numerous unknown parameters (leakage factors and Manning coefficients distributed along the river). These coefficients may have the same effects on the heads : for example, the decrease of the Manning's coefficients or the increase of the leakage coefficients raise the head in the river.

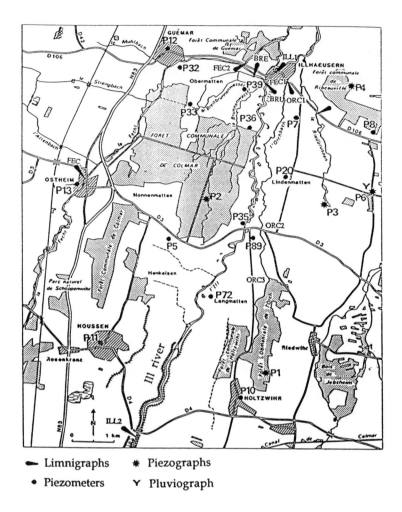

— Limnigraphs   ✳ Piezographs

• Piezometers   Y Pluviograph

Figure 1 : Measurement network near the river Ill.

The other approach was successful in simulating heads and the river discharge. Results in steady state are given in Esteves and Ackerer [10]. Results for transient simulation for a flood from 20/08/87 to 08/10/87 are shown in figure 2. The model was calibrated on steady state observations made in spring 1987, the same leakage and Manning's coefficients are used for the transient simulation. The variations of the river discharge are well reproduced by the model. The simulated head elevations are too sharp close to the river and too smooth far from the river. The water exchanges between the river and the aquifer at low water and high water are very important (fig 3). Sharp variations are due to the dams.

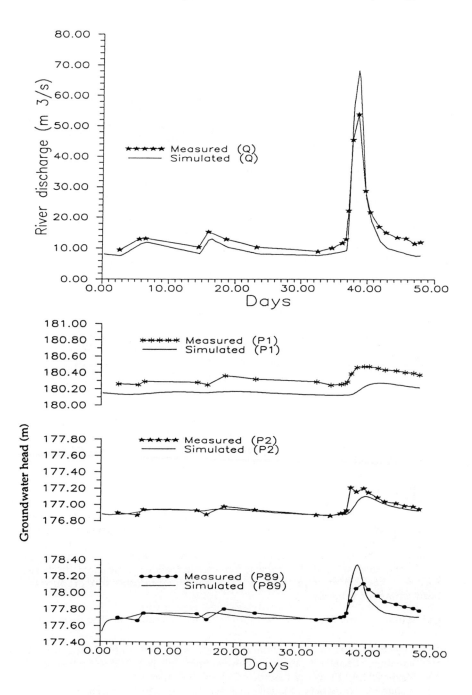

Figure 2 : Simulated and measured discharge and head variations.

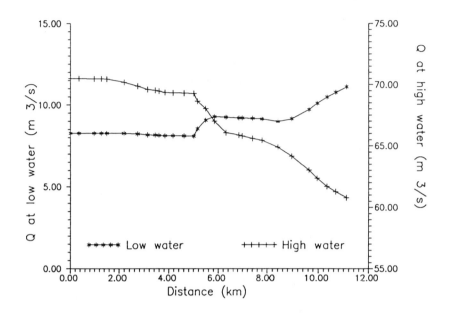

Figure 3 : Computed discharge along the river Ill.

CONCLUSIONS

The proposed coupled subsurface-surface flow model has been validated for a partially penetrating river with strong interactions with its alluvial aquifer. The classical approach in describing the exchanges between the aquifer and the river, often satisfactory for groundwater flow modelling, is not appropriate to our case study. The simulations underline that model calibration has to be based on comparisons between measured and calculated groundwater heads, surface water heads and discharge. If the proposed model managed to simulate discharge variations in the river, improvements have to be done in order to reproduce the head variations properly.

REFERENCES

[1]   Ligget J.A. and Cunge J.A. Numerical methods of solution of the unsteady flow equations. Chapter 4 , Unsteady flow in open channels, (Ed K. Mahmood, V. Yevjevich), Vol 1, Water Resources Pub., Fort Collins.
[2]   Kohane R. Numeriches Modell zur instationnären Hochwasserberechnung in Fliessgewässern mit geglierdertem Querschnitt. Programmdokumentation n° 89/15(HG105). Institut für Wasserbau, University of Stuttgart (FRG), 1989.

[3]   Pinder G.F. and Gray W.G., Finite element simulation in surface and subsurface hydrology. Academic Press, New-York, 295 pp., 1977.

[4]   Herbert R., Modelling partially penetrating rivers on aquifer models. Ground Water, Vol. 8, pp. 29-36, 1970.

[5]   Pinder G.F. and Sauer S.R., Numerical simulation of flow wave modification due to bank storage effects. Water Ressources Research, Vol. 7 (1), pp. 63-70, 1971.

[6]   Cunningham A.B. and Sinclair P.J., Application and analysis of a coupled surface and groundwater model. J. of Hydrology, Vol. 43, pp. 129-148, 1979.

[7]   Morel-Seytoux H.J., Illangasekare T., Peters G., Field verifications of the concept of reach transmissivity. In IAHS-AIHs Pub., n° 128, pp. 355-359, Proc. of the symp. on l'hydrologie des régions à faibles précipitations, Camberra, Australia, Déc. 1979.

[8]   Bouwer H., Theory of seepage from open channels. Advances in Hydrosciences, (Ed. V.T.Chow),Vol. 5, pp. 121-171, Academic Press, New-York, 1969.

[9]   Esteves M., Etude et modélisation des relations aquifere-rivière dans le Ried de Colmar (Ht-Rhin, France). PhD Thesis, University Louis Pasteur, Strasbourg, France, 233 p., 1989.

[10]   Esteves M. and Ackerer P., Etude et modélisation des échanges entre la rivière Ill et sa nappe alluviale dans le Ried de Colmar (France). In IAHR-AIRH PuB. (Ed.P. Dahlblom and G. Lindh), Vol. 1,pp. 113-120, Proc. Int Symp., Interaction between groundwater and surface water, Ystad, Sweden, june 1988.

Least Squares and the Vertically Averaged Flow Equations

L.R. Bentley(*), G.F. Pinder(**)

(*) Department of Civil Engineering and Operations Research, Princeton University, Princeton, NJ 08544, USA

(**) College of Engineering and Mathematics, University of Vermont, Burlington, VT 05405, USA

INTRODUCTION

Tracking, as required for Lagrangian procedures, requires an accurate velocity field. In general, the velocity field must be generated by numerically solving the partial differential equations that govern the flow. An algorithm based on the least squares mixed finite element method (LESFEM) is introduced.

VERTICALLY AVERAGED FLOW EQUATIONS

The governing equations over the domain Ω are conservation of mass,

$$S \frac{\partial h}{\partial t} + \frac{\partial q_x}{\partial x} + \frac{\partial q_y}{\partial y} + B (h - h_r) - R - Q_p \delta(\underline{x} - \underline{x}_p) = 0 \tag{1}$$

Darcy's law in the x-direction,

$$q_x + T_{xx} \frac{\partial h}{\partial x} + T_{xy} \frac{\partial h}{\partial y} = 0, \tag{2}$$

and Darcy's law in the y-direction,

$$q_y + T_{yx} \frac{\partial h}{\partial x} + T_{yy} \frac{\partial h}{\partial y} = 0, \tag{3}$$

where h is head, q_x and q_y are fluxes, S is the storage coefficient, B is the leakance, h_r is a leakage reference head, R is infiltration, Q_p is a pumping rate (positive injection) and T_{ij} are transmissivities.

Boundary conditions are:

$$\mathbf{BH}_i h + \mathbf{BX}_i q_x + \mathbf{BY}_i q_y = \bar{b}_i \quad \underline{x} \cap \delta\Omega_i, \tag{4}$$

where $\mathbf{BH}_i$, $\mathbf{BX}_i$, $\mathbf{BY}_i$ are boundary operators, $\delta\Omega_i$ is the segment of boundary associated with the ith type boundary condition, $\bar{b}_i$ is the specified boundary value.

GRID DEFINITION

A rectangular mesh of grid blocks is oriented along the x and y-axis. Grid blocks can be refined by divisions of integral powers of two (Figure 1). For example, a grid refinement of two means that each side of the grid block is split in the center, and four elements are created. When two grid blocks of different refinement share a side, "hanging nodes" are created. The hanging nodes have degrees of freedom which are active on only one side of the element boundary, and, as will be detailed later, they will require special treatment.

Figure 1 illustrates a circular boundary that has been approximated by a series of straight line segments. Penalty residuals[1], equation (9), are used to enforce boundary conditions.

Three trial functions are constructed with bilinear basis functions on the rectangular grid just described:

$$\hat{h} = H_J(t)\,\Phi_J(x,y)\,, \quad \hat{q}_x = QX_J(t)\,\Phi_J(x,y)\,, \quad \hat{q}_y = QY_J(t)\,\Phi_J(x,y)\,, \quad (5)$$

where $\Phi_J(x,y)$ are bilinear basis functions, the repeated J index implies a summation over all of the active nodes and $H_J(t)$, $QX_J(t)$ and $QY_J(t)$ are time varying values of the state variables at the Jth node.

LEAST SQUARES EQUATIONS

The trial functions, equation (5), are substituted into the conservation of mass equation, (1), the two Darcy law equations, (2) and (3), and the boundary condition equations, (4), to form residuals. After approximating the time derivative with an implicit finite difference, the conservation of mass residual is:

$$Rm = (\frac{S}{\Delta t}+B)\Phi_J(\underline{x})H_J + \frac{\partial \Phi_J(x)}{\partial x}QX_J + \frac{\partial \Phi_J(x)}{\partial y}QY_J$$
$$- \left[\frac{S}{\Delta t}\hat{h}^{n-1}+B\,h_r +R +Q_p\,\delta(\underline{x}-\underline{x}_p)\right], \quad (6)$$

where the coefficients H_J, QX_J and QY_J are the node values of the respective trial functions at the solution time level, n, and $\hat{h}^{n-1}$ is the value of the trial function at the previous time level.

The Darcy law residuals are:

$$Rq_x = \Phi_J(\underline{x})QX_J + \left[T_{xx}\frac{\partial \Phi_J(x)}{\partial x}+T_{xy}\frac{\partial \Phi_J(x)}{\partial y}\right]H_J \text{ , and} \quad (7)$$

$$Rq_y = \Phi_J(\underline{x})QY_J + \left[T_{yx}\frac{\partial \Phi_J(x)}{\partial x}+T_{yy}\frac{\partial \Phi_J(x)}{\partial y}\right]H_J \text{ .} \quad (8)$$

Penalty residuals are used to enforce boundary conditions:

$$Rb_i = H_J\,\mathbf{BH}_i\,\Phi_J(\underline{x})+QX_J\,\mathbf{BX}_i\,\Phi_J(\underline{x})+QY_J\,\mathbf{BY}_i\,\Phi_J(\underline{x})-\overline{b}_i \quad \underline{x}\cap\delta\Omega_i \text{ .} (9)$$

Finally, residuals are needed to enforce continuity of hydraulic head and the component of flux normal to the element boundary at hanging nodes, JC:

$$Rch_{JC} = \frac{H_{JC^+} + H_{JC^-}}{2} - H_{JC} , \tag{10a}$$

$$Rcx_{JC} = Wax_{jc} \left[\frac{QX_{JC^+} + QX_{JC^-}}{2} - QX_{JC} \right] , \tag{10b}$$

$$Rcy_{JC} = Way_{jc} \left[\frac{QY_{JC^+} + QY_{JC^-}}{2} - QY_{JC} \right] , \tag{10c}$$

where JC^+ and JC^- are the indices of the nodes adjacent to the hanging node on the connecting element boundary and Wax_{jc} and Way_{jc} activate or inactivate a flux connection residual depending on the orientation of the element boundary.

The inner products are defined:

$$(u, v) = \int_{\Omega} u\, v\, d\Omega , \qquad <u, v>_i = \int_{\delta\Omega_i} u\, v\, d\delta\Omega . \tag{11}$$

The weighted sum of the squares of the residuals is:

$$\varepsilon = Wm\,(Rm, Rm) + Wq\,(Rq_x, Rq_x) + Wq\,(Rq_y, Rq_y) +$$

$$Wb_I <Rb_i, Rb_i>_i + Wch\, Rch_{JC}\, Rch_{JC} +$$

$$Wcq\, Wax_{jc}\, Rcx_{JC}\, Rcx_{JC} + Wcq\, Way_{jc}\, Rcy_{JC}\, Rcy_{JC} , \tag{12}$$

where Ws are weights associated with a residual type and a repeated index implies summation. If there are M active nodes in the mesh, equation (12) contains $3M$ unknowns. A system of $3M$ equations is generated by taking the derivative of equation (12) with respect to each of the unknowns and setting it to zero:

$$\frac{\partial \varepsilon}{\partial H_I} = 0 , \quad \frac{\partial \varepsilon}{\partial QX_I} = 0 , \quad \frac{\partial \varepsilon}{\partial QY_I} = 0 \qquad I = 1, M . \tag{13}$$

Equation (13) is the set of discrete equations that must be solved in the least squares groundwater flow problem. The equations will always generate symmetric, positive-definite system matrices.

DISCUSSION

Parameters are specified at all nodes except hanging nodes. The values at hanging nodes are automatically set to the average of the values of the two adjacent nodes of the connection face. Interior to elements, the parameter values are interpolated using the bilinear basis. Thus, all parameters are C^0 continuous. Since fluxes are C^0 continuous, at least in the least squares sense, the resulting velocity field will also be C^0 continuous.

The direct application of equations (13) leads to insufficient fluxes when point sources exist. The poor global behaviour is a direct result of the trial space being unable to accomodate Darcy's law in the elements touching the source location. In order to avoid the problems just described, point sources are located within elements, and the Darcy's law is not written inside those elements. When a source element is identified, all the contributions associated with Darcy's law are ignored. In order to avoid tracking difficulties in Lagrangian formulations near point sources, a local singular interpolating function is added to the trial function after the solution has been computed.

An iterative matrix solver that uses compressed storage is required. A preconditionned conjugate gradient algorithm[2] was chosen as the LESFEM flow code solver. The algorithm uses a compressed array which has length equal to the rank of the system matrix, and width equal to the number of non-zero entries found in the row with the maximum number of non-zero entries.

The LESFEM flow algorithm displays sensitivity to the choice of weights. The algorithm may not converge or may converge to a poor solution if inappropriate weighting is used. When the residuals associated with equations (6) through (8) are greatly different in magnitude, the algorithm may produce poorly conditionned matrices. The conditionning can be significantly improved by scaling time so that the transmissivities are on the order of one. Further research is needed to find optimal weighting and scaling procedures.

PUMPING WELL IN A CIRCULAR AQUIFER

The following examples simulate a pumping well in a homogeneous confined circular aquifer of radius 1000 meters. The boundary of the domain and the grid used for the simulation are illustrated in Figure 1. A pumping well is located in the element just northeast of the center. A zero hydraulic head condition is applied along the straight line segments that approximate the circular boundary, and the initial head is zero everywhere. The residual weights were $Wm = 10^4$, $Wq = 1.0$, $Wb\,1 = 10.0$, $Wch = 1.0$ and $Wcq = 10^4$.

The LESFEM drawdown results at four locations are compared to the Theis solutions in Figure 2. Near the source, the LESFEM drawdowns show some departure at early time, but, as time progresses, the solutions match well in all regions. Finally, the LESFEM heads stop decreasing, because the constant head boundary begins to influence the solution.

The steady state result can be computed in one step by setting the storage term to zero. The steady state heads of the one step solution are contoured in Figure 1. The zero head contour falls directly on the imposed zero head boundary. Two LESFEM x-direction flux profiles are compared to the flux profiles derived from the Thiem solution in Figure 3. The first profile starts at the pumping well and strikes directly west, and the second starts at the pumping well and strikes southwest at 225°. The effect of adding the singular function to produce better tracking velocities in the source element is also shown in

Figure 3.

CONCLUSION

The least squares mixed finite element (LESFEM) groundwater flow algorithm has several positive attributes. A symmetric positive-definite sytem matrix is stored in compressed form and is solved with a preconditionned conjugate gradient algorithm. Irregular domains and local grid refinement are constructed on orthogonal meshes. Use of orthogonal meshes allows simple, automated mesh generation procedures. Similarly, combining the penalty method for enforcing boundary conditions with an orthogonal mesh allows simple, automated boundary specification. The continuous velocity fields are desireable for tracking. Finally, tracking is simplified on an orthogonal mesh.

The accuracy and operational advantages of LESFEM indicate that further investigation of the procedure is warranted. In particular, formal analysis of LESFEM remains to be done. Optimal values for residual weights and optimal length and time scales need to be investigated. Finally, a more effective preconditionner for the conjugate gradient solver needs to be implemented.

This work was supported by the Air Force Office of Scientific Research, Bolling AFB, D.C.

REFERENCES

1. J. P. Laible and G. F. Pinder, 'Least squares collocation of differential equations on irregularly shaped domains using orthogonal meshes,' NUMER. METH. PAR. DIF. EQ., 5: 347-361 (1989).
2. A. Peters, B. Romunde and F. Sartoretto, 'Vectorized implementation of some MCG codes for fe solution of large groundwater flow problems,' in PROC. INTL. CONF. ON COMP. METHS. IN FLOW ANALYSIS, Sept. 5-8, Okayama, 1: 123-130 (1980).

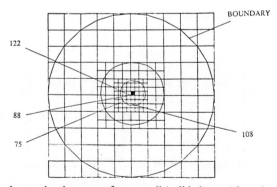

FIGURE 1 Steady state head contours from a well (solid element) in a circular aquifer.

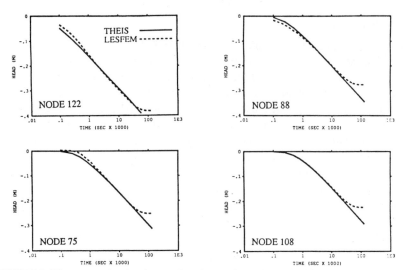

FIGURE 2 Time versus head at four nodes of analytic (solid) and LESFEM (dashed) solutions.

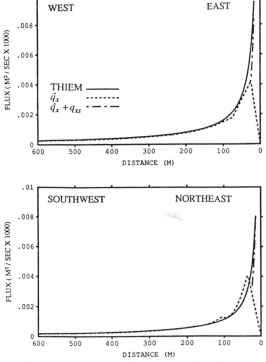

FIGURE 3 Steady state flux profiles of analytic (solid), LESFEM (dashed) and LESFEM with singular flux function (short-long dashed) solutions.

A-Posteriori Errors of Finite Element Models in Groundwater and Seepage Flow

U. Meissner(*), H. Wibbeler(**)

(*) Technische Hochschule Darmstadt, Institute for Applied Informatics and Numerical Methods in Civil Engineering, D-6100 Darmstadt, West Germany
(**) Universität Hannover, Sonderforschungsbereich 205, D-3000 Hannover, West Germany

ABSTRACT

The contribution supplies a consistent theory for the estimation of numerical errors. The presented theory is applied to the finite element analysis of groundwater and seepage flow. A least square variational principle is used to formulate a higher order approximation of finite element errors by means of the h- or p–version. The numerical method is applied to individual finite element patches in order to improve velocities and to compute mean errors within these subdomains. Numerical studies with the h–version are presented in comparison to an exact solution of a specific flow problem.

INTRODUCTION

The a–posteriori analysis of approximation errors from finite element computations is of great importance for practical applications in engineering. The reliability of numerical results depends on the magnitude of approximation errors. Therefore computer programms should supply the user with quantitative and qualitative informations about these errors and their distribution. Adaptive mesh refinement techniques can be utilized to achive optimal numerical solutions with a minimum of costs; Babuska and Gui (1986); Devloo Oden and Strouboulis (1987), Kikuchi (1986), Rank and Werner (1986). Advanced software products should be developed to process a–priori and a–posteriori informations, using new tools of knowledge based systems; Rank and Babuska (1987).

The mathematical theory for the estimation of approximation errors has been strongly improved by numerous contributions since the early publications of Babuska and Rheinboldt (1978). Zienkiewicz and Zhu (1987) proposed a simple error estimator in addition to more sophisticated error estimators and indicators published and applied earlier; Kelly, Gago and Zienkiewicz (1983); Babuska and Miller (1984). Ibid. comparative studies were carried out by definition of the effectivity index as a ratio of the predicted to the correct error values.

In this context the present formulation delivers a modification and extension of well–known theories. At first the principle of virtual work is straight forward derived from a least square variational principle, related to the commonly known energy norm; Meissner and Menzel (1989). Secondly the same principle is used to gain a higher order approximation of finite element errors in velocities. After a first finite element analysis a second patch–wise calculation is neccessary a–posteriori to compute an estimation of the errors by means of the h– or p–version. These numerical results are used to improve the original results and to calculate mean quantities of errors within each finite element patch.

GOVERNING EQUATIONS

Figure 1 illustrates the definitions for velocities,discharges,and hydaulic head to describe the two–dimensional flow problem.

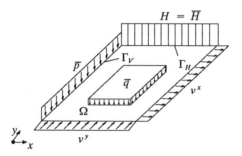

Figure 1: Definitions

The linear problem is described by the following differential equations:

Darcy's law

$$v^\alpha = -K^{\alpha\beta} H_{,\beta} \qquad in \ \Omega \tag{1}$$

with the permeabilty tensor $K^{\alpha\beta}$,

equation of continuity

$$v^\alpha|_\alpha - \bar{q} = 0 \qquad in \ \Omega \tag{2}$$

with a prescribed discharge $\bar{q}$.

To solve the problem the following boundary conditions have to be specified:

boundary conditions for the hydraulic head

$$H = \bar{H} \qquad on \ \Gamma_H \subseteq \Gamma \tag{3}$$

with a prescribed function $\bar{H}$,

boundary conditions for the velocities

$$v^\alpha n_\alpha + \bar{p} = 0 \qquad on \ \Gamma_v = \Gamma \setminus \Gamma_H \qquad (4)$$

with a prescribed line discharge $\bar{p}$, where n_α denotes the outward normal vector.

In the finite element method equations (1),(2) are valid for each subdomain and equations (3), (4) are to be modified slightly for the inclusion of inter-element boundary conditions. Equations (1), (3) hold exactly for hydraulic head models, whereas for equations (2), (4) only an approximate solution is constructed.

For analytical solutions the stream function formulation

$$v^\alpha = \epsilon^{\alpha\beta} \Psi_{,\beta} \qquad (5)$$

is useful, where $\epsilon^{\alpha\beta}$ denotes the ϵ-tensor, which transforms equations (1)–(2) into the problem

$$\Psi|_\alpha{}^\alpha = 0 \qquad (6)$$

for $K^{\alpha\beta} = const.$ and $\bar{q} \equiv 0$.

ANALYTICAL SOLUTION

Figure 2 illustrates a specific problem for which an analytical solution is given.

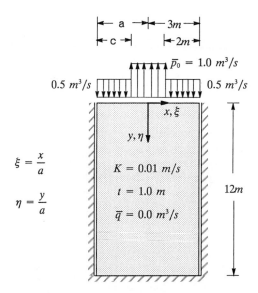

Figure 2: Semi–infinite Flow Problem

For this special case the line discharge is expanded by the Fourier–series

$$\bar{P}\,(\xi) \;=\; -\,\bar{p}_0 \sum_{n=1}^{\infty} a_n \, \cos \, (\beta_n \, \xi) \tag{7a}$$

with the Fourier–coefficients

$$a_n \;=\; -2 \, \cos \, \beta_n \, \sin \, (\beta_n \, \frac{c}{a}) \, / \, (\beta_n \, \frac{c}{a}) \;\; , \tag{7b}$$

$$\beta_n \;=\; n \, \pi \;\; , \tag{7c}$$

and the solution of equ. (6) is given by

$$\Psi \;=\; \bar{p}_0 \, a \sum_{n=1}^{\infty} \frac{a_n}{\beta_n} \, e^{-\beta_n \eta} \, \sin \, (\beta_n \, \xi) \tag{8}$$

This exact solution is further on taken as a reference for the comparison with approximate finite element solutions. Figures 6a–8a display the distribution of hydraulic head and velocities of the analytical solution on a coarse 6x8–grid.

RESIDUALS

In finite element models the hydraulic head H is approximated by polynomials $\hat{H}$ of order p within each element which satisfy the compatibility requirement (3). The approximation error is defined by

$$\tilde{H} \;=\; H - \hat{H} \;=\; O \, (h^{p+1}) \tag{9}$$

with a Taylor–series truncation error of order $O \, (h^{p+1})$ where h denotes the geometrical mesh refinement ratio. From equation (1) follows the approximation error of the velocity field

$$\tilde{v}^\alpha \;=\; v^\alpha - \hat{v}^\alpha \;=\; O \, (h^p) \qquad . \tag{10}$$

After any finite element computation the following residuals can be evaluated and displayed. By substitution of equ. (10) into equations (2) and (4) the discharge residual

$$\tilde{q} \;=\; -\hat{v}^\alpha|_\alpha + \bar{q} \;=\; O \, (h^{p-1}) \tag{11}$$

$$\;=\; \tilde{v}^\alpha|_\alpha \qquad\qquad in \; \Omega$$

and the inter–element flow residual

$$\tilde{p} \;=\; \hat{v}^\alpha \, n_\alpha + \bar{p} \;=\; O \, (h^p) \tag{12}$$

$$\;=\; -\tilde{v}^\alpha \, n_\alpha \qquad\qquad on \; \Gamma_v$$

are gained. Both types of residuals can be regarded as error discharges of the finite element system. Figures 9a and 9b display the distribution of the inter–element flow residual for two different finite element discretizations of the example.

LEAST SQUARE VARIATIONAL PRINCIPLE

The specified equations of the flow problem can be summarized mathematically in the following variational principle

$$\hat{E}\,(\hat{H}) \;=\; \frac{1}{2}\,\int\!\!\int_{\Sigma\Omega} (H_{,\alpha} - \hat{H}_{,\alpha})\,K^{\alpha\beta}\,(H_{,\beta} - \hat{H}_{,\beta})\,d\Omega \;\geq\; 0$$

$$=\; \frac{1}{2}\,\int\!\!\int_{\Sigma\Omega} \bar{H}_{,\alpha}\,K^{\alpha\beta}\,\bar{H}_{,\beta}\,d\Omega \;=\; O\,(h^p) \tag{13}$$

with the first variation

$$\delta\hat{E}\,(\hat{H}) \;=\; 0 \tag{14}$$

as a neccessary condition for the minimum of functional (13) which is always not negative as the permeability tensor $K^{\alpha\beta}$ is positive definite. In this formulation the integral terms are to be summed up over the finite elments involved as the symbols $\Sigma\Omega$, $\Sigma\Gamma$ indicate in the following.

The variation leads straight forward to an extended formulation of the well–known principle of virtual work by use of equations (2), (4)

$$\int_{\Sigma\Gamma_H} \delta\hat{H}\,(v^\alpha - \hat{v}^\alpha)\,n_\alpha\,d\Gamma \;-\; \int_{\Sigma\Gamma_v} \delta\hat{H}\,(\hat{v}^\alpha\,n_\alpha + \bar{p}\,)\,d\Gamma$$

$$-\int\!\!\int_{\Sigma\Omega} \delta\hat{H}\,(-\hat{v}^\alpha|_\alpha + \bar{q}\,)\,d\Omega \;=\; 0 \quad. \tag{15}$$

From this equation the system matrix relationship

$$\int_{\Sigma\Gamma_H} \delta\hat{H}\,v^\alpha\,n_\alpha\,d\Gamma \;=\; -\underline{\delta\hat{H}}\,(\,\underline{\hat{K}}\,\underline{\hat{H}} - \underline{\bar{Q}}\,) \tag{16}$$

is established by use of the approximation function $\hat{H}$, and the finite element solution can be calculated after the hydraulic head boundary condition (3)

$$\delta\hat{H} \;=\; \delta H \;=\; 0 \qquad on\;\Gamma_H \tag{17}$$

has been included.

Back–substitution of equ. (15) into the functional (13) leads to the energy norm $\| \hat{e} \|$

$$\frac{1}{2} \| \hat{e} \|^2 \; = \; \hat{E}_{\min} \; = \; \Pi_i^{\min} - \hat{\Pi}_i^{\min} \; \geq \; 0 \tag{18}$$

for the approximation error with the internal potential of the exact solution

$$\Pi_i^{\min} \; = \; \frac{1}{2} \int\limits_{\Sigma\,\Omega} \int H_{,\alpha} \; K^{\alpha\beta} \; H_{,\beta} \; d\Omega \tag{19}$$

and of the approximate solution

$$\hat{\Pi}_i^{\min} \; = \; \frac{1}{2} \int\limits_{\Sigma\,\Omega} \int \hat{H}_{,\alpha} \; K^{\alpha\beta} \; \hat{H}_{,\beta} \; d\Omega \quad . \tag{20}$$

Figures 6b–8b and 6c–8c display finite element solutions for two different meshes in comparison to the analytical solution of the example. Figure 10 illustrates the convergence of the approximate solutions to the exact solution by use of equ. (18).

The great advantage of the variational formulation is that in addition to the residuals (11) and (12) also the resultants of the approximation errors along boundaries Γ_H can be calculated a–posteriori from equ. (15) as sources $\tilde{T}$

$$\int\limits_{\Sigma\,\Gamma_H} \delta\hat{H} \; \bar{v}^\alpha \; n_\alpha \; d\Gamma \; = \; \int\limits_{\Sigma\,\Gamma_\nu} \delta\hat{H} \; \bar{p} \; d\Gamma \; + \; \int\limits_{\Sigma\,\Omega} \int \delta\hat{H} \; \bar{q} \; d\Omega \tag{21}$$

$$= \; - \delta\hat{H}^T \; (\; \underline{\hat{K}} \; \underline{\hat{H}} - \underline{\bar{Q}} \;) - \int\limits_{\Sigma\,\Gamma_H} \delta\hat{H} \; \hat{v}^\alpha \; n_\alpha \; d\Gamma$$

$$= \; - \delta\hat{H}^T \; \tilde{T}$$

at discrete nodal points using the matrix relationship (16) and equation (1). It is obvious that these fictive sources $\tilde{T}$ and the residuals $\bar{p}$ and $\tilde{q}$ must be in a state of self–equilibrium, which is not only valid for the whole assembled finite element system, but also for each single element and for any subdomain, e.g. individual finite element patches. Figure 3 illustrates this a–posteriori state of the error discharges.

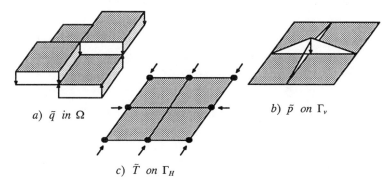

$a) \; \tilde{q} \; in \; \Omega$

$b) \; \tilde{p} \; on \; \Gamma_\nu$

$c) \; \tilde{T} \; on \; \Gamma_H$

Figure 3: Error Discharges at Patch

The discrete sources $\tilde{T}$ can be interpreted as the resultants of the weighted residuals $\tilde{q}$ and $\tilde{p}$ at the "outer" boundary Γ_H .

A-POSTERIORI ERROR ESTIMATOR

In absence of the exact solution for any arbitrary system the exact approximation error measured in the energy norm (18), (13) cannot be calculated. Therefore an additional higher order approximation with the h– or p–version is needed to achieve an estimation of the exact error . Figure 10 illustrates this goal and the use of an error estimator $\hat{\bar{E}}$, predicting the true error $\hat{\bar{E}}_{min}$ in energy.

In order to approximate the approximation error $\bar{H}_{,\alpha}$ a second functional, related to equations (13), (14), is defined

$$E \; (\hat{\bar{H}}) \;\; = \;\; \frac{1}{2} \int_\Sigma \int_\Omega (\bar{H}_{,\alpha} - \hat{\bar{H}}_{,\alpha}) \; K^{\alpha\beta} \; (\bar{H}_{,\beta} - \hat{\bar{H}}_{,\beta}) \; d\Omega \;\; \geq \;\; 0 \tag{22}$$

where $\hat{\bar{H}}$ represents a higher order contribution to equ. (9)

$$\tilde{\bar{H}} \;\; = \;\; \bar{H} - (\hat{H} + \hat{\bar{H}}) \;\; = \;\; \bar{H} - \hat{\bar{H}} \tag{23a}$$

with

$$\mathring{\hat{H}} \;\; = \;\; \hat{H} + \hat{\bar{H}} \; . \tag{23b}$$

Figure 4 illustrates two alternatives for constructing the higher order approximation $\mathring{\hat{H}}$ by a refinement of the original mesh used for the approximation $\hat{H}$. This technique is later on applied to individual patches which are cut out of the finite element system.

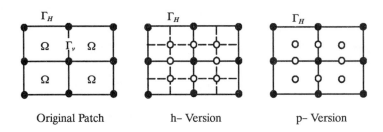

Figure 4: FEM Mesh Refinements

By the first variation

$$\delta E \left(\hat{\tilde{H}} \right) = 0 \tag{24}$$

the error between the original solution $\tilde{H}_{,\alpha}$ and the higher order approximation $\hat{\tilde{H}}_{,\alpha}$ is minimized in a least square sense, leading to the following equation by use of equations (11), (12)

$$\int\int_{\Sigma \Omega} \delta \hat{\tilde{H}}_{,\alpha} \, K^{\alpha\beta} \, \hat{\tilde{H}}_{,\beta} \, d\Omega \quad = \quad - \int_{\Sigma \Gamma_H} \delta \hat{\tilde{H}} \, \bar{v}^{\alpha} \, n_{\alpha} \, d\Gamma \tag{25}$$

$$+ \int_{\Sigma \Gamma_v} \delta \hat{\tilde{H}} \, \bar{p} \, d\Gamma + \int\int_{\Sigma \Omega} \delta \hat{\tilde{H}} \, \bar{q} \, d\Omega \quad .$$

The last two terms can be evaluated with the known residuals $\bar{q}$ and $\bar{p}$ of equations (12), (11) and the first term on the right-hand side represents the known fictive error sources of equ. (21) when $\delta \hat{\tilde{H}}$ is restricted to

$$\delta \hat{\tilde{H}} = \delta \tilde{H} = - \delta \hat{H} \qquad on \ \Gamma_H \tag{26}$$

by use of equ. (9).

In this way the new finite element relationship

$$\delta \hat{\tilde{H}}^T \, \underline{\hat{K}} \, \underline{\hat{H}} = \delta \hat{\tilde{H}}^T \, \underline{\tilde{Q}} \tag{27}$$

is set up which can be solved for the refined finite element system because $\underline{\tilde{Q}}$ includes only known quantities.

From the solution $\hat{\tilde{H}}$ an improvement of velocities is gained due to equ. (23b)

$$\overset{\circ}{v}{}^{\alpha} = \hat{v}^{\alpha} + \hat{\bar{v}}^{\alpha} \quad . \tag{28}$$

The quality of the improvement can be displayed in comparision to equations (11), (12) by

$$\mathring{\bar{q}} \;\; = \;\; -\mathring{v}^{\alpha} \,|_{\alpha} + \bar{q} \qquad\qquad in \;\; \Omega \tag{29}$$

and

$$\mathring{\bar{p}} \;\; = \;\; \mathring{v}^{\alpha} \; n_{\alpha} + \bar{p} \qquad\qquad on \;\; \Gamma_{v} \;\;. \tag{30}$$

The back–substitution of equ. (25) into equ. (22) leads to the energy norm $\| e \|$

$$\frac{1}{2} \, \| e \|^{2} \;\; = \;\; E_{min} \;\; = \;\; \hat{E}_{min} - \hat{\hat{E}} \;\; \geq \;\; 0 \tag{31}$$

for the residual approximation error with the error norm (18) and the error estimator

$$\hat{\hat{E}} \;\; = \;\; \frac{1}{2} \int\limits_{\Sigma\,\Omega}\!\!\int \hat{\hat{H}}_{,\alpha} \; K^{\alpha\beta} \; \hat{\hat{H}}_{,\beta} \; d\Omega \tag{32}$$

$$= \;\; \frac{1}{2} \; \hat{\hat{\underline{H}}}{}^{T} \; \hat{\underline{K}} \; \hat{\hat{\underline{H}}} \;\;.$$

Figure 10 illustrates the relationship between $\hat{E}$ the error norm of the original mesh, $\hat{\hat{E}}$ the error norm of a completely refined new finite element system, and $\hat{\hat{E}}$ the higher order approximation of $\hat{E}$. From the least square properties of the functionals (13), (22) it is obvious that $\hat{\hat{E}}$ represents a lower bound of $\hat{E}$ and approximates $\hat{\hat{E}}$

$$\hat{E}_{min} \;\; \geq \;\; \hat{\hat{E}} \;\; \approx \;\; \hat{\hat{E}}_{min} \;\; \geq \;\; 0 \;\;. \tag{33}$$

This means in consequence that the error estimator $\hat{\hat{E}}$ leads to a relevant estimation for the improved solution, but not for the solution $\hat{E}_{min}$ of the original system.

ERROR ESTIMATION WITHIN FINITE ELEMENT PATCHES

The numerical procedure of the algorithm estimating the approximation errors of the improved velocities (28) follows three steps:

1. Finite Element analysis of the original mesh $\qquad \hat{H}, \; \hat{v}^{\alpha}, \; (\hat{E}_{min})$,

2. Patch–wise finite element analysis of individual subdomains with the higher order approximation $\qquad\qquad\qquad\qquad\qquad \hat{\hat{H}}, \; \hat{\hat{v}}^{\alpha}, \; \hat{\hat{E}}, \; \mathring{\hat{v}}^{\alpha}$,

3. Computation of mean quantities for the estimation of errors within the subdomains.

The variational approach can be applied to any subdomain because the neccessary informations about the errors in equ. (25)

$$\bar{v}^\alpha \ on \ \Gamma_H \ , \ \ \bar{p} \ on \ \Gamma_v \ , \ \ \bar{q} \ in \ \Omega$$

can be calculated in any case. As the quality of the higher order solution $\hat{\hat{H}}$ depends very much on the quantity of informations about these errors , the minimum configuration to analyse seems to be the refined finite element patch and the maximum configuration is the refined total system. As the interest of the procedure is to gain an improvement of velocities (28) it is only neccessary to exclude all rigid body modes when solving equ. (25), because the solution $\hat{\hat{H}}_{,\alpha}$ approximates the gradient $\tilde{H}_{,\alpha}$ with respect to equ. (22). The procedure has great advantages for parallel computing because the analysis of individual finite element patches can be performed independently of each other.

For the efficiency of the algorithm the computation of the following mean quantities is proposed which are related to each finite element patch:

a) Energy density, equ. (32)

$$< \hat{\hat{\epsilon}} > \ = \ \hat{\hat{E}} \ / \int_{\Sigma\Omega} \int d\Omega \ , \tag{34}$$

b) Mean error of velocities

$$< \hat{\hat{v}} >^2 \ = \ 2 \hat{\hat{E}} \ / \int_{\Sigma\Omega} \int \| K_{\alpha\beta} \| \ d\Omega \quad , \tag{35}$$

c) Effectivity index

$$\hat{\Theta}^2 \ = \ < \hat{\hat{\epsilon}} > \ / \ < \overset{\circ}{\hat{\epsilon}} > \ , \tag{36}$$

$$< \overset{\circ}{\hat{\epsilon}} > \ = \ \overset{\circ}{\hat{E}} \ / \int_{\Sigma\Omega} \int d\Omega,$$

$$\overset{\circ}{\hat{E}} \ = \ \frac{1}{2} \int_{\Sigma\Omega} \int (v^\alpha - \overset{\circ}{\hat{v}}{}^\alpha) \ K_{\alpha\beta} \ (v^\beta - \overset{\circ}{\hat{v}}{}^\beta) \ d\Omega \ .$$

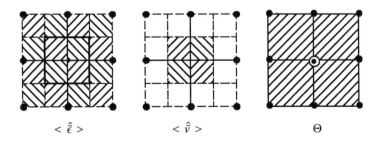

$$< \hat{\hat{\epsilon}} > \qquad\qquad < \hat{\hat{v}} > \qquad\qquad \Theta$$

Figure 5: Mean Quantities within Patches

For the representation of the following results Figure 5 illustrates the subdomains for which these mean quantities are evaluated within each patch. The mean error of velocities and the effectivity index are then associated to the centre node of the patch and their distribution is linearly interpolated.

NUMERICAL RESULTS

The patch–wise application of the h–version of the procedure is illustrated for the choosen example by a comparison of the exact solution and the approximate solution of different FEM discretizations (6x8 , 12x16–meshes).

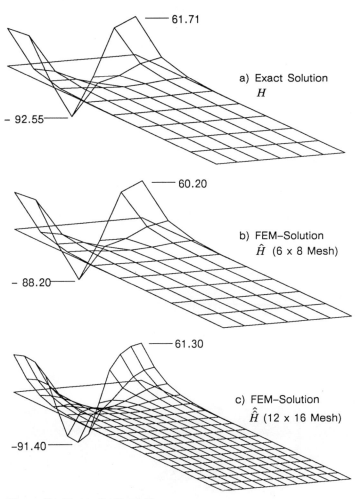

Figure 6: Hydraulic Head H

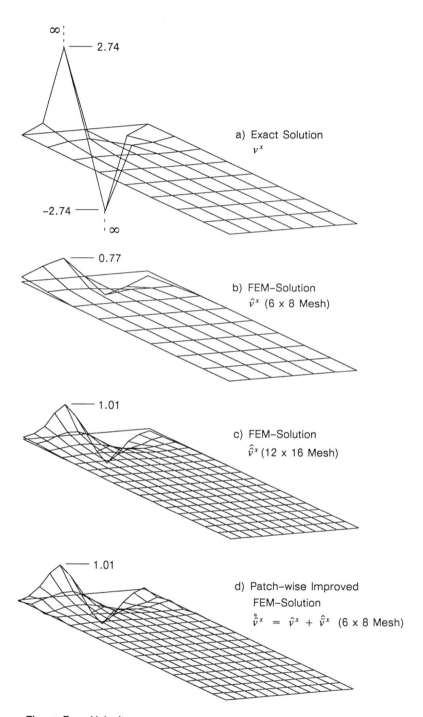

a) Exact Solution
v^x

b) FEM–Solution
$\hat{v}^x$ (6 x 8 Mesh)

c) FEM–Solution
$\hat{v}^x$ (12 x 16 Mesh)

d) Patch–wise Improved
FEM–Solution
$\overset{\circ}{\hat{v}}{}^x = \hat{v}^x + \tilde{\hat{v}}{}^x$ (6 x 8 Mesh)

Figure 7:   Velocity v^x

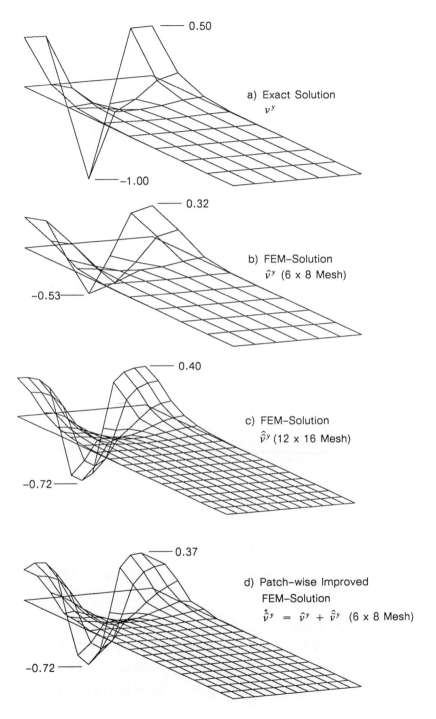

0.50

a) Exact Solution
v^y

−1.00

0.32

b) FEM−Solution
$\hat{v}^y$ (6 x 8 Mesh)

−0.53

0.40

c) FEM−Solution
$\hat{v}^y$ (12 x 16 Mesh)

−0.72

0.37

d) Patch−wise Improved
FEM−Solution
$\overset{\ast}{v}{}^y = \hat{v}^y + \hat{\bar{v}}^y$ (6 x 8 Mesh)

−0.72

Figure 8:   Velocity v^y

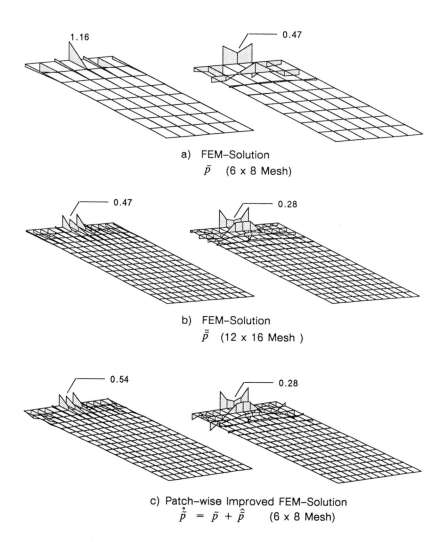

a)   FEM-Solution
$\tilde{p}$ (6 x 8 Mesh)

b)   FEM-Solution
$\bar{\tilde{p}}$ (12 x 16 Mesh)

c) Patch-wise Improved FEM-Solution
$\overset{\circ}{\tilde{p}} = \tilde{p} + \hat{\tilde{p}}$ (6 x 8 Mesh)

Figure 9 :   Residual Discharge p

Figures 7d–8d and 9c demonstrate the quality of results which is achieved by the patch–wise improvement of the 6x8–mesh results in comparison to the 12x16–solution. The additional computational effort leads to an improved solution for the velocities $\overset{\circ}{\tilde{v}}{}^{\alpha}$ and for the residuals $\overset{\circ}{\tilde{q}}$, $\overset{\circ}{\tilde{p}}$ which is very close to the solution $\bar{\tilde{q}}$, $\bar{\tilde{p}}$ of the refined total system.

Figure 10 illustrates the convergence of the FEM solutions $\hat{E}_{min}$, of the improved FEM solutions $\overset{\circ}{\hat{E}}$ and of the error estimator $\hat{\hat{E}}$ against the exact internal potential Π_i^{min} for different discretizations. It becomes obvious that the estimator $\hat{\hat{E}}$ supplies a good estimation of the improved solution $\overset{\circ}{\hat{E}}$ and the associated refined mesh solution $\overset{\circ}{\hat{E}}$.

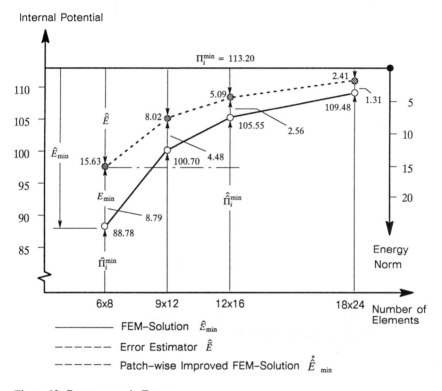

Figure 10: Convergence in Energy

The distribution of energy over the system is illustrated in Figures 11–13 by use of the patch mean values (34), (35), (36). Figure 11 underlines that also the energy density of the error estimator $< \hat{\hat{\epsilon}} >$ represents a good approximation of the exact error even in its distribution over the system. The same fact becomes evident from the effectivity index plotted in Figure 13. Its values greater than one express the upper bound quality of the error estimation for regions with reasonable high errors, Figures 11,12. It approaches zero in those parts of the system which are not strongly affected by the residuals of the finite element approximation, Figure 9.

Figure 12 is concerned with the distribution of mean errors in velocities. It expresses the same results as in Figure 11 in a slightly different manner which the engineer might be more interested in.

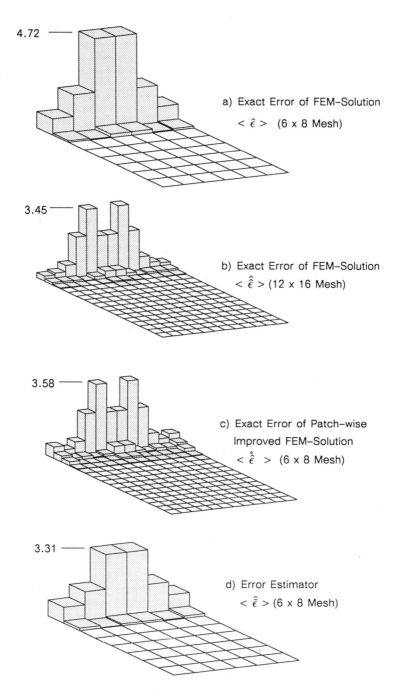

Figure 11 :   Error in Energy Density

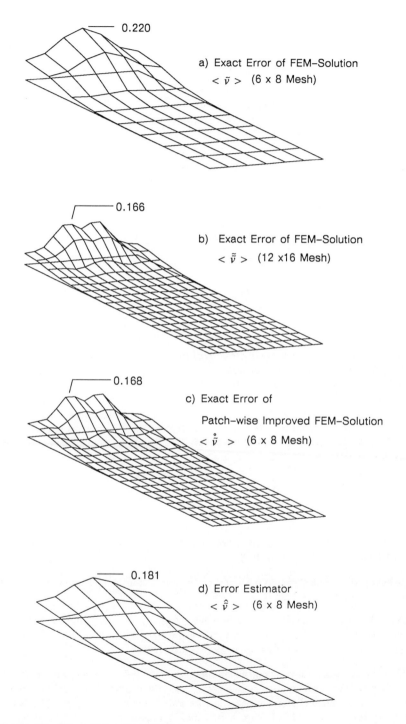

0.220

a) Exact Error of FEM−Solution
   $< \tilde{v} >$ (6 x 8 Mesh)

0.166

b) Exact Error of FEM−Solution
    $< \tilde{\tilde{v}} >$ (12 x16 Mesh)

0.168

c) Exact Error of
   Patch−wise Improved FEM−Solution
   $< \overset{\circ}{\tilde{v}} >$ (6 x 8 Mesh)

0.181

d) Error Estimator
   $< \hat{\tilde{v}} >$ (6 x 8 Mesh)

Figure 12: Mean Error of Velocities

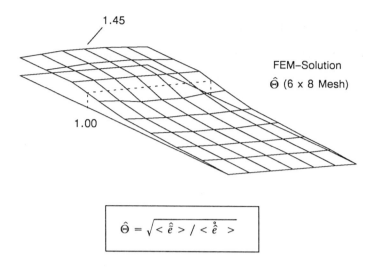

1.45

1.00

FEM–Solution
$\hat{\Theta}$ (6 x 8 Mesh)

$$\hat{\Theta} = \sqrt{<\hat{\hat{e}}> / <\overset{\circ}{\hat{e}}>}$$

Figure 13 :   Effectivity Index Θ

CONCLUSION

The proposed error estimator (32) was derived from a least square variational principle
(22), (24) including FEM approximation errors in hydraulic head gradients. Its reliability
was demonstrated by a typical example for which an analytical solution is available. The
patchwise procedure of its application can be extended to various families of finite elements. The patch strategy can be optimally used in algorithms of parallel computing. The
error estimator supplies a reliable estimation of the distribution of errors in energy and
velocities which is useful for adaptive mesh refinement techniques to get an optimal approximation of the exact solution. The developed theory is also applicable to different types and combinations of flow systems. This fact is very important for the use in general
purpose FEM program systems.

REFERENCES

I. Babuska, W. C. Rheinboldt: Error estimates for adaptive finite element computations.
SIAM J. Num. Anal. 15, p. 736 (1978).

I. Babuska, W. C. Rheinboldt: A posteriori error estimate for the finite element method.
Int. J. Num. Meth. Eng. 12, p. 1597 (1978).

D. W. Kelly, J. P. de S. R. Gago, O. C. Zienkiewicz: A posteriori error analysis and adaptive
processes in the finite element method: Part I – error analysis; Part II – adaptive mesh refinement. Int. J. Num. Meth. Eng. 19, p. 1593; p. 1621 (1983)

I. Babuska, A. Miller: The post processing approach in the finite element method–Part 1:
calculation of displacements, stresses and other higher derivations of the displacements;
Part 2: the calculation of stress intensity factors; Part 3: a posteriori error estimates and
adaptive mesh selection. Int. J. Num. meth. Eng. 20, p. 2085; p. 1111; p. 2311 (1984)

I. Babuska, W. Gui: Basic principles of feedback and adaptive approaches in the finite element method. Comp. Meth. Appl. Mech.&Eng. 55, p. 27 (1986).

N. Kikuchi: Adaptive grid–design methods for finite element analysis. Comp. Meth. Appl. Mechn. & Eng. 55, p. 129 (1986).

E. Rank, W. Werner: An adaptive finite element approach for the free surface seepage problem. Int. J. Num. Meth. Eng. 23, p. 1217 (1986)

P. Devloo, J. T. Oden, T. Strouboulis: Implementation of an adaptive refinement technique for the SUPG algorithm. Comp. Meth. Appl. Mech & Eng. 61, p. 339 (1987).

O. C. Zienkiewicz, J. Z. Zhu: A simple error estimator and adaptive procedure for practical engineering analysis. Int. J. Num. Meth. Eng. 24, p. 337 (1987)

E. Rank, I. Babuska: An expert system for the optimal mesh design in the hp–version of the finite element method. Int. J. Num. Meth. Eng. 24, p. 2087 (1987)

U. Meissner, A. Menzel: Die Methode der finiten Elemente. Springer–Verlag Berlin 1989.

Relative Efficiency of Four Parameter-Estimation Methods in Steady-State and Transient Ground-Water Flow Models

M.C. Hill

US Geological Survey, PO Box 25046, MS. 413, Lakewood, CO 80225, USA

ABSTRACT

The modified Gauss-Newton parameter-estimation method calculated parameters using less computer processing time than that required by three conjugate-direction methods in four test cases.

INTRODUCTION

Parameters in numerical ground-water flow models have been successfully estimated using nonlinear-optimization methods such as the modified Gauss-Newton (GN) method and conjugate-direction methods[17]. The relative efficiency of these methods, in terms of minimizing total computer processing time and storage, is important if large, transient problems are to be calibrated by these methods. Evaluation of relative efficiency is difficult because the calculations required by the methods may be executed many different ways and because the performance of the methods is problem-dependent. The efficiency of some parameter-estimation methods has been evaluated for transient, one-dimensional ground-water flow problems by Gavalas and others[3], and for steady-state, two-dimensional ground-water flow problems by Cooley[2]. Gavalas and others[3] did not use the more efficient Newton method[1] to calculate step size in their conjugate-direction method, so their results, which generally favor GN, are not conclusive. Cooley[2] reported that GN was much more efficient than the Fletcher-Reeves (FR) and quasi-Newton (QN) regression methods for four test cases, but the single decomposition of the direct solver he used to solve for hydraulic heads and sensitivities for each iteration may not be as effective in transient problems.

This paper investigates the relative efficiency of GN and three conjugate-direction parameter-estimation methods on two-dimensional, steady-state and transient ground-water flow test cases. The steady-state test cases are included to compare the performance of the algorithm with published examples. The three conjugate-direction methods are FR, QN, and combination Fletcher-Reeves quasi-Newton (FR-QN). All three are combined

with Newton's method of calculating step size. The numerical
ground-water flow model is described by McDonald and Harbaugh[4].

PARAMETER-ESTIMATION METHODS

The parameter-estimation methods used in this paper are
discussed briefly here; see the cited literature for more
information and additional references.

The objective function $(S(\underline{b}))$ minimized by all the regression
methods is:

$$S(\underline{b}) = (\underline{h}* - \underline{h})^T \underline{w} (\underline{h}* - \underline{h}) + (\underline{b}* - \underline{b})^T \underline{u} (\underline{b}* - \underline{b}) \qquad (1)$$

where T indicates the transpose of a vector; $\underline{h}*$ and $\underline{h}$ are
vectors with ND components equal to the observed and calculated
hydraulic heads, respectively; $\underline{b}*$ and $\underline{b}$ are vectors with NP
components equal to the prior estimates and calculated values of
the parameters, respectively; and $\underline{w}$ and $\underline{u}$ are known weight
matrices that indicate the accuracy and correlation of the
observed hydraulic heads and prior estimates on the parameters,
respectively. For this paper, $\underline{w}$ and $\underline{u}$ are diagonal matrices.

GN is discussed by Cooley[2] and Yeh[7]. In GN, parameter
estimation is accomplished by iteratively updating the
parameters using $\underline{d}$, the solution of the normal equations:

$$(\underline{X}^T \underline{w} \, \underline{X} + \underline{R}^T \underline{u} \, \underline{R} + m\underline{I}) \, \underline{d} = \underline{X}^T \underline{w} \, (\underline{h}* - \underline{h}) + \underline{R}^T \underline{u} \, (\underline{b}* - \underline{b}) \qquad (2)$$

where $\underline{X}$ is an ND by NP matrix of sensitivities (the derivative
of hydraulic head at observation locations with respect to the
parameters) evaluated for the most recent parameter estimates; $\underline{R}$
is an NP by NP matrix with diagonal elements equal to 1.0 for
parameters with prior estimates and all other entries equal to
0.0; m is a Marquardt parameter; and $\underline{I}$ is an NP by NP identity
matrix. To produce a better conditioned problem, Eq. (2) is
scaled using the diagonal elements of the matrix on the left-
hand side before being solved. Convergence is achieved when the
largest absolute value of the components of $\underline{d}$ is less than a
specified convergence criterion, which was equal to 0.01 in this
paper. To calculate hydraulic heads and sensitivities using the
numerical flow model, each GN iteration requires NP+1 solutions
of matrix equations of order NN, the number of nodes in the
grid.

All three conjugate-direction methods are discussed by Carrera
and Neuman[1], and FR and QN are discussed by Cooley[2]. In
conjugate-direction methods, parameter estimation is
accomplished using the gradient vector, which has NP components
equal to the derivatives of Eq. (1) with respect to the
parameters. The gradient vector is calculated efficiently using
the adjoint-state method[6] and is modified by approximations of
the inverse of the matrix on the left-hand side of Eq. (2) to
define the direction of parameter change, $\underline{d}'$. FR and QN differ
in how the matrix inverse is approximated. In the combined
method, FR is used in the first iterations, followed by QN. New
parameter estimates are calculated as $\underline{b}_{new} = \underline{b}_{old} + a\underline{d}'$, where a
is a step size calculated by Newton's method. To produce a

better posed problem, the estimated parameters are scaled as
defined by the user so all estimated parameters have similar
values. The performance of conjugate-direction methods is
strongly influenced by the scaling used, and optimal scaling is
determined by trial and error. Convergence is achieved when the
maximum absolute value of the components of the gradient vector
and of $a\underline{d}'$ are less than a specified convergence criterion,
which is equal to 0.01 in this paper. To calculate the
hydraulic head, the adjoint states, and a function required by
Newton's method, each iteration of a conjugate-direction
regression method requires three solutions of matrix equations
of order NN.

The computer processing time required by either GN or the
conjugate-direction methods mostly is spent solving the NP+1 or
the three matrix-equation solutions required at each iteration,
respectively, as discussed above. Although a GN iteration
requires more matrix-equation solutions than a conjugate-
direction iteration when the number of parameters is larger than
two, the conjugate-direction method is less effective at
locating the optimum parameter values; therefore, in groundwater flow problems, 5 to 10 times more iterations commonly are
required to achieve convergence[2]. If all the required matrix-
equation solutions used the same amount of computer processing
time, the most efficient regression method could be identified
by comparing the number of solutions per iteration multiplied by
the number of iterations required for convergence. However, if
a sequence of solutions is steady state or has the same time-
step size, the matrices involved in the solutions are identical,
and, if a direct solver is used, the entire sequence can be
solved using one matrix decomposition. Decomposition sometimes
used more than 75% of the computer time required by the direct
solver for the test cases in this paper, so using one
decomposition was very advantageous. In GN, the NP+1 solutions
at steady state and at each time step produce such sequences,
and the sequences are longer if the time-step size is constant.
In conjugate-direction methods, adjoint states are solved
backwards in time, so appropriate sequences occur only for
steady-state problems or when the flow problem is linear and the
time-step size is constant. The direct solver used in this work
is a D4 solver. Because round-off error prevents the D4 solver
from being used for problems with more than a few thousand grid
nodes, results produced using an iterative conjugate-gradient
solver with the modified incomplete Cholesky preconditioner
(PCG) with convergence criterion equal to 0.0001 also are
discussed.

The precision of the arrays in the FORTRAN algorithm can
significantly affect the performance of the parameter-estimation
methods[35]. In this paper, the following arrays are double-
precision: the solution of all matrix equations of order NN; the
left-hand-side matrix and $\underline{d}$ of equation (2); the approximated
inverse matrix of QN; and the gradient, an ND-length vector used
to calculate the step size, and $\underline{d}'$ of the conjugate-direction
methods. All other arrays are single-precision. Double-
precision scalar variables are used to improve accuracy where

possible. Lack of precision is expected to cause the most
problems in conjugate-direction regression.

TEST CASES

Test cases were chosen from the literature to verify the
performance of the regression methods and to keep the
descriptions included here brief. See the cited literature for
full descriptions of the test cases.

Test cases 1a and 1b are Cooley's[2] second test problem, which is
steady state, and a transient version of that test problem,
respectively. The aquifer fills most of a 13,200-foot by 14,200-
foot area and is simulated using a 17-row by 16-column grid. In
both test cases, 14 parameters are estimated; 9 have prior
estimates. In the steady-state test case, there are 32
hydraulic-head observations corrupted with $N(0,1)$ noise.
Transient flow conditions are created by turning the two wells
off and creating 3 more hydraulic-head observations at each of
10 locations used in the steady-state simulation. The three
observations occur at 1.52 days, 5.38 days, and 30.0 days; two
locations are in each of zones 1 and 3; and six, including the
two wells, are in zone 2. There are 15 time steps in the
simulation, and time-step size is increased by a factor of 1.2
each time step, starting with a value of 0.42 days. Values of
the storage coefficients for zones 1 through 3 are 0.00075,
0.0005, and 0.0006. Observations in the transient test case
were not corrupted with noise. The scaling used for conjugate-
direction methods is the same as described by Cooley[2].

Test cases 2a and 2b are Carrera and Neuman's[1] steady-state and
transient two-dimensional theoretical test cases, respectively.
The aquifer fills a 6-km square area and is simulated using an
8-row by 8-column grid. The estimated parameters are 9
transmissivity values, which are scaled by dividing by 10 and
taking their natural logs. Initial values were 1.1 and after
accounting for differences in scaling, are about 10% larger than
the initial values used by Carrera and Neuman[1]. Uncorrupted
observations were used in both test cases. A storage-
coefficient value of 0.0005 was used in this paper.

RESULTS

The results of the simulations are listed in table 1. When they
converged, all the parameter-estimation methods produced similar
final values of the parameters and objective function, so these
values are not included in the table. To put the array storage
requirements in perspective, the storage used when solving only
for hydraulic heads for the four test cases with the D4 solver
is 5732, 6004, 1328, and 1388. The FR-QN results listed are for
the FR iterations that produced fastest convergence, as
determined by trial and error. FR-QN results are not reported
for the transient problems because the long computer processing
times required by FR and QN indicate that even if FR-QN did
converge, the computer processing time would far exceed that
required by GN.

The number of iterations required by test cases 1a and 2a indicates that the model developed for this paper performs similarly to the model used by Cooley[2] and somewhat differently than the model used by Carrera and Neuman[1]. The difference in the iterations required for test case 2a could result from using different convergence criteria and initial parameter estimates and from differences in the precision of arrays in the two models.

TABLE 1: Results of Test Cases Using D4 and PCG Solvers
[*, parameter estimation did not converge in 100 iterations, the computer processing time for 50 iterations is listed;--, results not reported]

Method	Solver	Test Case			
		1a	1b	2a	2b
		Number of Iterations			
GN	D4	3 [1](3)	4	9	9
	PCG	3	4	9	9
FR	D4	* [1](*)	*	* [1](86)	*
	PCG	*	*	*	*
QN	D4	40 [1](37)	*	62 [1](53)	*
	PCG	40	*	44	*
FR-QN	D4	[2]25;40	--	[2]20;43 [1](-;51)	--
	PCG	[2]25;40	--	*	--
		Computer Processing Time, in Seconds[3]			
GN	D4	5	87	3	43
	PCG	29	239	20	163
FR	D4	36	380	10	105
	PCG	96	654	30	291
QN	D4	29	379	12	108
	PCG	70	612	25	249
FR-QN	D4	29	--	9	--
	PCG	72	--	--	--
		Array Storage Requirements[4]			
GN	D4	9892	13069	2942	4482
	PCG	8350	11527	3051	4591
FR	D4	9593	12573	2772	4439
	PCG	8051	11031	2881	4548
QN	D4	9805	12785	2862	4529
	PCG	8263	11243	2971	4638
FR-QN	D4	9805	12785	2862	4529
	PCG	8263	11243	2971	4638

[1] Numbers in parentheses are the iterations to convergence reported for 1a by Cooley[2], and for 2a by Carrera and Neuman[1].
[2] FR and total iterations.
[3] On a PR1ME 6655 (Use of the brand name in this paper is for identification purposes only and does not constitute endorsement by the U.S. Geological Survey).
[4] In equivalent single-precision words of central memory.

For simulations that did not converge in 100 iterations, the computer processing time for 50 iterations is listed for reference. The long computer processing times required in these simulations indicate that changes to induce convergence, such as

making more model arrays double precision, would not produce a
parameter-estimation method that was more efficient than GN.

The number of iterations required to achieve convergence with
conjugate-direction methods sometimes depended on the solver
used for the three matrix equations of order NN. For example,
for test case 2a, QN required 62 iterations using the D4 solver,
but only 44 using PCG. The variation in performance with the
solver may result from lack of double   precision in parts of
the solver and in the model in general. The number of
iterations required by GN was not affected by the solver.

DISCUSSION AND CONCLUSIONS

The results indicate that the modified Gauss-Newton method
estimates parameters using much less computer processing time
(less than 50% with the direct D4 solver; less than 80% with
PCG), and slightly more array storage (as much as 6%) than the
three conjugate-direction methods in four test cases that
represent steady-state or transient ground-water flow in two
two-dimensional theoretical aquifer systems. The modified
Gauss-Newton method also has the advantage of not requiring
user-defined scaling and was not as susceptible as the
conjugate-direction methods to numerical problems, such as
limited precision of model arrays. These results suggest that,
unless the small amount of extra array storage required is
unavailable, the modified Gauss-Newton method is more effective
for the problems considered. The fact that this conclusion is
valid when using the PCG solver suggests that the modified
Gauss-Newton method would also be more efficient for larger
problems than those studied.

REFERENCES CITED

1. Carrera, Jesus and Neuman, S.P. (1986) Estimation of Aquifer
   Parameters Under Transient and Steady State Conditions, Water
   Resources Research, 22(2):199-242.
2. Cooley, R.L. (1985) A Comparison of Several Methods of
   Solving Nonlinear Regression Groundwater Flow Problems, Water
   Resources Research, 21(10):1525-1538.
3. Gavalas, G.R., Shah, P.C., and Seinfeld, J.H. (1976)
   Reservoir History Matching by Baysian Estimation, Soc. Pet.
   Eng. J., 16(6):337-350.
4. McDonald, M.G. and Harbaugh, A.W. (1988) A Modular Three-
   Dimensional Finite-Difference Ground-Water Flow Model, U.S.
   Geological Survey Techniques of Water Resouces
   Investigations, Bk.6, Ch. Al, 548p.
5. Stewart, G.W. (1973) Introduction to Matrix Computations,
   Academic Press, 441p.
6. Townley, L.R. and Wilson, J.L. (1985) Computationally
   Efficient Algorithms for Parameter Estimation and Uncertainty
   Propagation in Numerical Models of Groundwater Flow, Water
   Resources Research, 21(12):1851-1860.
7. Yeh, W-G. (1986) Review of Parameter Identification
   Procedures in Groundwater Hydrology: the Inverse Problem,
   Water Resouces Research, 22(2):95-108.

Adaptive Multigrid Method for Fluid Flow in Porous Medium

L. Ferragut(*), F. Pétriz(**)

(*) Dpto. de Matemática Aplicada y Métodos Informáticos, ETSI Minas, UPM Madrid, Spain

(**) Dpto. de Matemática Aplicada, ETSI Industriales, U. de Zaragoza, Spain

ABSTRACT

In this paper we deal with the finite element h-adaptive method for typical second order elliptic problems which modelizes the flow in porous medium. We consider first a simple and reliable error estimator that leads naturally to the resolution of a complementary problem. The local error indicators allow us to design an adaptive strategy and the built multigrid structure is used to solve both, the original system of finite element equations and the complementary system. Finally we give a representative example to check the validity of the method.

STATEMENT OF THE PROBLEM

In this paper Ω shall be a given, bounded domain in $\mathbb{R}^d$ with boundary $\Gamma = \Gamma_o \cup \Gamma_1$. We have to solve the second order boundary value problem:

$$-\frac{\partial}{\partial x_i} \left(a_{ij} \frac{\partial u}{\partial x_j} \right) = f \qquad \text{in } \Omega \qquad (1)$$

$$u = g_o \qquad \text{on } \Gamma_o \qquad (2)$$

$$a_{ij} \frac{\partial u}{\partial x_j} n_i = g_1 \qquad \text{on } \Gamma_1 \qquad (3)$$

where the matrix a is symmetrical and positive. To this problem we shall associate the energy norm:

$$\| v \|_E = \left(\int_\Omega a_{ij} \frac{\partial v}{\partial x_i} \frac{\partial v}{\partial x_j} \right)^{1/2} \qquad (4)$$

FINITE ELEMENT APROXIMATION AND ERROR ESTIMATION

Let be V a finite dimensional space of peace-wise polynomes of degree k over each triangle T in a triangulation of Ω.

We search for uh such that $u_h - u_{ho} \in V$ and for all v in V verify:

$$\int_\Omega a_{ij} \frac{\partial uh}{\partial x_i} \frac{\partial v}{\partial x_j} = \int_\Omega fv + \int_{\Gamma_1} g_1 v \qquad (5)$$

where uho is a function of V such that uho = g_1 on Γ_1.

We look for an estimation of the error $\| u-uh \|_E$; consider the affine variety of vector functions:

$$\mathbb{V}^{f \cdot g_1} = \{ q; \text{ div } q + f = 0 , \int_{\Gamma_1} q \cdot n \, v = \int_{\Gamma_1} g_1 v \quad \forall v \}$$

Then we have the following error estimation which proove can be found in Ferragut, Pétriz and Thomas [1]:

For all $p \in \mathbb{V}^{f, g_1}$ $\| u_h - u \|_E =$

$$(\int_\Omega A_{ik}(p_i - a_{ij}\frac{\partial uh}{\partial x_j}) (p_k - a_{kl}\frac{\partial uh}{\partial x_l}))^{1/2} \qquad (6)$$

where A is the inverse matrix of a.

We look for a value of p the nearest, as possible, to the vector function ($a_{ij} \partial u/\partial x_j$), this leads to the introduction of the mixed formulation of the problem (1),(2),(3); find the couple (p,u) such that:

$$\int_\Omega A_{ij} p_i q_j + u \frac{\partial q_i}{\partial x_i} = \int_\Gamma q_i n_i g_o \qquad \forall q \qquad (7)$$

$$\int_\Omega v (\frac{\partial p_i}{\partial x_i} + f) = 0 \qquad \forall v \qquad (8)$$

$$\int_\Gamma p_i n_i \varphi = \int_{\Gamma_1} g_1 \varphi \qquad \forall \varphi \qquad (9)$$

In practice we approach the mixed problem by finite elements, following Thomas [2], or Raviart and Thomas [3].

REFINEMENT ALGORITHM

Taking the value:

$$\eta_T = (\int_T A_{ik}(p_i - a_{ij}\frac{\partial uh}{\partial x_j}) \ (p_k - a_{kl}\frac{\partial uh}{\partial x_l}))^{1/2} \qquad (10)$$

as error indicator, we refine the mesh according to the following strategy:

1. If $\eta_T \geq \gamma \ \eta_{opt}$ we refine by four elements, where γ is near the unity and $\eta_{opt}^2 = 1/N \sum \eta_T^2$; N is the actual number of elements before refinement.

2. If $\eta_T > \beta \ \eta_{opt}$ with $\beta \approx 0.5$ we refine by two elements.

With this two kind of subdivision we reach quicly a mesh with uniform error distribution. For the element subdivision, the algorithm of Rivara [4] has been employed.

MULTIGRID METHOD

To solve both the standard discrete variational problem (5) and the mixed variational problem (7),(8) and (9) we use the multigrid methods with the secuence of nested subspaces, corresponding to the nested grids generated by the refinement algorithm.

For the system (5) we use as smoothing method the diagonal preconditioned conjugate gradient method; the matrix of the change of finite element basis is used as restriction operator and the interpolation operator is used as prolongation.

On the other hand the mixed discrete system correspondig to (7),(8) and (9) could be written with matrix notation as:

$$A \ p \ + \ B^t \ u \ = \ b_1 \qquad (11)$$

$$B \ p \qquad\qquad = \ b_2 \qquad (12)$$

We have used a smoother which could be considered as the following Arrow-Hurwicz-Uzawa iteration, see Fortin and Glowinski [5]:

$$p^{n+1} = p^n - \omega S^{-1}(A_r p^n + B^t u^n - b_1 - rB^t b_2) \qquad (13)$$

$$u^{n+1} = u^n + \rho(Bp^{n+1} - b_2) \qquad (14)$$

where $A = A + rB^t B$, S is an auxiliary operator and ω an acceleration parameter; in our case ωS is the operator associated with a few steps of the diagonal preconditioned conjugate gradient method.

Concerning the system (11), (12) the restriction and prolongation operator are the L^2- projection and the injection respectively for the Lagrange multiplier u, as we do not have any continuity requirement for this variable. For the flows p we can construct easily the interpolation operator, when we split by his middle point one side of one element, the flow values of the two new sides are taken to be one half of the flow value of his father; the values trough a new side at the interior of a triangle can also be calculated using the divergence theorem. The restriction operator for the flows p to pass from a fine grid to a coarse one is straightforward; we have only to sum up in each side the flow values of the two daughter sides.

APLICATION

Let us consider the problem which modelizes the flow in a porous medium in the domain shown in figure 1, corresponding to the equations (1),(2),(3) with $f = 0$ with the boundary conditions specified in the figure and the conductivity matrix a being the identity. The segment AG represents a vertical plate. In the figure 2 the fifth grid adapted is sketched and the contours corresponding to the values ranging from 1.5 to 2 with a 0.05 interval. In table 1 the main features of the adopted multigrid adaptive process are presented. β stands for the rate of convergence which corresponds to the expression $\| error \| \leq C N^{-\beta}$, N being the number of degrees of freedom of the standar variational aproximated problem. For this case, the theorical value of β for a uniform refinement is 0.25. The improvement in the rate of convergence can be seen.

Regarding the multigrid algorithm, the tolerance in the solution of the two problems is of order 10^{-3}. For every grid the convergence has been reached for the number of iterations indicated in table 1. The number of iterations stabilizes on refining the grid.

In most cases,when dealing with fluid flow in porous
medium, the flows values obtained by the   resolution
of the mixed problem are   sougth, as   well   as   the
piezometric head u, this fact justify the use of the
present adaptive finite element procedure.

Table 1. Perfomances of the adaptive multigrid method

Mesh	Iter. s. p.	Iter. m. p.	Equat. p.head.	Equat. flows	Error	β
1	1	1	6	19	0.28	
						0.2685
2	2	1	21	58	0.20	
						0.2988
3	1	2	55	158	0.15	
						0.5197
4	1	2	120	345	0.10	
						0.3541
5	1	1	251	730	0.077	

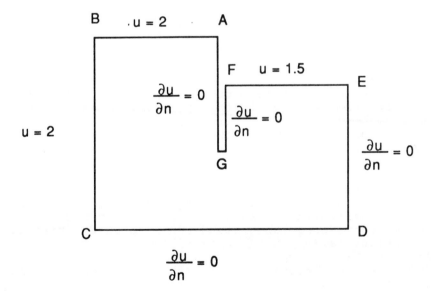

Fig.1 Flow problem in porous medium

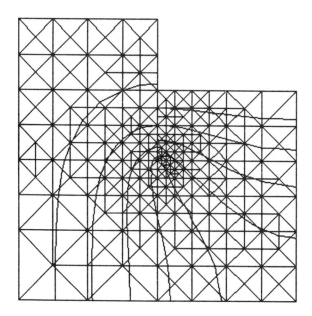

Fig.2 Fifth adapted grid and contours.

REFERENCES

1. Ferragut, L., Pétriz, F., and Thomas, J.M. Error
   estimation, mixed formulation and adaptive
   multigrid thecniques. To appear.

2. Thomas, J.M. Sur l'analyse numerique des methodes
   d'elements finis hybrides et mixtes. These détat.
   Université Pierre et Marie Curie, Paris, 1977.

3. Raviart, P.A. and Thomas, J.M. A mixed finite
   element method for second order elliptic problems.
   (Ed. Gallini, I., Magenes, E.), pp. 292 to 315,
   Lecture Notes in Math. 606. Springer Verlag,
   Berlin, 1977.

4. Rivara, C. Algorithm for refining triangular grids
   suitable for adaptive and multigrid techniques,
   Int. J. Num. Meth. in Eng. Vol.20, pp.
   745-756, 1984.

5. Fortin, M. and Glowinski R. Methodes de Lagrangien
   Augmenté, Dunod, Paris, 1982.

Triangular Finite Element Meshes and their Application in Ground-Water Research

J. Buys(*), J.F. Botha(*), H.J. Messerschmidt(**)
(*) Institute for Ground-water Studies and
(**) University of Orange Free State,
Bloemfontein, Republic of South Africa

ABSTRACT

An algorithm for the efficient construction of a triangular finite element mesh, will be discussed, showing how to handle information about any discontinuities like dykes or mountain ranges, in such a way that it can be incorporated in the final contour map. Since triangular finite element meshes are well-known in finite element models of ground-water resources, application of the algorithm is illustrated through two examples of different fields of interest: drawing of contour maps and the computation of ground-water velocities.

INTRODUCTION

Contour maps of a regionalised variable can be computed automatically in several ways. One method often used in commercial available packages, is to interpolate the observed values to a square mesh and use linear products of one-dimensional interpolation polynomials to construct the contour map. If the data are not closely spaced, which is usually the case in the earth sciences, the interpolation to a square grid can introduce considerable errors in the final map.

It is always possible to connect a set of points on a two-dimensional domain with a triangular mesh. Moreover, the edges of a triangular mesh can be more easily aligned with the linear structures, causing the discontinuities in the earth sciences. For example, it often happens that the ground-water levels across dykes and similar structures can differ by tens of metres. No contour map can reflect this phenomenon, unless special provision has been made in the basic algorithm to account for it. Fortunately, the positions of such discontinuities are mostly known. We, therefore, elected to base our contouring algorithm on a triangular mesh rather than a square one.

There exist various methods by which a triangular mesh can be constructed (Lyness & Asquith [1], McLain [2] and Watson [3]). However, not all of them can easily be computerised. Moreover, see Botha and Pinder [4], the accuracy of triangular interpolation is adversely affected if an angle of the triangle is small, as is the case in an elongated triangle. What one would like to have, is an algorithm that can construct a triangular mesh that is as nearly equiangular as possible. Such an algorithm, described by Green and Sibson [5], forms the basis of our algorithm.

TRIANGULATION

We use the recursive algorithm, described by Green & Sibson [5], to find a triangulation for a finite set of distinct points in two dimensions. This triangulation, called the Delaunay triangulation, is as near as possible equiangular (Sibson [6]). By making an addition to the algorithm, consisting of a pair of rules to handle any degeneracies (Buys, Botha & Messerschmidt [7]), it ensures that the subsequent triangulation is not ambiguous and fully triangulated. A degeneracy occurs when four, or more, points all lie on the same circle. In this form the algorithm can also be used to compute a triangular finite element mesh for the nodes of a rectangular grid, which is fully degenerate.

It is obvious that a discontinuity, with known position, can always be incorporated into the triangulation of the data points. The triangles, crossed by the discontinuity, are constructed between data points on opposite sides of the discontinuity. All contour lines which pass through these triangles, will consequently be influenced by values on opposite sides of the discontinuity. Since it is important that the contour lines stop directly at the discontinuity, each triangle crossed by a discontinuity is subdivided into two. The crossing points are added as extra points to the set of data points. When the end points of the discontinuity are also added, it yields a triangulation where the edges joining the extra points, form a boundary between the triangles on both sides of the discontinuity.

FINITE ELEMENT OUTPUT

The information about the triangles, or to borrow a term from the method of finite elements, is given in terms of the element topology - a list of elements and their three associated node numbers. In order to draw a complete continuous contour, rather than element by element, a list of the three adjacent elements is also needed (Lyness & Asquith [1]). This information, called the mesh map, together with the nodal information, is derived from the triangulation algorithm.

The mesh map plays an important role in the handling of discontinuities. Elements on the mesh edge do not have elements adjacent to all three edges. For these edges, the mesh map has empty entries. Elements with one or more edges next to a discontinuity, can be treated similarly. With this technique, the contour plotting algorithm will interrupt the contour line when it reaches this edge. Contour lines that go around the end points of the discontinuity are not influenced.

EXTRA POINTS

Elements that share an edge on the discontinuity, also share two extra points as nodal points. The extra points on the discontinuity, added to the triangulation, are not actual data points. In order to draw contours through these elements, it will be necessary to estimate z-values for each of these extra points, depending on the side of the discontinuity through which the contour line is followed. The data points on each side are used to estimate these values, one for the one side of the discontinuity and the other for the opposite side.

In order to select the right value for an extra point, when contours are followed through the elements, the elements next to a discontinuity must be associated to one and only one side of it. This is achieved by dividing the area into zones, grouping the data points and elements according to their position with respect to the discontinuity, and then attaching this information to the finite element output.

To estimate the z-values for the extra points, the elements in each zone are scanned in turn, finding all the extra points on the boundary of this zone. Then, for each of these points, the three nearest data points in the zone are selected and used to estimate the z-value. The z-values, associated with end points of singularities which cannot be associated with a particular side of a discontinuity, are estimated similarly, but now using the three nearest data points, irrespective of their zone.

An appealing feature of this method is that the zones are only used to estimate the z-values of the extra points, and to follow contours through the elements next to the discontinuity. It does not have any effect on the contour lines away from the discontinuities, even for applications where more than one discontinuity is present in the same region. The flow around the end points is also not affected.

COMPUTATION OF THE CONTOUR LINES

The contour plotting algorithm that processes the contour lines, must make special provision for the various z-values that are associated with the extra points, Whenever a z-value for an extra point is needed, the zone number of the element is used to collect the correct z-value. No other arrangements have to be made, because the edges of the elements that lie on the discontinuity, are already marked as mesh edges in the finite element output.

When calculating the piecewise linear segments, the open contour segments, starting at a mesh edge, are processed first, following by the closed contour segments (Lyness & Asquith [1]). Adjacent elements are found from the mesh map. If a contour passes exactly through a node, the contour value is decremented temporarily to find the next edge crossed by the contour segment. This assures that the line still passes through the node, without any problems in finding the adjacent element.

Curve fitting techniques can be used to smooth the piecewise linear representation produced. We have used a weighted least square cubic-spline approximation to smooth the contours (De Boor [8]). The coordinates on the element edges are used as knots, and normal linear interpolation is used to find enough coordinates at equal intervals between the knots. If 'meandering' of contours, in coarse mesh regions, occurs, more knots are introduced. This is done by finding more coordinates between the knots and specify some of these also as knots. The number of knots can be increased until no further 'meandering' occurs.

VELOCITY COMPUTATION

The triangulation computed for a finite set of distinct points, can also be used to compute the ground-water velocities for each element. It can then be represented graphically by means of a vector, showing the direction and relative magnitude, with respect to each other. Over each triangle, a plane surface is used to approximate the z-values; the gradient of this plane is then taken as the gradient. Given the values of porosity and conductivity, the fluid velocity can be calculated from Darcy's law (Pinder, Celia & Gray [9]). The resulting velocity vector is drawn at the centroid of the triangle.

ILLUSTRATION ON GROUND-WATER DATA

As an example, we have chosen a region where the presence of the two geological dykes affects the flow of ground water in the region. The effect of including the dykes into the contouring algorithm can be judged by comparing the contours in Figures 1 and 2. Of particular importance is the difference in the positions of contour lines on opposite sides of the dykes.

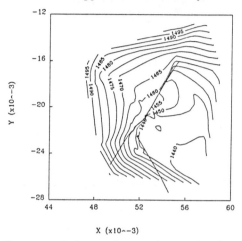

Figure 1   Contours of observed ground-water levels without taking notice of the presence of dykes.

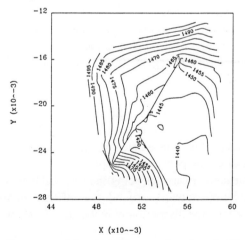

Figure 2   Contours of the ground-water levels as in Figure 1, but this time taking the position of dykes into account.

Figure 3 shows the velocity vectors computed for an area surrounding a pan used to infiltrate partially purified water into a phreatic aquifer.

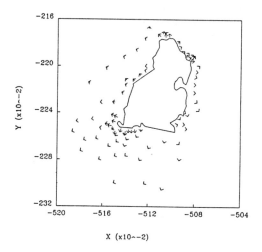

Figure 3   Velocity vectors computed for the ground-water levels of an area surrounding a pan.

CONCLUSION

The method described in this paper, combines the advantage of drawing contours from an irregular triangular grid with the advantage of adding information about discontinuities in the region. It also provides a method for drawing very accurate contour lines that can clearly reflect the influence of geographical discontinuities in the region, if the position of the discontinuities is added.

A disadvantage of the package in its present form is the absence of a facility to interpolate to points inside the triangles (Watson & Philip [10]). Such a facility would be very useful in finite element work and will be added shortly to the package. Otherwise the package has proved its worth in numerous applications. Two particular examples being the establishment of an efficient ground-water observation network around the infiltration pan shown in Figure 3 and in delineating the discontinuity in the aquifer shown in Figures 1 and 2.

REFERENCES

1.   Lyness, J.F. and Asquith, L.A. A Simple Contour Plotting Program for Finite Element Output, Software for Engineering Problems. (Ed.

Adey, R.A.), Computational Mechanics, Southampton, 1985, pp. 86–94.

2. McLain, D.H. Drawing Contours from arbitrary Data Points. Computer Journal, Vol. 17, 1974, pp. 318-324.

3. Watson, D.F. ACORD - Automatic Contouring of Raw Data. Computers and Geosciences, Vol. 8, 1982, pp. 97-101.

4. Botha, J.F. and Pinder, G.F. Fundamental Concepts in the Numerical Solution of Differential Euqations. John Wiley and Sons, New York, 1983.

5. Green, P.J. and Sibson, R. Computing Dirichlet Tessellations in the Plane. Computer Journal, Vol. 21, 1978, pp. 168-173.

6. Sibson, R. Locally Equiangular Triangulation. Computer Journal, Vol. 21, 1978, pp. 243-245.

7. Buys, J., Botha, J.F. and Messerschmidt, H.J. Coping with Degeneracy in the Computation of Dirichlet Tessellations. South African Computer Journal, 1990 (to appear).

8. De Boor, C. A Practical Guide to Splines. Springer-Verlag, New York, 1978.

9. Pinder, G.F., Celia, M. and Gray, W.G. Velocity Calculation from Randomly Lacated Hydraulic Heads. GROUND WATER, Vol. 19, 1981, pp. 262-264.

10. Watson, D.F. and Philip, G.M. Triangle Based Interpolation. Mathematical Geology, Vol. 16, 1984, pp. 779-795.

Comparison of Gradient Algorithms for Groundwater Flow Simulation using the Integrated Finite Difference Method

M. Ferraresi

Facoltà di Ingegneria, Università di Bologna, I-40136 Bologna, Italy

ABSTRACT

The Integrated Finite Difference approach to groundwater flow simulation lends itself to explicit matrix formulations analogous to those obtained for pipe network problems. The circumstance suggests the opportunity of considering the classes of gradient algorithms more frequently used in hydraulic network analysis also in the field of flow through porous media. The global gradient technique is highly commendable when dealing with unsteady conditions, but also in steady state problems it seems more suitable in comparison with the loop gradient approach and the linear theory method.

INTRODUCTION

Partial differential equations are at the heart of most computer simulations of continuous physical systems, such as those encountered in water resources analysis. The numerical solution of Cauchy and boundary value problems is usually achieved by transforming the partial differential equations first into sets of ordinary ones through spatial integration and then into systems of algebraic equations by considering finite time increments.

After the completion of the first stage, all phenomena formally exhibit the same mathematical structure, irrespective of their original dimension, so that the algorithms particularly devised for one–dimensional processes can be considered also for the solution of two– and three–dimensional problems. In this framework, the simulation of surface or groundwater hydrological systems through the Integrated Finite Difference (IFD) method may take advantage of the gradient techniques applied in hydraulic network analysis.

HYDRAULIC NETWORKS AND THE IFD REPRESENTATION

The analysis of water distribution networks consists of determining both the flow rate along every hydraulic connection (open channel or pressure pipe) and the piezometric head at every node, given the geometry and the physical properties of the system, together with adequate initial and boundary conditions. The integration in space of the partial differential mass conservation and flow equations yields the following system of ordinary differential equations in matrix form (e.g. Ferraresi et al. [1]):

$$\mathbf{C}_d^{-1}\mathbf{q} + \mathbf{L}\,\mathbf{h} = -\mathbf{Y}_d\,d\mathbf{q}/dt \qquad (1a)$$

$$\mathbf{L}'\mathbf{q} = -\mathbf{g} + \mathbf{S}_d\,d\mathbf{h}/dt \qquad (1b)$$

where upper and lower case bold letters denote matrices and vectors respectively, the subscript d means matrix diagonalization and the apex ' indicates matrix transposition. $\mathbf{C}$ is the diagonal matrix of connection conductivities, of size $nc \times nc$ if nc is the number of connections in the network; $\mathbf{q}$ and $\mathbf{h}$ are the vectors of the nc unknown discharges along connections and of the nn unknown nodal heads, being nn the number of nodes. $\mathbf{Y}$ and $\mathbf{S}$ are diagonal matrices representing the inertia along connections and the storage at nodes, while $\mathbf{g}$ is the vector of flow rate demand at nodes. $\mathbf{L}$ is a matrix of size $nc \times nn$, describing the network layout in terms of connection-to-node topology, whose elements l_{ij} can take only three values: 0 if connection i does not link node j, 1 or -1 if the flow of connection i enters or leaves node j respectively. The elements of $\mathbf{C}$, $\mathbf{Y}$ and $\mathbf{S}$ in general depend on $\mathbf{q}$, $\mathbf{h}$ and time t, but the flow rates $\mathbf{q}$, still function of time, are constant along each connection. According to this representation, the hydraulic system can store water only at its nodes, while the inertial effects belong uniquely to the connections.

An analogous formulation can be obtained by spatially integrating the fundamental groundwater equation:

$$div\,(\,\mathbf{K}\,\vec{grad}\,h\,) + g^v \;=\; S^v\,\frac{\partial h}{\partial t} \tag{2}$$

over elemental sub-regions Ω into which the total flow domain is discretized, following the IFD method (Narasimhan and Witherspoon [2]):

$$\int_\Omega \mathbf{K}\,\vec{grad}\,h\,\vec{n}\,d\Omega + g^v\,V \;=\; S^v\,V\,\frac{\partial h}{\partial t} \tag{3}$$

where the integral on the left–hand side represents the summation of fluxes across the surface Ω of the element, whose volume is V; $\vec{n}$ is the unit vector normal to Ω and super-script v denotes quantities per unit volume. If the element interfaces are perpendicular to the lines joining adjacent nodes, the IFD approximation for Equation (3) at node j can be written:

$$\Sigma_i\,\Omega_{ij}\,K_{ij}\,(h_i - h_j)\,/\,d_{ij} + g^v_j\,V_j \;=\; S^v_j\,V_j\,\Delta h_j\,/\,\Delta t \tag{4}$$

where the summation is extended over all nodes i linked with j, d_{ij} being their distances and Ω_{ij} the relative interface areas. In the IFD approximation (4) of Equation (3), K_{ij} must be regarded as the diagonal element of tensor $\mathbf{K}$ multiplying the normal piezometric gradient. Off–diagonal anisotropic contribution is taken into account by considering the gradient tangential to the interface between two adjacent nodes. This gradient can be computed by fitting a piecewise piezometric surface, with the same procedure adopted in the Finite Element method, to the triangular elements used in the discretization procedure according to the Thiessen polygon methodology. The approximate solution is given in terms of a linear combination of interpolation functions exactly reproducing nodal point values and providing head values at every location within each triangle as a function of space coordinates only. With the assumption of linear spatial interpolation, Equation (4) for two–dimensional problems can be written as:

$$\Sigma_i\,\Omega_{ij}\,K_{ij}^{n,n}\,(h_i - h_j)\,/\,d_{ij} + \Sigma_i\,p_{ij}\,K_{ij}^{n,t}\,(h_r - h_l)_{ij} + g_j \;=\; S_j\,\Delta h_j\,/\,\Delta t \tag{5}$$

where $K^{n,n}$ and $K^{n,t}$ are the hydraulic conductivities contributing to the normal flow rate for normal and tangential gradients respectively, p_{ij} is the aquifer thickness, $(h_r - h_l)_{ij}$ is the fluid potential difference between right and left reference points for connection ij, i.e. nodes linked with both node i and node j.

In order to represent the anisotropy effect, a new matrix $\mathbf{L_t}$ of size $nc \times nn$ must be introduced, whose elements $l_{t,ij}$ take on value 0 if node j is not linked with both the nodes identifying connection i, values 1 or -1 otherwise, the node j being at the right or at the left of connection i respectively. Calling $\mathbf{L_n}$ the layout matrix previously defined as $\mathbf{L}$ and observing that $h_i - h_j = \mathbf{L_n}\mathbf{h}$ and $h_r - h_l = \mathbf{L_t}\mathbf{h}$, according to the IFD methodology the head loss – flow relationship can be expressed as follows:

$$\mathbf{q} + (\,\mathbf{C_n}\,\mathbf{L_n} + \mathbf{C_t}\,\mathbf{L_t}\,)\,\mathbf{h} \;=\; 0 \tag{6}$$

where the conductivity matrices $\mathbf{C_n}$ and $\mathbf{C_t}$, both diagonal and of size $nc \times nc$, depend on the aquifer structure and hydrodynamics. For confined conditions and darcyan flow:

$$\mathbf{C_n} = \left[\frac{K^{n,n}\,w\,p}{d}\right]_d \quad;\quad \mathbf{C_t} = \left[K^{n,t}\,p\right]_d \tag{7a}$$

w and p being the width and the thickness of the connection. For unconfined conditions and darcyan flow:

$$\mathbf{C}_n = \left[\frac{K^{n,n}\,w}{2\,d}\right]_d [\,|\mathbf{L}_n|\,(\mathbf{h}-\mathbf{z})]_d \quad ; \quad \mathbf{C}_t = \left[\frac{K^{n,t}}{2}\right]_d [\,|\mathbf{L}_n|\,(\mathbf{h}-\mathbf{z})]_d \qquad (7b)$$

where $|\mathbf{L}_n|$ is the matrix of the absolute values of the elements of $\mathbf{L}_n$ and $\mathbf{z}$ is the vector of the aquifer bottom elevation at the nn nodes. For non–darcyan flow, a simplified form of the conductivity matrices is:

$$\mathbf{C}_n = \left[\alpha_n\,|\mathbf{q}|^{1-u}\right]_d \quad ; \quad \mathbf{C}_t = \left[\alpha_t|\mathbf{q}|^{1-u}\right]_d \qquad (7c)$$

based on the hypotheses that the head losses can be expressed according to the Hazen-Williams friction formula and that both conductivities be proportional to $\mathbf{q}$ instead of its normal and tangential components respectively. These assumptions do not represent practical restrictions and can be easily modified. Equation (5) can be written as follows:

$$\mathbf{L}_n'\,\mathbf{q} - \mathbf{S}\,\mathbf{h} = -\mathbf{g} \qquad (8)$$

where the matrix $\mathbf{S}$ and the vector $\mathbf{g}$ now include respectively the time step Δt and the contribution to Δh of the head at the previous time step. Since time derivatives of groundwater flow velocities are negligible, Equations (6) and (8) are formally analogous with (1a) and (1b). The overall system of connection and nodal equations can be written in partitioned matrix form as:

$$\begin{bmatrix} \mathbf{I} & (\mathbf{C}_n\mathbf{L}_n + \mathbf{C}_t\mathbf{L}_t) \\ \mathbf{L}_n' & -\mathbf{S} \end{bmatrix} \begin{bmatrix} \mathbf{q} \\ \mathbf{h} \end{bmatrix} = \begin{bmatrix} \mathbf{0} \\ -\mathbf{g} \end{bmatrix} \qquad (9)$$

In isotropic conditions the similarity to Equations (1a) and (1b) is complete, if $\mathbf{C} = \mathbf{C}_n$ and $\mathbf{L} = \mathbf{L}_n$:

$$\begin{bmatrix} \mathbf{C}^{-1} & \mathbf{L} \\ \mathbf{L}' & -\mathbf{S} \end{bmatrix} \begin{bmatrix} \mathbf{q} \\ \mathbf{h} \end{bmatrix} = \begin{bmatrix} \mathbf{0} \\ -\mathbf{g} \end{bmatrix} \qquad (10)$$

GRADIENT ALGORITHMS

The direct solution of the problem is possible when the Equations (9) are linear, i.e. only for darcyan flow and confined aquifer conditions, with $\mathbf{S}$ not dependent on $\mathbf{h}$. In this case the flow rates and the piezometric heads can be computed as:

$$\begin{bmatrix} \mathbf{q} \\ \mathbf{h} \end{bmatrix} = \begin{bmatrix} \mathbf{A}_1 & \mathbf{A}_2 \\ \mathbf{A}_3 & \mathbf{A}_4 \end{bmatrix} \begin{bmatrix} \mathbf{0} \\ -\mathbf{g} \end{bmatrix} \qquad (11)$$

where the matrix inversion is performed analytically, yielding:

$$\mathbf{A}_1 = \mathbf{I} - (\mathbf{C}_n\mathbf{L}_n + \mathbf{C}_t\mathbf{L}_t)\,[\mathbf{L}_n'(\mathbf{C}_n\mathbf{L}_n + \mathbf{C}_t\mathbf{L}_t) + \mathbf{S}]^{-1}\mathbf{L}_n' \qquad (12a)$$

$$\mathbf{A}_2 = (\mathbf{C}_n\mathbf{L}_n + \mathbf{C}_t\mathbf{L}_t)\,[\mathbf{L}_n'(\mathbf{C}_n\mathbf{L}_n + \mathbf{C}_t\mathbf{L}_t) + \mathbf{S}]^{-1} \qquad (12b)$$

$$\mathbf{A}_3 = [\mathbf{L}_n'(\mathbf{C}_n\mathbf{L}_n + \mathbf{C}_t\mathbf{L}_t) + \mathbf{S}]^{-1}\mathbf{L}_n' \qquad (12c)$$

$$\mathbf{A}_4 = -[\mathbf{L}_n'(\mathbf{C}_n\mathbf{L}_n + \mathbf{C}_t\mathbf{L}_t) + \mathbf{S}]^{-1} \qquad (12d)$$

The solution is immediately available as:

$$\mathbf{q}(t) = -(\mathbf{C}_n\mathbf{L}_n + \mathbf{C}_t\mathbf{L}_t)\,\mathbf{h}(t) \quad ; \quad \mathbf{h}(t) = [\mathbf{L}_n'(\mathbf{C}_n\mathbf{L}_n + \mathbf{C}_t\mathbf{L}_t) + \mathbf{S}]^{-1}\mathbf{g}(t) \qquad (13)$$

When the aquifer is phreatic or the groundwater flow in non–darcyan, iterative methods must be applied: the problem is solved by successive linear approximations, obtained neglecting quadratic and higher order terms in the Taylor expansion of the non–linear equations (e.g. Artina and Todini [3]). Writing the Equation (9) as:

$$f[\mathbf{q}(t), \mathbf{h}(t)] = 0 \tag{14}$$

f being a given mathematical function, the Taylor expansion yields:

$$f[\mathbf{q}^I(t), \mathbf{h}^I(t)] + J[\mathbf{q}^I(t), \mathbf{h}^I(t)]\begin{bmatrix} d\mathbf{q}^I \\ d\mathbf{h}^I \end{bmatrix} \simeq f[\mathbf{q}^{\#}(t), \mathbf{h}^{\#}(t)] \tag{15}$$

where J is the Jacobian matrix of f, $d\mathbf{q}$ and $d\mathbf{h}$ are the total differentials of the unknowns, superscript I indicates the iteration, $\mathbf{q}^{\#}$ and $\mathbf{h}^{\#}$ are the values satisfying Equation (14). From Equation (15) the $I + 1$ iteration is computed as follows:

$$\begin{bmatrix} \mathbf{q}^{I+1} \\ \mathbf{h}^{I+1} \end{bmatrix} = \begin{bmatrix} \mathbf{q}^I \\ \mathbf{h}^I \end{bmatrix} + \begin{bmatrix} d\mathbf{q}^I \\ d\mathbf{h}^I \end{bmatrix} = \begin{bmatrix} \mathbf{q}^I \\ \mathbf{h}^I \end{bmatrix} - J^{-1}[\mathbf{q}^I(t), \mathbf{h}^I(t)] f[\mathbf{q}^I(t), \mathbf{h}^I(t)] \tag{16}$$

Applying Equation (16) to (9) with conductivity matrices expressed by Equation (7b), i.e. for unconfined, darcyan flow, the following recursive algorithm is found:

$$\mathbf{q}^{I+1} = -\mathbf{E}\mathbf{h}^{I+1} \; ; \; \mathbf{h}^{I+1} = [\mathbf{L}_n'\mathbf{E} + \mathbf{S}]^{-1}\mathbf{g} \tag{17}$$

$\mathbf{E}$ is an $nc \times nn$ matrix function of $\mathbf{h}^I$:

$$\mathbf{E} = [\mathbf{B}_n(\mathbf{L}_n\mathbf{h}^I)_d + \mathbf{B}_t(\mathbf{L}_t\mathbf{h}^I)_d]|\mathbf{L}_n| + (\mathbf{C}_n\mathbf{L}_n + \mathbf{C}_t\mathbf{L}_t) \tag{18}$$

and $\mathbf{B}_n$ and $\mathbf{B}_t$ are the diagonal matrices at the right hand side of Equations (7b), premultiplying the diagonal matrix of the aquifer average thickness along connections. Applying analogously Equation (16) to (9) with conductivity matrices expressed by Equation (7c), i.e. for non–darcyan flow, the iterative solution algorithm takes the form:

$$\mathbf{q}^{I+1} = [\mathbf{A}_2^*\mathbf{L}_n' - \mathbf{I}](\mathbf{F}^{-1} - \mathbf{I})\mathbf{q}^I - \mathbf{A}_2^*\mathbf{g} \tag{19a}$$

$$\mathbf{h}^{I+1} = \mathbf{A}_4^*\mathbf{L}_n'(\mathbf{F}^{-1} - \mathbf{I})\mathbf{q}^I - \mathbf{A}_4^*\mathbf{g} \tag{19b}$$

The matrix $\mathbf{F}$ is easily computed by differentiation of Equation (9) as:

$$\mathbf{F} = \mathbf{I} + (1 - u)_d[\mathbf{q}]_d^{-1}[\mathbf{C}_n(\mathbf{L}_n\mathbf{h})_d + \mathbf{C}_t(\mathbf{L}_t\mathbf{h})_d] \tag{20}$$

The matrices $\mathbf{A}^*$ are derived from the analytical inversion of the Jacobian matrix of the problem; in particular:

$$\mathbf{A}_2^* = \mathbf{F}^{-1}(\mathbf{C}_n\mathbf{L}_n + \mathbf{C}_t\mathbf{L}_t)[\mathbf{L}_n'\mathbf{F}^{-1}(\mathbf{C}_n\mathbf{L}_n + \mathbf{C}_t\mathbf{L}_t) + \mathbf{S}]^{-1} \tag{21a}$$

$$\mathbf{A}_4^* = -[\mathbf{L}_n'\mathbf{F}^{-1}(\mathbf{C}_n\mathbf{L}_n + \mathbf{C}_t\mathbf{L}_t) + \mathbf{S}]^{-1} \tag{21b}$$

The gradient algorithms of Equations (17) and (19) are analogous to the iterative technique proposed by Todini and Pilati [4] for the analysis of pipe networks and called global Newton–Raphson. The solution is achieved by iteratively solving a system of linearized equations of size equal to the number of nodes nn, instead of dealing with one of size $(nc + nn)$. In steady state conditions, the same recursive algorithms hold, provided that $\mathbf{S}$ is neglected in all matrices $\mathbf{A}^*$. This solution scheme corresponds to the algorithm called nodal Newton–Raphson in pipe network analysis.

Two commonly used gradient techniques for the solution of hydraulic networks, the linear theory method and the loop Newton–Raphson algorithm, can also be derived from System (9), by considering linear combinations of its head loss – flow equations.

The linear theory method is particularly devised for steady state problems: piezometric heads are eliminated as direct unknowns by means of the loop incidence matrix $\mathbf{M}$ of size $nc \times nl$, where the number of independent loops nl can always be made equal to $nc - nn$ by adding dummy connections to the network. Following Todini and Pilati [4], the elements $m_{i,k}$ of $\mathbf{M}$ take on value 0 if connection i does not belong to loop k, values 1 or -1 otherwise, the flow along connction i being in the positive or in the negative direction of loop k respectively. If isotropic conditions hold, with $\mathbf{L} = \mathbf{L}_n$ and $\mathbf{C} = \mathbf{C}_n$, the relationships $\mathbf{M'L} = \mathbf{L'M} = 0$ are immediately found. Equation (9) is then equivalent to:

$$\begin{bmatrix} \mathbf{M'C^{-1}} & 0 \\ \mathbf{R} & \mathbf{L^*} \\ \mathbf{L'} & -\mathbf{S} \end{bmatrix} \begin{bmatrix} \mathbf{q} \\ \mathbf{h} \end{bmatrix} = \begin{bmatrix} 0 \\ -\mathbf{g} \end{bmatrix} \tag{22}$$

where $\mathbf{R}$ is an $nn \times nc$ matrix of flow resistence along nn connections linearly independent from the first nl relationships and $\mathbf{L^*}$ is the corresponding layout matrix: both $\mathbf{R}$ and $\mathbf{L^*}$ are subsets of $\mathbf{C^{-1}}$ and $\mathbf{L}$ respectively. The Equation (22) offers in general no practical advantage over (9), but in steady state conditions it can be reduced to:

$$\begin{bmatrix} \mathbf{M'C^{-1}} \\ \mathbf{L'} \end{bmatrix} [\mathbf{q}] = \begin{bmatrix} 0 \\ -\mathbf{g} \end{bmatrix} \tag{23}$$

where the unknowns $\mathbf{h}$ no longer appear if $\mathbf{C}$ does not depend on them. The iterative algorithm for $\mathbf{q}$ entails now the inversion of a matrix of size nc. The solution is not unique in terms of heads: at least one reference value $\mathbf{h}_0$ must be provided as a boundary condition.

The solution space is further reduced in the loop Newton–Raphson method, where the corrections $\Delta \mathbf{q}$ along the connections are computed in terms of the flow rate corrections $\Delta \mathbf{q}_l$ for each loop:

$$\Delta \mathbf{q} = \mathbf{M} \Delta \mathbf{q}_l \tag{24}$$

Provided that also at the initial iteration step the flow rates $\mathbf{q}$ satisfy the nodal conditions of mass conservation, the Equation (9) can be written as:

$$\mathbf{M'C^{-1}q} = 0 \tag{25}$$

whose recursive solution algorithm is:

$$\mathbf{q}^{I+1} = [\mathbf{I} - \mathbf{M}(\mathbf{M'C_q^{-1}M})^{-1}\mathbf{M'C^{-1}}]\mathbf{q}^I \tag{26}$$

where $\mathbf{C}_q$ is the derivative of $(\mathbf{C^{-1}q})$ with respect to $\mathbf{q}$. The size of the problem to be solved is now nl, if $\mathbf{C}$ does not depend on $\mathbf{h}$. Again at least one reference value $\mathbf{h}_0$ must be provided as a boundary condition.

COMPARISON AND CONCLUSIONS

The linear theory method and the loop gradient algorithm can be regarded as different formulations of the global gradient approach, the difference resting on the linear combinations of the head loss – flow equations considered in the solution system. Their practical application is however restricted to those problems where the nodal piezometric heads can be discarded as direct unknowns, i.e. in steady state, isotropic conditions with the friction and storage coefficients independent from $\mathbf{h}$. Otherwise some unnecessary complications would be introduced in the computation of the flow rates $\mathbf{q}$, namely the need of solving a system of linearized equations of size nn in addition to the original one, whose size is nc for the linear theory method and nl for the loop gradient. The global gradient scheme entails the iterative solution of a system of size nn in every condition. In addition to these computations, every algorithm requires the scalar projection of linear combinations of the results into the space of the remaining unknowns by means of algebraic transformations.

For regular rectangular grids nc is about twice nn, while nl is about equal to nn, but irregular IFD meshes are not seldom found where nc is three or four times greater than nn and consequently also nl noticeably exceeds the number of nodes. The linear theory method is far below the other two techniques from the point of view of size and sparseness of the coefficient matrix to be inverted, which is not symmetrical even for isotropic media. On the other hand the loop gradient algorithm requires an initial solution in terms of flow rates satisfying the nodal mass balance; this solution can be easily computed, assuming for instance unit gradients along every connection, but it may affect the convergence rate of the recursive procedure.

The global gradient method does not need a loop or path definition, as is the case of the other two techniques, and even though this task can be performed automatically (Todini and Pilati [4]) the general input preparation is simpler. The conclusion that its inherent advantages make the global approach preferable to the linear theory and the loop gradient algorithms, reached by Salgado et al. [5] for pressure pipe network analysis are substantially corroborated also in groundwater flow simulation, with the additional merit of its ability to deal with unrestricted structural and hydrodynamic aquifer conditions.

The gradient algorithms above described were applied to various test problems of groundwater flow. The analysis was focussed essentially on the algorithm performance: for fairness of comparison, the time integration scheme, the method of solution of the resulting algebraic systems and the initial balanced discharge distribution were assumed common to all tests.

The number of iterations required to achieve the solution did not differ significantly, but noticeable differences were ascertained in terms of computational time, owing to the different size of the problem and favouring he global gradient approach. Slower convergence was found when strong heterogeneity was assumed, either in terms of hydraulic conductivity or of piezometric gradient steepness, induced by irregular withdrawals distribution. This circumstance seems to affect the linear theory method and the loop gradient technique more than the global gradient algorithm, providing further reasons for preferring the latter approach.

REFERENCES

[1] Ferraresi, M., Franchini, M., Lamberti, P., Pilati, S. and Todini, E. A numerical method for the solution of a class of hydraulic network problems, Quaderni Istituto di Costruzioni Idrauliche, 1986.

[2] Narasimhan, T.N. and Witherspoon, P.A. An integrated finite difference method for analyzing fluid flow in porous media. Water Resources Research, 12(1), p. 57–64.

[3] Artina, S. and Todini, E. On the convergence of solution methods for water supply distribution networks, Quaderni Istituto di Costruzioni Idrauliche, 1990

[4] Todini, E. and Pilati, S. A gradient algorithm for the analysis of pipe networks, in Computer Applications in Water Supply (Ed. Coulbeck,B. and Orr,C.H.), pp. 1–20. Research Studies Press Ltd. and Wiley & Sons, Leicester, 1988

[5] Salgado, R., Todini, E. and O'Connell, P.E. Comparison of the gradient method with some traditional methods for the analysis of water supply distribution networks, in Computer Applications in Water Supply (Ed. Coulbeck,B. and Orr,C.H.), pp. 38–62. Research Studies Press Ltd. and Wiley & Sons, Leicester, 1988

A Direct Computation of the Permeability of Three-Dimensional Porous Media

S. Succi(*), A. Cancelliere(*)(**), C. Chang(***), E. Foti(**), M. Gramignani(**), D. Rothman(***)

(*) IBM European Center for Scientific and Engineering Computing, Via del Giorgione 159, 00147 Rome, Italy

(**) Department of Hydraulic Engineering, University of Catania, Viale A. Doria 6, Catania, Italy

(***) Department of Earth, Atmospheric and Planetary Sciences, Massachussets Institute of Technology, Cambridge, MA 02139, USA

Abstract

We present a series of high-resolution numerical simulations of flows in porous media with the Lattice Boltzmann Method (LBM). Quantitative evaluations of medium's permeability as a function of porosity are presented.

Introduction

In the recent years a new computational technique based on a special type of lattice Boltzmann equation has been proposed as an alternative tool to investigate complex hydrodynamical flows in two and three dimensions. This technique, intimately related to lattice gas automata, has been introduced to overcome two major drawbacks of boolean lattice gas simulations: statistical noise and need of huge look-up tables to evolve in time the automaton state. Except for the possibility of performing boolean (hence "exact" in a digital computer) simulations, all the remaining advantages of lattice gas computing are retained, namely:

1) ideal amenability to vector/parallel processing;

2) ease of implementation of complex boundary conditions.

Because of these properties, LBM is indeed an excellent candidate to determine the transport properties of complex flows. Numerical investigations on the permeability of two-dimensional porous media using a cellular automata model have first been performed by Rothman [1] ; Succi et al. [2] extended these studies to three-dimensional flows through simple porous media in which the rock geometry was modelled by cubes. In this paper we present an application of the LBM to the determination of the permeability of a three-dimensional porous medium in which the rock geometry is modelled, more realistically, by penetrable spheres.

The Lattice Boltzmann Equation

The microscopic dynamics of Lattice Gas Automata (LGA) is described in terms of a set of boolean variables $n_i(\bar{x}, t)(i = 1,....,b)$, at the sites $\bar{x}$ of a regular lattice in a D dimensional space. In the following, the time t is considered to be a discretized variable: $t = 1,2,3.....$ Let $\bar{c}_i$ be the vector describing the single cell of the lattice. The meaning of $n_i(\bar{x}, t)$ is the following: if $n_i = 0$, there is no particle moving in the $\bar{c}_i$ direction at $\bar{x}$; if $n_i = 1$, there is one particle of unit mass moving with velocity $\bar{c}_i$. The vector $\bar{c}_i$ completely describes the geometry of the system. One can show that LGA converges to the Navier-Stokes limit, if the tensor $T_{klmn} \equiv \sum c_{ik} c_{il} c_{im} c_{in}$ $(k, l, m, n = 1, D)$ is isotropic. Thus the choice of the lattice geometry is crucial. Unfortunately there is no 3D lattice which satisfies the above condition. Frisch and al. [3] have shown that one can use a 4D face centered-hypercube (FCHC) lattice ($b = 24$) projected onto a three-dimensional subspace. The dynamics of the system is described by collision rules among the whole set of $n_i(\bar{x}, t)$ for a given lattice site. Because the n_i are boolean variables, the collision rules must be designed in such a way that no more than one particle can have the "velocity" $\bar{c}_i$ at the same location $\bar{x}$ (exclusion principle). This requirement leads to a Fermi-Dirac equilibrium distribution. Next, mass and momentum must be conserved during a collision. In general the set of the collision rules is defined by a look-up table with 2^b-bits entries where 2^b is the number of possible states at a given site. Because the viscosity of the fluid is controlled by the collision rules, these must be defined in an optimal way, in order to obtain the most efficient LGA for fluid flows. This choice of the optimal collision rules leads to a linear programming problem which may be difficult to be solved for large values of D, i.e. for the FCHC lattice in 4D. The hydrodynamical fields of LGA are a suitable space time averaging of the original boolean variables. In particular, the density and the momentum are given by:

$$\rho = <\sum_i n_i(\bar{x}, t)> \tag{1}$$

$$\rho\bar{v} = <\sum_i n_i(\bar{x}, t)\,\bar{c}_i> \tag{2}$$

where $< \dots >$ stands for some kind of average. The equations for the variables defined by (1) and (2) converge to the Navier-Stokes equations. This has been verified in two and three dimensional numerical experiments. In the Boltzmann approach, the occupation numbers n_i are replaced by the corresponding mean populations $N_i = < n_i >$. This means that instead of following each particle in detail, one is only interested in the story of an average particle. By supplementing this averaging procedure with some further statistical assumption (molecular chaos), a Lattice Boltzmann Equation can be derived in the following form:

$$S_i N_i \equiv N_i(\bar{x} + \bar{c}_i, t + 1) - N_i(\bar{x}, t) = \Omega_i(N) \tag{3}$$

where Ω_i represents the change in N_i as due to collision. The advantage of this equation over the boolean one is that the functions N_i (which are now real variables) are smooth functions of space and time because the single particle fluctuations have been averaged out by the very definition of the mean population. Another practical advantage is that the 3D hydrodynamics can be simulated without using the huge look-up table required for the boolean version. On the other hand, it is also clear that all the physics related to particle correlations is lost. One usually linearizes the collision operator Ω_i around the uniform steady state $N_i = d$, d being the density per link of the discrete fluid. Equation (3) then yields:

$$S_i N_i \equiv \sum_{j=1}^{b} \Omega_{ij}(N_j - N_j^{eq}) \tag{4}$$

where Ω_{ij} is $\partial\Omega_i/\partial N_j$ and the equilibrium distribution N_i^{eq} is expanded to the second order in the local velocity field u. Equation (3) is taken as a starting point for our experiments.

Implementation of the code

Fluid flow through a porous medium is simulated via a three step process: a translation, where the particles move from one site to the next along each direction, a reflection, where particles colliding with solid sites are reflected with opposite speed, and a collision step, where particles at each site are redistributed according to rules subject to the constraints of mass and momentum conservation. The implementation of these processes leads to three basic routines MOVE, REFLEC, COL for the translation, reflection and collision steps respectively. The MOVE routine involves addresses shifting which implies only load/store operations and should therefore take a negligible amount of CPU time. However this routine can become unexpectedly time consuming if data are not accessed in a proper way. Therefore, in order to maximize the efficiency of the 3090 high-speed buffer (64 Kbytes of fast memory), a stride-one data access has been implemented; furthermore by exploiting the Vector Facility with an appropriate Do-Loops nesting, a bar speed up of about a factor 2 was obtained. This led to a well-optimized routine which takes 10% of total CPU time at each step. The collision step, which represents the computational kernel of any lattice gas code, was implemented in the COL routine. Because of its "local" nature, this process is very suitable for parallelization and vectorization since the updating of one site requires variables related only to that site. This routine requires about 300 floating point operations per site but this value could be significantly reduced by eliminating the quadratic terms in the equilibrium distribution which are presumably not influent for Reynolds number as small as those encountered in the study of flows through porous media. The REFLEC routine is necessary in order to treat the geometrical complexity of a porous medium. A non-slip condition at wall sites (i.e. zero velocity on the surface of solids) was imposed by reflecting back to the fluid, with opposite speeds, particles colliding with solid sites. In our scheme, wall sites are fictious ones laying in between solid sites and fluid sites. This increases the complexity of the code but, on the other hand, it guarantees a better fulfillment of the non-slip boundary condition. In order to speed up the code, a topological distinction among solid sites was introduced by distinguishing between "border" sites (i.e. solid sites next to fluid ones) and "inner" sites which are sites completely surrounded by solid ones. The topological information corresponding to the solid sites has been encoded into a number of arrays which serve as (indirect) addresses to perform the reflections only at the border sites. Since only border sites are involved in this routine, its efficiency is a function of the complexity of the medium implemented. The percentage of CPU time per step is within the range of 10%-20%.

Results

Calculations were carried out on an IBM 3090/VF 600 E multiprocessor running under MVS/XA operating system. Two classes of microgeometries have been investigated, using spheres of radius 8 lattice units and 4 lattice units respectively. The latters were implemented both in a 64 cubic lattice and in a 128 cubic one. The spheres are randomly placed on the lattice with their centers at distances of multiple of the radius, so that they can overlap. Several values of porosity can be obtained by varying the number of spheres. See for instance Fig. 1 which shows a transversal section of the porous flow at a solid fraction of .3 in a 128 cubic lattice.

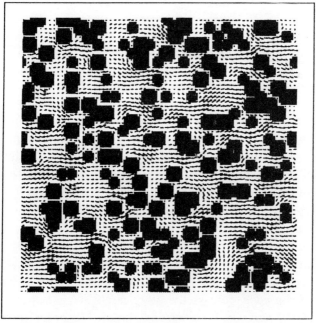

Figure 1. Flow on the X-Y plane in a 128 cubic lattice

A constant pressure gradient along the x direction is imposed and the total velocity is obtained, once the flow reaches a steady state, by summing the x components of velocities at each non solid site in the lattice.   The mean velocity per particle in the x direction can be calculated by averaging over the sum of the particles at each site. The permeability can then be calculated from the mean velocity $\bar{v}_x$, since the viscosity and the pressure gradient are known. The permeability k can be expressed as:

$$k = q \times \frac{\mu}{grad\ P} \qquad (5)$$

where q is the volumetric flow rate per unit area $= \phi\ \bar{v}_x$ (ϕ being the porosity) and $\mu = 0.258$ is the dynamic viscosity. The pressure gradient is expressed in terms of a volumic force f which is added and subtracted to all directions projecting onto x all over non solid sites; in formulae $grad\ P = 12 \times f \times \sqrt{2}$.   The mean particle velocity is defined as follows:

$$\bar{v}_x = \frac{\sum_i v_{xi}}{\sum_i N_i} = \frac{\sum_i v_{xi}}{N \times \phi \times \rho \times \sqrt{2}} \qquad (6)$$

where the sum $\sum v_{xi}$ is extended over all non-solid sites, N is the number of sites, $\rho = .328 \times 24$, and $\sqrt{2}$ is a normalization factor due to the shape of the lattice.   For each microgeometry generated, the value of permeability has been computed. In Fig. 2 we represent the permeability k, normalized to the square of the sphere's radius (r), as a function of the solid fraction $s = (solid\ volume/total\ volume)$.   The continuous curve plotted in Fig. 2 represents a theoretical upper bound evaluated by Weissberg and Prager using variational arguments [4] .

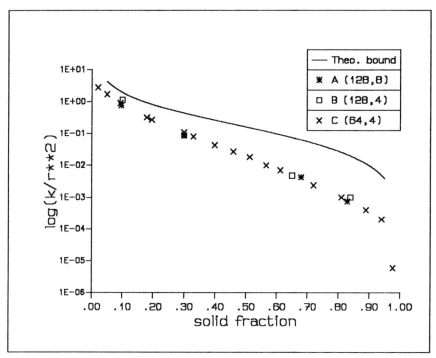

Figure 2. Normalized permeabilities as a function of solid fraction

From this figure we see that, indeed, for each of the three microgeometries A (128,8), B (128,4), C (64,4), the numerical results lay below the theoretical upper bound, as they should. We also note that the permeabilities calculated in the three cases A, B, C, match within less than a 30% factor which is a nice indication that, probably, the continuum limit has been reached by the numerical simulations. In the region of high solid fractions, our results could in principle be influenced by spurious Knudsen layer effects because, as the pore size becomes smaller than a few lattice units, the lattice Boltzmann gas does not behave like a fluid but rather as an ensemble of "free particles" [3]. Despite of this, even in this region, the results don't exhibit any evident anomaly, which might be due to some sort of statistical compensation of the errors. These problems are currently under investigation.

Acknowledgements

A.C., E.F., M.G. acknowledge financial support by the IBM European Center for Scientific and Engineering Computing, where all the numerical calculations have been performed.

References

[1]   Frisch U., d'Humieres D., Hasslacher B., Lallemand P., Pomeau Y. and Rivet J.P., *Complex Systems,* n. 1 648 (1987).

[2]   Rothman D., *Geophysics,* n.4 509 (1988).

[3]   Succi S., Foti E., Higuera F., *Europhysics Letters,* n. 10   433 (1989)

[4]   Weissberg H. L., Prager S., *Phys. Fluids,* 13 2958 (1970)

SECTION 2 - UNSATURATED GROUNDWATER FLOW

A Finite Element Adapted Matricial Programming Language (FEMPL) for Modelling Variably Saturated Flow in Porous Media with Galerkin-type Schemes

G. Gaillard, J. Bovet

Institute of Soil and Water Management and Department of Mathematics, Swiss Federal Institute of Technology, CH-1015 Lausanne, Switzerland

ABSTRACT

A Finite Element adapted Matricial Programming Language (FEMPL) has been developed in order to facilitate the implementation of Galerkin-type schemes. Owing to its specific characteristics (extension of the multi-entries array concept) and to an appropriate set of matricial and finite element operations, FEMPL provides brevity and efficiency in the programming task. Examples are given in the frame of transfer modelling in variably saturated porous media with applications to infiltration and redistribution in homogeneous and stratified soil columns.

INTRODUCTION

Since Neuman [9] and Duguid and Reeves' [5] early works, the finite element method (FEM) has become a worthwhile tool for modelling water and mass flow in variably saturated porous media. Except for ordinary cases with constant boundary conditions and uniform soil texture, where the finite difference method (FDM) is advantageous, more elaborated numerical procedures are required, among which the FEM, with its well-known mathematical basis and a widespread use, is very profitable [1, 8]. FEM is helpful not only for bi- and tridimensional systems, where irregular shapes of flow domain and complexities in prescription of boundary conditions can occur, but also for one-dimensional situations for instance in studying effects of heterogeneities on water movement or in describing advance rate of an infiltration front near the soil surface.

However, when using a standard programming language (FORTRAN, etc.), resulting schemes are difficult to implement and require considerable efforts and expensive time, particularly for a non-specialist. Although proper softwares can be developed or re-used to obtain satisfactory specific codes, it is difficult for the user to modify or to improve it and still keep a complete control on the whole program. So, it appeared easier and more elegant to avoid these drawbacks by developing a specific matricial-type programming language which includes instructions appropriate to the FE schemes. For these reasons, it has been called FEMPL (Finite Element adapted Matricial Programming Language).

This paper presents the principal characteristics of FEMPL with programming examples in the frame of water transfer in variably saturated porous media. Some applications to infiltration and redistribution in homogeneous and stratified soil columns are also shown.

CHARACTERISTICS OF FEMPL

The objective of FEMPL is to simplify programming of FE algorithms by providing some specific simple routines. This results in a significant reduction of program length and developing time. Consequently, legibility is improved, and all kinds of further modifications can be easily introduced. Although we only describe specific characteristics of FEMPL, it must be kept in mind that it is a comprehensive programming language including all the necessary standard instructions [3].

Basic concepts

The FEMPL syntax is based on the so-called "structures", which consist in a generalization of the multi-entries array concept. A structure is a multi-indexed array where the indices (expressed below between brackets) denote particular finite element related items, such as "element" (e), "node" (n), "mesh point" (s), "Gauss point" (g), "dimension" (d) or "variable" (v). A structure can be a vector, a matrix or a specified arranged set of matrices and vectors. For instance, the index characterizations (e, g, d, d), (e, n, n) and (s, s) can denote respectively a set of tensors (d, d) expressed for Gauss integration points (g) of each element (e), a set of elementary contributions (n, n) calculated for each element (e) and a matrix after assembly (s, s) according to the global coordinates system (see examples in the next paragraph).

Owing to this basic concept, numerical operations on total or specified partial structures can be directly formulated without syntax complexities provided indices compatibility is satisfied.

Numerical, matricial and finite element operations

In addition to the arithmetic operations and the conventional mathematical functions (exponential, trigonometric, etc.), FEMPL includes matricial procedures (expressed below in bold type) : multiplication *****, determinant **DET**, inversion **INV**, transposition **TRS**, concatenation **CNC**, extraction **EXT**, permutation **PER**, etc. Practical algorithms, as Gauss-type numerical integration **SUM** and solution of linear equations system **SOL** are also implemented. Finally, FE related operations are included : assembly **ASS** and disassembly **DAS**. All together these instructions allow powerful programming with very few instruction lines, as shown in the examples of the next paragraph.

Implementation

In its actual and experimental version, FEMPL has been interfaced with worksheet (in this case LOTUS-123 with implementation of a FE library) for pre- and postprocessing operations. The compiler has been written in TURBO/PASCAL for PC [2]. However, FEMPL, in its design, is a programming language independent of the type of implementation. This is a stand-alone system, like FORTRAN, PASCAL, BASIC or ADA.

APPLICATION OF FEMPL TO FINITE ELEMENT SIMULATION OF TRANSFER IN POROUS MEDIA

Several programs were developed with FEMPL to simulate water and mass transfer in saturated and variably saturated porous media. Since FEMPL is consistent with the matricial notation, significant advantage can be taken from the similarity of the encountered numerical schemes, by simple recovery or rapid creation of short programming modules. Moreover, because of its specific multi-indices characterization, FEMPL allows immediate transposition from one- to two-dimensional simulation schemes, provided a necessary re-formulation of the boundary conditions is defined (in this case it is possible to get two or more coordinates systems to overlay).

Example of programming

For illustration purposes, we present the core of a FE code for water transfer in variably saturated porous media [6]. Variational principle and Galerkin approximation are applied to the well-known Richard's model, with FD time discretization [1, 8, 9]. Knowing the water head $\{\psi\}^k$ at time t_k, $\{\psi\}^{k+1}$ at time t_{k+1} is obtained after solving the following system of non-linear equations [8: eq. 4.7.2.8 and following] :

$$[E]^{k+\varepsilon} \{\psi\}^{k+1} = \{G\}^{k+\varepsilon} \tag{1}$$

with

$$[E]^{k+\varepsilon} = \varepsilon [A]^{k+\varepsilon} + \frac{[B]^{k+\varepsilon}}{\Delta t} \tag{2}$$

and

$$\{G\}^{k+\varepsilon} = \{F\}^{k+\varepsilon} + \left((\varepsilon-1) [A]^{k+\varepsilon} + \frac{[B]^{k+\varepsilon}}{\Delta t} \right) \{\psi\}^k \tag{3}$$

where

$$A_{IJ}^{k+\varepsilon} = \sum_e \int_{R^e} \hat{K}_{ij}^{k+\varepsilon} \frac{\partial N_I}{\partial x_i} \frac{\partial N_J}{\partial x_j} dR \tag{4}$$

$$B_{IJ}^{k+\varepsilon} = \sum_e \int_{R^e} \hat{c}^{k+\varepsilon} N_I N_J dR \tag{5}$$

$$F_I^{k+\varepsilon} = - \sum_e \left(\int_{R^e} \hat{K}_{ij}^{k+\varepsilon} \frac{\partial N_I}{\partial x_i} e_j dR + \int_{B_2^e} N_I q dB \right) \tag{6}$$

where square brackets, braces, indices and exponents denote respectively matrices (s,s), vectors (with s components), component references and time t, ε is a factor defined in the interval [0,1] such that $\{\psi\}^{k+\varepsilon} = \varepsilon \{\psi\}^k + (1 - \varepsilon) \{\psi\}^{k+1}$, Δt is the time interval, N_I are the shape functions defined on elements R^e, subdomains of the whole domain R whose boundary fraction B_2 supports Neumann-type boundary conditions, $\hat{K}_{ij}^{k+\varepsilon}$ and $\hat{c}^{k+\varepsilon}$ are the approximated hydrodynamic characteristics, q is the Darcy flux and $\sum_e$ denotes the assembly operation.

Expressing (4), (5) and (6) according to the matricial notation [11], for instance:

$$\hat{K}_{ij}^{k+\varepsilon} \frac{\partial N_I}{\partial x_i} \frac{\partial N_J}{\partial x_j} <=> \{\partial_x N_I\}^T [K]^{k+\varepsilon} \{\partial_x N_J\} \tag{7}$$

The corresponding coding is directly (numerical integration according to Gauss procedure) [6] :

AI = **TRS** (DXNG) * (TKG ∧ CINTG) * DXNG.   (8)
A = **ASS** (**SUM** (AI, g)).

BI = CINTG ∧ CEMG ∧ NGNG.   (9)
B = **ASS** (**SUM** (BI, g)).

FI = **EXT** ((TKG ∧ CINTG) * DXNG, e, g, 2, n).   (10)
F = **ASS** (**SUM** (FI, g)).

After prescription of the pertinent boundary conditions (Dirichlet's type : D; Neumann's type : N) :

E = (EPS ∧ A) + ((1 / DTZERO) ∧ B).   (11)
G = ((((EPS – 1) ∧ A) + ((1 / DTZERO) ∧ B)) * HWZERO) – (N + F).   (12)
HWPSUN = **SOL** (E, G, D).   (13)

where (the indices are denoted between square brackets) :

A, B, E [s, s], AI, BI [e, g, n, n]   matrices and integrands defined in (2), (4) and (5)
D, F, G, N [s], FI [e, g, d, n]   vectors and integrand defined in (3), (6) and above
CEMG [e, g]   capillary capacity at Gauss points
CINTG [e, g]   Gauss integration coefficient

DTZERO [ ]   time interval Δt
DXNG [e, g, d, n]   derivative of the element shape functions N_I at Gauss points

EPS [ ]   factor ε
HWZERO and HWPSUN [s]   water head respectively at times t_k and t_{k+1}
NGNG [g, n, n]   product of the element shape functions N_I at Gauss points
TKG [e, g, d, d,]   tensor of hydraulic conductivity at Gauss points

^   arithmetic multiplication
*   matricial multiplication

ASS   assembly function
EXT   extraction function
SOL   linear equations system solution function
SUM   integration function (Gauss procedure)
TRS   transposition function

Each FEMPL function modifies the index characterization, for example, one proceeds in equation (8) from [e, g, n, n] for AI to [s, s] for A, i. e. from local to global coordinates systems.

It is easy to note the efficiency of this type of coding as soon as a matricial notation is used. No particular limitation is encountered. Similar results are obtained for other matricial FE calculations. For example, FEMPL expression of rigidity matrix (8) is rigorously the same whether the dimension is 1 or 2; moreover, with proper conversion from hydraulic conductivity to hydrodynamic dispersion function, equation (8) can be recovered without modifications for mass transfer simulation. Practically no effort is required from the user in this case. An interesting advantage is that very little time and effort must be devoted to these standard operations; therefore the user can emphasize on more important parts of the program, e.g. optimization of the algorithmic structure, implementation of efficient convergence procedures, etc. Besides, one is able to develop rapidly a common basis of transfer simulation around a simple central core of coding.

Application : infiltration and redistribution in homogeneous and stratified soil columns
Several programs were developed to simulate successfully saturated and unsaturated infiltration, redistribution and drainage in different homogeneous and stratified soil columns. First, we present the simulations of the infiltration experiments of Touma and Vauclin [12, 6]. Then, simulation of constant flux infiltration and subsequent redistribution in a stratified soil column is shown [4]. The non-linearities are dealt according to an iterative procedure of readjustment of hydrodynamic functions in the middle of the time interval until satisfactory convergence is achieved. The time stepping is optimized following the readaptation method described by Gureghian [7].

In Fig. 1 simulated and measured moisture profiles of saturated (2.3 cm water head at soil surface) and unsaturated (constant flux of 8.3 cm/h) infiltrations in a coarse sand [12] are presented. The same hydrodynamic analytical functions were used. The results are quite satisfactory for this case characterized by a sharp interface of saturation [6].

An infiltration at constant flux (1.43 10^{-5} m/s) was performed on a two-layered soil column (medium above coarse sands) of 3 m length and 0.95 m diameter. As soon the water front had reached the layer interface, redistribution was allowed with no flux at the surface. The results of the simulation are presented in Fig. 2 together with the corresponding moisture

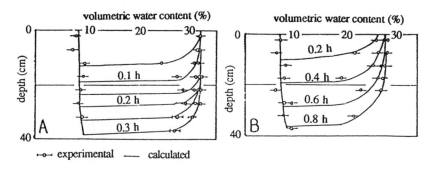

Fig. 1   Simulated and measured moisture profiles of saturated (1A) and unsaturated (1B) infiltrations. Experiments of Touma and Vauclin [12, Fig. 5a and 8]

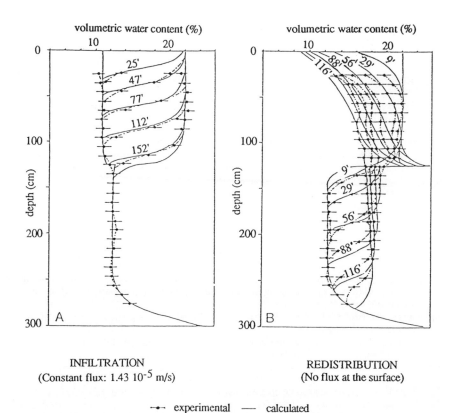

Fig. 2   Simulated and measured moisture profiles of unsaturated infiltration (2A) followed by redistribution (2B). Experiments on a two-layered soil column of Christe and Gaillard [4]

measurements. The use of the neutron-probe technique explains the alteration of the profiles at the soil interface as well as the lack of measurements at the surface. The results are considered satisfactory, the global behaviour being correctly reproduced in the two layers and the moisture profile near the interface satisfying the pertinent requirements [4, 10].

CONCLUSION

In its present state of development and implementation, FEMPL is more appropriate for research needs (test of algorithms, simulation of experiments in laboratory systems, etc.) than for large field-scale three-dimensional modelling. This is mainly due to the non-implementation of specific procedures for solving memory-size problems and reducing time execution. But owing to very short program development and moderate simulation times, FEMPL appears to be a very useful tool of FE modelling, especially for people who do not have time to deal with complex coding requirements. Moreover, because of its structure, adaptation to software development does not modify at all the programs, but only the FEMPL compiler. Further adaptations are now under development for an extended use of FEMPL.

ACKNOLEDGMENTS

This work was supported by the "Fonds National Suisse de la Recherche Scientifique, Projet Milieux non saturés: Transferts et comportement".
Thanks to J.-D. Favrod and G. de los Cobos for their important contribution in conception and development of FEMPL.

REFERENCES

1. Becker, E.B., Carey, G.F. and Oden, J.T. Finite elements : an introduction, Vol. 1, Prentice-Hall Inc., Englewood Cliffs, New Jersey, 1981.
2. Bovet, J. et de los Cobos, G. Le LPM - Langage de Programmation Matriciel : Concept et mode d'emploi, 19 pp., EPF-IGRHAM, Lausanne, 1989.
3. Bovet, J., de los Cobos, G. et Gaillard, G. Le LPM - Langage de programmation matriciel : Syntaxe et programmation, 40 pp., EPF-IGRHAM, Lausanne, 1989.
4. Christe, R. et Gaillard, G. Infiltration non saturante et redistribution dans une colonne de sol stratifiée : simulation expérimentale et modélisation par éléments finis. Bull. CHYN, Vol. 9, in press, 1990.
5. Duguid, J. O. and Reeves, M. Material transport in porous media: A finite element Galerkin model, Oak Ridge National Laboratory, Oak Ridge, Tennessee, 1976.
6. Gaillard, G. Simulation par éléments finis du mouvement de l'eau en milieu poreux déformable variablement saturé - Cas unidimensionnel vertical stratifié - Le programme VARSAT1, 31 pp., EPF-IGRHAM, Lausanne, 1989.
7. Gureghian, A.B. A two-dimensional finite-element solution scheme for the saturated-unsaturated flow with applications to flow through ditch-drained soils, J. Hydrol., Vol. 50, pp. 333-353, 1981.
8. Huyakorn, P.S. and Pinder, G.F. Computational methods in subsurface flow, Academic Press Inc., New-York, 1983.
9. Neuman, S.P. Saturated-unsaturated seepage by finite elements, J. Hydr. Div. ASCE, HY12, Vol. 99, 2233-2250, 1973.
10. Poulovassilis, A. Steady-state potential and moisture profiles in layered porous media, Soil Sci. 107(1), pp. 47-52, 1969.
11. Segerlind, L. J. Applied finite element analysis, John Wiley & Sons, New York, 1976.
12. Touma, J. and Vauclin, M. Experimental and numerical analysis of two-phase infiltration in a partially saturated soil, Transport in Porous Media, Vol. 1, pp. 27-55, 1986.

Numerical Methods for Nonlinear Flows in Porous Media

M.A. Celia, E.T. Boutoulas, P. Binning
Water Resources Program, Dept. of Civil Engineering and Operations Research, Princeton University, Princeton, NJ 08544, USA

INTRODUCTION

Nonlinear flows in porous media occur in many water resources problems. One of the most important and challenging cases is multiphase flow, wherein more than one fluid occupies the pore space of the medium. Examples include infiltration or redistribution of water in unsaturated soils, and contamination of groundwater by non-aqueous phase liquids (NAPL's). In such cases the fluid mass balance equations are coupled and highly nonlinear. Design and implementation of robust and efficient numerical methods is an on-going research challenge.

This paper addresses some of the important numerical issues by looking at the model single-equation representation of unsaturated flow, the Richards equation. Numerical methods are presented that produce superior results over a wide range of soil conditions. These techniques are used to demonstrate what we call the hierarchical philosophy of modeling flow and transport in porous media. An example of infiltration into an initially very dry soil, followed by redistribution, is used to demonstrate this philosophy. Extension of these concepts to the fully multiphase case is also discussed.

COMPUTATIONAL APPROACH

A number of important aspects of numerical modeling of non-linear flows in porous media can be illustrated by focusing on the unsaturated zone, with Richards equation used to describe the movement of liquid. Richards equation may be written as

$$\frac{\partial \theta}{\partial t} - \underset{\sim}{\nabla} \cdot K \underset{\sim}{\nabla} h - \frac{\partial K}{\partial z} = 0, \qquad \begin{array}{c} \underset{\sim}{x} \in \Omega \\ t > 0 \end{array} \qquad (1)$$

where θ denotes moisture content, h is pressure head, K is unsaturated hydraulic conductivity, z is the vertical direction (positive upward), and the medium is assumed to be isotropic. Equation (1) is referred to as the mixed form of Richards equation.   When numerically solving Richards equation, the first important choice that must be made is which form of the equation to use as the governing partial differential equation. This question was addressed in detail in [4], where it was shown that the best choice is the mixed form of Richards equation, because it maintains the advantages of both the θ-based and h-based forms without suffering from their respective limitations [4].

The appropriate linearization of this nonlinear equation has been presented elsewhere ([1],[4]) and will only be discussed briefly here.   The procedure is initialized by a time discretization of equation (1) using, say, a backward Euler method. Because the coefficient K is nonlinear, a simple iteration strategy is then introduced such that the evaluations of θ and h at the new time level (denoted by θ^{n+1}, h^{n+1}) are also at the new iteration level (denoted by $\theta^{n+1,m+1}$ and $h^{n+1,m+1}$), while K is lagged by an iteration, $K^{n+1,m}$.   The key to the method is to then expand θ at the new iteration level in terms of h, using the fact that θ is a function of h.   Thus

$$\theta^{n+1,m+1} = \theta^{n+1,m} + \frac{d\theta}{dh}\bigg|^{n+1,m}(h^{n+1,m+1}-h^{n+1,m}) + O((\delta h)^2) \quad (2)$$

Use of equation (2) (ignoring the $O((\delta h)^2)$ terms) allows θ at the unknown iteration to be replaced by θ at a known iteration and h at the unknown iteration level.   Therefore the only unknown in the resulting equation is $h^{n+1,m+1}$, and appropriate numerical approximation in space (say by finite elements) allows an updated estimate of h to be determined.   The resulting approximation may be written as follows,

$$\frac{C^{n+1,m}}{\Delta t}(\delta h) - \underset{\sim}{\nabla} \cdot K^{n+1,m}\underset{\sim}{\nabla}(\delta h)$$

$$= \underset{\sim}{\nabla} \cdot K^{n+1,m}\underset{\sim}{\nabla}h^{n+1,m} + \frac{\partial K^{n+1,m}}{\partial z} - \left(\frac{\theta^{n+1,m}-\theta^n}{\Delta t}\right) \quad (3)$$

where $\delta h \equiv h^{n+1,m+1}-h^{n+1,m}$ is the increment in iteration and $C \equiv d\theta/dh$ is the specific moisture capacity.   Notice that upon convergence, both the right and left sides of equation (3) go to zero.   Because the right side is a simple difference approximation to the mass balance statement (the governing equation), it can be shown [4] that this approximation possesses the conservative property, thereby numerically conserving mass.   In contrast, it can be noted that the usual h-based equation leads to poor mass balance results, especially for infiltration into initially very dry soils, because the analogous right side term is not of this form [4].

Mass conservation alone is not sufficient to guarantee good numerical solutions. This is especially true in very dry soils, where the nonlinearity is very strong. In this case, finite element approximations should employ mass lumping to avoid oscillatory behavior around steep moisture fronts. Mass lumping assures that the numerical approximation obeys a maximum principle (which is also obeyed by the differential equation) ([2],[4]). Distributed mass matrices allow violation of this principle, thereby leading to oscillatory solutions. Such oscillations can lead to nonconvergence of the iterative scheme [4].

A final consideration in solving Richards equation is the solution of the resulting algebraic equations. In multiple spatial dimensions, direct solution is computationally prohibitive and iterative solvers are required. Recent work [3] has shown that the incomplete Cholesky pre-conditioned conjugate gradient (ICCG) method is superior to traditional iterative solvers, and leads to very efficient solutions for large multi-dimensional problems. Both theoretical and numerical results support this contention [3]. This solver is used in all unsaturated flow results reported below.

EXTENSION TO TWO-PHASE FLOW

The unsaturated flow equation, which assumes constant air pressure, is a simplification of the fully multiphase case, and so the results cited above serve as an excellent guide in designing approximations for two-phase flow systems. For example, a mixed equation form of the governing equations leads to a mass-conservative approximation, while mass lumping eliminates oscillations at sharp infiltration fronts. This information is being used to design two-phase flow simulators. In addition, a symmetric matrix structure will be maintained so that efficient iterative matrix solvers such as ICCG can be used. We expect to report on our computational experiences at the meeting.

APPLICATION

The numerical methods discussed above have been used to simulate an unsaturated zone field test being carried out in New Mexico. The example serves to highlight the general modeling philosophy that is being followed, and to indicate the types of problems the methods are capable of solving. The experimental design is described in [5]. The test consisted of infiltration at a constant rate (approximately 2 cm/day) along a strip 4 meters wide and 10 meters long. The infiltration was imposed for 82 days, after which the infiltration was stopped and redistribution allowed to occur. The system was modeled as a two-dimensional vertical section beneath the center of the strip. A schematic of the system is shown in Figure 1. The simulation is numerically challenging because the soil was initially very dry, with initial water tensions estimated to be of the order of 10,000. centimeters.

The approach to modeling this problem was based on what we call a hierarchical philosophy. The basic idea is to begin with a simple description of the system, and to hierarchically add physical complexity. For example, in the New Mexico case study an early estimate of the material relationship between θ and h was available. Thus a homogeneous medium was assumed, and a K vs. h relationship was inferred from the θ vs. h relaiton via the procedure of van Genuchten [6]. Based on this data, simulations were performed to estimate the infiltration and redistribution phases of the experiment. A typical result is shown in Figure 2. Next, as some data was collected on the spatial variability of both the θ vs. h curves and the saturated conductivity, more spatial complexity was added to the model. For example, Figures 3 and 4 show solutions based on different levels of complexity. Figure 3 is a simulation result with K_{sat} spatially variable (with given correlation structure), all other parameters assumed uniform. Figure 4 uses both K_{sat} and the exponent n in the van Genuchten parameterization of conductivity [6] spatially variable but uncorrelated, all other parameters held uniform. The two results are seen to be significantly different, highlighting the importance of spatial variability in nonlinear physics. For a system that has natural spatial variability, these simple results indicate that homogeneous solutions are both qualitatively and quantitatively incorrect. The results also point out the importance of scale in defining parameters and in simulating nonlinear processes in natural porous media. A variety of other simulation scenarios have been investigated, and some of these will be discussed in our presentation. It should be noted that the simulations reported here used approximately 5,000 nodes, and all simulations were carried out on desktop workstations (mostly on a microVax II).

Modeling of the New Mexico test site has shown that simulation results for an entire hierarchy of system descriptions are crucial to understanding nonlinear physics at scales of practical importance. Robust and very efficient numerical methods, such as those described herein, are a necessary prerequisite for such a simulation approach.

ACKNOWLEDGEMENTS

This work was supported in part by the U.S. Nuclear Regulatory Commission under Contract Number NRC-04-88-074, by the National Science Foundation under Grant 8657419-CES, and by the U.S. Environmental Protection Agency under Agreement CR-814946.

Although the research described in this article has been funded in part by the U.S.E.P.A., it has not been subjected to Agency review and therefore does not necessarily reflect the views of the Agency and no official endorsement should be inferred.

REFERENCES

[1] Allen, M.B. and C. Murphy, "A finite element collocation method for variably saturated flow in porous media," *Num. Meth. PDE's*, 3, 229-239, 1985.

[2] Bouloutas, E.T. "Improved numerical methods for simulation of fluid flow and contaminant transport in partially saturated porous media," PHD Thesis, MIT, 1989.

[3] Bouloutas, E.T. and M.A. Celia, "Fast iterative solvers for linear systems arising in the simulation of unsaturated flow systems," subm. to *Adv. Water Res.*, 1990.

[4] Celia, M.A., E.T. Bouloutas and R.L. Zarba, "A general mass-conservative numerical solution for unsaturated flow," to appear, *Water Resources Research*, 1990.

[5] Wierenga, P.J., D. Hudson, J. Vinson, and R.G. Hills, "Las Cruces trench experiment database: Experiment I," Publication in progress, 1990.

[6] van Genuchten, M.Th., "A closed-form equation for predicting the hydraulic conductivity of unsaturated soils," *Soil Sci. Soc. Am. J.*, 44, 892-898, 1980.

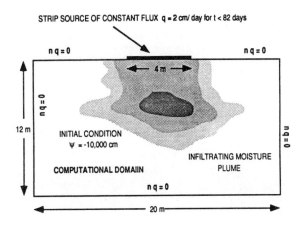

FIGURE 1 - Schematic of physical system to be simulated.

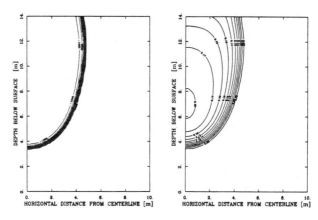

FIGURE 2 - Simulation results (t=523 days) for homogeneous case, showing symmetric half-plane solutions for both pressure head (left) and moisture content (right).

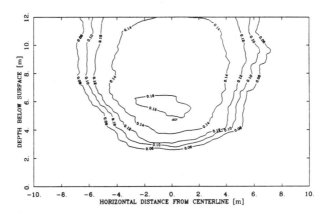

FIGURE 3 - Simulation results for moisture content (t=1629 days) using variable K_{sat}.

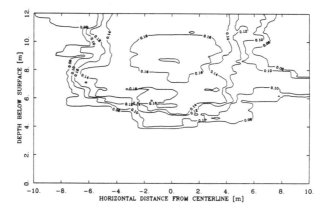

FIGURE 4 - Simulation results (t=1033 days) using variable K_{sat} and n, uncorrelated.

Numerical Analysis of Nonlinear Unsaturated Flow Equations

R. Ababou

Southwest Research Institute, Center for Nuclear Waste Regulatory Analyses, San Antonio, TX 78228, USA

ABSTRACT:

The numerical behavior of the nonlinear unsaturated flow equation is examined analytically for an implicit finite difference scheme. The governing equation combines nonlinear diffusion and convection operators, and is characterized by a simple Peclet number. Numerical errors are investigated using truncation error analysis, frozen stability analysis, and functional analysis of the nonlinear mapping associated with Picard iterations. These approaches shed light on different but complementary aspects of the same numerical problem.

INTRODUCTION:

Flow in unsaturated porous media is governed by a strongly nonlinear diffusion type equation with a nonlinear, forced convection term due to gravity. These features make it particularly difficult to solve by any means. Analytical solutions have been and are still providing valuable insights for certain classes of flows, but most realistic problems have to be solved numerically: see [1] and [2] for high-resolution supercomputer simulations of 3-dimensional, transient unsaturated flow in randomly heterogeneous and stratified media.

Our experience is that the strong nonlinearity of unsaturated flow usually causes considerable numerical difficulties and requires trial-and-error adjustments of mesh size, time step, relaxation parameters, tolerance criteria, and adaptive controls. To improve the efficiency of future numerical algorithms will require some understanding of how the specific features of the unsaturated flow equation contribute to numerical errors. Exploring this question constitutes the main purpose of this paper.

UNSATURATED FLOW EQUATIONS:

For transient flow in variably saturated porous media, a mixed variable formulation of the governing equation is obtained by combining the mass conservation equation $(\partial\theta/\partial t = -\nabla.Q)$ with the Darcy-Buckingham equation $(Q=-K\nabla H)$:

$$\partial\theta(h,x)/\partial t = \nabla(\ K(h,x)\ (\nabla h + g)\ ) \qquad (1)$$

where $H=h+g.x$ is the hydraulic potential, h is pressure head, and g is the cosine vector aligned with the acceleration of gravity and equal to $(0,0,-1)$ if the third axis is vertical pointing downwards. Defining the specific moisture capacity $C=\partial\theta/\partial h$ yields the pressure-based Richards equation:

$$C(h,x)\ \partial h/\partial t = \nabla(\ K(h,x)\ (\nabla h + g)\ ) \qquad (2)$$

Introducing a nonlinear moisture diffusivity $D=K/C$, assuming a spatially homogeneous moisture retention curve $\theta(h)$, and $h \leq 0$ everywhere, yields the moisture-based version of eq.(1):

$$\partial\theta/\partial t = \nabla(\ D(\theta,x)\ \nabla\theta + g\ K(\theta,x)\ ) \qquad (3)$$

In the detailed 3-dimensional simulations of [1] and [2], the finite difference method was applied to the mixed form (1), which is more mass conservative than eq.(2) and is not limited to negative pressures as eq.(3). For convenience, however, we will use the standard Richards equation (2) for the numerical analyses to be developed in this paper.

The nonlinear gravity term containing (g) acts as forced convection, in competition with the diffusion term represented by the elliptic operator $\nabla(K\nabla h)$. For homogeneous media, eqs.(2) and (3) can be reformulated as follows, respectively:

$$C\ (\partial h/\partial t + V.\nabla h) = \nabla(K\nabla h) \qquad (2)'$$

and:

$$\partial\theta/\partial t + V.\nabla\theta = \nabla(D\nabla\theta) \qquad (3)'$$

where:

$$V = -\ (\partial K/\partial\theta)\ g \qquad (4)$$

The vector V represents the velocity of pressure or moisture disturbances, in the absence of the right-hand side diffusive terms. Note that moisture and pressure waves propagate at the same speed in homogeneous media. The convective-diffusive form taken by equations (2-3) suggests that a Peclet number might be used to characterize convection versus diffusion effects.

A Peclet number emerges quite naturally from the Kirchhoff transform formulation [1]. This transform is valid for the class of heterogeneous media possessing a homogeneous relative conductivity curve $Kr(h)$, i.e. with a separable conductivity curve $K(h,x)=Ks(x)Kr(h)$. But we focus here on the more restricted case where both the saturated and relative conductivities are homogeneous, i.e. with a homogeneous conductivity curve K(h)=KsKr(h). The Kirchhoff transform is then defined by:

$$\phi(h) = \int_{-\infty}^{h} K(h') \, dh' \qquad (6)$$

Substituting eq.(6) in eq.(2) and using chain rules yields:

$$\partial\phi/\partial t + V.\nabla\phi = D \, \nabla^2\phi \qquad (7)$$

Here again, the wave velocity V is given by equation (4), or equivalently by $V = -\alpha Dg$, where $\alpha=\partial\ell nK/\partial h$ is the slope of the log-conductivity/pressure curve.

Given the form of the Kirchoff equation (7), the Peclet vector Pe=$V\ell$/D emerges as a relative measure of convective versus diffusive moisture transport over the chosen length scale (ℓ). Furthermore, substituting the above expression for V gives a very simple expression for the Peclet vector:

$$Pe = -\alpha\ell g \qquad (8)$$

which reveals the special role played by the α-parameter. Since transforms like $\theta(h)$ or $\phi(h)$ do not fundamentally alter the ratio of convection versus diffusion coefficients, this Peclet vector characterizes the transport of pressure as well as moisture, Kirchoff potential, or any other quantity that can be related to pressure in a one-to-one fashion.

TRUNCATION ERROR ANALYSIS:

In this section, we evaluate truncation error as a function of mesh size and time step for a nonlinear, implicit finite difference discretization of the Richards equation. Note that truncation analysis compares the discrete and differential operators, but is not concerned with the numerical errors incurred by the dependent variable itself, or with the space-time propagation of such errors (stability), or with the additional errors incurred while solving the nonlinear discretized system (linearization). Our purpose here is to identify potential sources of inaccuracies by looking at the leading order terms of truncation error.

Consider the Richards equation (2), to be solved for a 1-dimensional homogeneous medium using a fully implicit nonlinear finite difference discretization (Euler backwards in time, 2-point centered in space). The differential and discretized equations are, respectively:

$$\mathcal{L}(h) = -C(h)\,h_t + (\,K(h)\,(h_x + g)\,)_x = 0 \tag{9.a}$$

$$L(h_i^{n+1}) = -C_i^{n+1}\,\frac{h_i^{n+1} - h_i^n}{\Delta t} + \frac{K_{i+1/2}^{n+1}}{\Delta x}\left[\frac{h_{i+1}^{n+1} - h_i^{n+1}}{\Delta x} + g\right]$$
$$-\frac{K_{i-1/2}^{n+1}}{\Delta x}\left[\frac{h_i^{n+1} - h_{i-1}^{n+1}}{\Delta x} + g\right] = 0 \tag{9.b}$$

where g is a cosine representing gravity, with g=0 if x is horizontal and g=-1 if x is vertical downwards. The coefficients of the discrete operator are fully nonlinear, being expressed at the current time step (n+1). The mid-nodal conductivities are approximated by a geometric weighting:

$$K_{i+1/2} = \left[\,K(h_i)\,K(h_{i+1})\,\right]^{1/2} \tag{10}$$

In the case of an exponential K(h) curve, this scheme weights pressures arithmetically, and it yields the exact midnodal conductivity in zones of spatially constant pressure gradient.

The truncation error $E(h)=L(h)-\mathcal{L}(h)$ at the nodes of the space-time mesh was calculated in [1] using intermediate results from [3]. We choose here to express the final result in terms of both flux (Q) and pressure (h) as follows:

$$E(h) \approx \Delta t\left[\frac{\partial}{\partial x}[\,K\,\frac{\partial}{\partial t}(\,\frac{\partial h}{\partial x}\,)\,] - \frac{C}{2}\,\frac{\partial^2 h}{\partial t^2}\right] \tag{11}$$
$$+ \frac{\Delta x^2}{24}\left[\frac{\partial}{\partial x}(\,K\,\frac{\partial^3 h}{\partial x^3}\,) - \frac{\partial^3 Q}{\partial x^3}\right] - \frac{\Delta x^2}{8}\left[\alpha\,\frac{\partial}{\partial x}(\,Q\,\frac{\partial^2 h}{\partial x^2}\,)\right]$$

where $Q=-K(\partial h/\partial x+g)$, and $\alpha=\partial \ell nK/\partial h$. It is interesting to note that one recovers the linear heat equation by letting K=1, C=1, and $\alpha=0$ in (9-11). Inserting in eq.(11) the identities:

$$h_t = h_{xx} \quad\text{and}\quad h_{tt} = -Q_{xxx} = h_{xxxx}\,,$$

leads to the verification of a well known result [4]: the order of accuracy of the linear heat equation increases from $O(\Delta t)+O(\Delta x^2)$ to $O(\Delta t^2)+O(\Delta x^4)$ with the choice $\Delta t/\Delta x^2 = 1/6$.

Let us now discuss the implications of (11) in the fully nonlinear case. The $O(\Delta t)$ term, due to temporal discretization errors, appears to be controlled by the rate of change of the pressure gradient and by the second order time-derivative of pressure. The first of the two $O(\Delta x\Delta x)$ terms is due to spatial discretization errors other than midnodal conductivity weighting. The second $O(\Delta x\Delta x)$ term is due solely to errors in evaluating midnodal conductivities by the geometric rule (10), and vanishes in regions of spatially constant pressure gradient, as expected in the case of exponential K(h).

Equation (11) simplifies considerably in the steady state case, since the 1-dimensional flux Q becomes constant in both space and time. The result suggests that, even in the transient case, spatial errors are dominated by the rate of change of pressure curvature with depth, which can become quite large near sharp wetting fronts above and below the inflexion point. This type of information may be used for designing optimal adaptive grid procedures.

STABILITY ANALYSIS:

To complement the previous truncation error analysis, we now examine how numerical errors propagate as a function of time. In addition, we hope to capture at least some of the additional error amplification effects due to inexact treatment of nonlinearity. Our approach is to analyze the stability of a linearized version of the finite difference problem, such that all nonlinear coefficients are evaluated from the solution at the previous time step (no iterations). The unstable effect of linearization is partially taken into account by unfreezing the nonlinear convective coefficient, while other coefficients remain frozen.

We focus once more on the case of 1-dimensional homogeneous media as in eqs.(9). Consider the following linearized form of the finite difference system:

$$C_i^n \frac{h_i^{n+1}-h_i^n}{\Delta t_n} = \frac{1}{\Delta x}\left[K_{i+\frac{1}{2}}^n\left[\frac{h_{i+1}^{n+1}-h_i^{n+1}}{\Delta x}+g\right]-K_{i-\frac{1}{2}}^n\left[\frac{h_i^{n+1}-h_{i-1}^{n+1}}{\Delta x}+g\right]\right] \quad (12)$$

where the superscript (n+1) indicates the current time level. The form of this finite difference system suggests that, while the nonlinear diffusion operator is treated implicitly, the nonlinear gravity term $g(K[i+1/2]-K[i-1/2])/\Delta x$ is treated explicitly since it is entirely evaluated at the previous time level. Based on this remark, we will now examine how this discrepancy affects the numerical stability of the solution.

The proposed method is to develop a Fourier stability analysis of equation (12) with partially frozen coefficients. This is analogous to the usual frozen coefficients analysis as described in [4], except that the nonlinearity of the gravity term is taken into account via the quasilinear approximation:

$$g\frac{K_{i+\frac{1}{2}}^n - K_{i-\frac{1}{2}}^n}{\Delta x} = g\frac{K_i^n}{\Delta x}\frac{\alpha}{2}(h_{i+1}^n - h_{i-1}^n) + O(\Delta x^2) \quad (13)$$

where again $\alpha=d\ell nK/dh$. The leading term on the right-hand side is expected to be a reasonable approximation of the left-hand side if the quantity $\alpha|h[i+1]-h[i]| \approx \alpha\Delta x|\partial h/\partial x|$ is on the order of unity or less. At any rate, even rough indications on the numerical stability of the nonlinear unsaturated flow system will be useful given the lack of theoretical results in this area. With this provision, substituting (13) into (12) yields the following mixed implicit/explicit scheme:

$$-\tilde{D}_{i-\frac{1}{2}} h_{i-1}^{n+1} + (1+\tilde{D}_{i-\frac{1}{2}}+\tilde{D}_{i+\frac{1}{2}}) h_i^{n+1} - \tilde{D}_{i+\frac{1}{2}} h_{i+1}^{n+1} \simeq$$

$$-\frac{1}{2} g \, \alpha\Delta x \, \tilde{D}_i \, h_{i-1}^n + h_i^n + \frac{1}{2} g \, \alpha\Delta x \, \tilde{D}_i \, h_{i+1}^n \tag{14.a}$$

where $\tilde{D}$ is the dimensionless diffusion coefficient:

$$\tilde{D}_{i(\pm\frac{1}{2})} = \frac{K_{i(\pm\frac{1}{2})}}{C_i} \frac{\Delta t}{\Delta x^2} \tag{14.b}$$

The stability of equation (14) with frozen diffusion coefficients can be studied in the standard way using Fourier stability analysis [4]. This leads to a complex amplification factor, ρ, characterizing the growth rate of numerical errors in time:

$$\rho \simeq \frac{1 + j \, \alpha g \Delta x \, \tilde{D}_i \, \sin(k\Delta x)}{1+(\tilde{D}_{i+\frac{1}{2}}+\tilde{D}_{i-\frac{1}{2}})(1-\cos(k\Delta x)) - j(\tilde{D}_{i+\frac{1}{2}}-\tilde{D}_{i-\frac{1}{2}})\sin(k\Delta x)} \tag{15}$$

where j is the square-root of -1, and k is a Fourier mode or wavenumber taking discrete values: $k \in \{\pi/L, \cdots, n\pi/L\}$.

Requiring $|\rho| \lesssim 1$ in equation (15) finally leads to the necessary and sufficient stability condition:

$$Pe = |\alpha g \Delta x| \leq 2 \sqrt{1+(2 \frac{K_i}{C_i} \frac{\Delta t}{\Delta x^2})^{-1}} \tag{16}$$

where Pe represents the grid Peclet number [see eq.(8)]. If the Peclet number is less than 2, then the stability condition is always satisfied irrespective of the time step size. On the other hand, if the Peclet number is greater than 2, stability requires a stringent constraint on the time step size. To summarize, the stability condition is:

$$\text{either:} \quad Pe = |\alpha g \Delta x| \leq 2,$$

$$\text{else:} \quad Pe = |\alpha g \Delta x| \geq 2 \quad \text{and} \quad \frac{K_i}{C_i} \frac{\Delta t}{\Delta z^2} \leq \frac{2}{(Pe-2)(Pe+2)} \tag{17}$$

Recall that the Peclet number was defined as a convection to diffusion ratio [see discussion above eq.(8)]. When the grid Peclet number of eqs.(16-17) is much smaller than unity, pressure disturbances appearing at any node are smeared out by diffusion before reaching the next node (stable case).

Finally, the effects of heterogeneity can be analyzed in a qualitative manner as follows. Assume for instance that the a-parameter of the exponential conductivity curve is spatially variable. The local Peclet number is therefore also spatially variable. Assuming (roughly) that the previous stability analysis still holds locally, we see from equation (17) that instabilities must be triggered in zones of coarse porosity where a takes large values. Equation (15) shows that the most unstable Fourier modes are those with largest wavenumbers, having fluctuation scales comparable to mesh size. And, equation (16) indicates that such instabilities will grow faster where moisture diffusivity is high, e.g. in wet zones.

In order to minimize the chances of explosive error amplification, it seems reasonable to require that the vertical mesh size be a fraction of the average length scale $1/a$, which typically lies in the range 10-100 cm for sandy to clayey soils. This guideline was used to design large scale numerical experiments of unsaturated flow in [1] and [2].

CONVERGENCE ANALYSIS OF NONLINEAR PICARD ITERATIONS:

In practice, an iterative scheme such as Picard or Newton must be used to iteratively linearize and solve the nonlinear algebraic system at each time step. For instance, a modified Picard scheme that preserves the symmetry of the system was used in [1,2]. In this section, we show how the convergence of the Picard scheme can be investigated by applying functional analysis methods [5,6,7] to the nonlinear mapping associated with the iteration scheme. The proposed approach is to apply the Picard method directly to the partial differential equation of unsaturated flow, and to examine the convergence properties of the resulting iteration scheme, a priori independent of discretization.

For illustration here, we will restrict our analysis to the special case of steady unsaturated flow in a spatially homogeneous 1-dimensional medium, for which an exact solution can be derived by direct integration. Assume that $K(h)$ is exponential with exponent $a=\partial \ln K \partial h$ and that the x-axis is vertical downwards. Define the dimensionless variables:

$$\xi = x/L, \quad \Psi = -h/L, \quad a = a/L, \quad q = Q/Ks, \quad \text{and} \quad k = K/Ks,$$

where Ψ is the dimensionless suction head, always positive in unsaturated media. Our model problem is steady infiltration or evaporation in a vertical column extending, say, from soil surface at z=0 (ξ=0) to a water table or other boundary at z=L (ξ=1). This can be formulated as the boundary value problem:

$$\mathcal{L}(\Psi) = (\ k(\Psi)\ (\Psi_\xi + 1)\ )_\xi = 0$$

$$\Psi(0)=\Psi_0\ ;\quad \Psi(1)=\Psi_1$$

(18)

A straightforward integration of (18) yields the conductivity profile $k(\xi)$, which itself can be used to obtain the suction profile $\Psi(\xi)$. The complete solution is given by:

$$k(\Psi(\xi)) = q + (k_1-q)\ \exp\{a(\xi-1)\}$$ (19.a)

$$q = \{k_1-k_0\exp(a)\}/\{1-\exp(a)\}$$ (19.b)

$$k(\Psi)=\exp(-a\Psi),\ k_0=k(\Psi_0),\ k_1=k(\Psi_1),$$ (19.c)

$$q \leq \{1-k_1\exp(-a)\}/\{1-\exp(-a)\}$$ (19.d)

where the constant dimensionless flux q is either positive (downwards) or negative (upwards). Note that the solution is only valid for boundary conditions such that $\Psi(\xi) \geq 0$ on the $[0,1]$ interval. This requires satisfying $\Psi_0 \geq 0$, $\Psi_1 \geq 0$, and the inequality (19.d) which boils down to $q \leq 1$ if $a \gg 1$.

Let us now apply a Picard scheme with relaxation parameter ω to iteratively solve (18). The solution is constructed by way of an iterated mapping:

$$(\ k(\Psi^n)\ (\Psi^{n+1}-\Psi^n)_\xi\ )_\xi\ =\ -\ \omega\ \mathcal{L}(\Psi^n)$$

$$\Psi^{n+1}(\xi) - \Psi^n(\xi) = 0\quad \text{at}\quad \xi = 0\ \text{and}\ \xi = 1$$

(20)

The residual operator $\mathcal{L}(\Psi)$ is the same as the one defined in equation (18), and n is the iteration counter. The Dirichlet conditions are implemented exactly at each iteration, since they are linear.

At each iteration, the iterated mapping of equation (20) is a boundary value problem that is directly integrable in terms of the incremental suction $\Psi[n+1]-\Psi[n]$. One obtains after some manipulations:

$$\Psi^{n+1} = \{\ (1-\omega)\ \mathcal{U}(.) + \omega\ \mathcal{P}(.)\ \}\ \Psi^n$$ (21.a)

where $\mathcal{U}$ is the identity operator satisfying $\mathcal{U}(\Psi)=\Psi$, and $\mathcal{P}$ is the Picard iteration operator defined by:

$$\mathcal{P}(\Psi^n) = \{\ (\Psi_1-\Psi_0+1)\ \mathcal{F}(\Psi^n) + (\Psi_0- \xi)\ \}$$ (21.b)

where $\mathcal{F}$ is the ratio of two integral operators:

$$F(\Psi^n) = \left[\int_0^{\xi} k(\Psi^n(s))^{-1}ds\right] / \left[\int_0^1 k(\Psi^n(s))^{-1}ds\right] \qquad (21.c)$$

Each operator $\mathcal{U}(.)$, $\mathcal{P}(.)$, and $\mathcal{F}(.)$ maps onto itself the space of continuous functions $\Psi(\xi)$ defined on the interval $[0,1]$, and $(21.a)$ yields $\Psi[n+1](\xi) = \Psi[n](\xi)$ at $\xi=0$ and $\xi=1$.

The convergence properties of the Picard scheme are most directly related to the properties of the iteration operator $(21.a)$ which maps the old solution $\Psi[n]$ into the new solution $\Psi[n+1]$. Without going into details, let us point out that the contraction mapping theorem and Ostrowski's local convergence theorem [7] can be used to test the conditions under which convergence occurs, and to estimate convergence rate in some function space norm. Work along these lines is ongoing. The case of transient flow will be developed by applying the same principles to the semi-discretized differential flow equation.

REFERENCES:

1.   Ababou R. (1988): Three-Dimensional Flow in Random Porous Media, Ph.D. thesis, M.I.T., Cambridge, MA 02139, U.S.A., pp. 833, [Chapters 5 & 7].

2.   Ababou R., and Gelhar L. W. (1988): A High-Resolution Finite Difference Simulator for 3D Unsaturated Flow in Heterogeneous Media, in: Computational Methods in Water Resources (VIIth Internat. Conf.), M. Celia et. al. ed., Elsevier & Comput. Mech. Publi., Vol. 1, 173-178.

3.   Vauclin M., Haverkamp R., and Vachaud G. (1979): Résolution Numérique d'une Equation de Diffusion Nonlinéaire, P.U.G., Grenoble, France, pp. 183.

4.   Ames W. F. (1977): Numerical Methods for Partial Differential Equations, Academic Press, New York, pp. 365, [Chapter 2].

5.   Curtain R. F. and Pritchard A. J. (1977): Functional Analysis in Modern Applied Mathematics, Academic Press, New York, pp. 339.

6.   Milne R. D. (1980): Applied Functional Analysis: An Introductory Treatment, Pitman, Boston.

7.   Ortega J. M. and Rheinboldt W. C. (1971): Iterative Solution of Nonlinear Equations in Several Variables, Academic Press, New York, pp. 572, [5.1,7.1,10.1,12.1].

Time-Discretization Strategies for the Numerical Solution of the Nonlinear Richard's Equation

C. Paniconi, A.A. Aldama, E.F. Wood

Department of Civil Engineering and Operations Research, Princeton University, Princeton, New Jersey, USA

Abstract

The transient, nonlinear Richards' equation describing flow in partially saturated porous media is solved numerically using several time-discretization strategies. The aim of the research is to compare the computational efficiency of conventional iterative schemes with several newer non-iterative formulations. The Implicit Factored method in particular appears to be an attractive alternative because of its high accuracy and two-level nature. The performance of the different schemes is evaluated over a range of accuracy levels, using both the standard form of Richards' equation and a modified form with a forcing term added (allowing comparisons with an exact solution). The limitations and advantages of the different time-discretization approaches are discussed.

Introduction

The governing equation for one-dimensional flow in partially saturated porous media, Richards' equation, is obtained by combining Darcy's law with the continuity equation. Expressing this equation with pressure head ψ as the dependent variable, with t representing time, and with the vertical coordinate z taken positive upward yields

$$S(\psi)\frac{\partial \psi}{\partial t} = \frac{\partial}{\partial z}\left(K_s K_r(\psi)\left(\frac{\partial \psi}{\partial z} + 1\right)\right) \tag{1}$$

where $S(\psi) = d\theta/d\psi$ is the specific soil moisture capacity, θ is the volumetric moisture content, and the hydraulic conductivity $K(\psi)$ is expressed as a product of the conductivity at saturation, K_s and the relative conductivity, $K_r(\psi)$.

Equation (1) is highly nonlinear due to pressure head dependencies in the moisture capacity and conductivity terms, the latter term contributing a nonlinearity to both a diffusion-type term and the gravitational gradient term $\partial K/\partial z$. In solving (1) numerically, implicit time discretizations are used, ensuring stability of the overall scheme. The nonlinear nature of the equation, however, requires that the implicit scheme be solved iteratively, unless the nonlinear components are linearized in some manner. The most common approach has been to solve (1) numerically using either the Picard or Newton iteration methods.

In this paper we compare the performance of six time-discretization strategies for solving (1): the Newton and Picard iterative schemes; non-iterative versions of the Newton and Picard methods; a three-level non-iterative scheme; and a linearization method applied to a variant of (1). The order of accuracy of the different schemes is compared, and the schemes are evaluated on the basis of computer time expended to achieve a given level of accuracy. A forcing term is introduced into Richards' equation so that the six strategies can also be compared against an exact solution. A more detailed development and discussion of the work described here will be found in Paniconi [8].

Time-Discretization Schemes

We apply a finite element Galerkin discretization in space, with linear basis functions, to transform (1) into the system of ordinary differential equations

$$\mathbf{A}(\Psi)\Psi + \mathbf{F}(\Psi)\frac{d\Psi}{dt} = \mathbf{q}(t) - \mathbf{b}(\Psi) \tag{2}$$

where Ψ is the vector of pressure head nodal values, $\mathbf{q}$ is the specified boundary flux vector and the nonlinear terms $\mathbf{A}, \mathbf{F}, \mathbf{b}$ are functions of $K_r(\psi)$ $(\mathbf{A}, \mathbf{b})$ and $S(\psi)$ $(\mathbf{F})$.

Newton and Picard Methods   The $O(\Delta t^2)$ Crank-Nicolson scheme is used to discretize (2) over time, giving

$$\mathbf{f}(\Psi^{k+1}) \equiv \tfrac{1}{2}\mathbf{A}^{k+\frac{1}{2}}(\Psi^{k+1} + \Psi^k) + \mathbf{F}^{k+\frac{1}{2}}\frac{\Psi^{k+1} - \Psi^k}{\Delta t} - \mathbf{q}^{k+\frac{1}{2}} + \mathbf{b}^{k+\frac{1}{2}} = 0 \tag{3}$$

where $\mathbf{A}^{k+\frac{1}{2}} \equiv \mathbf{A}(\Psi^{k+\frac{1}{2}}) = \mathbf{A}(\tfrac{1}{2}\Psi^{k+1} + \tfrac{1}{2}\Psi^k)$ and likewise for the other nonlinear terms. We express the Newton iteration scheme for system (3) as

$$\mathbf{f}'(\Psi^{k+1,(m)})\,\mathbf{h} = -\mathbf{f}(\Psi^{k+1,(m)}) \tag{4}$$

where $\mathbf{h} \equiv \Psi^{k+1,(m+1)} - \Psi^{k+1,(m)}$ and the Jacobian for the system is

$$f'_{ij} = \tfrac{1}{2}A_{ij} + \frac{1}{\Delta t}F_{ij} + \sum_s \frac{\partial A_{is}}{\partial \psi_j^{k+1}}\psi_s^{k+\frac{1}{2}} + \frac{1}{\Delta t}\sum_s \frac{\partial F_{is}}{\partial \psi_j^{k+1}}(\psi_s^{k+1} - \psi_s^k) + \frac{\partial b_i}{\partial \psi_j^{k+1}} \tag{5}$$

The Picard iteration scheme can be written as

$$(\tfrac{1}{2}\mathbf{A}^{k+\frac{1}{2},(m)} + \frac{1}{\Delta t}\mathbf{F}^{k+\frac{1}{2},(m)})\,\mathbf{h} = -\mathbf{f}(\Psi^{k+1,(m)}) \tag{6}$$

In (4) and (6) superscript k is a time step index and superscript (m) an iteration index.

Non-Iterative Newton and Picard Methods   The non-iterative Newton and Picard schemes are derived by taking a single iteration with the Newton or Picard methods, using as initial guess the solution from the previous time step. That is, we take $\Psi^{k+1} = \Psi^{k+1,(1)}$ and $\Psi^{k+1,(0)} = \Psi^k$. With these assumptions the Newton scheme (4) is reduced to the non-iterative scheme

$$\left[\frac{1}{\Delta t}\mathbf{F}^k + \tfrac{1}{2}\mathbf{A}'^k + \tfrac{1}{2}\mathbf{A}^k + \tfrac{1}{2}\mathbf{B}'^k\right](\Psi^{k+1} - \Psi^k) = -\mathbf{A}^k\Psi^k - \mathbf{b}^k + \mathbf{q}^{k+\frac{1}{2}} \tag{7}$$

where $A'_{ij} \equiv \sum_s(\partial A_{is}/\partial \psi_j)\psi_s$ and $B'_{ij} \equiv \partial b_i/\partial \psi_j$. The Picard scheme (6) reduces to

$$\left[\frac{1}{\Delta t}\mathbf{F}^k + \tfrac{1}{2}\mathbf{A}^k\right](\Psi^{k+1} - \Psi^k) = -\mathbf{A}^k\Psi^k - \mathbf{b}^k + \mathbf{q}^{k+\frac{1}{2}} \tag{8}$$

We will at times refer to (7) and (8) as the linearized Newton and Picard schemes, respectively, since these non-iterative schemes can also be derived by linearizing the Crank-Nicolson system (3), in a procedure similar to that used to obtain the Implicit Factored scheme presented below. This derivation is useful to illustrate that the non-iterative strategies (7) and (8) are $O(\Delta t)$ accurate, compared to $O(\Delta t^2)$ accuracy for their iterative counterparts (4) and (6).

Implicit Factored Method The Implicit Factored method is a linearization procedure introduced by Beam and Warming [1] and Briley and McDonald [2] which can be used to transform an implicit $O(\Delta t^2)$ scheme into a non-iterative scheme with the same level of accuracy. There is a complication in applying the Implicit Factored method to Richards' equation in the form (1), as the nonlinearity in the time derivative term is of the form $f(\varphi)(\partial\varphi/\partial t)$ rather than $\partial g(\varphi)/\partial t$. Linearizing such a term to $O(\Delta t^2)$ accuracy will require higher order Jacobian terms or three time levels. In order to avoid this complication, we can re-cast Richards' equation with soil moisture content θ as the dependent variable, or we can maintain a head-based formulation and write Richards' equation in the form

$$\frac{\partial\psi}{\partial t} = H(\psi)\frac{\partial}{\partial z}\left(K_s K_r(\psi)\left(\frac{\partial\psi}{\partial z}+1\right)\right) \tag{9}$$

where $H(\psi) \equiv 1/S(\psi) = (d\theta/d\psi)^{-1}$.

Applying a finite element Galerkin discretization in space to (9) produces the system of ordinary differential equations

$$\mathbf{G}(\mathbf{\Psi})\mathbf{\Psi} + \mathbf{F}\frac{d\mathbf{\Psi}}{dt} = \mathbf{q}(t,\mathbf{\Psi}) - \mathbf{b}(\mathbf{\Psi}) \tag{10}$$

where $\mathbf{G} \equiv \mathbf{A} + \mathbf{C} + \mathbf{D}$ and with the system matrices and vectors, over local subdomain element $\Omega^{(e)}$, expressed as

$$\mathbf{A}^{(e)} = K_s^{(e)}\int_{\Omega^{(e)}} H^{(e)}K_r^{(e)}\mathbf{N}_{,z}^{(e)T}\mathbf{N}_{,z}^{(e)}\,dz\;;\quad \mathbf{b}^{(e)} = K_s^{(e)}\int_{\Omega^{(e)}} H^{(e)}K_r^{(e)}\mathbf{N}_{,z}^{(e)T}\,dz$$

$$\mathbf{C}^{(e)} = K_s^{(e)}\int_{\Omega^{(e)}} H'^{(e)}K_r^{(e)}\mathbf{N}^{(e)T}\mathbf{N}_{,z}^{(e)T}\mathbf{N}_{,z}^{(e)}\mathbf{\Psi}^{(e)}\,dz\;;\quad \mathbf{F}^{(e)} = \int_{\Omega^{(e)}} \mathbf{N}^{(e)T}\mathbf{N}^{(e)}\,dz \tag{11}$$

$$\mathbf{D}^{(e)} = K_s^{(e)}\int_{\Omega^{(e)}} H'^{(e)}K_r^{(e)}\mathbf{N}^{(e)T}\mathbf{N}_{,z}^{(e)T}\,dz\;;\quad \mathbf{q}^{(e)} = \pm\left[H^{(e)}q_{r_N}\mathbf{N}^{(e)T}\right]\Big|_{\Gamma_N}$$

In (11), $H' = dH/d\psi$, $\mathbf{N}$ is the vector of basis functions, $\mathbf{N}^T$ is the transpose of $\mathbf{N}$, $\mathbf{N}_{,z} = d\mathbf{N}/dz$, and q_{r_N} is a specified Darcy flux on the natural boundary Γ_N.

In comparing (2) and (10) we note that the finite element discretization of (9) introduces two additional system matrices $\mathbf{C}$ and $\mathbf{D}$, matrix $\mathbf{F}$ is now independent of $\mathbf{\Psi}$, and the boundary flux vector $\mathbf{q}$ now has a nonlinear dependency. The Crank-Nicolson scheme is used to discretize (9) in time, and Taylor series expansions are used to linearize the nonlinear terms to $O(\Delta t^2)$ accuracy. This results in the $O(\Delta t^2)$ Implicit Factored scheme, which we express as

$$\mathbf{U}^k\left(\mathbf{\Psi}^{k+1} - \mathbf{\Psi}^k\right) = \mathbf{v}^k \tag{12}$$

where

$$\mathbf{U}^k \equiv \frac{1}{\Delta t}\mathbf{F} + \tfrac{1}{2}\left[\mathbf{G}^k + \mathbf{G}'^k + \mathbf{B}'^k - \mathbf{Q}_M^k\right]\;;\quad \mathbf{v}^k \equiv \mathbf{q}_v^k - \mathbf{b}^k - \mathbf{G}^k\mathbf{\Psi}^k \tag{13}$$

$$(Q_M^k)_{ij} \equiv \begin{cases} q_{\Gamma_N}^{k+\frac{1}{2}} H'(\psi_i^k), & \text{if } i = j \text{ and } i \in \Gamma_N \\ 0, & \text{if } i \neq j \text{ or } i \notin \Gamma_N \end{cases} \qquad (14)$$

$$(q_V^k)_i \equiv \begin{cases} q_{\Gamma_N}^{k+\frac{1}{2}} H(\psi_i^k), & \text{if } i \in \Gamma_N \\ 0, & \text{if } i \notin \Gamma_N \end{cases} \qquad (15)$$

and where $G'_{ij} \equiv \sum_s (\partial G_{is}/\partial \psi_j)\psi_s$ and $B'_{ij} \equiv \partial b_i/\partial \psi_j$.

Three-Level Lees Scheme A non-iterative $O(\Delta t^2)$ accurate scheme involving three time levels was introduced by Lees [7] and Douglas [5] for discretizing equations of the form $b(u)u_t = (a(u)u_x)_x + f(u, u_x)$. Applying the scheme to Richards' equation (1) spatially discretized as (2), we obtain the three-level Lees scheme

$$\mathbf{F}(\mathbf{\Psi}^k)\frac{\mathbf{\Psi}^{k+1} - \mathbf{\Psi}^{k-1}}{2\Delta t} + \frac{1}{3}\mathbf{A}(\mathbf{\Psi}^k)(\mathbf{\Psi}^{k+1} + \mathbf{\Psi}^k + \mathbf{\Psi}^{k-1}) = \mathbf{q}(t^k) - \mathbf{b}(\mathbf{\Psi}^k) \qquad (16)$$

which we write in the form

$$[\frac{1}{2\Delta t}\mathbf{F}^k + \frac{1}{3}\mathbf{A}^k]\mathbf{\Psi}^{k+1} = \frac{1}{2\Delta t}\mathbf{F}^k\,\mathbf{\Psi}^{k-1} - \frac{1}{3}\mathbf{A}^k[\mathbf{\Psi}^k + \mathbf{\Psi}^{k-1}] - \mathbf{b}^k + \mathbf{q}^k \qquad (17)$$

The three-level Lees scheme is attractive in that it is a non-iterative, $O(\Delta t^2)$ accurate method which does not involve the computation of Jacobian matrices. However, the scheme requires an additional level of storage (though only for heads) and its three-level nature makes dynamic timestep control less straightforward. In addition, an analysis by Varah [9] of the Lees scheme applied to Burgers' equation, and implementation of the scheme to solve porous media flow problems (Culham and Varga [4], Paniconi [8]) suggests that the scheme may generate nonlinear instabilities and oscillations.

Comparison of Time-Discretization Schemes

Infiltration, drainage, and evaporation simulations were run under various boundary and initial condition scenarios, using a range of soil properties. We used the relations developed by van Genuchten and Nielsen [6] to describe the soil water retention and conductivity curves $\theta(\psi)$ and $K_r(\psi)$, with modifications as described below to accomodate the Implicit Factored scheme. For each test problem, all six schemes were run repeatedly, using a different, fixed timestep size Δt for each run. All test problems used a mesh size Δz fine enough to ensure that spatial discretization errors did not dominate over time discretization errors. Error norms were computed by comparing the numerical solution $\hat{\psi}$ at some fixed midnodal point z_{mid} in the column with a base case, or 'exact' solution ψ_{exact}. That is

$$\varepsilon(\Delta t) = \max_t \left| \hat{\psi}(z_{mid}, t; \Delta t) - \psi_{exact}(z_{mid}, t) \right|, \qquad t = \Delta t, 2\Delta t, \ldots, T \qquad (18)$$

The amount of CPU used for each run was recorded, and for the iterative schemes, the number of iterations as well. The test problems were run on a CDC Cyber 205.

Two types of base case solutions were used for the test problems: an exact solution obtained from a variant of Richards' equation (described below), and a numerical solution of the original Richards' equation solved using the iterative Newton scheme with a very fine mesh and timestep size.

Extended van Genuchten Moisture Curves In order to avoid a singularity in form (9) of Richards' equation, the soil water retention function of van Genuchten and Nielsen [6] was extended by introducing parameters S_s and ψ_o:

$$\theta(\psi) = \begin{cases} \theta_r + (\theta_s - \theta_r)[1 + \beta]^{-m}, & \psi \leq \psi_o \\ \theta_r + (\theta_s - \theta_r)[1 + \beta_o]^{-m} + S_s(\psi - \psi_o), & \psi \geq \psi_o \end{cases} \qquad (19)$$

where $\beta \equiv (\psi/\psi_s)^n$, $\beta_o \equiv \beta(\psi_o) = (\psi_o/\psi_s)^n$, $m = 1 - 1/n$, and n, θ_r, θ_s, ψ_s are fitting parameters. S_s provides the option of having a nonzero specific soil moisture capacity in the saturated and near-saturated region, and ψ_o can be calculated after specifying an S_s value by imposing a continuity requirement on $\theta(\psi)$ and $d\theta/d\psi$.

Exact Solution with Forcing Function   A forcing term $f(z,t)$ is introduced into Richards' equation (1)

$$S(\psi)\frac{\partial \psi}{\partial t} = \frac{\partial}{\partial z}\Big(K_s K_r(\psi)\big(\frac{\partial \psi}{\partial z} + 1\big)\Big) + f(z,t) \tag{20}$$

or, for the Implicit Factored scheme,

$$\frac{\partial \psi}{\partial t} = \frac{1}{S(\psi)}\frac{\partial}{\partial z}\Big(K_s K_r(\psi)\big(\frac{\partial \psi}{\partial z} + 1\big)\Big) + f_{IF}(z,t) \tag{21}$$

with $f_{IF}(z,t) = f(z,t)/S(\psi)$. Adapting one of the exact solutions presented by Brutsaert [3] to an infiltration problem with extended van Genuchten moisture curves, we get

$$\psi_{exact} = \psi(z,t) = \psi_s \left[(1 + \beta_o)\, \exp\big(\frac{L - z}{ma\sqrt{t + b}}\big) - 1\right]^{1/n} \tag{22}$$

where L is the height of the soil column and a, b are constants. The forcing functions are backed out of (20) and (21), assuming (22), and incorporated into the finite element codes.

Discussion of Results   We present, in Figure 1, the results from two test problems, plotted in CPU seconds as a function of error norm, or accuracy level. Table 1 gives the parameter values used in these runs. The initial condition and the boundary condition at the base node for the first test problem are determined by the exact solution, which was generated using $a = 10.0, b = 0.8$. For the second problem, the initial condition was hydrostatic equilibrium with a base node BC of $\psi = 0.0$, and the base case solution was generated using a mesh size $\Delta z = 0.01$ and a time step size $\Delta t = 0.0025$.

Table 1.   Parameter Values for Test Problems 1 and 2 (units: m, hr)

Test Problem	Base Case	θ_r	θ_s	ψ_s	n	S_s	ψ_o
1	Exact	0.15	0.38	-1.2	4.0	0.01	-0.3117
2	Newton	0.08	0.45	-3.0	3.0	0.001	-0.1911

Test Problem	Δt range	T	K_s	Surface BC	L	Δz	z_{mid}
1	$[0.01, 25.0]$	100.0	0.0004	$\psi = \psi_o$	5.0	0.005	2.5
2	$[0.005, 4.0]$	32.0	5.0	$q = t/64$	10.0	0.1	8.0

In Figure 1a, where the base case solution is exact, the Implicit Factored scheme is the most efficient over all levels of accuracy. The Picard, Newton, and Lees schemes are of comparable efficiency throughout, while the $O(\Delta t)$ accurate linearized Newton and Picard schemes perform poorly when high accuracy requirements are imposed. In Figure 1a, an error norm of 0.006 corresponds to an error of approximately 1% in pressure heads.

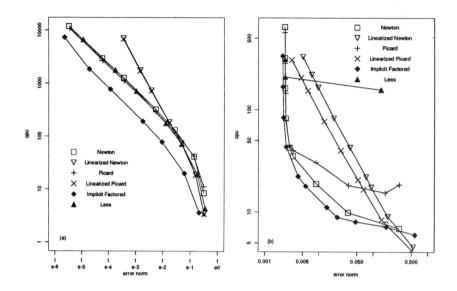

Figure 1. Evaluation of time-discretization strategies: (a) Exact solution used as base case; (b) Fine grid Newton used as base case.

In many problems where Richards' equation is solved without an explicit forcing term, and where boundary conditions are not time-varying, the only forcing occurs during early stages of the simulation, when the solution must adapt to the imposed boundary conditions. In the absence of other complicating factors, such as dry soil initial conditions, highly nonlinear moisture curves, or alternating episodes of infiltration, drainage, and evaporation, we often find in such problems that the Newton and Picard schemes are inefficient during the first few time steps, requiring many iterations to converge, but that after this initial difficulty both methods converge in 2 or 3 iterations per time step, making these methods equally or more efficient than the three-level Lees scheme and the linearized (Implicit Factored and non-iterative Newton and Picard) schemes. Figure 1b shows the results of a test problem where the surface boundary flux is time-varying and the soil is initially quite dry near the surface. In this plot, an error norm of 0.07 corresponds to an error of approximately 1% in pressure heads. For this problem, the Picard scheme required many more iterations to converge than the Newton scheme, the Lees scheme was unstable except at very small time steps, and the non-iterative Picard and Newton schemes were inefficient at error norms below 2%. We see that the Implicit Factored and iterative Newton schemes are the most efficient for this test problem, with the Implicit Factored scheme being slightly more efficient.

Conclusions

We have compared the accuracy and computational efficiency of six time-integration strategies. In our numerical tests we observed the theoretically expected rates of convergence for each of the six schemes. Between the iterative Newton and Picard schemes, we find the Newton method to be more efficient in problems involving highly nonlinear soil properties or time-varying forcing terms. Of the four non-iterative schemes investigated as alternatives to the conventional Newton and Picard methods for solving the nonlinear Richards' equation, we find that the $O(\Delta t^2)$ accurate Implicit Factored scheme shows the most promise. The main limitation on this scheme is that it requires nonsingular moisture characteristics in order to handle problems involving flow in saturated or near-saturated soils. For the van Genuchten moisture curves used in this work, removing this singularity resulted in a reciprocal moisture capacity function with very steep gradients in its higher derivatives, and these gradients caused the Implicit Factored scheme to perform poorly in near-saturated regions. Another drawback is the complexity of the finite element system resulting from Richards' equation in the form (9), with two additional, nonsymmetric system matrices and a nonlinear flux component. These shortcomings may make it worthwhile to investigate an Implicit Factored scheme implementation for the moisture content based form of Richards' equation or for the mixed formulation of the unsaturated flow problem.

References

1. BEAM, R. M. AND WARMING, R. F. An implicit finite-difference algorithm for hyperbolic systems in conservation-law form. *J. Comput. Phys.* 22 (1976), 87–110.

2. BRILEY, W. R. AND McDONALD, H. Solution of the multidimensional compressible Navier-Stokes equations by a generalized implicit method. *J. Comput. Phys.* 24 (1977), 372–397.

3. BRUTSAERT, W. Some exact solutions for nonlinear desorptive diffusion. *J. Appl. Math. Phys.* 33 (1982), 540–546.

4. CULHAM, W. E. AND VARGA, R. S. Numerical methods for time-dependent, nonlinear boundary value problems. *Soc. of Pet. Eng. J.* 11 (1971), 374–388.

5. DOUGLAS, J. *A survey of numerical methods for parabolic differential equations.* In *Advances in Computers, Vol. 2*, F. L. Alt, Ed. Academic Press, New York, 1961, 1–54.

6. VAN GENUCHTEN, M. T. AND NIELSEN, D. R. On describing and predicting the hydraulic properties of unsaturated soils. *Ann. Geophys.* 35 (1985), 615–628.

7. LEES, M. A linear three-level difference scheme for quasilinear parabolic equations. *Math. Comp.* 20 (1966), 516–522.

8. PANICONI, C. Ph.D. dissertation , Princeton University, Princeton, N. J., in preparation.

9. VARAH, J. M. Stability restrictions on second order, three level finite difference schemes for parabolic equations. *SIAM J. Numer. Anal.* 172 (1980), 300–309.

3-D Modelling of Coupled Groundwater Flow and Transport within Saturated and Unsaturated Zones

J.M. Usseglio-Polatera(*), A. Aboujaoudé(*), P. Molinaro(**), R. Rangogni(**)

(*) Laboratoire d'Hydraulique de France (LHF), BP 172X, 38042 Grenoble, France
(**) ENEL/CRIS, Via Ornato 90/14, 20162 Milano, Italy

ABSTRACT

The paper presents the numerical investigations carried out during the development of 2-D and 3-D codes implementing DEDALE modelling system for coupled groundwater flow and solute transport within saturated and unsaturated soils. DEDALE is based on a fractional step (process splitting) approach involving finite difference discretisations.

OBJECTIVES

In groundwater contamination problems, flow and transport equations must be coupled together when the fluid density is varying with the contaminant concentration. This is the case with salt intrusion problems in coastal aquifers. The 2-D and 3-D numerical modelling of these processes generates computational difficulties especially when both saturated and unsaturated zones are concerned and when hydraulic singularities (wells, drain-pipes, trenches, impervious walls) and intricate topologies of the computational domain have to be taken into account.

The major purpose of this paper is to provide an overview and an analysis of the experience carried out jointly by ENEL/CRIS (Italy), LHF and SOGREAH (France) during a collaboration in the development and the validation of the modelling system DEDALE implemented in 2-D and 3-D versions.

DEDALE is based on a fractional step (process splitting) approach [6] involving finite difference schemes. A 2-D vertical version has been developed and validated and is fully operational now [5]. A 3-D version is under development and validation. The authors place the emphasis on the evolution in the computational approach between the 2-D and 3-D versions.

GOVERNING EQUATIONS

It is assumed that Darcy's formulation applies. A Forchheimer formulation has been introduced in the 2-D vertical version in order to account for some turbulence effects (flow modelling in breakwaters under short waves) but, as far as the computational method is concerned, due to the linearization of the quadratic term, it makes no difference with Darcy's formulation. Therefore DEDALE solves the following equations [1]:

Darcy's law with variable density:

$$\vec{q} = -K\left[\overrightarrow{gradh} + (h+z)\frac{\overrightarrow{grad\,\rho}}{\rho}\right] \qquad (1)$$

Fluid mass balance equation:

$$\frac{\partial\,(\rho\theta)}{\partial t} + div\,(\rho\vec{q}) = 0 \qquad (2)$$

Contaminant mass balance equation (Advection-Diffusion):

$$\frac{\partial\,(\rho\theta\,C)}{\partial t} + div\,[\rho C\,\vec{q} - \rho\theta\,D\,\overrightarrow{grad}\,C] = 0 \qquad (3)$$

Fluid density variation with respect to contaminant content: $\rho = f_1\,(C)$ $\qquad$ (4)

Hydraulic conductivity variation with respect to water content: $K = f_2\,(\theta)$ $\qquad$ (5)

Capillary pressure variation with respect to water content: $h + z = f_3\,(\theta)$ $\qquad$ (6)

with:
$\vec{q}$ flow velocity components (q is the velocity vector)
h   piezometric head
C   contaminant concentration
ρ fluid density (with contaminant)
θ water content
K   hydraulic conductivity
z   elevation with respect to a given datum (z is assumed to increase downwards)
g   gravity
D   diffusion tensor
p   pressure [p = ρg (h+z)]

NUMERICAL SCHEME FOR THE 2-D VERTICAL VERSION

According to the process splitting technique [8], within every time increment, three steps are solved in succession on a conventionally space-staggered grid : flow, contaminant advection and contaminant diffusion. This leads to a partially coupled solution [7].

First step - Flow Calculation
The first step involves a combination of equations (1) and (2) leading to a new estimate of piezometric heads:

$$\frac{\partial\,(\rho\theta\,)}{\partial t} - div\,[\rho K\,\overrightarrow{gradh} + K\,(h+z)\,\overrightarrow{grad\rho}] = 0 \qquad (7)$$

Note that $\vec{q}$ have been eliminated and that its calculation is not required to solve the problem. Note also that K and θ are related by (5).
Equation (7) can be transformed into:

$$\rho\gamma\,(h)\,\frac{\partial h}{\partial t} - div\,[\rho K\,(h)\,\overrightarrow{grad}\,h + K(h)\,(h+z)\,\overrightarrow{grad}\,\rho] + \theta\,(h)\,\frac{\partial\rho}{\partial t} = 0 \qquad (8)$$

where $\gamma(h) = \partial\theta/\partial h$ is the specific water capacity.

This equation with h is solved using a variant of ADI (Alternate Direction Implicit) methods. An iterative procedure is implemented in order to check the precision with a criterion to be compared to the maximum variation of h between two successive iterations. K(h), θ(h) and ρ are discretised explicitly for the sake of linearisation.

Note that γ(h) is updated at each iteration. Hence, since coefficients are variable, a conventional conjugate gradient method could not have been used.

The method ends up with the inversion of tridiagonal matrices; this is done through a Thomas double sweep algorithm. θ can then be calculated from (6).

Second step - Contaminant Advection

$$\frac{\partial C}{\partial t} + \frac{\overrightarrow{q}}{\theta} \cdot \overrightarrow{grad}\, C = 0 \tag{9}$$

This step is solved by means of the method of characteristics in two dimensions. The calculation of the characteristic line is based on a high order Runge-Kutta method. Each basic time step is split up in order to have nearly regular trajectory segments. The value of C at the foot of the characteristic line is estimated through cubic interpolation except near steep fronts where linear interpolation is adopted. ρ can then be updated, based on Eq (4).

Third step - Contaminant Diffusion

$$\frac{\partial C}{\partial t} - \frac{1}{\rho\theta}\, div\, [\, \rho\, \theta\, D\, \overrightarrow{grad}\, C] = 0 \tag{10}$$

This equation is linearised, assuming that ρ is known from the previous step. After discretization, this equation is split up into both plane directions and the system is solved through an iterative method based on a variant of conjugate gradient with coordination in space [2]. The discretization ends up, after space splitting, with the inversion of tridiagonal matrices and the Thomas double sweep algorithm is implemented again.

NUMERICAL SCHEME FOR THE 3-D VERSION

The numerical scheme has been significantly modified for the 3-D version, mainly the transport/dispersion part. As a matter of fact, numerical modelling of complex real world problems is always a trade-off between accuracy and manageability and it is well known that the implementation of numerical methods which are theoretically the most accurate may lead to severe degradations of this accuracy when complex geometries including various singularities are dealt with. Accordingly, the 3-D algorithm has been simplified and involves only two fractional steps within each time increment:

First step - Flow calculation
The same variant of ADI method than in the 2-D version is used to solve Eq. (8) including the iterative procedure. The updating of .γ(h) at each iteration may induce a too slow convergence and oscillations. In this respect, γ(h) is updated as follows:

$$\gamma\,(h)_k^{n+\frac{1}{2}} = \frac{1}{2}\,\gamma\,(h)^n + \frac{1}{k-1}\sum_{j=1}^{k-1}\gamma_j\,(h)$$

where k is the number of the current iteration and n the index of the current time step.

<u>Second step - Contaminant Advection-Diffusion</u>
The method of characteristics is very accurate provided that the calculation of the characteristic line and the interpolation at the foot of this characteristic line are reliable enough. In 3-D, with a computational domain full of singularities (wells, drain-pipes) inducing locally very unsteady evolutions, within an intricate geometry or when large time steps are used near steady state, these requirements are not easily fulfilled and the theoretical advantages of the method are significantly hindered.

A numerical method able to manage the respective influences of advection and diffusion through the Peclet number has been selected: the "Power Law Scheme", proposed by S. V. Patankar [4] and based on a weighting of the upwind scheme and of the centered scheme depending on the Peclet number.

SPECIFIC BOUNDARY CONDITIONS

DEDALE can account for prescribed head or flux and concentration, or impervious boundaries. In addition, specific boundary conditions are allowed:
- *Seepage surface*: the boundary condition is calculated automatically by the code during the iterative procedure. It may be impermeability (in the unsaturated zone), prescribed head or zero-pressure (in the saturated zone). This condition is required for the simulation of large wells and trenches.
- *Wells*: the internal boundary condition may be prescribed head or prescribed discharge (pumped or injected). Given the initial level, the well geometry, the prescribed discharge and the computed flow from/to the aquifer, a specific module calculates the water level inside the well at every time step through an iterative procedure. In 3-D, wells can also be accounted for as local source/sink terms at given grid points.
- *Trenches*: a specific procedure is implemented in 2-D based on the same principles than wells but without the axial symmetry; in 3-D, the treatment is the same as for wells since the difference is only topological.
- *Drain-pipes*: a drain-pipe is considered as a zero-pressure internal boundary conditions and is dealt with automatically. A drain-pipe within the unsaturated zone is ignored.
- *Semi-pervious* or *impervious thin layers*: they are introduced through an exchange coefficient depending of the thin layer hydraulic conductivity and thickness. They do not require any particular grid refinement.
- *Non-diffusive boundary-condition for concentration*: at inflow boundaries, the contaminant concentration must be prescribed. At outflow boundaries, the concentration is calculated in the advection step following the characteristic line coming from inside the model. Zero diffusion is then assumed within the boundary cell. This procedure eliminates most numerical diffusion near the boundary but work is under way to eliminate the artificial diffusion which may result from the estimate of $\overrightarrow{\text{grad}}\ \rho$ in the flow step near outflow boundaries.

SOIL CHARACTERISTICS

The soil characteristics required for computations are as follows[3]:
- capillary tension $\Psi\ (\theta)$ as a function of soil water content θ;
- hydraulic conductivity $K(\theta)$ as a function of soil water content θ.

They may be input by the user analytically or point by point. The user must be very careful over the definition of the curves, especially on the derivability of $\Psi(\theta)$. They have a strong influence on flow within the unsaturated zone and on the ability of the iterative procedure to converge rapidly at every time step.
Fig.1 gives examples of typical curves.

Each couple of curves K (θ), Ψ (θ) is assigned to the grid points included within a 3-D block defined by the user. This makes it possible to account for any heterogeneity. Different curves may be given for horizontal and vertical directions in order to account for anisotropy.

VALIDATION AND APPLICATION EXAMPLES

The 2-D version of DEDALE has been validated extensively on the reference problems available in the literature and versus the results obtained by other computer codes [5, 6]. It has been used for several real world applications, namely modelling of progressive salt intrusion generated by intensive pumping in coastal aquifers (Fig.2) and modelling of flow/pressure evolution through breakwaters at sea under the influence of short waves (Fig.3). An extensive program of validation of the 3-D version has been set up and the results will be presented in a following paper.

Acknowledgments
The choice of relevant numerical methods has been made in collaboration with Dr. Michel Bonneton, senior technical adviser at LHF.

REFERENCES

1.   Bear, J. and Verruijt, A., Modelling Groundwater Flow and Pollution , D. Reidel Publishing Company, 1987.

2.   Benque, J.P., Cunge, J.A., Feuillet, J., Hauguel, A. and Holly, F.M., New Method for Tidal Current Computation, ASCE Journal of Waterway, Port, Coastal and Ocean Division, Vol. 108, WW3, August 1982.

3.   Bonneton, M., Jardin, P., Lavedan, G., Moullard, P.Y., The MINOS Model, 2-D Representation of Subsurface Flow in Saturated and Unsaturated Zones, Proceedings of 21st IAHR Congress, Melbourne, Australia, August 1985.

4.   Patankar, S.V., Numerical Heat Transfer and Fluid Flow, Series in Computational Methods in Mechanics and Thermal Sciences, Mc Graw Hill Publishers, 1980.

5.   Rangogni, R., Un caso di verficica del codice DEDALE, Internal report (in Italian) ENEL-DSR-CRIS No 3728, February 1989.

6.   Rangogni, R., Calcoli di filtrazione per la simulazione di un dewatering: Confronto tra schemi 2D e 3D, Internal report (in Italian) ENEL-DSR-CRIS No 3754, April 1989

7.   Usseglio-Polatera, J.M. and Jardin, P., Software Environment for Transport and Dispersion of Contaminants within Saturated and Unsaturated Zones, Poster Paper to the International Symposium on Contaminant Transport in Groundwater, Stuttgart, F.R.G., April 1989.

8.   Usseglio-Polatera, J.M. and Chenin-Mordojovich, M.I., Fractional Steps and Process Splitting Methods for Industrial Codes, Proceedings of the VII Inter,. Conf. on Computational Methods in Water Resources, MIT, Cambridge, USA, June 1988.

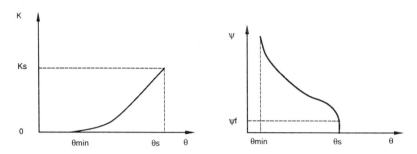

Fig. 1 Examples of curves K(θ) and ψ(θ)

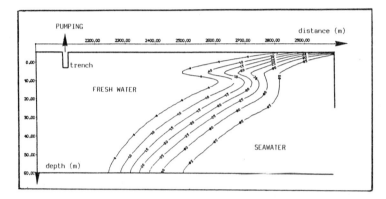

Fig. 2 Application of DEDALE to the salt intrusion in a coastal aquifer with hererogeneous soils due to pumping (Iso-concentrations in salt)

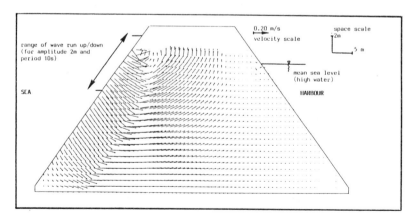

Fig. 3 Application of DEDALE to the calculation of unsteady flow through a breakwater under short waves (Darcy's velocity field at tidal high water and near minimum run down of waves)

A Method to Fit the Soil Hydraulic Curves in Models of Flow in Unsaturated Soils

D.R. Hampton

Department of Geology, Western Michigan University, Kalamazoo, MI 49008, USA

INTRODUCTION

Modeling fluid flow through unsaturated porous media is complicated by the need to describe the soil hydraulic curves. These curves define the relationships between water content and matric head, and between one of these and hydraulic conductivity.

Several methods are used to model these curves in unsaturated media. One common method is to fit an analytical curve (e.g., Brooks-Corey or van Genuchten) to available soil hydraulic data. However, analytical curves do not in general adequately describe the hydraulic curves for dry soils or dual porosity media, and often fit the measured data in the wet range only qualitatively.

Another method is to input data to a curve-fitting routine, such as linear or cubic-spline interpolation. Linear interpolation is inadequate since the derivatives of the curves are undefined at the data points. The derivatives of the curves are very important since they appear in the common form of the Richards equation used to describe flow through unsaturated media (see below). If cubic spline input points are irregularly spaced, as would be expected given the rapid variation of head and conductivity with water content, the resulting curve may not be smooth or even monotonic. Hence, the derivatives can become meaningless.

A model of coupled heat and water flow in saturated/unsaturated soils (Hampton [1]) was used to test the performance of three curve representations: analytical formulas, a cubic spline, and a curve-fitting routine called MONDER developed by Fritsch and Carlson [2]. MONDER produces a smooth, monotonic curve for any monotonic data set.

Two test cases were examined in the course of validating the computer model, HTRANS. The first was Philip's [3] quasi-analytical solution for infiltration without ponding in Yolo light clay. The second was a laboratory study of the water content and temperature changes in a heated box of clay soil. The soil water characteristic curve for this clay suggests that it was a dual-porosity soil.

INFILTRATION INTO YOLO LIGHT CLAY

To test the unsaturated flow component of HTRANS, one-dimensional infiltration into a vertical soil column was simulated. The problem was to calculate water contents at various times and depths in a homogeneous, vertical soil column which is infinitely long while water is infiltrating at the top. This may be described by the Richards equation:

$$\frac{\partial \theta_w}{\partial t} = C_w \frac{\partial \psi}{\partial t} = \frac{\partial}{\partial z}\left[K_{wz}\frac{\partial \psi}{\partial z}\right] + \frac{\partial K_{wz}}{\partial z} \tag{1}$$

where θ_w is volumetric water content,
ψ^w is matric (pressure) head,
C_w is the slope of the ψ versus θ_w curve, and
K_{wz} = hydraulic conductivity in the z direction.

Philip [3] developed a quasi-analytical solution to Equation 1 for a certain set of boundary conditions. Philip's method gives the depth which reaches any given water content, as a function of time. Haverkamp et al [4] compared their model with Philip's results using the analytical formulas below for the hydraulic properties of Yolo light clay:

$$K_w = K_{sat}/[1 + (|\psi|)^{1.77}/124.6] \tag{2}$$

where $K_{sat} = 1.23 \times 10^{-5}$ cm/sec, and

$$\theta_w = \theta_{wr} + 739 \ (\phi - \theta_{wr})/[739 + (\ln|\psi|)^4] \tag{3}$$

where $\phi = 0.495$ and $\theta_{wr} = 0.124$. For $\psi \geq -1$ cm, $\theta_w = \phi$.

A 40 cm deep "infinite" soil column was discretized as a finite element mesh with 80 6-node triangular elements. The time derivative was finite differenced, allowing one to choose Crank-Nicolson ($\alpha = 0.5$), fully implicit ($\alpha = 1$) or in between these two. The Crank-Nicolson method was found to be more accurate and was used below unless otherwise noted.

The hydraulic functions given in [4], Equations 2 and 3, were used to obtain the results in Figure 1a. HTRANS can use analytical functions for the hydraulic properties. The wetting front calculated by HTRANS did not penetrate as deep as the quasi-analytical results at low and high water contents, while there is some overshoot in the middle. However, these results match those reported in [4].

Equation 2 does not match Philip's K data around $\psi = -100$ cm. Equation 2 yields slightly higher conductivities than Philip's data in the wet and dry ranges, while it underestimates K for volumetric water contents between 0.33 and 0.39 by a factor of 1.3 to 2.2. This may explain why the wetting profile shown in Fig. 1a does not match Philip's solution.

HTRANS matched Philip's solution more closely when the hydraulic conductivity curve used corresponded more closely with Philip's data. The points from three K curves were provided as input data for three different HTRANS runs using its cubic spline routines. As expected, the results obtained using points taken from Eq. 2 as cubic spline input data coincide with the results shown in Fig. 1a. When Philip's 22 data points for K were used, the predicted wetting front was at the right depth, but was unacceptably rough. Water content did not decrease monotonically with increasing depth. Hence a third curve was created by smoothing Philip's K curve for water contents between 0.32 and 0.4. Taking cubic spline points from the "smoothed Philip" curve, HTRANS calculated a wetting profile which agrees better with Philip's solution and is nearly monotonic (Figure 1b).

The roughness of the wetting profile in Fig. 1b compared to the smoothness of the wetting front in Fig. 1a illustrates the need for smooth interpolations of the K - ψ and the ψ - θ functions. In this case the roughness is entirely due to the choice of K curve since Eq. 3 was used to describe the ψ - θ curve for all runs. Cubic spline functions can produce smooth curves for these properties if the input data are smooth and if sufficient points are used. The roughness of the wetting front in Fig. 1b may be due to roughness in the K curve produced using the smoothed Philip data points as cubic spline input data. However, attempts to generate a smoother wetting front using closer or different K data points for the cubic spline were unsuccessful. A smooth wetting front resulted when fully-implicit time integration was used, but then the front was retarded by 1 to 3 cm relative to the fronts in Figure 1.

The problem was solved again using MONDER. The wetting fronts were slightly smoother and better-looking when MONDER was used. To make the wetting front reasonably smooth, the time integration level used was $\alpha = 0.6$ (more implicit than Crank-Nicolson).

MODELING DUAL-POROSITY AND/OR DRY MEDIA

HTRANS was also used to calculate the temperature and moisture contents in a disk-shaped box of clay with a heating element in the center (Hampton [1]). The clay was initially relatively dry, with a water content below the wilting point. Water contents were measured by gamma-ray attenuation during the experiment and gravimetrically at the end.

Water content versus matric head data for the clay soil suggest that it is a dual porosity medium (Figure 2). This may be attributed to the way soil was prepared. The clay was originally blocky and hard, but was pulverized and passed through #6 and #10 sieves (U.S. Standard sieve sizes). The resulting soil behaved like a sand whose grains were microporous aggregates.

Attempts to simulate the evolution of water content and temperature with time in this experiment consistently met with failure until MONDER was used. When cubic splines were used for the water content vs matric head curve, often the model would become unstable whenever the water content attained a low value at a point. Both cubic splines and MONDER appeared to adequately follow Figure 2; these two curves if superimposed could be distinguished only at very low water contents. However, examining the slopes of the two curves interpolated from the same input data points (Figure 3) shows that MONDER generates a smoother curve whose slope is relatively well-behaved. The instability at low water contents when a cubic spline was used was apparently due to a reversal in the water content versus matric head curve at a water content of 0.075, in which the slope of the curve actually became negative!

CONCLUSIONS

A new method for interpolating soil hydraulic curves from measured data points has been suggested and evaluated. It appears to have promise, especially in dry or dual porosity media where these curves are hard to fit.

Prior to simulating flow through unsaturated media, it would be prudent to carefully examine the smoothness of the curves fit to the data relating matric head, hydraulic conductivity and water content. As aptly stated by Fritsch (1981, written communication), "$\psi(\theta)$ apparently has vertical slopes near the ends of the θ-range. This can cause severe numerical problems!"

ACKNOWLEDGEMENTS

I thank Fred Fritsch and R. E. Carlson for sharing MONDER and Pete Haaf for computational help.

REFERENCES

1. Hampton, D.R. Coupled Heat and Fluid Flow in Saturated-Unsaturated Compressible Porous Media. Ph.D. dissertation, Colorado State University, Dept. of Civil Engineering, Ft. Collins, CO 80523, U.S.A., pp. 293, 1989.
2. Fritsch, F.N. and Carlson, R.E. Monotone Piecewise Cubic Interpolation, SIAM J. Numerical Analysis, Vol. 17, pp. 238-246, 1980.
3. Philip, J.R. The Theory of Infiltration: 1., Soil Science, Vol. 83, pp. 345-357, 1957.
4. Haverkamp, R., Vauclin, M., Touma, J., Wierenga, P.J. and Vachaud, G. A Comparison of Numerical Simulation Models for One-Dimensional Infiltration, Soil Sci. Soc. Amer. J., Vol. 41, pp. 285-294, 1977.

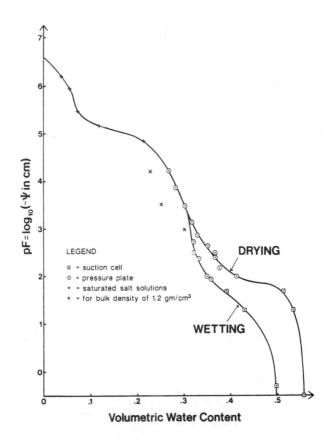

Fig.2 Water characteristic curve for clay soil

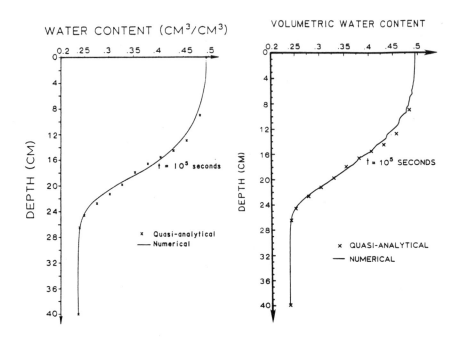

(a) Haverkamp's K   (b) Smoothed Philip K

Figure 1   Infiltration into Yolo light clay

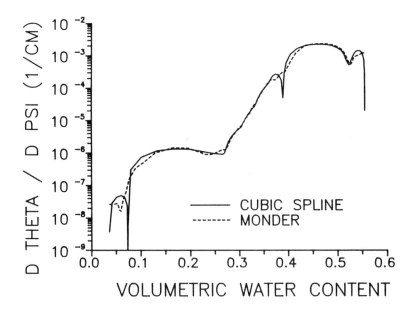

Fig.3 Slopes of interpolated characteristic curve

A Computational Investigation of the Effects of Heterogenity on the Capillary Pressure-Saturation Relation

L.A. Ferrand, M.A. Celia

Water Resources Program, Dept. of Civil Engineering and Operations Research, Princeton University, Princeton, NJ 08544, USA

ABSTRACT

Capillary drainage in porous media can be simulated using a computational model based on pore scale physics. This model can be used to generate the capillary pressure – saturation relation for heterogeneous as well as homogeneous media. Results generated for media consisting of alternating layers of fine and coarse material are shown to differ significantly from curves calculated as simple weighted averages of the results for the individual materials.

INTRODUCTION

Models of multiphase fluid displacement in porous media are used to predict the flow of water in the unsaturated zone as well as the transport of immiscible contaminants in groundwater. These models require specification of the material–dependent relation between fluid contents and fluid pressures. As an alternative to exhaustive laboratory measurements, it is possible to approach parameter identification from a more theoretical point of view. Well–defined pore-scale physics, based on fluid-fluid and fluid-solid interfacial behavior, can be used to simulate fluid displacement in a lattice model of pore space. This model can be used to generate the porous–medium–scale capillary pressure – saturation (P_c-S) relation for arbitrary porous media.

THE COMPUTATIONAL MODEL FOR IMMISCIBLE FLUID DISPLACEMENT

In the computational model, pore space is conceptualized as a cubic lattice of hollow, spherical sites which represent pore bodies. Each site is connected to six neighbors by hollow cylindrical bonds or pore throats. Site and bond radii are chosen at random from separate probability distributions. Figure 1.1 is a schematic of a cross–section through one layer of the lattice. Figure 1.2 shows the structure of the three-dimensional lattice. Once the geometry of the pore space is fixed, the distribution of fluids within the lattice is determined by application of the Young–Laplace equation which relates capillary pressure, surface tension, contact angle and the mean radius of curvature of the fluid-fluid interface.

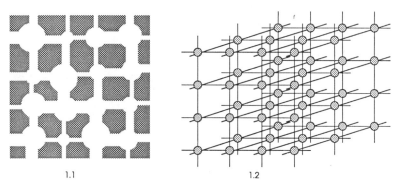

Figure 1. Three-dimensional lattice model of pore space (from Ferrand and Celia (1989b)). 1.1: Section through a single layer. 1.2: Cubic lattice structure.

Each primary drainage simulation mimics an experiment performed in a suction or pressure cell. All sites and bonds within the lattice are initially filled with wetting fluid. Sites on one face of the lattice are assumed to be in contact with a reservoir of nonwetting fluid; sites on the opposite face are in contact with a reservoir of wetting fluid.

Like an experimental sample, the computational lattice is subjected to incremental increases of capillary pressure. The movement of fluid-fluid interfaces through the lattice in response to each change is calculated using the Young–Laplace equation. Pores with the largest radii are drained first, followed by those of successively smaller radii as capillary pressure increases. Because sites must have radii larger than all adjoining bonds, drainage patterns are controlled only by the distribution of bond radii.

Once fluid distributions have been determined for a given value of capillary pressure, a global wetting fluid saturation can be calculated by averaging over the pore space of the computational lattice. Repeated application of this procedure gives a series of data points which are analogous to the P_c-S curves found in laboratory experiments. Details of the drainage algorithm and calibration of the model can be found in Ferrand and Celia (1989a,b).

APPLICATION TO HETEROGENEOUS MEDIA

Given a computational model which simulates capillary fluid displacement in a physically meaningful way, we can use it to predict the effects of medium heterogeneity on P_c-S relations.

A simple heterogeneous medium might consist of layers, either vertical (perpendicular to fluid reservoir faces) or horizontal (parallel to fluid reservoir faces) of two different materials. In a real soil these might be sand and clay layers.   We simulate this type of medium by assigning bond radii from lognormal probability distributions with different means in specified sections of the same lattice.

Figure 2 illustrates the layering patterns used in the computational experiments reported here.   Cases A and B represent homogeneous samples: A is a "fine" material (mean bond radius (μ) = 0.07mm), material B is "coarse" (μ = 0.14mm). Samples C, D and E are made up of two layers (one fine and one coarse) of the same thickness.   Samples F, G, H and I consist of five layers of varying thickness.   In each computational drainage experiment, the wetting fluid reservoir is applied at the bottom face while the nonwetting fluid reservoir is applied at the top face.

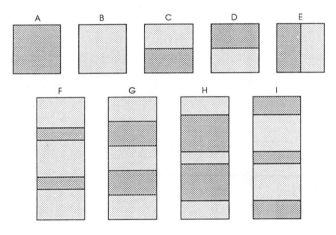

Figure 2.   Layering patterns for heterogeneous media.   Light areas indicate coarse material (mean bond radius (μ)=0.14mm), dark areas are fine material (μ=0.07mm).   Lattice dimensions: A,B,C,D and E: 29x25x25 sites, F,G,H and I: 61X25x25 sites.

Figure 3 illustrates drainage curves generated by the model for two-layered samples.   In this figure and in the figures which follow, each curve represents a mean of five Monte Carlo simulations for the specified distribution.   Open circles (A,B) represent curves generated for homogeneous media. Open triangles represent the mean of homogeneous curves.   Solid triangles represent P_c-S curves generated by the computational model for sample C (Figure 3.1), sample D (Figure 3.2) and sample E (Figure 3.3).   Simulated curves are fairly well approximated by the mean curves for the vertically layered case

(E) and for the horizontally layered case in which the coarse
material overlays the fine material (C). The approximation is
very poor for the horizontally layered case in which the fine

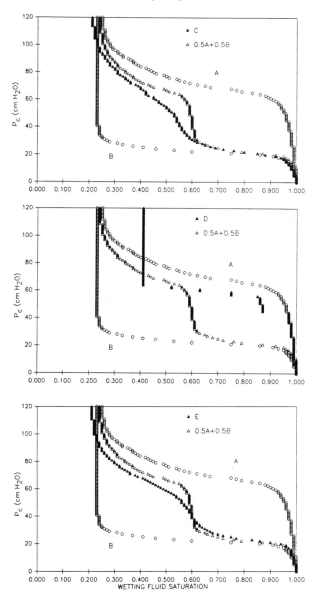

Figure 3. P_c-S drainage curves generated on 29x25x25
homogeneous (open symbols) and heterogeneous (solid symbols)
lattices. Each curve represents a mean of five Monte Carlo
simulations. 3.1: Samples A,B and C. 3.2: Samples A,B and D.
3.3: Samples A,B and E.

material overlays the coarse (D). The behavior observed in
this case is consistent with the fact that the entry pressure
for the fine material is higher than the capillary pressure at

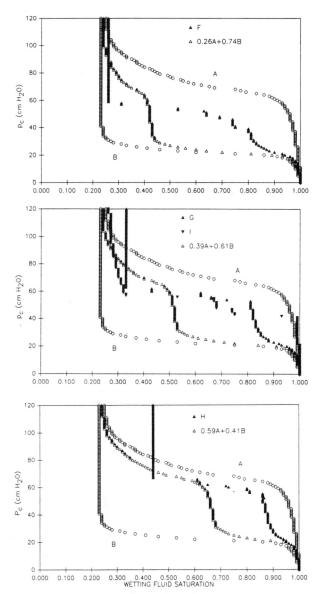

Figure 4. P_c-S drainage curves generated on 29x25x25
homogeneous (open symbols) and 61x25x25 heterogeneous (solid
symbol) lattices.   Each curve represents a mean of five Monte
Carlo simulations.   4.1: Samples A,B and F.   4.2: Samples A,B,G
and I.   4.3: Samples A,B and H.

which the coarse material is drained to its residual value.   No drainage can occur in the lower layer until the invading nonwetting fluid establishes a continuous pathway through the fine layer. As soon as this occurs, the coarse layer fully drains.   Because this eliminates continuous pathways between wetting fluid in the upper layer and the wetting fluid reservoir at the bottom, the wetting fluid remaining in the upper layer is trapped and we see much higher residuals than in homogeneous cases.   We can expect to see similar effects in more complex heterogeneous media.

Figure 4 illustrates drainage curves generated by the model for multilayered samples.   Again, open circles (A,B) represent curves generated for homogeneous media.   Open triangles represent a weighted average of homogeneous curves. Weights are calculated based on the relative volume of each type of material in the lattice.   Solid triangles represent P_c-S curves generated by the model for sample F (Figure 4.1), samples G and I (Figure 4.2) and sample H (Figure 4.3).   In all cases, the weighted average curves are a poor approximation to the simulated curves.   Note, in cases F, G and H, which have similar layering patterns, that the degree to which predicted residual saturations diverge from homogeneous residuals increases as the relative volume of fine material increases.

Figure 4.2 also provides a comparison between drainage curves for two samples which have identical relative volumes of fine and coarse materials and different layering structures. While these curves nearly coincide in the range where most drainage occurs, both entry pressures and residual saturations differ significantly.   These example calculations lead us to conclude that the particular structure of heterogeneities may need to be taken into account in predicting constitutive relations in heterogeneous media.   Simple averages, based on the amount of each material present, may seriously misrepresent the real behavior of heterogeneous systems.

ACKNOWLEDGEMENTS

This work was supported in part by the U.S. Geological Survey under Grants 14–08–0001–G1473 and 14–08–0001–G1747, and by the National Science Foundation under grant 8657419–CES.

REFERENCES

[1] Ferrand, L.A. and M.A. Celia, "Development of a three-dimensional network model for quasi-static immiscible displacement," in Contaminant Transport in Groundwater, H.E. Kobus and W. Kinzelbach (eds.), A.A. Balkema, Rotterdam, 1989a.
[2] Ferrand, L.A. and M.A. Celia, "A percolation-based model for drainage in porous media," submitted to Water Resour. Res., 1989b.

Comparison of $P,H,$ and R-version Adaptive Finite Element Solutions for Unsaturated Flow in Porous Media

J.R. Lang, L.M. Abriola, A. Gamliel

Department of Civil Engineering, The University of Michigan, Ann Arbor, MI, 48109-2125, USA

ABSTRACT

Three one-dimensional self-adaptive Galerkin finite element models for the solution of the non-linear Richards equation for transient unsaturated flow in porous media are presented. The presented models are examples of *p*-, *h*-, and *r*-versions of the finite element method. Comparisons are made between solutions obtained using these approaches and a non-self-adaptive finite element model. Local and global errors of the numerical solutions are measured in two different norms. Advantages and disadvantages of each self-adaptive method are briefly discussed.

INTRODUCTION

The development of self-adaptive numerical models has received much recent attention in the literature. Herein the term 'self-adaptive,' is used to describe models which improve the accuracy of an approximated solution in response to feedback from less accurate solutions.

Self-adaptive methods are computationally desirable for many important environmental problems which typically involve the propagation of a front in time. Such methods are computationally efficient for these types of problems since a relatively coarse approximation is generally adequate in regions away from the front, while a fine approximation is required at the front to provide sufficient resolution. In this paper, one dimensional solutions of the non-linear equation governing unsaturated flow are computed employing three self-adaptive finite element models. Although many models have been developed for the solution of unsaturated flow problems (see [1]), very few have been self-adaptive [2][3]. In this work, the performance of three distinct approaches to solution refinement are compared, the so-called *h*-, *r*-, and *p*-versions.

THE MATHEMATICAL MODEL

The governing equation for vertical isothermal flow of water in a rigid unsaturated soil matrix is given by [4]:

$$C \frac{\partial h}{\partial t} = \frac{\partial}{\partial z} \left(K k_{rw} \left(\frac{\partial h}{\partial z} - 1 \right) \right) \tag{1}$$

where: $C = n \, (ds/dh)$ is the soil moisture capacity,
          n is the matrix porosity,
          h is the suction head,
          s is the water saturation,
          z is the vertical coordinate (directed downwards),
          t is the time coordinate,
          K is the saturated hydraulic conductivity,
          k_{rw} is the relative permeability of the water phase.

Since the soil moisture capacity, C, and the relative permeability, k_{rw}, are functions of h, Equation (1) is a nonlinear partial differential equation in the suction head, h.

To solve Equation (1) numerically, Galerkin's method is applied to the governing equation. For both the *h*- and *r*-version methods, the traditional chapeau basis functions are employed. The *p*-version method uses a family of hierarchic Lagrangian basis functions whose linear member is the chapeau function. Since k_{rw} and ds/dh are functions of h, these coefficients are also expanded in terms of the basis functions in each scheme. The two parameters, K and n, are assumed piecewise constant over each element.

Applying Green's theorem to the flux term in the weighted residual equation and discretizing the time derivative using a backwards difference approximation yields a fully implicit system of non-linear algebraic equations:

$$[M]^{t+\Delta t} \{h\}^{t+\Delta t} = \{r\}^{t+\Delta t} \tag{2}$$

A Newton-Raphson scheme is coupled with a Thomas algorithm to solve this system. The convergence criterion used for all simulations presented herein is

$$|h^{k+1}-h^k|_\infty / |h^{k+1}|_\infty \le e_r \tag{3}$$

where the superscript on h indicates the iteration level during a given time step. All the simulations presented in this paper use a value of 10^{-4} for e_r. Details pertaining to individual adaptive solution schemes are highlighted below. It should be noted that all solutions were obtained in incremental form to minimize truncation error.

SELF-ADAPTIVE ALGORITHMS

In *p*-version self-adaptive algorithms, solution refinement is achieved by adjusting the order of interpolation on a fixed grid discretization. This method was originally developed to solve elliptic problems in mechanics [5][6]. In this work, the *p*-version scheme uses members of a family of C^0 hierarchic Lagrangian basis functions [5]. Only the linear and quadratic members of the hierarchic basis function set are employed. The three degrees of freedom for a given quadratic element are two nodal variables and a so called 'nodeless' variable which is the second derivative of the function at the element midpoint. If the global finite element matrix is formed using linear basis functions, selective enrichment of the solution is achieved by adding rows and columns containing the new terms resulting from the addition of quadratic terms to the trial function. Thus, previously formed matrices using lower order basis functions are imbedded in matrices formed with higher order basis functions, and the computational effort spent in forming the original global matrix with linear basis functions is retained. In addition, the initial solution for h, obtained from iterations over the original global matrix equation, can serve directly as a first estimate for iterations over the enriched mesh and no interpolation is required.

The *p*-version algorithm consists of an iteration solution scheme patterned after a block Gauss-Seidel approach [5]. First, the linear finite element problem (2) is solved to convergence. Next, selected elements are enriched with quadratic terms and an expanded matrix formed. Following enrichment, the original converged solution is used as a first estimate for the enriched grid solution. Further details on the algorithm may be found in [2][7].

An element is enriched if the change in the saturation per unit length in the i<u>th</u> element exceeds the criterion, e_q:

$$|s_{i+1}-s_i|/\Delta z \le e_q \tag{4}$$

Here the subscript on s indicates the node number. This criterion is evaluated using the saturations obtained from the initial linear problem (2). An e_q value of 10^{-5} was used for all the simulations contained in this paper.

In an h-version method the solution is refined by adding new nodes in desired locations. This increases the number of degrees of freedom, while retaining the original order of interpolation. The approach has been widely used for the solution of many different classes of problems. The h-version method used in this paper is loosely based upon a scheme presented by Ewing, et al. [8]. Equation (2) is formulated and solved to convergence on the initial grid. Selected elements are then identified and divided into two equal elements using the refinement criterion (4). Thus, one additional degree of freedom is added to each refined element. Equation (2) is then reformulated on the new grid and solved to convergence as was done on the original grid. In contrast to the p-version simulator, the refined matrix equation is solved using initial head values from the beginning of the time step since experience showed that convergence was not achieved when the converged unknowns from the iterations over the unrefined grid were used as estimates for the unknowns on the refined grid. The h-version scheme requires that previous time level values of the nodal variables be interpolated from the previous values associated with the unrefined grid. This interpolation is performed linearly. Once an element had been refined, however, the values from the previous time step are available on subsequent time steps and no interpolation is necessary.

R-version adaptivity differs from the other methods discussed in this paper in that a *fixed* number of degrees of freedom are distributed in a fashion to maximize accuracy. The order of interpolation is retained. Redistribution of the fixed number of degrees of freedom is accomplished by decreasing the spacing between nodes where large gradients and curvatures in the unknown variable exist and increasing the spacing where the gradients and curvatures are small. The method presented in this paper was originally developed for the solution of nonlinear multiphase flow problems [9]. This algorithm solves for the nodal locations and corresponding h values in a sequential manner. To obtain nodal distribution equations, discrete equidistribution criteria for gradient [10] and curvature [11] are formulated. These criteria are then combined to generate the following set of nonlinear algebraic equations in the nodal positions (x_i):

$$x_{i-2} \{ -B_1[(m_{i-1} - m_{i-2})^2 + \varepsilon_2]^{1/2} \} +$$
$$x_{i-1} \{ -[m_{i-1}^2 + \varepsilon_1]^{1/2} + (B_1 - B_2) [(m_i - m_{i-1})^2 + \varepsilon_2]^{1/2} \} +$$
$$x_i \{ [m_{i-1}^2 + \varepsilon_1]^{1/2} + [m_i^2 + \varepsilon_1]^{1/2}$$
$$+ B_1[(m_{i-1} - m_{i-2})^2 + \varepsilon_2]^{1/2} + B_2[(m_{i+1} - m_i)^2 + \varepsilon_2]^{1/2} \} +$$
$$x_{i+1} \{ -[m_i^2 + \varepsilon_1]^{1/2} + (B_2 - B_1) [(m_i - m_{i-1})^2 + \varepsilon_2]^{1/2} \} +$$
$$x_{i+2} \{ -B_2[(m_{i+1} - m_i)^2 + \varepsilon_2]^{1/2} \} \tag{5}$$

Here m_i is the piecewise linear approximation of the gradient in the unknown in element i. B_1 and B_2 are curvature weighting parameters and ε_1 and ε_2 are known as the artificial repulsive force and artificial viscosity respectively. The last two parameters are commonly employed to retain a few grid points at positions of small or vanishing gradient and curvature [12]. In this work, ε_1 and ε_2 are computed using the following expressions:

$$\varepsilon_1 = (c_1 \times m_{i\ max})^2 \tag{6a}$$
$$\varepsilon_2 = (c_2 \times m_{i\ max})^2 \tag{6b}$$

Here, $m_{i\ max}$ is the maximum element wise gradient in the suction head and c_1 and c_2 are constant input parameters.

In the r-version self-adaptive algorithm Equation (5) is first solved to convergence using a banded matrix solver. Convergence for the grid is attained when the largest change in the nodal positions over the last iteration is less than one tenth of the minimum element size. After the new grid is established, the nodal variables associated with the new nodal locations are interpolated linearly from the nodal variables associated with the old nodal locations. Then matrix equation (2) is formulated using linear basis functions and solved to convergence. A new grid is generated when the length of the element containing the maximum gradient in the unknown is greater than 2.0 times the length of the smallest element. Thus, a new grid may not be generated every time step. Further details on this moving grid algorithm may be found in [5][13][14] along with guidelines for the selection of parameters appearing in (5).

SIMULATIONS

The finite element models discussed above were used to simulate water infiltration into a vertical column of homogeneous sand. The following expressions were employed for the saturation and the relative permeability:

$$k_{rw} = A / (A + |h|^B) \tag{7a}$$
$$s = a(s_s - s_r)/(a + |h|^b) + s_r \tag{7b}$$

Here A, B, a, and b are empirical constants and s_s and s_r are the saturated and residual water levels in the medium. Table 1 contains the values of the various constants and numerical criteria as well as the boundary and initial conditions used for the numerical simulations presented in this paper.

Table 1. Parameters for example simulations

$n = 0.30$	$K = 34.0$ cm/h	$s_s = 0.95$
$s_r = 0.25$	$A = 1.175 \times 10^6$	$B = 4.74$
$a = 1.611 \times 10^6$	$b = 3.96$	

Boundary and initial conditions:

$s_w = 0.333$ at	$t = 0$	$0 < z < 500$ cm
$s_w = 0.890$ at	$z = 0$	$t \geq 0$
$s_w = 0.333$ at	$z = 500$ cm	$t \geq 0$

Moving grid parameters:

$c_2 = 0.1$	$c_2 = 0.0$	$B_1 = 1.05 \times 10^{-7}$
$B_2 = 1.0 \times 10^{-7}$		

A finite element code, validated by comparison with an analytical solution [15] is used in common by all self-adaptive algorithms. Time step size is controlled by the number of iterations over Equation (2) required to achieve convergence. If the solution requires less than three iterations at the previous time level, the time step is multiplied by 1.5 to obtain the new time step size. A small time step was used at the start of most simulations to account for the discontinuous initial conditions. All simulations used a value of 5×10^{-5} h for the initial time step. Solutions were computed with each method, p-, h-, r- and non-adaptive versions, for a variety of discretizations. Figure 1 presents a representative result from a 51 node discretization ($\Delta z = 10$ cm).

As a measure of the relative accuracies of different solutions, two approximate error measures were computed. These are the discrete L_2 and L_∞ norms. Both are determined relative to a fine mesh numerical approximation of the true solution (1000 linear elements, $\Delta z = 0.5$ cm). Error measures and computational times for the simulations shown in Figure 1 are given in Table 2. Here computational effort is reported as the number of cpu seconds required to execute the program. I/O time was not included. A Sun SPARCstation1 was used for all simulations.

DISCUSSION

All self-adaptive schemes resulted in a dramatic improvement in solution accuracy over a non-adaptive approach. The r-version was found to consistently out-perform the h- and p-versions in accuracy measures and was quite competitive in CPU time. The p-version was usually more accurate than the h-version (especially at relatively coarse discretizations) and was substantially faster in most applications. Convergence of the h-scheme tended to require more iterations than the p-version approach and frequent reevaluation of the Jacobian in the Newton-Raphson scheme. In addition, h-version solutions tended to be extremely responsive to the number of refined elements. Error norms varied in excess of one order of magnitude as e_q was decreased until a value of 10E-5 was reached. The p-version did not exhibit such extreme behavior and was relatively insensitive to the degree of enrichment.

Although it would appear from these results that the r-version offers uncontested computational advantage for a desired level of accuracy, it should be noted that this approach requires the determination of several parameters to which the solution is extremely sensitive. This selection must be done in a quasi-trial and error process which requires an experienced user. The p- and h- versions, on the other hand, are relatively easy to use since they have few adjustable parameters.   In addition, extensions of these two approaches to higher dimensions appear more straightforward

Figure 1:   51 node comparison

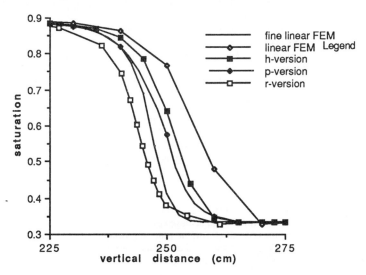

Table 2. Comparison of FEM Models

FEM type	NDF	Error measures L_2	L_∞	CPU time (seconds)
Non-adaptive	51	118.8	25.0	14.5
P-version	62	46.4	12.1	22.5
H-version	62	63.8	16.25	36.1
R-version	51	33.5	8.7	36.2

ACKNOWLEDGEMENTS

This work was supported, in part, by the National Science Foundation under grant ECE-8451469.

REFERENCES

1. van Genuchten, M. Th., Progress in Unsaturated Flow and Transport Modeling, Reviews of Geophysics, 25, 135-140, 1987.
2. Abriola, L.M., Finite Element Solution of the Unsaturated Flow Equation using Hierarchic Basis Functions, Proceedings of the Sixth International Conference on Finite Elements in Water Resources, Lisbon, Portugal, 125-133, Springer-Verlag, 1986.
3. Sorek, S. and C. Braester, An Adaptive Eulerian-Lagrangian Approach for the Numerical Simulation of Unsaturated Flow, Proceedings of the Sixth International Conference on Finite Elements in Water Resources, Lisbon, Portugal, 87-100, Springer-Verlag, 1986.
4. Huyakorn, P.S. and G. F. Pinder, Computational Methods in Subsurface Flow, Academic Press, New York, pp. 146-150, 1983.
5. A. Peano, Hierarchies of Conforming Finite Elements for Plane Elasticity and Plate Bending, Computers and Mathematics with Applications, 2, 211-224, 1976.
6. Zienkiewicz, O.C., J. P. De S. R. Gago and D. W. Kelly, The Hierarchical Concept in Finite Element Analysis, Computers and Structures, 16, 53-65, 1983.
7. Abriola, L.M. and J.R. Lang, Self-Adaptive Hierarchic Finite Element Solution of the One-dimensional Unsaturated Flow Equation, International Journal for Numerical Methods in Fluids, in press, 1990.
8. Ewing, R.E., Adaptive Grid Refinement Methods for Time Dependent Flow Problems, Communications in Applied Numerical Methods, 3, 351-358, 1987.
9. Gamliel, A., Simulation of Immiscible Multiphase Flow in Porous Media Using a Moving Grid Finitie Element Method, Ph.D. Thesis, Dept of Civil Engineering, Univ. of Michigan, 1989.
10. Thompson, J.F., A Survey of Dynamically-Adaptive Grids in the Numerical Solution of Partial Differential Equations, Applied Numerical Mathematics, 1, 3-27, 1985.
11. Mosher, C.M., A Variable Node Finite Element Method, J. Computational Physics, 57(2), 157-187, 1985.
12. Miller, K. and R. Miller, Moving Finite Elements. I, SIAM J. Numerical Analysis, 18, 1019-1032, 1981.
13. Gamliel, A. and L.M. Abriola, A Moving Grid Solution for the Coupled Nonlinear Equations Governing Multiphase Flow in Porous Media: 1. Model Development, in review.
14. Gamliel, A. and L.M. Abriola, A Moving Grid Solution for the Coupled Nonlinear Equations Governing Multiphase Flow in Porous Media: 1. Example Simulations and Sensitivity Analysis, in review.
15. Haverkamp, R., M. Vauclin, J. Touma, P. J. Wierenga and G. Vachaud, A Comparison of Numerical Simulation Models for One-dimensional Infiltration, Soil Science Society of America Journal 41, 285-294, 1977.

SECTION 3 - MULTIPHASE FLOW

Multiphase Flow Simulation in Groundwater Hydrology and Petroleum Engineering

R.E. Ewing(*), M.A. Celia(**)
(*) Institute for Scientific Computation, University of Wyoming, Laramie, WY 82071, USA (**) Department of Civil Engineering and Operations Research, Princeton University, Princeton, NJ 08544, USA

INTRODUCTION

Numerical simulation of the flow of multiphase and/or multicomponent fluids in porous media is important in many branches of science and engineering. For example, in the areas of soil science, agricultural engineering, and groundwater hydrology, movement of fluids and their dissolved components in both saturated and unsaturated soils is an important environmental consideration. In petroleum engineering, improved recovery of oil is based on simulation of multiphase, multicomponent fluid transport in deep rocks. In both the groundwater contamination and petroleum recovery problems, mass transfer across phase boundaries may also be an important consideration.

The underlying physics of the groundwater and petroleum problems are very similar. However, for a variety of reasons, the analysis and solution of these problems in the fields of groundwater hydrology and petroleum engineering have historically been independent. In an initial attempt to bridge the schism between these two fields, this paper focuses on analysis of the unsaturated flow and petroleum recovery problems in a common framework, within which both similarities and differences may be illuminated.

For the present study, we assume there are two fluids flowing simultaneously in a porous solid. In unsaturated flow, these fluids are water (w) and air (a); in the petroleum problem the fluids are assumed to be water and oil (o). Relevant material properties, including the capillary pressure-saturation and relative permeability-saturation relations, are assumed to be known.

Statements of mass conversion for each phase, coupled with the multiphase version of Darcy's equation, lead to a set of nonlinear parabolic partial differential equations that govern the movement of the fluids. For multiphase flow systems, the nonlinear nature of these equations leads to solutions that are significantly different from those to the analogous linear equations. These nonlinear effects are important and must be understood. The approach taken herein is to rewrite the governing equations in the form of a nonlinear advection-diffusion equation and to infer qualitative and quantitative information about solution behavior from these equations. In addition, the advection-diffusion equation form leads naturally to an Eulerian-Lagrangian Localized Adjoint Method (ELLAM) (Celia et al. [4]) for numerical solution. The ELLAM technique is a generalized characteristic method that provides proper treatment of boundary conditions and thereby possesses the conservative property. This paper focuses on both the nonlinear physics of multiphase systems and the associated numerical solution methods.

MULTIPHASE FLOW EQUATIONS

We first make some assumptions to determine a set of model equations which possess most of the key properties of general multiphase flow equations. We try to state these assumptions explicitly and perhaps to indicate some of the difficulties involved in generalizations of these assumptions.

Although the compressibility of the solid matrix and the fluids, especially air, may be important in certain cases, we ignore these effects in the present treatment. We thus assume that both the fluids and the media for the model equations treated below are incompressible. This assumption implies that the density of each phase is constant in space and time.

The equations describing two-phase, immiscible, incompressible displacement in a porous medium are given by

$$\Phi \frac{\partial S_w}{\partial t} - \nabla \cdot \left(k \frac{k_{rw}}{\mu_w} (\nabla p_w + \rho_w g \nabla z) \right) = F_w, \quad x \in \Omega, t \in J, \quad (1)$$

$$\Phi \frac{\partial S_i}{\partial t} - \nabla \cdot \left(k \frac{k_{ri}}{\mu_i} (\nabla p_i + \rho_i g \nabla z) \right) = F_i, \quad x \in \Omega, t \in J, \quad (2)$$

for $i = a, o$, denoting air or oil, respectively. Here $S_i, k_{ri}, \mu_i, p_i, \rho_i, and F_i, i = w, a, o$, are the saturations, relative permeabilities, viscosities, pressures, densities, and source rates of the respective phases. Φ, k and g are the porosity, intrinsic or absolute permeability, and gravity terms and z, Ω, and J are the vertical direction (assumed positive upward) and the spatial and temporal domains of the problem. $p_{ciw} = p_i - p_w$ is the capillary pressure with $i = a$ or o.

The standard assumption made in unsaturated flow in soils is that the air phase is infinitely mobile with an essentially constant pressure, equal to atmospheric, and that Equations (1) and (2) can be reduced to a single, classical Richards' equation for the water phase (Hillel [11]). Three standard forms of the unsaturated flow equation, the "h-based" form, the "θ-based" form, and the "mixed" form are given below:

$$C(h)\frac{\partial h}{\partial t} - \nabla \cdot K(h)\nabla h - \frac{\partial K}{\partial z} = 0 \,, \tag{3}$$

$$\frac{\partial \theta}{\partial t} - \nabla \cdot D(\theta)\nabla \theta - \frac{\partial K}{\partial z} = 0 \,, \tag{4}$$

$$\frac{\partial \theta}{\partial t} - \nabla \cdot K(h)\nabla h - \frac{\partial K}{\partial z} = 0 \,, \tag{5}$$

where θ is the moisture content (defined by $\theta = \Phi S_w$) and h is the water head (defined by $h = p_w/\rho_w g$). $C(h) \equiv \frac{d\theta}{dh}$ is the specific moisture capacity function, $K(h)$ is the unsaturated hydraulic conductivity, and $D(\theta) \equiv \frac{K(\theta)}{C(\theta)}$ is the unsaturated diffusivity. In stating the above governing equations, we also assume that the relationships between θ and h (S_w and p_c) and K and θ (k_{ri} and S_w) are known explicitly from capillary pressure and relative permeability curves.

Although each of the model Equations (1)–(5) are formally parabolic, their properties are not well understood due to the nonlinearities involved. For unsaturated flow, we know (Celia et al. [1]) that the formulation given by Equation (3) can give problems in mass-balance calculations. Also, Equation (4) breaks down as θ tends to its saturated value, since in backing θ out of a capillary pressure curve, the term $D(\theta)$ loses its meaning. Thus, arguments have been made (Celia et al. [1]) that the mixed form of Richards' equation is to be preferred.

In petroleum engineering applications, the system (1)–(2) is turned into a new system by summing Equations (1) and (2) and using this sum as one of the governing equations. The summation eliminates the time derivatives because $S_w + S_i = 1$, and the system is assumed to be incompressible. In addition, the pressure of each fluid is replaced by an equivalent expression involving the average of the two fluid pressures and the capillary pressure. Coupling the summed equation with the water phase equation leads to the following system.

$$\nabla \cdot V_t = F_w + F_i \,, \tag{6}$$

$$V_t = -k\lambda(S_w)(\nabla p + \bar{\rho}g\nabla z) \,, \tag{7}$$

$$\Phi\frac{\partial S_w}{\partial t} + \nabla \cdot (\lambda_{fw}V_t) - \nabla \cdot (D\nabla S_w) = F_w \,, \tag{8}$$

where V_t and p are the velocity and pressure of the total fluid, $\bar{\rho}$ is the "superficial" average fluid density, λ and λ_{fw} are the total and fractional mobilities, and D is a diffusion term; they are given by the following definitions

$$\lambda = \left(\frac{k_{rw}}{\mu_w} + \frac{k_{ri}}{\mu_i} \right), \quad \lambda_{fw} = \frac{k_{rw}}{\lambda \mu_w}, \quad \bar{\rho} = \frac{\rho_i + \rho_w}{2}, \tag{9}$$

$$p = \frac{p_i + p_w}{2} + \frac{1}{2} \int_0^{p_c} (\lambda_{fi} - \lambda_{fw}) d\eta \tag{10}$$

$$D = k\lambda(S_w)\lambda_{fw} \left(\frac{dp_c}{dS_w} \nabla S_w + \frac{\rho_i - \rho_w}{2} g \nabla z \right). \tag{11}$$

Notice that because the saturation S_w is the dependent variable in Equation (8), the equation suffers the same limitations as the θ-based form of Richards' Equation (4). However, it is a very instructive form and provides insight into solution behavior when two mobile fluid phases are present. In particular, the fractional flow function $\lambda_{fw}(S_w)$, defined in Equation (9), provides important qualitative information about aspects such as the shape and speed of infiltrating fronts.

EFFECTS OF NONLINEAR FLUX FUNCTIONS IN ADVECTIVE-DIFFUSION EQUATIONS

From looking at the linear advection-diffusion equation with a small diffusion coefficient ($\epsilon \ll 1$)

$$au_t + bu_x - \epsilon u_{xx} = 0, \tag{12}$$

we note that information basically follows the directions of the secondary characteristics (Celia and Gray [2]) with a speed determined by b/a. In the simple nonlinear form given by Burger's equation

$$u_t + uu_x - \epsilon u_{xx} \equiv u_t + \frac{1}{2}(u^2)_x - \epsilon u_{xx} = 0, \tag{13}$$

we note that since the flux function $u^2/2$ is concave up, the speed of information is given by u, the derivative of the flux function; thus, points with higher values of u travel faster than those with lower values. This causes a monotonically decreasing initial condition to sharpen up to form a narrow (width determined by ϵ) front which then moves at the speed determined by the top value of u. If ϵ were zero, non-unique solutions would exist, and the true physical shock would be given via an entropy condition (or a Rankine-Hugoniot condition). The non-convexity of the flux function causes the sharp front to form and persist in the presence of diffusion; it also determines the speed of propagation of the front.

For standard petroleum engineering problems, the flux function, given in Equation (8) by $\lambda_{fw} V_t$ forms a non-convex "S-shaped curve" like the fractional flow curve given in Figure 1 and produces characteristic flow properties. Given a monotonically decreasing initial condition, the convex region will sharpen to form a front, the concave region will spread out in a rarefaction wave and the front will move roughly at the speed given by the slope of the line originating at the origin and tangent to the fractional flow function multiplied by the total velocity. The shape of the front is determined by the shape of the diffusion term $D(s)$ and is usually diffused on both sides of the front.

The moving fronts generated by water infiltration through unsaturated soils have somewhat different properties. A typical fractional flow curve for an air-water system is given in Figure 2. It is very small for a large portion of the saturation range, rises rapidly for high saturations, and does not have the convex region at the top. Due to the shape, there should be no rarefaction portion of the saturation curves, as there is in petroleum applications. Since the fractional flow function is also a multiplier in the diffusion term, D, there is no diffusion for lower saturations at the base of the infiltration fronts, but there is significant diffusion near the top saturations at the head of the fronts. We see this behavior in the computed profiles in Figure 3. Also, the speed of the front is essentially constant and is determined by the slope of the fractional flow curve at the saturation of the top leading edge of the front, multiplied by the total fluid velocity.

OPERATOR SPLITTING AND ELLAM

Motivated by the concepts discussed in the last section, we can use the fact that we can determine the speed of the front to develop very efficient computational procedures. This idea has already been applied very effectively in petroleum applications (Espedal and Ewing [8], Dahle et al. [6,7]). In these problems, we take the information from the flux function that predicts the velocity and linearize about that value, retaining the convex part of the fractional flow to balance the diffusion term. See Dahle et al. [6] for details. A modified method of characteristics then combines the accumulation and the linearized flux function to determine a directional derivative along the characteristics of the flow and to move the front along these characteristics. For miscible displacement in porous media, when the flux function is already linear with respect to the velocity, this symmetrizes the operator and allows the use of very large time steps with stable and accurate results. For the multiphase problems with the nonlinear flux function, the operator is not linearized fully. The convex part of the flux function still yields a nonsymmetric term that must be treated by Petrov-Galerkin methods or some other upstream-weighting technique.

In Herrera and Ewing [10], these operator splitting, modified method of characteristic ideas were generalized to two-phase flow in porous media through the ELLAM ideas. Since the ELLAM ideas (Celia et al. [4]) will be discussed in several papers in the proceedings and the related talks at the conference, we will not go into details about the formulation, but refer the reader to the accompanying papers of Celia, Herrera, and Russell.

We next consider ELLAM methods for the advection-diffusion formulation of unsaturated flow problems. From the proceeding section, we note that given an initial condition, if the fractional flow function is strictly concave, as λ_{fw} in Figure 2, then a front will form and move with a constant velocity, but with significant diffusion at the high saturations. Thus, if we control our numerical scheme to time step along the characteristics of the flow with added diffusion in the regions of higher saturation, the method should work extremely well.

Because the multiphase flow equations may be viewed as a nonlinear advection-diffusion equation, characteristic methods become a natural choice for numerical simulations. Recent work of, for example, Espedal and Ewing [8], has used an operator-splitting concept to arrive at a characteristic method of solution. To illustrate the general concept, consider the following split equations:

$$\Phi \frac{\partial \bar{S}}{\partial t} + \frac{d}{dS} f^m(\bar{S}) \cdot \nabla \bar{S} \equiv \Phi \frac{d}{d\tau} \bar{S} = 0 , \tag{14}$$

$$\Phi \frac{\partial S_w}{\partial t} + \nabla \cdot (b^m(S_w)S_w) - \epsilon \nabla \cdot (D(S_w)\bar{V}S_w) = F_w , \tag{15}$$

for $t_m \leq t \leq t_{m+1}$. The splitting of the fractional flow function into two parts $f^m(S_w) + b(S_w)S_w$ is constructed such that $f^m(S_w)$ is linear in the shock region (Espedal and Ewing [8]). We note that Equation (14) produces the same unique physical solution as Equation (8) if D and F_w in Equation (8) were zero provided that an entropy condition were imposed in both cases. Thus, we will always have a unique solution to Equation (14) and, as in the miscible case, by using a characteristic method to treat Equation (14), we can use large Δt without loss of stability or accuracy. Note that Equation (15) is not a symmetric operator, and Petrov-Galerkin with some upstream-weighted test function may be required for stability.

A similar solution procedure may be derived in the ELLAM context, with the added feature of proper boundary condition treatment and therefore improved mass balance (Celia and Zisman [5]). Following the ideas of the split equations above, the ELLAM might use test functions that satisfy the constant-coefficient advection part (Equation (14)) and the remaining spatial part analogous to the spatial derivatives in Equation (15), viz.,

$$-b^m \nabla W - \epsilon \nabla D \nabla W = 0 . \tag{16}$$

Because these operators have variable coefficients, the concepts described for ELLAM in Herrera and Ewing [10] and the paper in these proceedings by Russell must be applied. We also note that the test functions satisfying Equation (16) are similar to those termed Optimal Test Functions by Celia and Herrera [3].

COMPUTATIONAL EXAMPLES FOR UNSATURATED FLOW

In order to illustrate the importance of the advection-diffusion formulation given by Equation (8) for unsaturated flow, we consider some experimental data presented earlier (Touma and Vauclin [12]), for two-phase infiltration. The relative permeability data from that experiment were used to obtain the fractional flow curves presented in Figure 2. Then the *a priori* analysis based upon these fractional flow curves was used to predict the behavior of the wetting fronts given in Figure 3. The constant speed of the front motivates the use of ELLAM techniques, which can take full advantage of this property to use very long time steps in a stable and accurate fashion.

The computations presented in Figure 3 were not made with ELLAM techniques but with a code described in the literature (Celia et al. [4]) using the "mixed" form of Richards' equation discussed above. Although the code computed the frontal advance very accurately, due to the linearization and stability criteria, there is a limitation on the time step used for these computations. ELLAM should have no such restriction for these problems.

Because a "mixed" form of Richards' equation was used, there was no *a priori* intuition of how the nonlinearities in the measured data would affect the shape or speed of the infiltration fronts. By considering the form of the model given by Equation (8), we were able to assess the properties of the fronts before the computations were made. The convexity of the fractional flow curve for water given in Figure 2 predicted that the front would be very sharp, would have no characteristic rarefaction form common to petroleum applications, and would move at a constant speed, given by the slope of the line emanating from the origin and tangent to the fractional flow function multiplied by the total velocity at the relevant saturation. The form of the diffusion term from Equation (8) can also be obtained *a priori* and illustrates why there is no diffusion at the base of the front, but why diffusion is present at the top of the front.

Because this analysis provides important *a priori* information about soultion behavior, improved numerical solution methods may be devised. In this case, ELLAM methods appear to be the natural choice, given the explicit information regarding frontal behavior provided from the fractional flow curves.

We realize that ELLAM is not the complete solution. Because infiltrating fronts can be very sharp, and remain that way for long periods because of the nonlinear effects described above, fine spatial resolution is still required in the vicinity of the front. Thus some combination of local refinement and ELLAM is warranted. Again, the *a priori* information can be used to significant advantage in the refinement scheme, since the frontal velocity would be well estimated. A combination of ELLAM, which alleviates Courant-Friedrichs-Levy conditions on time step, and local grid refinement, which provides required resolution of the steep front, appears to be a very promising approach to pursue.

We also realize that the advantages of these approaches will be compromised when looking at multiphase fluid displacement in heterogeneous formations. For these cases, we are currently investigating possible combinations of the mixed equation solutions of Celia et al. [1], the ELLAM concepts described above, and the possibility of using mixed finite element methods (Ewing et al. [9]). By combining knowledge about the nonlinear physics of multiphase systems with appropriate numerical methods that naturally accomodate the dominant physical processes involved, we hope to develop a reliabale and efficient simulator (or set of simulators) to model the general multiphase flow problem. These simulators are intended to allow realistic treatment of important formation heterogeneities, be equally applicable to hydrologic and petroleum problems, incorporate mass exchange between phases, and solve the concommitant miscible transport problem. This paper represents our initial efforts in this pursuit.

CONCLUSIONS

The problem of multiphase flow in porous media is very important in both the hydrologic and petroleum communities. While historically there has been little interaction between the two fields, the strong similarities between the respective physical systems invites collaboration and sharing of information, with concommitant mutual benefit. An introductory comparative analysis of the unsaturated flow and oil recovery problems has been undertaken. This paper describes some of this effort, focusing on the importance of nonlinear physics and the implication for design of numerical simulators. In particular, the problem of infiltration of water into an unsaturated soil (Touma and Vauclin [12]) was examined, using analysis techniques rooted in petroleum engineering. This analysis led to characteristic methods as a potentially powerful solution method, which leads naturally to an ELLAM formulation. The advantages, as well as anticipated difficulties, of the ELLAM method were discussed. Overall, we believe that only through thorough understanding of nonlinear physics, nonlinear mathematics, and numerical analysis can effective solutions for general multiphase problems be achieved.

ACKNOWLEDGMENTS

This work was supported in part by the National Science Foundation under Grants 8657419-CES and RII-8610680, by the Office of Naval Research under Contract No. 0014-88-K-0370, and by the U.S. Environmental Protection Agency under Assistance Agreement CR 814945. Although the research described in this article has been funded in part by the U.S.E.P.A., it has not been subjected to Agency review and therefore does not necessarily reflect the views of the Agency and no official endorsement should be inferred.

REFERENCES

1. Celia, M. A., Bouloutas, E. T., and Zarba, R. L. A General Mass-Conservative Numerical Solution for the Unsaturated Flow Equation. Water Resources Research, submitted.

2. Celia, M. A. and Gray, W. G. Fundamental Concepts for Applied Numerical Simulation. (to be published by Prentice-Hall), 1990.

3. Celia, M. A. and Herrera, I. Solution of General Ordinary Differential Equations by a Unified Theory Approach. Numerical Methods for Partial Differential Equations, 3, 117–129, 1987.

4. Celia, M. A. , Russell, T. F., Herrera, I. , and Ewing, R. E. An Eulerian-Lagrangian Localized Adjoint Method for the Advection Diffusion Equation. Advances in Water Resources, to appear.

5. Celia, M. A. and Zisman, S. An Eulerian-Lagrangian Localized Adjoint Method for Reactive Transport in Groundwater. Proceedings Seventh International Conference on Computational Methods in Water Resources, Venice, Italy, this volume.

6. Dahle, H. K., Espedal, M. S., and Ewing, R. E. Characteristic Petrov-Galerkin Subdomain Methods for Convection Diffusion Problems. Numerical Simulation in Oil Recovery (M. F. Wheeler, editor), IMA, Springer-Verlag, Berlin, 2:77–88, 1988.

7. Dahle, H. K. , Espedal, M. S., Ewing, R. E., and Saevareid, O. Characteristic Adaptive Subdomain Methods for Reservoir Flow Problems. Numerical Methods for Partial Differential Equations, to appear.

8. Espedal, M. S. and Ewing, R. E. Characteristic Petrov-Galerkin Subdomain Methods for Two-Phase Immiscible Flow. Computer Methods in Applied Mechanics and Engineering, 64, 113–135, 1987.

9. Ewing, R. E., Koebbe, J. V. , Gonzales, R. and Wheeler, M. F. Mixed Finite Element Methods for Accurate Fluid Velocities. Finite Elements in Fluids, Wiley, New York, 2, 233-243, 1985.

10. Herrera, I. and Ewing, R. E. Localized Adjoint Methods: Applications to Multiphase Flow Problems. Proceedings Fifth Wyoming Enhanced Oil Recovery Symposium, May 10–11, 1989, Casper, Wyoming, 155–173, 1990.

11. Hillel, D. Fundamentals of Soil Physics. Academic Press, New York, 1980.

12. Touma, J. and Vauclin, M. Experimental and Numerical Analysis of Two-Phase Infiltration in a Partially Saturated Soil. Transport in Porous Media, 1, 1:27–55, 1986.

TOUMA AND VAUCLIN DATA, MU(AIR)=MU(WATER)

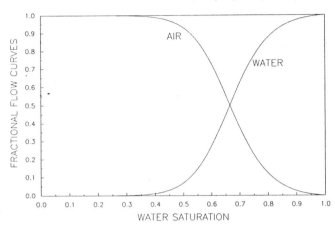

Figure 1.

DATA OF TOUMA AND VAUCLIN (1986)

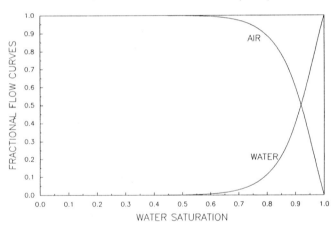

Figure 2.

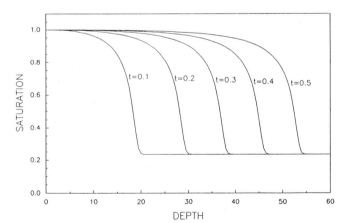

Figure 3.

A Collocation Based Parallel Algorithm to Solve Immiscible Two Phase Flow in Porous Media

J.F. Guarnaccia(*), G.F. Pinder(**)
(*) Department of Civil Engineering and Operations Research, Princeton University, Princeton, NJ 08544, (currently at the College of Eng. and Math., University of Vermont, Burlington, VT 05405, USA)
(**) College of Eng. and Math., University of Vermont, Burlington, VT 05405, USA

Introduction

Numerical solution of immiscible two phase flow problems is a computationally intensive task. Not only must one solve coupled nonlinear equations, but due to stability and accuracy considerations, one must use fine time and space discretizations. In response to growing demands for faster computers, developers have designed a class of machines called 'parallel processing computers.' These computers in effect link together more than one processor such that intercommunication and data transfer can occur. The result is that if a program can be reduced to a series of independent tasks, one can have more than one processor working on the problem, and thus reduce the overall turnaround time. It is up to the code developer to write algorithms which will efficiently utilize the number of processors available, thus the motivation in the development of a parallel collocation based alternating direction algorithm to solve immiscible two phase flow problems.

Multiphase Equation Development:

The mass balance equations describing the simultaneous flow of two immiscible phases, one wetting (w, eg. water) and one nonwetting (n, eg. oil), under isothermal, incompressible (constant denisties) conditions are written as:

$$\phi \frac{\partial S_w}{\partial t} - \frac{\partial}{\partial x_i}\left[\frac{k_{ij}\, k_{rw}}{\mu_w}\left(\frac{\partial p_w}{\partial x_j} - \rho_w g \frac{\partial z}{\partial x_j}\right)\right] = 0 \qquad \text{1a}$$

$$\phi \frac{\partial S_w}{\partial t} + \frac{\partial}{\partial x_i}\left[\frac{k_{ij}\, k_m}{\mu_n}\left(\frac{\partial p_n}{\partial x_j} - \rho_n g \frac{\partial z}{\partial x_j}\right)\right] = 0 \qquad \text{1b}$$

where porosity is assumed time invariant, and the subscripts i and j are used for summation convention. All other terms are defined in the nomenclature section. In writing equations (1) we have used the constitutive relation:
$$S_w + S_n = 1.$$
We now assume the following functional dependencies:
$$S_w = S_w(p_c); \quad p_c = p_c(S_w) = (p_n - p_w); \quad \text{and } k_{rw} = k_{rw}(S_w).$$
Given these relations, equations (1) form a set of coupled nonlinear equations which in general must be solved using an iterative numerical scheme.

Numerical Solution Algorithm:

Equations (1) are solved using a new collocation based alternating direction method amenable to parallel processing. To begin, the time derivative is approximated using an implicit backward differencing scheme, and the equations are linearized by lagging the nonlinear terms by an iteration (successive substitution). This time step/iteration scheme results in the following approximation to the accumulation term for each phase equation:

$$\frac{\partial S_w}{\partial t} = S_w' \frac{\partial p_c}{\partial t} = S_w'\left(\frac{\partial p_n}{\partial t} - \frac{\partial p_w}{\partial t}\right) \approx \frac{S_w^{'n+1,\,m}}{\Delta t}\left[\left(p_n^{n+1,\,m+1} - p_n^n\right) - \left(p_w^{n+1,\,m+1} - p_w^n\right)\right] \qquad 2$$

where the superscripts 'n' and 'm' represent time step and iteration respectively, and $S_w' = \partial S_w/\partial p_c$.

To linearize the flow term we lag the relative permeability by an iteration. This yields:

$$\frac{\partial}{\partial x_i}\left[\frac{k_{ij}\,k_{r\alpha}\!\left(S_w^{n+1}\right)}{\mu_\alpha}\left(\frac{\partial p_\alpha^{n+1}}{\partial x_j} - \rho_\alpha g \frac{\partial z}{\partial x_j}\right)\right] \approx \frac{\partial}{\partial x_i}\left[\lambda_\alpha\!\left(S_w^{n+1,m}\right)\left(\frac{\partial p_\alpha^{n+1,m+1}}{\partial x_j} - \rho_\alpha g \frac{\partial z}{\partial x_j}\right)\right], \; \alpha = w, n \qquad 3$$

where for convenience we introduce the coefficient $\lambda_\alpha = k_{ij}k_{r\alpha}/\mu_\alpha$.

The balance equations have been written such that the phase pressures are the dependent variables. Therefore, let us define the iterative increment for each phase:

$$\delta p_\alpha^{n+1,\,m+1} = \left(p_\alpha^{n+1,\,m+1} - p_\alpha^{n+1,\,m}\right), \qquad \alpha = w, n. \qquad 4$$

Substitution of equations (2) through (4) into equations (1) yields a set of linear partial differential equations in terms of the iterative increments δp_w and δp_n:

$$\frac{\phi}{\Delta t}\,S_w^{'n+1,\,m}\begin{bmatrix}1 & -1\\ -1 & 1\end{bmatrix}\begin{Bmatrix}\delta p_w^{n+1,\,m+1}\\ \delta p_n^{n+1,\,m+1}\end{Bmatrix} + \begin{bmatrix}\frac{\partial}{\partial x_i}\left[\lambda_w^{n+1,m}\left(\frac{\partial\,\delta p_w^{n+1,m+1}}{\partial x_j} - \rho_w g \frac{\partial z}{\partial x_j}\right)\right]\\[6pt] \frac{\partial}{\partial x_i}\left[\lambda_n^{n+1,m}\left(\frac{\partial\,\delta p_n^{n+1,m+1}}{\partial x_j} - \rho_n g \frac{\partial z}{\partial x_j}\right)\right]\end{bmatrix} =$$

$$\frac{\phi}{\Delta t}\,S_w^{'n+1,\,m}\begin{bmatrix}-1 & 1\\ 1 & -1\end{bmatrix}\begin{Bmatrix}\left(p_w^{n+1,\,m} - p_w^n\right)\\ p_n^{n+1,\,m} - p_n^n\end{Bmatrix} - \begin{bmatrix}\frac{\partial}{\partial x_i}\left[\lambda_w^{n+1,m}\left(\frac{\partial p_w^{n+1,m}}{\partial x_j} - \rho_w g \frac{\partial z}{\partial x_j}\right)\right]\\[6pt] \frac{\partial}{\partial x_i}\left[\lambda_n^{n+1,m}\left(\frac{\partial p_n^{n+1,m}}{\partial x_j} - \rho_n g \frac{\partial z}{\partial x_j}\right)\right]\end{bmatrix} \qquad 5$$

The continuous domain on which the problem is defined is discretized into a finite number of rectangular elements and nodes. The dependent variables in (5) are approximated in space using a linear combination of bicubic Hermite polynomial basis functions. For two-dimensional problems this results in four degrees of freedom at each node for each variable (the function, the two spatial derivatives and the cross derivative). The other variables which are space dependent are approximated using bilinear Lagrange basis functions ($k_{r\alpha}$ and S_w').

The discretized equation (5) is reduced to a set of linear algebraic equations by employing the orthogonal collocation method; a method of weighted residuals where the weighting function is the Dirac delta function. This is equivalent to driving the residual to zero at specified points in the domain which are denoted as collocation points. Thus no formal integrations are required, and generation of the system matrix is computationally analogous to the finite difference method.

Orthogonal collocation results when the equations are written at the four gauss quadrature points in each element.

The solution of the system matrix created by using the above discretization is performed using an alternating direction technique (AD). This technique is depicted diagrammatically in Figure 1. Rows and columns of the mesh are isolated by projecting the unknowns associated with them to the new iterate by using a linear projection. This has the effect of spatially decoupling the problem. One first performs a 'horizontal sweep' by concurrently sweeping through all the rows of the mesh. Information is not made available as it is obtained until all rows are calculated. In this way the scheme mimics a Jacobi block iterative matrix solver. Having just calculated the 'm+1' iterate level, one performs the same operations on the columns of the mesh ('vertical sweep') to obtain the 'm+2' solution. By alternating solution directions boundary conditions are best propagated.

As described, this AD scheme has several important attributes. First, for each direction all the one-dimensional sweeps can be performed concurrently, and therefore, the method is well suited for parallel processing; second, it can directly accommodate cross-derivative terms of the permeability tensor as no operator splitting is required; third there is only a linear relation between mesh size and computational effort; and fourth, it is easy to implement. For a more detailed description of the AD algorithm see Guarnaccia and Pinder (1989).

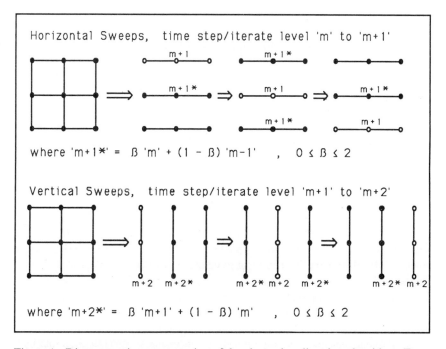

Figure 1 - Diagrammatic representation of the alternating direction algorithm. For 'horizontal sweeps' the solution at the 'm+1' iterate is obtained by three independent operations. The rows with the open nodes are evaluated using projected information from adjacent rows. The 'vertical sweeps' operate likewise on columns to obtain the 'm+2' iterate. ß is the projection parameter.

Example Simulation:

We have chosen to model an immiscible flood of trichloroethylene (TCE) in water. TCE is a denser than water non-aqueous phase liquid that has received a lot of research attention over the past several years. As a result there is much data, both numerical and experimental, with which to compare. The particular problem chosen was borrowed from the work of Abriola (1983). Due to space constraints the reader is referred to this work for details regarding specific aquifer and fluid parameters and $k_{r\alpha}$, S_w, and p_c functional relationships. Suffice it to say that the fluid properties are based on experimental results while the porous medium is idealistically homogeneous. The problem domain is depicted in Figure 2. The aquifer is confined and initially saturated with water at static equilibrium. In addition there is a pressure drop from right to left yielding a mild water flow in that direction. Pressure constraints enforce the condition at the nodes labeled 'TCE saturated' (these nodes are held at residual water saturation, 0.306).

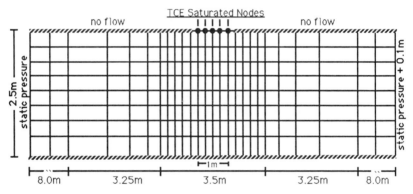

Figure 2 - Diagrammatic representation of the TCE infiltration simulation, showing mesh geometry and boundary conditions.

Results for two simulation times are plotted in Figures 3 and 4. Overall mass balance for the simulation was 5 percent. The simulation was run with a maximum time step of 20 seconds and time step size governed by number of iterations (increase < 8 iterations, decrease > 8 iterations). An iteration is obtained after the two dimensional sweeps are completed. An average time step for the entire simulation was approximately 15 seconds based on the convergence criteria: $\| \delta p_\alpha^{n+1,m+1} \|_2 \leq 1.0$. This criteria gave the best mass balance for the number of iterations required for convergence.

Abriola's (1983) finite difference solutions obtained for the same problem were of generally the same shape but the saturation front was much more dispersed. Her mesh, however, was much less refined than the one used here. When this code was used to simulate a column experiment referenced by Abriola (1983), a sharp front solution was observed and the calculated breakthrough time matched the experimental. While one cannot be sure of the 'correctness' of these 2-dimensional solutions one can say that they look 'reasonable.' Note that in Figure 4 the TCE front has hit the bottom and is not only spreading laterally, but has trapped some water directly under the source.

The model was run on an IRIS-4D parallel processing computer. Speedups of 3.8 were obtained when four processors were utilized. This high level of efficiency results from the fact that on the order of 99 percent of the calculations are performed in parallel mode.

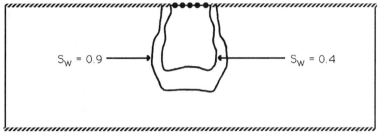

Figure 3 - Solution after 5,000 seconds. Note that the residual saturation of water is 0.306.

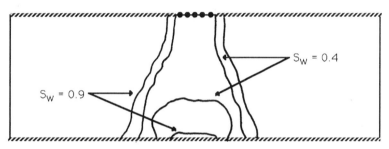

Figure 4 - Solution after 10,000 seconds. Note that water is trapped under front.

Conclusion

The collocation based AD algorithm described herein has been shown to be simple to implement, robust in its simulation versatility, and efficient in a parallel computation environment.

Nomenclature

g	=	gravitational coefficient	k_{ij} =	local intrinsic permeability tensor
k_r	=	relative permeability	p =	pressure
p_c	=	capillary pressure	S =	phase volume saturation
t	=	time	v_i =	phase average velocity
z	=	scalar potential associated with gravity		
λ	=	transmissibility tensor	μ =	dynamic viscosity
ρ	=	phase average density	ϕ =	porosity

ACKNOWLEDGMENT - The authors wish to acknowledge the financial support of the U.S. Environmental Protection Agency assistance I.D. No. CR-814946-01-1 and the IBM Corporation.
DISCLAIMER- Although the research described in this article has been funded in part by the U.S. Environmental Protection Agency under assistance I.D. No. CR-814946-01-1, it has not been subjected to agency review and therefore does not

necessarily reflect the views of the Agency and no official endorsement should be inferred.

References

Abriola, L.M., <u>Multiphase Migration of Organic Compounds in a Porous Medium</u>: <u>A Mathematical Model</u>, Lecture Notes in Engineering, **8**, Springer Verlag, Berlin, 1984, pages 112-142, 158-174.

Guarnaccia, J.F. and G.F. Pinder, <u>A Parallel Collocation Based Algorithm for the Generalized Transport Equation</u>, in *Applications of Supercomputers in Engineering: Fluid Flow and Stress Analysis*, Proceedings of the First International Conference, Edited by C. A. Brebbia and A. Peters, Southampton, UK, September 1989, pages 67-78.

Numerical Simulation of Three-Phase Multi-Dimensional Flow in Porous Media

B.E. Sleep, J.F. Sykes

Department of Civil Engineering, University of Waterloo, Waterloo, Ontario, Canada N2L 3G1

ABSTRACT

This study presents a three-phase, multidimensional finite difference model for flow in porous media incorporating a dynamic air phase. The numerical model is verified against existing analytical solutions for one-dimensional, two-phase infiltration with gravity. Simulations are presented to demonstrate the effect of a dynamic air phase in three-phase flow.

INTRODUCTION

Although three-phase flow problems have been addressed in the petroleum reservoir engineering literature for many years, treatment of three-phase flow systems of interest to contaminant hydrogeologists has been limited. Abriola and Pinder (1985), Faust (1985), Parker et. al. (1987) and Forsyth (1988) have developed three-phase models for flow in porous media. These investigators assumed that the air phase was passive, prescribing uniform air phase pressures.

Simulation of three-phase flow in porous media requires the solution of a coupled set of three highly nonlinear equations for the water, air and nonaqueous liquid (NAPL) phases. Solution of these equations in more than one dimension, for practical sized domains is only feasible with the use of iterative solution methods. Behie and Forsyth (1984) used the incomplete lower-upper (ILU) factorization method with the orthomin acceleration method for black oil and thermal enhanced oil recovery simulation.

This paper presents a finite difference model for three-phase (air-water-NAPL), three-dimensional flow. The model is verified with an analytical solution for two-phase water-NAPL and water-air flow with gravity and capillarity. The impact of a dynamic gas phase is illustrated with numerical simulations in one and two dimensions.

EQUATIONS OF THREE-PHASE FLOW

The equations describing the flow of three immiscible fluid phases in porous media may be written as

$$\frac{\partial}{\partial x_i}\left[\frac{\rho_l k_{ij} k_{rl}}{\mu_l}\left(\frac{\partial p_l}{\partial x_j} + \rho_l g \frac{\partial z}{\partial x_j}\right)\right] + q_l - \frac{\partial}{\partial t}(\theta \rho_l S_l) \quad (1)$$

where l represents the water (w), air (a), or oil (o) phases; ρ_l is the density, k_{ij} is the permeability tensor; k_{rl} is the relative permeability of phase l; p_l is pressure; θ is porosity; S_l is fluid saturation; and q_l represents sinks and sources of fluid.

These equations are subject to the constraints:

$$S_w + S_o + S_g - 1.0 \; ; \; p_w - p_{caw}(S_w) \; ;$$
$$p_o - p_w - p_{cow}(S_w) \; ; \; p_g - p_o - p_{cgo}(S_t) \quad (2)$$

where $S_t = S_w + S_o$. The capillary pressure equations are based on the assumption that in the three-phase system oil is of intermediate wettability relative to air and water and that there are only oil-water and air-oil interfaces. When it is assumed that the gas phase pressure is uniformly atmospheric the equation for air flow is not required, and S_t is a function of the oil phase pressure.

NUMERICAL APPROACH

Block-centered finite differences are used for the spatial discretization of Equation 1 for each of the three phases. The equation for a grid block of volume V_i, and time increment Δt from $N-1$ to N is

$$\frac{V_i}{\Delta t}\left[(\theta S_l \rho_l)^N - (\theta S_l \rho_l)^{N-1}\right]$$

$$- \frac{V_i}{\Delta x_i}[T_l \rho_l]^N_{i+1/2}\left[p^N_{l,i+1} - p^N_{l,i} + \rho^N_{l,i+1/2} g(z_{i+1} - z_i)\right] \quad (3)$$

$$- [T_l \rho_l]^N_{i-1/2}\left[p^N_{l,i} - p^N_{l,i-1} + \rho^N_{l,i-1/2} g(z_i - z_{i-1})\right] + q^N_l$$

plus similar flux terms for the y and z directions. Δx_i is the x dimension of the ith finite difference block and

$$[T]^N_{i\pm1/2} = \left(\frac{2}{\Delta x_{i\pm1} + \Delta x_i}\right)\left(\frac{k_{i\pm1/2}\ k^N_{rl,u}}{\mu_l}\right) \quad (4)$$

Implicit time stepping is used to ensure solution stability. Relative permeabilities are upstream weighted. Gas phase densities are assumed to be ideal. Equation (3), in combination with constitutive relationships for capillary pressures, relative permeabilities, and densities, is solved for p_w, S_w, and S_t for the dynamic air phase case, and for p_w and S_w for the passive air phase case.

Application of the Newton-Raphson technique to Equation (3) results in a sparse, nonsymmetric system of linear equations with three equations for each finite difference node. The linear equations are solved using block ILU second degree factorization with orthomin acceleration (Behie and Vinsome, 1982). An additional reduction of computational effort is achieved by using an alternating diagonal ordering.

MODEL VERIFICATION

Rogers' (1983) solution for vertical displacement of a wetting fluid by a nonwetting fluid was modified to describe non-wetting fluid displacing wetting fluid for model verification.

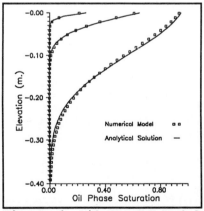

Figure 1 Oil-Water Model Verification: 1,10,60 min

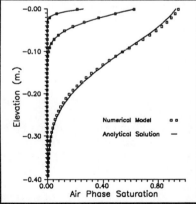

Figure 2 Air-Water Model Verification: 1,10,60 min

The relative permeability - saturation relationships used are

$$k_{rw} = S_w - S_{rw}; \quad k_{rnw} = 1 - S_w - S_{rnw} \tag{5}$$

where S_{rw} and S_{rnw} are the irreducible saturations of the wetting and nonwetting phases. The capillary pressure function, allowing the analytical solution, is of the form

$$P_c(S_w) = a \ln \left[(S_w - S_{rw})^b (S_w - 1 + S_{rnw})^{-c} (S_w + d)^e \right] + f \tag{6}$$

a, b and f are arbitrary constants, and c,d and e are functions of viscosities and irreducible saturations.

The results of simulations of oil and air displacing water for the numerical and analytical models, in Figures 1 and 2, show the excellent agreement between the solutions. In both cases wetting and nonwetting fluid viscosities were 1 cP and 0.8 cP, respectively. The large air phase viscosity was required to meet capillary presssure function constraints of the analytical solution. Fluid densities were 1000, 1200, and 1.3 kg/m³ for water, oil and air phases. The permeability and infiltration rate were 1×10^{-12} m² and 1×10^{-5} m/s. S_{rw} and S_{rnw} were zero.

THE EFFECT OF FINITE AIR PHASE MOBILITY

Figure 3 shows a simulation of oil influx under ponded conditions into a one-dimensional column. Oil flux rates are much less when air mobility is finite. The oil infiltrates as a very sharp front and moves to greater depths when air entrapment is ignored. The Brooks-Corey model (Corey, 1986) for capillary pressure and relative permeability relationships were used. The displacement pressures were 0.3, 0.1 and 0.15 m. for the air-water, oil-water and air-oil capillary pressures. The pore size distribution parameter was set at 1.8. The oil pressure was set at 0.04 m of water at the ground surface while water pressure was set as atmospheric at the bottom of the domain. Oil density and viscosity were 1200 kg/m₃ and .9 cP, and permeability was 1x10-12 m².

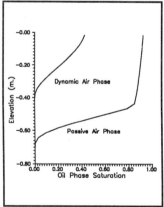

Figure 3 1-D Column Dynamic vs. Passive Air Phase (1 hr.)

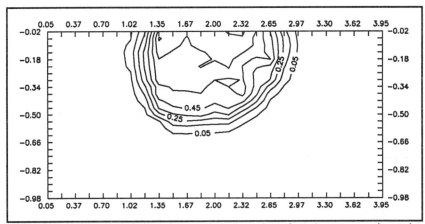

Figure 4 2-D Simulation, Random k Field, Dynamic Air Phase (15 min.)

Oil infiltration into a two-dimensional domain of 40 x .1 m width by 25 x 0.04 m. depth was simulated. A random permeability field with mean ln(k) of -25.24 and variance of 0.3 was generated with the turning bands method. A set of random entry pressures was generated using the Leverett scaling function (Corey, 1986) to scale the pressures to the distribution of permeabilities. The pore size distribution index was not varied. Oil pressure was set to 0.01 m. of water in 10 blocks (x = 1.5 to x = 2.5) at the ground surface. Air pressure was atmospheric in blocks with no ponded oil. Water pressure was atmospheric at the bottom of the domain.

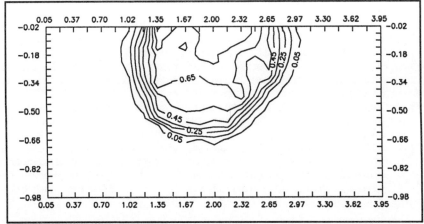

Figure 5 2-D Simulation, Random k Field, Passive Air Phase (15 min.)

Simulation results are presented in Figures 4 and 5. The maximum oil saturations were 0.85 and 0.76 for the passive and dynamic air phase cases. Total infiltration of oil was reduced from 0.226 m³ to 0.157 m³ as a result of the finite air phase mobility.

Total computation time for 35 time steps for the two-dimensional, dynamic air phase simulation was about 200 min. on a Sun Sparcstation 1. Approximately 65% of this time was used by the ILU-orthomin solver. For the passive air simulation approximately 120 min. were required, with 43% of this time spent solving the linear equations. An average of 15 orthomin iterations were required for the dynamic air case, compared to 8 for the passsive case.

CONCLUSIONS

Simulation of multidimensional three-phase flow in porous media with a dynamic air phase is feasible with the use of iterative solution methods. The inclusion of a dynamic air phase may be necessary to accurately determine the rate of infiltration of immiscible fluids in the vadose zone in some situations.

REFERENCES

1.   Abriola, L. M., and G. F. Pinder, A multiphase approach to the modeling of porous media contamination by organic compounds, Numerical simulation, *Water Resour. Res.*, **21**(1), 19-26, 1986.
2.   Behie, A., and P. A. Forsyth, Incomplete factorization methods for fully implicit simulation of enhanced oil recovery, *SIAM J. Sci. Stat. Comp.*, **5**, 543-561, 1984,
3.   Behie, A., and P. K. W. Vinsome, Block iterative methods for fully implicit reservoir simulation, *Soc. Pet. Eng.*, **22**, 658-668, 1982.
4.   Corey, A. T., Mechanics of Immiscible Fluids in Porous Media, Water Resources Publ., Colorado,1986.
5.   Faust, C. R., Transport of immiscible fluids within and below the unsaturated zone, *Water Resour. Res.*, **21**(4),1985.
6.   Forsyth, P. A., Simulation of nonaqueous phase groundwater contamination, *Adv. Water Resour.*, **11**(2), 1988.
7.   Rogers, C., M. P. Stallybrass, and C. L. Clements, On two phase infiltration under gravity and with boundary infiltration: application of a Backlund transformation, *Nonlinear Anal., Theory, Methods, and Applic.*, **7**, 785-799, 1983.

A Compositional Model for Simulating Multiphase Flow, Transport and Mass Transfer in Groundwater Systems

A.S. Mayer, C.T. Miller

Department of Environmental Sciences and Engineering, CB# 7400, 105 Rosenau Hall, University of North Carolina, Chapel Hill, NC 27599-7400, USA

INTRODUCTION

Mathematical models of multiphase flow and species transport in subsurface environments generally employ equilibrium partitioning relationships to describe interphase mass exchange (Sleep and Sykes[1]). However, the equilibrium approach for describing interphase mass exchange in groundwater systems requires validation through theoretical, laboratory, and field analyses. The present work concerns the development of a one-dimensional, three-fluid multiphase flow and single-species transport model. The numerical model incorporates nonequilibrium mass exchange relationships among a nonaqueous phase liquid (NAPL), a water phase, a gas phase, and a solid phase.

MASS BALANCE EQUATIONS

The mass balance equation describing bulk flow of individual phases can be written as (Abriola and Pinder[2]):

$$\sum_i \left(\frac{\partial}{\partial t} \left(\theta_\alpha \rho_\alpha \omega_\alpha^i \right) + \nabla \cdot \left(\theta_\alpha \rho_\alpha \omega_\alpha^i \mathbf{v}_\alpha \right) + \nabla \cdot \mathbf{J}_\alpha^i \right) = \sum_i \left(I_\alpha^i + \Gamma_\alpha^i + R_\alpha^i \right) \quad (1)$$

subject to the constraints of

$$\sum_i \omega_\alpha^i = 1, \quad \sum_\alpha \theta_\alpha = 1, \quad \sum_\alpha S_\alpha = 1, \quad \sum_\alpha I_\alpha^i = 0, \quad \sum_i \nabla \cdot \mathbf{J}_\alpha^i = 0 \quad (2)$$

where i denotes a species index, α denotes a phase index, θ_α = volumetric fraction, S_α = degree of saturation, ρ_α = density $[ML^{-3}]$, ω_α^i = mass fraction, $\mathbf{v}_\alpha$ = velocity $[LT^{-1}]$, $\mathbf{J}_\alpha^i$ = dispersive flux $[ML^{-3}T^{-1}]$, I_α^i = interphase exchange term $[ML^{-3}T^{-1}]$, Γ_α^i = source/sink term $[ML^{-3}T^{-1}]$, and R_α^i = reactive term $[ML^{-3}T^{-1}]$.

Flow equations can be derived by evaluating Equation (1) for a given α, yielding:

$$\frac{\partial}{\partial t} \left(\theta_\alpha \rho_\alpha \right) + \nabla \cdot \left(\rho_\alpha \frac{-\mathbf{k} k_{r\alpha}}{\mu_\alpha} \cdot \left(\nabla P_\alpha - \rho_\alpha \mathbf{g} \right) \right) = I_\alpha \quad (3)$$

where the phase velocity, $\mathbf{v}_\alpha$, has been replaced with a modified form of Darcy's law, the source/sink and reactive terms have been neglected, and $\mathbf{k}$ = intrinsic permeability tensor $[L^2]$, $k_{r\alpha}$ = relative permeability, μ_α = fluid dynamic viscosity $[ML^{-1}T^{-1}]$, P_α = fluid pressure $[ML^{-1}T^{-2}]$, and $\mathbf{g}$ = gravity vector $[LT^{-2}]$. Three mobile phases are described with flow equations—a water phase ($\alpha = w$), a NAPL ($\alpha = n$), and a gas phase ($\alpha = g$).

The mass balance equation for the transport of individual species can be written as

$$\frac{\partial}{\partial t}\left(\theta_\alpha \rho_\alpha \omega_\alpha^i\right) + \nabla \cdot \left(\theta_\alpha \rho_\alpha \omega_\alpha^i \mathbf{v}_\alpha\right) + \nabla \cdot \mathbf{J}_\alpha^i = I_\alpha^i + \Gamma_\alpha^i + R_\alpha^i \tag{4}$$

subject to Equation (2).

The dispersive term can be written as

$$\nabla \cdot \mathbf{J}_\alpha^i = -n\nabla \cdot \left(\rho_\alpha S_\alpha \mathbf{D}_\alpha^i \nabla \cdot \omega_\alpha^i\right) \tag{5}$$

where n = porosity and $\mathbf{D}_\alpha^i$ = hydrodynamic dispersion tensor $[L^2T^{-1}]$, consisting of a molecular diffusion and mechanical dispersion term.

The interphase mass exchange term for the exchange of mass from the β phase to the α phase is given by:

$$I_\alpha^i = \theta_\alpha \rho_\alpha K_{\alpha,\beta}(\omega_\alpha^{i*} - \omega_\alpha^i) \tag{6}$$

where $K_{\alpha,\beta}$ = mass transfer rate coefficient $[T^{-1}]$ and ω_α^{i*} = equilibrium value of mass fraction. The transport of the chemical species i is described with three transport equations—a water phase ($\alpha = w$), a gas phase ($\alpha = g$), and a solid phase ($\alpha = s$).

EQUATION FORMULATION

Equations (1) to (6) may be expanded to formulate a set of five coupled equations approximating the simultaneous flow of water, a NAPL, and a gas phase, and the transport of a single solute in the water and gas phases. The solid-phase mass fraction is formulated in an additional equation that is uncoupled from the other five flow and transport equations. The following equations summarize the governing flow and transport in a single dimension that is aligned with the gravity vector.

Water flow

$$n\left(S_w \beta_w - \frac{\partial S_w}{\partial P_{nw}}\right)\frac{\partial P_w}{\partial t} + n\frac{\partial S_w}{\partial P_{nw}}\frac{\partial P_n}{\partial t} - \rho_w g \frac{\partial K_w}{\partial P_{nw}}\left(\frac{\partial P_w}{\partial x} - \frac{\partial P_n}{\partial x}\right)$$

$$+\frac{\partial K_w}{\partial P_{nw}}\left[\left(\frac{\partial P_w}{\partial x}\right)^2 - \frac{\partial P_n}{\partial x}\frac{\partial P_w}{\partial x}\right] - K_w\frac{\partial^2 P_w}{\partial x^2}$$

$$+n\left(S_w K_{w,n} + \frac{\rho_g}{\rho_w}(1 - S_t)K_{g,w}H\right)\omega_w^i + \frac{\rho_s}{\rho_w}(1 - n)K_{s,w}[K_f - (\omega_s^i)''](\omega_w^i)^\eta \tag{7}$$

$$-n\left(\frac{\rho_g}{\rho_w}(1 - S_t)K_{g,w}\omega_g^i + S_w K_{w,n}\omega_{w,n}^{i*}\right) + \frac{\rho_s}{\rho_w}(1 - n)K_{s,w}(\omega_s^i)' = 0$$

where: g = gravitational constant $[LT^{-2}]$, H = Henry's Law constant, k = intrinsic permeability $[L^2]$, k_{rw} = relative permeability of water phase, K_f = Freundlich coefficient, $K_w = kk_{rw}/\mu_w$ = conductivity of water phase $[M^{-1}L^3T]$, $K_{g,w}$ = gas-water phase mass-transfer rate coefficient $[T^{-1}]$, $K_{s,w}$ = solid-water phase mass-transfer rate coefficient $[T^{-1}]$, $K_{w,n}$ = NAPL-water phase mass-transfer rate coefficient $[T^{-1}]$, P_n = pressure of NAPL phase $[ML^{-1}T^{-2}]$, P_w = pressure of water phase $[ML^{-1}T^{-2}]$,

P_{nw} = NAPL-water phase capillary pressure $[ML^{-1}T^{-2}]$, S_w = saturation of water phase, S_t = saturation of total liquid phases, β_w = compressibility of water phase α $[LT^2M^{-1}]$, η = (Freundlich exponent)$^{-1}$, μ_w = dynamic viscosity of water phase $[ML^{-1}T^{-1}]$, ρ_g = density of gas phase $[ML^{-3}]$, ρ_s = density of solid phase $[ML^{-3}]$, ρ_w = density of water phase $[ML^{-3}]$, ω_g^i = mass fraction of species i in gas phase, ω_w^i = mass fraction of species i in water phase, $(\omega_s^i)'$ = lumped coefficient for solid phase mass fraction, $(\omega_s^i)''$ = lumped coefficient for solid phase mass fraction, and $\omega_{w,n}^{i*}$ = equilibrium mass fraction of species i in water phase.

NAPL flow

$$
n\left[\left((S_t - S_w)\beta_n - \frac{\partial S_w}{\partial P_{nw}} - \frac{\partial S_t}{\partial P_{gn}}\right)\frac{\partial P_n}{\partial t} + \frac{\partial S_w}{\partial P_{nw}}\frac{\partial P_w}{\partial t} + \frac{\partial S_t}{\partial P_{gn}}\frac{\partial P_g}{\partial t}\right]
$$

$$
-\rho_n g\left[\frac{\partial K_n}{\partial P_{nw}}\left(\frac{\partial P_w}{\partial x} - \frac{\partial P_n}{\partial x}\right) - \frac{\partial K_n}{\partial P_{gn}}\left(\frac{\partial P_g}{\partial x} - \frac{\partial P_n}{\partial x}\right)\right]
$$

$$
+\left(\frac{\partial K_n}{\partial P_{nw}}\frac{\partial P_w}{\partial x} - \frac{\partial K_n}{\partial P_{gn}}\frac{\partial P_g}{\partial x}\right)\frac{\partial P_n}{\partial x} + \left(\frac{\partial K_n}{\partial P_{gn}} - \frac{\partial K_n}{\partial P_{nw}}\right)\left(\frac{\partial P_n}{\partial x}\right)^2 \quad (8)
$$

$$
-K_n\frac{\partial^2 P_n}{\partial x^2} - \frac{n}{\rho_n}\left[\rho_w S_w K_{w,n}\omega_w^i + \rho_g(1 - S_t)K_{g,n}\omega_g^i\right]
$$

$$
+\frac{n}{\rho_n}\left[\rho_w S_w K_{w,n}\omega_{w,n}^{i*} + (1 - S_t)K_{g,n}c_{gn}\right] = 0
$$

where $c_{gn} = P_v M_{g,n}/RT$ $[M^{-1}L^3]$, k_{rn} = relative permeability of NAPL, $K_n = kk_{rn}/\mu_n$ = conductivity of NAPL $[M^{-1}L^3T]$, $K_{g,n}$ = gas phase-NAPL mass-transfer rate coefficient $[T^{-1}]$, $M_{g,n}$ = molecular weight of gas phase component $[M\text{mol}^{-1}]$, P_g = pressure of gas phase $[ML^{-1}T^{-2}]$, P_{gn} = gas phase-NAPL capillary pressure $[ML^{-1}T^{-2}]$, P_v = saturated vapor pressure $[ML^{-1}T^{-2}]$, R = ideal gas constant $[ML^2T^{-2}\text{mol}^{-1}\text{o}^{-1}]$, T = temperature $[\text{o}]$, β_n = compressibility of NAPL $[LT^2M^{-1}]$, μ_n = dynamic viscosity of NAPL $[ML^{-1}T^{-1}]$, and ρ_n = density of NAPL $[ML^{-3}]$.

Gas flow

$$
n\left[\left((1 - S_t)\frac{\partial\rho_g}{\partial P_g} - \rho_g\frac{\partial S_t}{\partial P_{gn}}\right)\frac{\partial P_g}{\partial t} + \rho_g\frac{\partial S_t}{\partial P_{gn}}\frac{\partial P_n}{\partial t} + (1 - S_t)\frac{\partial\rho_g}{\partial\omega_g^i}\frac{\partial\omega_g^i}{\partial t}\right]
$$

$$
+g\left(K_g\frac{\partial\rho_g^2}{\partial P_g} + \rho_g^2\frac{\partial K_g}{\partial P_{gn}}\right)\frac{\partial P_g}{\partial x} - \left(K_g\frac{\partial\rho_g}{\partial P_g} + \rho_g\frac{\partial K_g}{\partial P_{gn}}\right)\left(\frac{\partial P_g}{\partial x}\right)^2 - \rho_g K_g\frac{\partial^2 P_g}{\partial x^2} \quad (9)
$$

$$
-\left(K_g\frac{\partial\rho_g}{\partial\omega_g^i}\frac{\partial\omega_g^i}{\partial x} - \rho_g\frac{\partial K_g}{\partial P_{gn}}\frac{\partial P_n}{\partial x}\right)\frac{\partial P_g}{\partial x} - g\left(\rho_g^2\frac{\partial K_g}{\partial P_{gn}}\frac{\partial P_n}{\partial x} - K_g\frac{\partial\rho_g^2}{\partial\omega_g^i}\frac{\partial\omega_g^i}{\partial x}\right)
$$

$$
+n(1 - S_t)\{\rho_g[(K_{g,w} + K_{g,n})\omega_g^i - K_{g,w}H\omega_w^i] - K_{g,n}c_{g,n}\} = 0
$$

where k_{rg} = relative permeability of gas phase, $K_g = kk_{rg}/\mu_g$ = conductivity of gas phase $[M^{-1}L^3T]$, and μ_g = dynamic viscosity of gas phase $[ML^{-1}T^{-1}]$.

Water transport

$$
n \left\{ S_w \frac{\partial \omega_w^i}{\partial t} + \left[\left(S_w \beta_w - \frac{\partial S_w}{\partial P_{nw}} \right) \frac{\partial P_w}{\partial t} + \frac{\partial S_w}{\partial P_{nw}} \frac{\partial P_n}{\partial t} \right] \omega_w^i \right\}
$$

$$
- \left(\rho_w g \frac{\partial K_w}{\partial P_{nw}} \omega_w^i + K_w \frac{\partial \omega_w^i}{\partial x} \right) \frac{\partial P_w}{\partial x} + \frac{\partial K_w}{\partial P_{nw}} \left(\frac{\partial P_w}{\partial x} \right)^2 \omega_w^i
$$

$$
- \left[K_w \frac{\partial^2 P_w}{\partial x^2} + \left(\frac{\partial K_w}{\partial P_{nw}} \frac{\partial P_w}{\partial x} - \rho_w g \frac{\partial K_w}{\partial P_{nw}} \right) \frac{\partial P_n}{\partial x} \right] \omega_w^i
$$

$$
+ \rho_w K_w g \frac{\partial \omega_w^i}{\partial x} + n \left(S_w K_{w,n} + \frac{\rho_g}{\rho_w} (1 - S_t) K_{g,w} H \right) \omega_w^i
$$

$$
+ \frac{\rho_s}{\rho_w} (1 - n) K_{s,w} [K_f - (\omega_s^i)''] (\omega_w^i)^\eta - \frac{\rho_g}{\rho_w} n (1 - S_t) K_{g,w} \omega_g^i
$$

$$
- n S_w K_{w,n} \omega_{w,n}^{i*} + \frac{\rho_s}{\rho_w} (1 - n) K_{s,w} (\omega_s^i)' - n S_w D_w^i \frac{\partial^2 \omega_w^i}{\partial x^2} = 0
$$

(10)

where D_w^i = a dispersion coefficient for the water phase $[L^2 T^{-1}]$.

Gas transport

$$
n(1 - S_t) \left(\rho_g + \frac{\partial \rho_g}{\partial \omega_g^i} \omega_g^i \right) \frac{\partial \omega_g^i}{\partial t} - n \left(\rho_g \frac{\partial S_t}{\partial P_{gn}} - (1 - S_t) \frac{\partial \rho_g}{\partial P_g} \right) \frac{\partial P_g}{\partial t} \omega_g^i
$$

$$
+ \left[n \rho_g \frac{\partial S_t}{\partial P_{gn}} \frac{\partial P_n}{\partial t} + g \left(K_g \frac{\partial \rho_g^2}{\partial P_g} + \rho_g^2 \frac{\partial K_g}{\partial P_{gn}} \right) \frac{\partial P_g}{\partial x} \right] \omega_g^i
$$

$$
- \left[\left(K_g \frac{\partial \rho_g}{\partial P_g} + \rho_g \frac{\partial K_g}{\partial P_{gn}} \right) \left(\frac{\partial P_g}{\partial x} \right)^2 + \rho_g K_g \frac{\partial^2 P_g}{\partial x^2} \right] \omega_g^i
$$

$$
- K_g \left(\frac{\partial \rho_g}{\partial \omega_g^i} \omega_g^i + \rho_g \right) \frac{\partial \omega_g^i}{\partial x} \frac{\partial P_g}{\partial x} + \frac{\partial K_g}{\partial P_{gn}} \left(\rho_g \frac{\partial P_g}{\partial x} - \rho_g^2 g \right) \frac{\partial P_n}{\partial x} \omega_g^i
$$

$$
+ K_g g \left(\frac{\partial \rho_g^2}{\partial \omega_g^i} \omega_g^i + \rho_g^2 \right) \frac{\partial \omega_g^i}{\partial x} + \rho_g n (1 - S_t) (K_{g,w} + K_{g,n}) \omega_g^i
$$

$$
- n(1 - S_t) [\rho_g K_{g,w} H \omega_w^i + K_{g,n} c_{g,n}] - n S_g \rho_g D_g^i \frac{\partial^2 \omega_g^i}{\partial x^2} = 0
$$

(11)

where D_g^i = a dispersion coefficient for gas phase $[L^2 T^{-1}]$.

Solid phase transport

$$
\frac{\partial \omega_s^i}{\partial t} - (K_{s,w}) \omega_s^i + (K_f K_{s,w}) (\omega_w^i)^\eta = 0
$$

(12)

where ω_s^i = solid phase mass fraction.

The solid phase transport equation is approximated with an implicit finite-difference equation. The solid phase mass fraction at the new time step is solved for and is substituted into the water phase flow and transport equations. This procedure eliminates the need to include a solid phase transport equation, but introduces the $(\omega_s^i)'$ and $(\omega_s^i)''$ lumped coefficients into the water phase flow and transport equations.

CONSTITUTIVE RELATIONS

The general equations summarized by Equations (7) to (12) require constitutive relations for closure. These relations include: $S_\alpha = f(P_g, P_n, P_w)$, $k_{r\alpha} = f(P_g, P_n, P_w)$, and $K_{\alpha,\beta} = f(P_g, P_n, P_w)$, when formulated in terms of the primary multiphase variables—P_α and ω_α^i. The constitutive relations used to provide closure follow formulations summarized by Parker et al.[3] for S_α and $k_{r\alpha}$, while the $K_{w,n}$ relation follows from Miller et al.[4], and other $K_{\alpha,\beta}$ terms were assumed constant.

COMPUTATIONAL APPROACH

A Petrov-Galerkin finite element method (Westerink and Shea[5]) was used to approximate Equations (7) to (12), while the implicit finite-difference method was used to resolve the time derivatives. A Picard iteration scheme was used to resolve the nonlinearities in the solution. As an alternate approach to solving the nonlinear system of equations, a Newton-Raphson scheme is currently being tested; the relative computational efficiency of the two schemes will be compared in a future publication.

RESULTS

The model was validated by comparison to two analytical solutions: a single-phase flow solution shown in Figure 1(a) (Carslaw and Jaeger[6]), and a transport solution that includes interphase mass transfer from a NAPL to a water phase shown in Figure 1(b) (Miller et al.[4]). The flow validation was performed for the case of an instantaneous change in a Dirichlet boundary condition, for a finite length (L) domain. The transport validation was performed for the case of advective and dispersive transport, with water as the single mobile fluid phase. Relevant parameters are summarized on the figures.

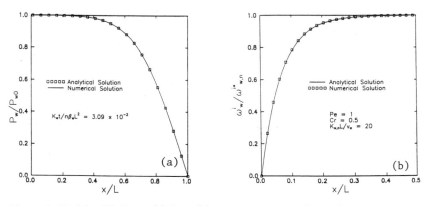

Figure 1. Model validations: (a) flow, (b) mass transport and transfer.

The results of an example simulation are shown in Figures 2(a) and 2(b). This example describes the imbibition of a NAPL into a vertically-oriented column that is at a residual water-phase saturation of 0.20. The upper boundary conditions are Dirichlet for all dependent variables: $P_{nw} = 4 \times 10^6$ g/(cm sec^2), $P_{gn} = 1 \times 10^6$ g/(cm sec^2), and $\omega_w^i = \omega_w^{i*}$. The lower boundary conditions are: Neumann—no flow—for the fluid phases, and Neumann—no dispersive flux—for the solute species equations. Initial conditions are: $P_{nw} = 4.9 \times 10^6$ g/(cm sec^2), $P_{gn} = 1.0 \times 10^2$ g/(cm sec^2), $\omega_w^i = 0$. Medium and fluid phase properties are as described by Abriola and Pinder[7], while constitutive pressure-saturation-permeability relations are as described by Parker et al.[3] for a sandy soil. NAPL-water mass transfer was approximated using $Sh = 100\theta_n Re^{0.35}$ (Miller et al.[4]), where Sh is the Sherwood number, and Re is the Reynolds number. A dispersion coefficient of 9.1×10^{-3} cm^2/sec was used.

Figure 2(a) shows the propagation of a NAPL front at three separate times, while Figure 2(b) shows the evolution of the water-phase mass fraction for the same times. These results show that the NAPL front is relatively steep, the rate of advance of the NAPL front decreases as a function of time, and the mass fraction is significantly below the equilibrium value—over the entire simulation. Thus, in this case, the equilibrium approach is not valid for describing NAPL-water phase mass transfer.

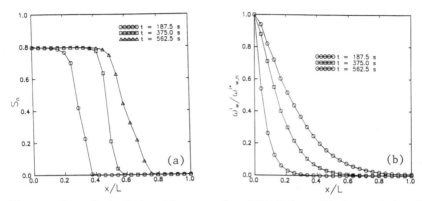

Figure 2. Example simulation: (a) propagation of NAPL front, (b) time evolution of water-phase mass fraction.

REFERENCES

1. Sleep, B.E. and Sykes, J.F. Modeling the Transport of Volatile Organics in Variably Saturated Systems, Water Resources Research, Vol. 25, pp. 81–92, 1989.

2. Abriola, L.M. and Pinder, G.F. A Multiphase Approach to Modeling of Porous Media Contamination by Organic Compounds 1. Equation Development, Water Resources Research, Vol. 21, pp. 11–18, 1985.

3. Parker, J.C., Lenhard, J. and Kuppusamy, T. Physics of Immiscible Flow in Porous Media, R.S. Kerr Environmental Research Laboratory, U.S. Environmental Protection Agency, Ada, OK, 1987.

4. Miller, C.T., Poirier-McNeill, M.M. and Mayer, A.S. Dissolution of Trapped Nonaqueous Phase Liquids: Mass Transfer Characteristics, submitted to Water Resources Research, 1990.

5. Westerink, J.J. and Shea, D. Consistent Higher Degree Petrov-Galerkin Methods for the Solution of the Transient Convection-Diffusion Equation, International Journal for Numerical Methods in Engineering, Vol. 28, pp. 1077–1101, 1989.

6. Carslaw, H.S. and Jaeger, J.C. Conduction of Heat in Solids, Oxford University Press, Oxford, UK, 1984.

7. Abriola, L.M. and Pinder, G.F. A Multiphase Approach to Modeling of Porous Media Contamination by Organic Compounds 2. Numerical Simulation, Water Resources Research, Vol. 21, pp. 11–18, 1985.

SECTION 4 - FLOW IN FRACTURED MEDIA

Particle Tracking in Three-Dimensional Groundwater Modelling

J. Trösch, A. von Känel
Laboratory of Hydraulics, Hydrology and Glaciology, Federal Institute of Technology, 8092 Zurich, Switzerland

INTRODUCTION

Particle tracking is a method for the calculation of streamlines used in the field of radioactive waste repositories. Traveltime and critical pathway of an imaginary particle released from the repository is determined. This pathline may be used for sophisticated one-dimensional models for the interaction of contaminants with the rock matrix.

The particle tracking is based on a three-dimensional finite element model for saturated flow. In three dimensions the streamlines cannot be calculated as orthogonal solution of the potential flow problem as in two-dimensional models. The solution has to be found by integration of the velocities over the pathline.

As the program is used for modelling flow in fractured rock, three-dimensional elements (bricks, prisms, tetraeders and pyramids) are used for the discretisation of the matrix, combined with two- and one-dimensional elements for fractures and wells.

The streamlines are integrated in the local coordinate system. This has major advantages as only few time consuming transformations of coordinates from global to local space are necessary and the element boundaries are planes. Problems may arise from an inadequate discretisation when local minima occur within elements or between adjacent elements. Special care has to be taken in these cases, to ensure that the streamlines are not trapped.

BASIC EQUATIONS

The continuity equation and Darcy's law give the wellknown equation for saturated groundwater flow:

$$\frac{\partial}{\partial x_i} k_{ij} \frac{\partial h}{\partial x_j} + q = 0 \quad ; \quad i, j = 1 \ldots 3 \tag{1}$$

and appropriate boundary conditions. Here x_i are the cartesian coordinates, k_{ij} the permeability tensor, h the head and q a source- or sink-term. The effective flow velocity v_i is derived directly from Darcy's law and the porosity ϵ of the rock matrix as

$$v_i = - \frac{k_{ij}}{\epsilon} \frac{\partial h}{\partial x_j} . \tag{2}$$

The definition of the streamline is

$$\frac{dx_i}{dt} = v_i(x,t) \tag{3}$$

Integration of this equation, combined with (2) gives the streamline and the elapsed time.

Eq.(1) is solved with the wellknown Galerkin finite element approximation with quadratic isoparametric elements.

For the solution of (3) in the local coordinate system s_i the inverse of the transformation has to be known. In the case of a 3-d element this is straightforward with the inverse of the Jacobian matrix D_{ij},

$$\frac{\partial s_i}{\partial x_j} = D_{ij}^{-1} = \left[\frac{\partial x_j}{\partial s_i}\right]^{-1} \quad , \quad i,j = 1 \ldots 3 \tag{4}$$

If the elements are 2-d or 1-d in a 3-d domain a metric tensor g_{im} is used

$$g_{im} = \frac{\partial x_k}{\partial s_i} \cdot \frac{\partial x_1}{\partial s_m} \cdot \delta_{k1} \tag{5}$$

with δ_{k1} the Kronecker matrix and the transformation (4) is in the form

$$\frac{\partial s_i}{\partial x_k} = g_{im}\delta_{k1} \frac{\partial x_1}{\partial x_m} = D_{ik}^{-1} \tag{6}$$

Further on the transformation of the vektor is needed, as for instance

$$\frac{\partial x_k}{\partial t} = \frac{\partial x_k}{\partial s_i} \cdot \frac{\partial s_i}{\partial t} \tag{7}$$

Introducing these transformations above in (3) and (2) gives

$$\frac{\partial s_i}{\partial t} = - \frac{1}{\epsilon} \frac{\partial s_i}{\partial x_k} K_{kl} \frac{\partial s_m}{\partial x_l} \cdot \frac{\partial h}{\partial s_m} \tag{8}$$

This equation is integrated with an Eulerian integration. Optionally a Runge-Kutta integration may be chosen.

IMPLEMENTATION

The element that contains the given starting point is searched with a Newton-Raphson step for a couple of nearest elements, until the local coordinates of the starting point lie within the element. As each element is a polyeder in the local space this check is easy. The point has to be always on the positive side of all halfspaces defined by the planes of the polyeder.

The timestep chosen depends on the velocity, resulting in a constant length: Thus it ensures that the number of steps is restricted and the length increment remains reasonable even with extreme velocities (e.g. near wells). If the streamline leaves one element the neighbour is found via a connection matrix, storing for every element the adjacent ones.

Special attention is given to 2-d and 1-d elements. 2-d elements lie normally between two 3-d elements. Once the track is in such an element, the velocity has no component normal to the element. To have an opportunity to leave this element, the velocities in the adjacent 3-d elements are calculated. If a velocity is pointing out of the fracture and if it is the largest one, the track continues in this 3-d element. Otherwise at the next step this check is repeated.

At the edges of the elements oszillations often occur. This is due to insufficient discretisation and because the velocity is not continous at element boundaries. This oszillations may give significant errors in traveltime. To circumvent this problem, all velocities in the elements concerned are calculated. If only 3-d elements are concerned, the element with the largest velocity is chosen, otherwise projections of the velocities on the edge are used.

Even with these sophisticated methods some streamlines are trapped within local minima. Only better discretisation could help in these cases.

EXAMPLES

As an example a dipoltest calculation is shown (Fig. 1). Some tracer is injected in the borehole at left and extracted in the other one. Part of tracer is flowing to an experimental tunnel at the top. Some shortcommings of the discretisation are seen within the three rectangular domains.

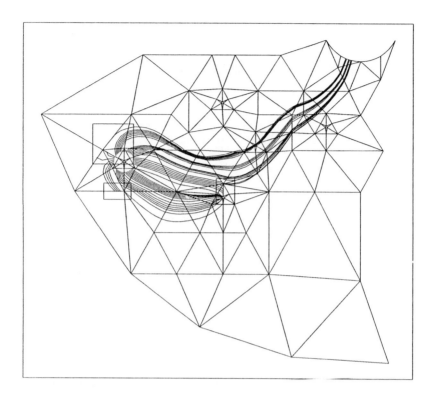

Fig. 1. Particle tracking for a dipol test in a fracture zone.

CONCLUSIONS

The method presented gives most of time good results, as can be
shown in comparisons with analytical solutions. If the porous
matrix is anisotropic, the method of projection to prevent oszillations may give erroneous results. In these cases a finer discretisation will be needed.

ACKNOWLEDGEMENTS

This project was partly founded by NAGRA, the National Coopera-
tive for the Storage of Radioactive Waste, Switzerland.

REFERENCE

von Känel, A. (1989). Particle Tracking - Advektiver Transport im
Grundwasser. Internal Report of the Laboratory of Hydraulics,
Hydrology and Glaciology, ETH Zurich.

Transport in Fractured Rock - Particle Tracking in Stochastically Generated Fracture Networks

J. Wollrath, W. Zielke

Institut für Strömungsmechanik und Elektronisches Rechnen im Bauwesen, Universität Hannover, D-3000 Hannover, FRG

ABSTRACT

The paper describes the algorithm of generating a fracture network used as an input for a flow and transport model. It also presents the method of particle tracking used to calculate flow paths and travel times.

Statistical distributions for fracture length, fracture aperture and fracture orientation are obtained based on available or synthetical data of fracture geometry and orientation. These distributions are used to generate two-dimensional networks representing the fractured rock. Observed single fractures can be used to condition the stochastically generated network.

The particle tracking algorithm uses the flow velocities calculated for the fracture network with a finite element model, developed at the Institute of Fluid Mechanics. The decision which fracture a particle passes through after an intersection of fractures, is based on a stochastical process which considers the flow rate of each fracture.

INTRODUCTION

The development of methods to predict flow and transport processes in fractured rock have received a strong impact from the need to store chemical or radioactive waste in deep rock caverns or salt mines. To analyze the safety of such deposits it is necessary to study transport paths and travel times in the case of a leakage of the deposit.

Often it is neccessary to consider the large influence of fractures in the safety studies. Starting with known data about fractures it is possible in some cases to substitute a model which consists of a discrete fracture network for a continuum model. The advantage of such an equivalent continuum model is that it can be handled in a way well known from groundwater pollution modelling.

The tools described in this paper are able to determine parameters like permeability, porosity and diffusion coefficients needed in an equivalent continuum model. The obtained breakthrough curves of a pollutant calculated by the discrete fracture model and the equivalent continuum model must by identical. This can be used to fit the parameters for the equivalent continuum model.

GENERATION OF A FRACTURE NETWORK

The calculation of flow through a discrete fracture network ordinarily needs exact knowledge of the number, the arrangement and the dimensions of all fractures in the modelling area. In the nature fractures exist in all orders of magnitude. The scale ranges from micro fissures to large geologic faults. Therefore, the modeller has to decide which scales of fractures are relevant for his problem. Furthermore, it is not possible to get the exact number, arrangement and dimensions of each fracture through observations and in-situ tests, but statistical distributions for this parameters could be obtained (e.g. Baecher et al. [2]).

The flow through a fracture network is threedimensional, but for basic calculations it is sufficient to do calculations in two dimensions. Detailed analysis requires to take into account the third dimension. This paper shows the generation of a fracture network in two dimensions. If one assumes that the main part of discharge takes place in distinct flow paths (channeling, Tsang et al. [5]) it is principally possible to take into account the third dimension in the algorithm shown in this paper.

For the twodimensional model a network of line elementes (fracture network) is generated in a square region (model region, Long [3]). The arrangement of the fractures in the model region is arbitrary. The network could be divided into several groups (see fig. 1c). Information of known single fractures or fracture distributions is used to condition the fracture network (Andersson et al. [1], see fig. 1a and 1b). The rock matrix between the fractures is assumed to be impermeable. If the direction of flow is from left to right, one can remove fracture clusters which are not connected to the left and to the right boundary (see fig. 1e). Also, the dead ends of the fractures can be removed (see fig. 1f). To investigate the direction dependency of the flow and transport parameters a selected window can be rotated (see fig. 1g).

PARTICLE TRACKING

The generated fracture network is used as an input for the flow model of the finite element program system ROCKFLOW ([4], Wollrath and Zielke [6]) to calculate the distribution of piezometric heads over the entire domain and the flow rates and velocities in each fracture.

Setting out a particle which represents a certain amount of pollutant at an arbitrary node of the fracture network the way and the time the particle needs to travel along this way are calculated using the before calculated velocities. At each node a decision has to be made which way (fracture) the particle takes after passing the node. The flow rate of all fractures (elements) in which the velocity vector points away from the considered node are summed and normalized to one. The flow rates of the individual elements are converted respectively. With the aid of a random number in the interval [0,1] the decision is made which fracture the particle passes next.

EXAMPLE USING DATA OF THE GRIMSEL ROCK LABORATORY

Fracture data from the analysis of boreholes of about 500 m length and several water injection tests which are made in the Grimsel rock laboratory, Switzerland are used to obtain fracture density (= number of fractures per unit volume) and statistical distributions for frature aperture and orientation. The model is fitted to a permeability of about $k_f = 10^{-6} \frac{m}{s}$ varying the fracture length. Fig. 2 shows the 70 m · 70 m model region consisting of about 1300 fracture with 9200 nodes and 17400 elements.

Fig. 3 shows the distribution of 500 particles started at the center of the left boundary after 10 days. In fig. 4 the ways the particles travel are plotted. The thickness of the lines are a measure for the number of particles passing the particular element. The breakthrough curve as a measure for the time the particles need to travel from left right is shown in fig. 5.

SUMMARY

Two parts of the program system ROCKFLOW are described as tools which can be used to calculate flow and transport processes in fractured rock. The twodimesional fracture generator creates a network of line elements used as the flow model. The calculated velocities in conjunction with a stochastical process are the basis for the particle tracking algorithm to simulate the transport behavior of a fracture network. Data of the Grimsel rock laboratory are used to show the application of the method.

ACKNOWLEDGEMENT

The work was initiated by the Federal Institute for Geoscience and Natural Resources (BGR) and is sponsored by the Federal Ministry of Research and Technology (BMFT) under contract 02 U 5937 3.

REFERENCES

[1] Andersson, J., A. M. Shapiro, and J. Bear. A stochastic model of a fractured rock conditioned by measured information. *Water Resources Research*, 20(1), 1984.

[2] Baecher, G. B., N. A. Lanney, and H. H. Einstein. Statistical description of rock properties and sampling. In *Proc. 18th U.S. Symp. Rock Mech.*, 1977.

[3] Long, J. C. S. *Investigation of Equivalent Porous Medium Permeability in Networks of Discontinuous Fractures*. PhD thesis, Lawrence Berkeley Laboratory, University of California, 1983.

[4] ROCKFLOW. *Theorie und Benutzeranleitung zum Programmsystem ROCK-FLOW*
Teil 1: Wollrath, J.: DURST/SM86 – Strömungsmodell für inkompressible Fluide;
Teil 2: Kröhn, K.-P.: DURST/TM88 – Transportmodell für inkompressible Fluide (Tracertransport);
Teil 3: Helmig, R., H. Shao: GM88 – Strömungsmodell für kompressible Fluide (Gasmodell);
Teil 4: Kröhn, K.-P.: TD88 – Transportmodell für kompressible Fluide
Teil 5: Wollrath, J.: DURST/GP87 – Graphikpaket;
Teil 6: Wollrath, J.: KG88 – Kluftgenerator;
Teil 7: Wollrath, J.: PT89 – Partikel-Tracking für Kluftnetze.
Institut für Strömungsmechanik und Elektronisches Rechnen im Bauwesen, Universität Hannover, 1986-89.

[5] Tsang, Y. W., C. F. Tsang, F. V. Hale, L. Moreno, and I. Neretnieks. *Channeling Characteristics of Flow and Solute Transport through a Rough-Surfaced Fracture*. Report No. LBL–23195, Lawrence Berkeley Laboratory, University of California, Berkeley, California, 1987.

[6] Wollrath, J. and W. Zielke. FE-Simulation von Strömungen im klüftigen Gestein. *Deutsche Gewässerkundliche Mitteilungen*, 34(1), 1990.

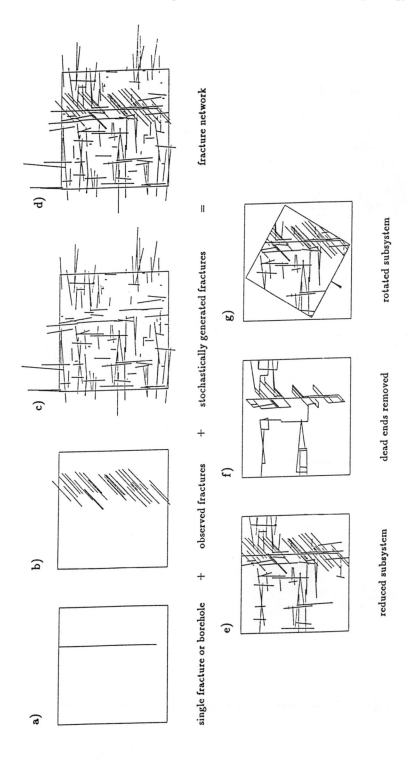

Figure 1: Sceme of the fracture network generator

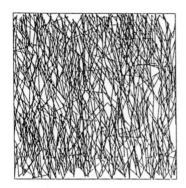

Figure 2: Fracture network of the
Grimsel model

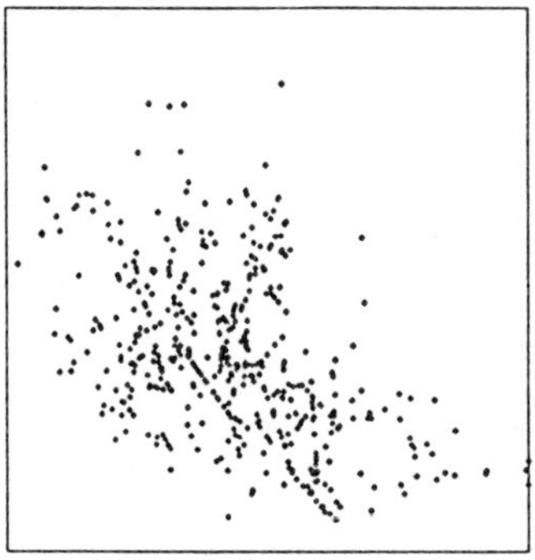

Luuuuul 10 M

Figure 3: Distribution of particles
started at the same time
after 10 days

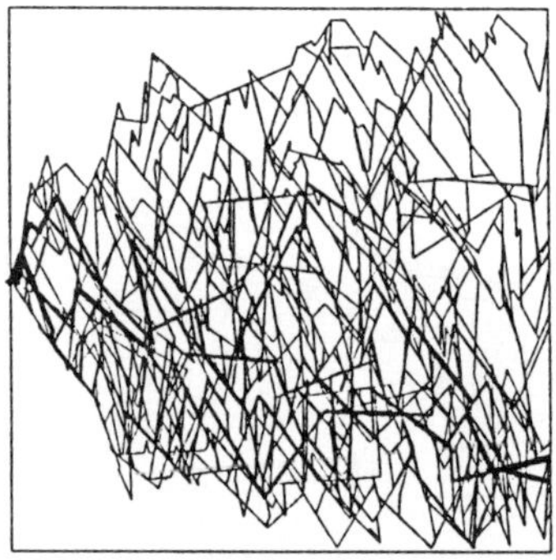

Figure 4: Particle tracks

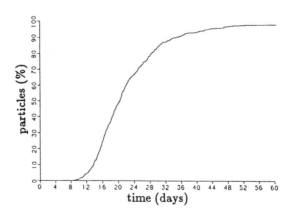

Figure 5: Breakthrough curve measured at the right boundary

An Efficient Semi-Analytical Method for Numerical Modeling of Flow and Solute Transport in Fractured Media

J. Birkhölzer(*), G. Rouvé(*), K. Pruess(**), J. Noorishad(**)

(*) Institut für Wasserbau und Wasserwirtschaft, Aachen University of Technology, Mies-van-der-Rohe-Str. 1, D-5100 Aachen, West Germany

(**) Earth Sciences Division, Lawrence Berkeley Laboratory, Berkeley, CA 94720, USA

ABSTRACT

A new semianalytical approach is presented for modeling flow and transport in fractured reservoirs. Based on the double porosity concept the model is capable of simulating interactions between rock matrix and fractures without noticeable increase in computer time and storage compared to single continuum calculations. Global flow in the fracture domain is described by a Galerkin Finite Element Method, whereas the transfer from matrix to fractures is calculated analytically by means of trial functions approximating the pressure or concentration distribution in the matrix blocks.

INTRODUCTION

In naturally fractured reservoirs most of the fluid is located in low-permeable matrix blocks; however, most of the fluid mobility occurs in a small volume of high-permeable, interconnected fractures. The matrix blocks act as sources (sinks) that feed (drain) the adjacent fractures. Considering solute transport in such a system advection and dispersion are relevant in the fracture network, whereas diffusion is the dominant process in the rock matrix. Despite of major advances in the past, numerical modeling of this behavior is still a difficult problem. Discrete fracture approaches are not applicable in field studies because of the enormous amount of input data and accompanying computer work. Porous media approximations have been shown to be inadequate in many cases since the "retarding" influence of the matrix is not taken into account. To overcome this drawback Barenblatt et al. [1] proposed the socalled "double porosity" model. Global flow is considered to take place in the fracture network while the rock matrix exchanges fluid and solute locally with the fissures.

Many workers have made the assumption of treating the interflow between fractures and matrix as quasisteady, with flow rate being proportional to the average pressure/concentration in

matrix and fractures. This simplifies the mathematics but does
not correctly represent the transient response in the matrix. In
the "nonsteady-state" approach flow inside the blocks and at the
block surface is approximated in spatial detail. The transient
leakage is calculated out at the gradient at the block surface
after solving the mass conservation equation in individual
blocks of given geometry.

Due to the fact that analytical solutions are limited to
simple cases quite a lot of numerical models have been developed
basing on the "nonsteady-state" method. Two different approaches
have been used: (1) Both fractures and matrix are discretized
simultaneously and solved in one differential equation system;
eg.[4, 5]. (2) Fracture network and matrix blocks are discretized as two different systems and coupled via leakage term; eg.
[6, 7, 8]. Although the latter approach is more efficient it
still involves a remarkable increase in computing time and storage compared to single continuum calculations.

In this paper we develop a new method for the simulation of
flow and transport in fractured media. It is conceptually similar to the second approach. Global flow/transport in the fracture network is approximated by a Galerkin Finite Element solution. The governing equation for the matrix, however, is described by semianalytical means, using simple trial functions for
the pressure and concentration distribution in the blocks. The
method is an adaption of a technique developed by Vinsome and
Westerveld [9] for describing heat exchange between a surface
with time-varying temperature and a semi-infinite half-space.
Pruess and Wu [10] followed this approach and presented a numerical solution scheme that incorporates heat and fluid exchange
between fracture and matrix blocks of arbitrary shape; this is
extended here to fluid and solute exchange including retardation
and radionuclide decay.

GOVERNING EQUATIONS

Flow and transport in a fractured reservoir can be described by
two coupled equations: two-dimensional in space for the fractures and one-dimensional for the matrix blocks. The theoretical
development of the governing equations and their solution is
presented in this paper only for the transport processes since
the semianalytical treatment of the fracture/ matrix leakage is
immediately applicable on flow problems. Following assumptions
are made: (1) The fracture network can be treated as a continuum. (2) Two-dimensional transport in the fracture ntwork is dominated by advection and hydrodynamic dispersion. (3) The permeability of the porous matrix is very low and transport in the
matrix will be mainly by molecular diffusion. Under these assumptions the mass conservation equation for the fracture network can be written in the form

$$n \frac{\partial}{\partial t} (C_f) + \frac{\partial}{\partial x_i} (q_i C_f) - \frac{\partial}{\partial x_i} (n D_{ij} \frac{\partial C_f}{\partial x_i}) - (1-n) M = 0 \quad (1)$$

C_f concentration of solute in fractured medium

D_{ij} hydrodynamic dispersion tensor, L^2/T

q_i Darcy velocities in the fractured medium, L/T

n fracture porosity (volume of fractures per unit volume of porous medium)

M diffusive flux from matrix to fracture per unit volume of matrix, $M/L^3/T$.

In the nonsteady-state approach the leakage term M in (1) depends on the concentration distribution in individual matrix blocks, which is evaluated out of the one-dimensional mass balance equation for a volume element $dV = A(z)dz$:

$$\phi \, R \, \frac{\partial C'}{\partial t} - D' \, \frac{\partial^2 C'}{\partial z^2} - D' \, \frac{\partial C'}{\partial z} \, \frac{\partial \ln (A(z))}{\partial z} + \lambda \, \phi \, R \, C' = 0 \qquad (2)$$

ϕ is the porosity in the matrix block, D' is the molecular diffusion coefficient, C' is the concentration in the matrix block, λ is a first order decay constant and R describes the retardation of a solute in the matrix, caused by adsorption/desorption processes. $A(z)$ is the interface area for flow in the matrix blocks at a distance z from the surface.

SOLUTION APPROACH

THE SEMIANALYTICAL METHOD
The matrix diffusion equation is solved by a semianalytical method, first presented by Vinsome and Westerveld [9] for heat transfer problems. The basic idea was to represent the temperature profile in the rock matrix by means of a reasonably flexible function containing a few parameters which can be evaluated in a simple and fast manner. Adapted to solute transport the trial function chosen by Vinsome and Westerveld can be written as follows:

$$C'(z,t) - C_i = (C_f - C_i + pz + qz^2) \, e^{-z/d}, \quad d : 0.5 \, (\frac{D' \, t}{R \, \phi})^{0.5} \qquad (3)$$

Here z is the distance from the boundary, perpendicular to the rock fracture interface, p and q are time-varying best fit parameters, d is the penetration depth for diffusion into the rock, C_f is the time-varying concentration at the interface and C_i is the initial condition (assumed to be uniform).
Due to the spatial concentration distribution in the entire fractured reservoir, each node in the FE-grid representing the global flow system is associated to a concentration profile in the adjacent matrix blocks. The coefficients p and q will be different for each grid node; they are determined concurrently with the global simulation from simple physical principles.

Inserting the semianalytical representation of C' and considering the mass balance at the block surface ($z=0$) equation (2) becomes:

$$\frac{R \, \phi \, (C_f - C_f^0)}{D \, \Delta t} = \frac{C_f - C_i}{d^2} - \frac{2p}{d} +$$

$$+ 2q + \frac{d\ln(A(z))}{dz} \Big|_{z=0} (p \, \frac{C_f - C_i}{d}) - \frac{\lambda \, R \, \phi}{D'} C_f. \qquad (4)$$

The time derivative is replaced by a FD-approximation. C_f and C_f^0 are the concentrations in the fractures at beginning and end of time step Δt, respectively. The condition that the rate of change of mass content in a block is equal to the solute transfer at the surface gives:

$$\frac{d}{dt} \int_V \phi\, R\, (C'(z,t) - C_i)\, dV = - D' \left.\frac{\partial C'}{\partial z}\right|_{z=0} A(0) +$$

$$- \lambda \int_V \phi\, R\, (C'(z,t) - C_i)\, dV \qquad (5)$$

V is the matrix block volume and A(0) is the block surface area. Using former assumptions the three-dimensional integration can be reduced to a onedimensional perpendicular to the block surface. The integral term in (5) becomes

$$I(t) = \int_V (C'(z,t) - C_i)\, dV = \int_0^{L/2} (C'(z,t - C_i)\, A(z)\, dz \qquad (6)$$

Introducing a FD approximation for the time derivative and evaluating the spatial derivative from (3) equation (5) finally becomes:

$$I(t + \Delta t) - I(t) + \lambda\, \Delta t\, I(t + \Delta t) = \frac{D'\,\Delta t}{R\,\phi} \left(\frac{C_f - C_i}{d} - p\right) \qquad (7)$$

Eqs. (4) and (7) (after inserting (3) into (6)) represent two linear equations for the unknown time-dependent parameters p and q. The volumetric rate of mass transfer from the blocks to the fractures is evaluated by applying Fick's law at the surface of the matrix blocks:

$$M = - \frac{A(0)}{V}\, D' \left.\frac{\partial C'}{\partial z}\right|_{z=0} = \frac{A(0)}{V}\, D' \left(\frac{C_f - C_i}{d} - p\right) \qquad (8)$$

The parameters p and q have to be evaluated at each time step for each node of the global Finite Element grid. Nevertheless the computer storage needed for this approach is quite small. Only the mass content integral (6) has to be stored for the next time step.

Note that advective/dispersive transport in the matrix can be readily incorporated into the semianalytical solution scheme as it is presented here. The fracture/matrix transfer can be evaluated using similar physical principles; the resulting equations, however, become more complicated.

DIRECT SOLUTION SCHEME

The transport processes in fractures and matrix are evaluated by two sets of differential equations which are coupled via leakage term M. A number of authors propose iterative solution techniques. With the semianalytical approach, however, a direct solution procedure can easily be employed. For each time step the parameters p and q and the mass transfer term M (eq. (8)) are evaluated in linear dependence of the unknown fracture concentration. Inserting (8) in (1) gives an equation system which is directly solved for the current concentration in the fracture domain by applying a standard Galerkin Finite Element formulation. Once the nodal concentrations have been obtained the values of I(t), p, q, M and C'(z,t) are determined by backward substitution.

PROXIMITY FUNCTION

The "nonsteady-state" treatment of the fracture/matrix leakage
requires detailed information about the matrix block shape. Many
workers used simple idealizations of the block geometry, eg.
prismatic blocks or spherical blocks. However, typical REVs of a
fractured reservoir contain a quite complex topological structure with a finite number of fractures and matrix blocks of different shape. Following the concept of "Proximity Functions"
proposed by Pruess, Karasaki [3] it is possible to represent all
the matrix blocks of a given reservoir subdomain by a function
Prox(z) which expresses the total fraction of matrix volume
within a distance z from the fractures (Prox(z) = V(z)/V with V:
total rock mass in the REV). The interface area for diffusive
flux at distance z is simply the derivative of the proximity
function. Using this concept the equations (4), (7) and (8) are
no longer restricted to individual blocks. They can easily be
extended to represent the total number of matrix blocks in the
REVs of a fractured aquifer.

For regularly shaped blocks the proximity functions can be
written down in analytical form. In naturally fractured reservoirs with irregular distributed fractures, they are approximated by Monte-Carlo-Studies using statistically generated fracture networks. In both cases the function and its derivative can
be written as polynomials in z. Thus, the integral term in (7)
can be easily evaluated by elementary means.

ONE-DIMENSIONAL FINITE ELEMENT APPROACH

To test the accuracy of the semianalytical solution for more
complex problems where analytical solutions are not available, a
second model was employed using a Galerkin Finite Element approximation of the one-dimensional diffusion equation (2) in the
matrix blocks. The spatial integration of the element matrices
is performed exactly without averaging A(z) to obtain improved
accuracy. This is important especially near the fracture/matrix
boundary where steep gradients may occur. Nevertheless, the integrals can still be evaluated by elementary means. Again a direct solution scheme is employed to solve the coupled fracture
and matrix equations. See details in Huyakorn et al. [8].

VERIFICATION

The method was verified by applying the model to a number of
test cases for both flow and transport problems and comparing
the results with available analytical solutions. One example is
selected to illustrate the accuracy and validity of the approach. It concerns longitudinal transport along a discrete
fracture and transverse diffusion into the adjacent matrix
blocks. An analytical solution has been developed by Tang et al.
[2]. A schematic description of the problem case, values of physical properties and boundary conditions are depicted in figure
1. In this simple case of parallel fractures the proximity func-

tion comes out as Prox(z) = z/1.2. Physical properties as well
as spatial and temporal discretization are similar to an example
calculated in Huyakorn et al. [8]. Concentration distributions
along the fracture and into the matrix for the semianalytical
and Tang's analytical solution are shown in figure 2. Excellent
agreement is achieved even for early time steps when steep gradients occur at the fracture/matrix boundary.

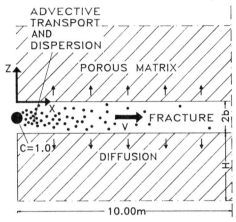

Fracture properties:

Aperture	2b	$= 10^{-4}$ m
Fracture Velocity	V	$= 0.01$ m/d
Fracture porosity	n	$= 0.5 \ 10^{-4}/1.2$
Darcy Velocity	q	$= n \cdot 0.01$ m/d
Longitudinal Dispersion	D_{xx}	$= 0.5138 \ 10^{-2}$ m^2/d

Matrix properties:

Matrix porosity	ϕ	$= 0.01$
Diffusion	D'	$= 1.382 \cdot 10^{-7}$ m^2/d
Proximity Function	Prox(z)	$= V(z)/v = z/1.2$
Halfwidth of Block	H	$= 1.20$ m

Radionuclide Properties: Initial and Boundary condition:

$\lambda = 0.154 \ 10^{-3} d^{-1}$, R = 1   C (x,0) = C'(z,0) = 0, C (0,t) = 1.0

Fig. 1: Schematic description of the test example

   To compare the accuracy of both approaches the same transport problem was also solved with the one-dimensional Finite
Element formulation. The discretization into the matrix was
achieved using five one-dimensional elements with nodal spacing
$\Delta z = 0.1$; 0.16; 0.24; 0.30 and 0.40. Results are also plotted
in figure 2. For early times the Finite Element solution lags
slightly behind the analytical. A much better agreement was obtained using a refined discretization in the matrix blocks. Simulations with a five-element matrix grid require about the same
computer time as the semianalytical approach, both need about
10% more than single continuum calculations. Refined discretization, however, leads to a significant increase in CPU time. Note

that the semianalytical approach requires an additional computer
storage of "knode" values compared to single continuum calculations.(knode: number of nodes of the global grid associated with
double porosity behavior). Using a 1-D FE approximation, however, a storage of "knode * kelem" is needed (kelem: average
number of 1-D elements in the matrix).

   It should be mentioned that several test examples were performed using complex proximity functions representing naturally
fractured formations. No differences showed up between the semianalytical and the 1-D FE approach if, in the latter case, a sufficient number of elements was used.

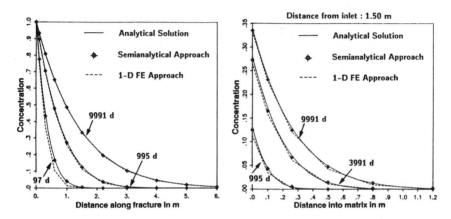

Fig. 2: Concentration distribution along the fracture and into
         the matrix

APPLICATION

A hypothetical, yet realistic, field-scale study is performed to
demonstrate the utility of the double porosity approach. A high-
permeable fault zone is located in a confined fractured aquifer
of 10.0m thickness. An isotropic permeability is assumed in the
fractured reservoir containing three sets of parallel equidi-
stant fractures. The matrix blocks are impermeable cubes with a
side length of L=2.4m. A single continuum steady-state flow
field simulation was performed using DIRICHLET-type boundary
conditions on the left and right boundary of the model region.
The velocity field is taken as input for a transient solute
transport calculation. A waste repository, located along the
y,z-plane, contaminates the fractured reservoir. A schematic description of the problem case, values of physical properties and
boundary conditions are given in figure 3. Also included in this
figure are contour lines of the piezometric head. To demonstrate
the influence of fracture/ matrix diffusion two simulations were
performed assuming single continuum and double porosity behaviour, respectively. The concentration distribution after 150 days
is plotted in figure 4. The solute is completely drained into
the fault zone and very fast carried by advection-dominated

transport into the far away part of the model region. Comparing the results of single and double porosity calculations indicates that matrix diffusion may act as a safety mechanism in contamination problems, slowing down the spreading of a contaminant in the fractured reservoir. Note that both the semianalytical and the 1-D FE model were applied to simulate the fracture/ matrix leakage. Excellent agreement was obtained between both solutions.

CONCLUSION

An efficient method is presented to simulate flow and transport in fractured reservoirs with "non-steady state double porosity" behavior. The approach uses simple trial functions to represent flow inside matrix blocks of arbitrary shape. Space discretization errors in the matrix are avoided. A direct solution scheme is employed to solve the two coupled equations for the fracture network and the matrix blocks. The method was verified by applying the model to a number of test cases and comparing the results with available analytical solutions. To check the accuracy for more complex problems a second model was developed which evaluates the interflow rate using a 1-D FE approximation for flow and transport in matrix blocks. Application was made to a hypothetical field scale study. The calculations show excellent agreement between both models; however, a remarkable saving of computer time and storage is achieved with the semianalytical method.

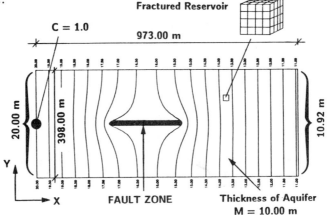

Fault zone properties:

$2b = 2.0m$, $k_f = 5.0m/s$
$\alpha_L = 100.0m$

Matrix properties:

$\phi = 0.01$, $D' = 0.3 \ 10^{-8} \ m^2/s$, $L = 2.4m$
$Prox(z) = 2.5 \ z - 2.083 \ z^2 + 0.579 \ z^3$

Fractured medium properties:

$k_f = 0.4 \ 10^{-8} \ m/s$, $n = 0.01$
$\alpha_L^f = 20.0m$, $\alpha_T = 5.0m$

Fig. 3: Schematic description of the field-scale study

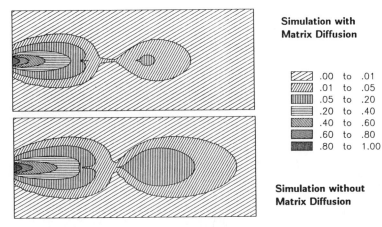

Simulation with
Matrix Diffusion

	.00 to	.01
	.01 to	.05
	.05 to	.20
	.20 to	.40
	.40 to	.60
	.60 to	.80
	.80 to	1.00

Simulation without
Matrix Diffusion

Fig. 4: Concentration distribution in the fractured reservoir

REFERENCES

[1] BARENBLATT, G.I.; I.P. ZHELTOV; and N. KOCHINA (1960): Basic Concepts in the Theory of Seepage of Homogeneous Liquids in Fissured Rocks. Prikl. Mat. Mekh., Vol. 24, pp. 852-864.

[2] TANG, D.H.; E.O. FRIND and E.A. SUDICKY (1981): Contaminant Transport in Fractured Porous Media: Analytical Solution for a Single Fracture. Wat. Resourc. Res., Vol. 17, pp. 555-564.

[3] PRUESS, K. and K. KARASAKI (1982): Proximity Functions for Modeling Fluid and Heat Flow in Reservoirs with Stochastic Fracture Distributions. Proceedings Eighth Workshop Geothermal Reservoir Engineering, Stanford CA.

[4] NOORISHAD, J. and M. MEHRAN (1982): An Upstream Finite Element Method for Solution of Transient Transport Equations in Fractured Porous Media. Water Resources Res., Vol. 18, pp. 588 - 596.

[5] PRUESS, K. and T.N. NARASIMHAN (1985): A Practical Method for Modeling Fluid and Heat Flow in Fractured Porous Media. Soc. Pet. Eng. J., Vol. 25, pp. 14 - 26.

[6] BIBBY, R. (1981): Mass Transport of Solutes in Dual - Porosity Media. Water Resources Res., Vol. 17, pp. 1075 - 1081.

[7] HUYAKORN, P.S.; B.H. LESTER and C.R. FAUST (1983): Finite Element Techniques for Modeling Groundwater Flow in Fractured Aquifers. Water Resources Res., Vol. 19, pp. 1019-1035.

[8] HUYAKORN, P.S.; B.H. LESTER and J.W. MERCER (1983): An Efficient Finite Element Technique for Modeling Transport in Fractured Porous Media: 1. Single Species Transport. Water Resources Res., Vol. 19, pp. 841 - 854.

[9] VINSOME, P.K.W. and J. WESTERVELD (1985): A Simple Method for Predicting Cap and Base Rock Heat Losses in Thermal Reservoir Simulations. J. Canadian Pet. Tech., Vol. 21, pp. 1861 - 1874.

[10] PRUESS, K. and Y.S. WU (1989): A New Semianalytical Method for Numerical Simulation of Fluid and Heat Flow in Fractured Reservoirs. Proceedings SPE Symposium on Reservoir Simulation, Houston TX.

Modelling Transport in Discrete Fracture Systems with a Finite Element Scheme of Increased Consistency

K.P. Kröhn, W. Zielke
Institut für Strömungsmechanik und Elektronisches Rechnen im Bauwesen, Universität Hannover, D-3000 Hannover 1, FRG

ABSTRACT

A finite element formulation with higher order accuracy for advection dominated transport problems in discrete fracture systems is described. The propagation of a cone shaped plume in a real fracture system was successfully simulated to demonstrate the performance of the improved scheme.

INTRODUCTION

The somewhat limited knowledge of flow and transport phenomena in fractured porous media has recently been enhanced by many field experiments. Mainly small domains are often investigated because of the large time scales involved, even if experimental velocities far exceed the natural fluid velocities.

The finite element program system ROCKFLOW has been developed to enable these processes to be calculated in deterministically described fractured rock. For this purpose, the program contains one-, two- and three-dimensional elements which can be combined arbitrarily in space. Since the standard finite element procedure tends to create intolerable oscillations in the solution for highly advective transport problems, the element formulation was modified to improve the solution.

A test case for this modification was defined and the ability of ROCKFLOW to deal with fractured structures was verified. The field data used were provided by the German Bundesanstalt für Geowissenschaften und Rohstoffe (Federal Institute for Geosciences and Natural Resources) BGR, who have conducted a number of experiments in crystalline rock formations at the Grimsel Rock Laboratory in northern Switzerland in cooperation with the Swiss Nationale Gesellschaft für die Lagerung radioaktiver Abfälle (National Company for depositing radioactive waste) NAGRA.

IMPROVED FINITE ELEMENT FORMULATION

Fluid flow and tracer transport described by the continuity equation

$$S_o \frac{\partial h}{\partial t} - \nabla * v = 0 \tag{1}$$

and the transport equation

$$n \frac{\partial c}{\partial t} + v * \nabla c - n \nabla * (D * \nabla c) + n \lambda c + q(c - c^*) = 0 \tag{2}$$

can be calculated by the finite element method with linear shape and test functions.

Gärtner [1] and Leismann [2] applied an analysis of consistency, which is a known technique in the finite difference method, to the equations relating to a Galerkin finite element formulation of the purely advective transport equation for one-dimensional and two-dimensional triangular elements. Following this procedure for isoparametric plane and hexahedral spatial elements and leaving the collocation point Θ for the time integration open, this time—consuming analysis yields two error terms. If incorporated in the finite element scheme with opposite signs, these new terms remove the error concerned without introducing new errors. The modified transport equation is of third order accuracy in time compared to the second order accuracy of the standard Crank-Nicolson-scheme and has the form

$$\frac{\partial c}{\partial t} + A(c) + (\Theta - \frac{1}{2}) \Delta t \left(A(c) \right)^2 - (\frac{1}{6} - \Theta + \Theta^2) \Delta t^2 \left(A(\frac{\partial c}{\partial t}) \right)^2 \tag{3}$$

with the differential operator A defined as

$$A(c) = v^T * (\mathbf{grad}\ c) \tag{4}$$

Leismann [2] showed that approximation of additional terms does not interfere with the improved scheme. Some basic comparisons for the improved accuracy are given by Kröhn, Wollrath and Zielke [4].

DESCRIPTION OF A GEOLOGICAL FRACTURE SYSTEM

Initial results of the experiments performed by the BGR at the Grimsel site indicated two planelike zones of high permeability which were caused by a locally increased number of small fissures. Each zone was considered to be a "macro fracture" and will be referred to as a "fracture" in the following. The strike and dip of the fractures were different and their locations were close enough to penetrate each other within the observed domain.

The surrounding granite rock had a permeability which was at least three orders of magnitude lower than in the fracture. Therefore the granite can be considered as impermeable compared to the fractures. The same applies to an even greater extend to the lamprophyre formation which constituted the vertical limits of the fracture system. Although an upper boundary could not be identified, pressure measurements indicated a hydraulic connection to the mountain surface. A lower boundary was not detected either.

One of the boreholes drilled with a dip of 30^0 connected both fractures. This hole penetrated the first fracture in the upper third (point I) and the second fracture in the lower third (point II). Two litres of water per minute discharged from the hole when the fracture system was hydraulically disturbed by the hole only.

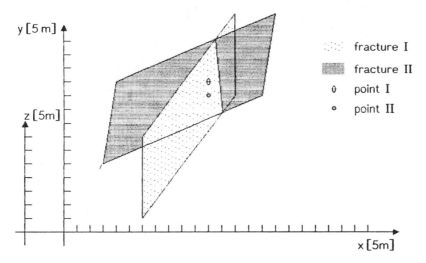

Figure 1   Geometry of the fracture system

NUMERICAL MODEL

While the vertical boundaries could be determined from geological maps, the upper and lower boundaries had to be chosen arbitrarily. For the purpose in question, horizontal cuts through the fracture system provided boundaries at a distance of at least 10 meters from the penetration points between the fractures and the borehole. The upper boundary was assumed to be open to fluid flow while zero flow conditions applied to all other boundaries of the system. According to the data recorded by the BGR, the fracture permeability was determined to be 9×10^{-5} m/s and the aperture 2 mm. Since the main fluid flow took place in the fractures, the rock matrix was neglected in the simulation.

The hydraulic system described was modelled with two-dimensional elements for the fractures and one-dimensional elements for the borehole. Both fractures had a rectangular shape and a sidelength of about 60 meters in the horizontal and 30 meters in the vertical direction. Fracture I was vertically orientated in space while fracture II had a dip of 80^0. The sloping intersection line between the fractures therefore resulted in an irregular element grid. The element mesh consisting of 2262 two-dimensional elements was designed so as to provide a fine grid spacing along the anticipated pathway of the plume and a coarser grid elsewhere

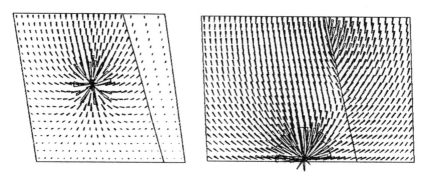

Figure 2 Cutout regions of the 2-D velocity field: fracture I (left) and fracture II (right)

A conelike concentration distribution with a radius of 5.5 meters was taken as an initial condition. Using elements with a typical side length of about one meter, the cone radius covered only six elements. The cone centre was located in fracture II in such a way that it travelled on curved streamlines to the intersection, passed the penetration line, underwent complete mixing with the fresh water and finally accelerated to the borehole where it left the system.

Apart from the initial plume there was no concentration within the domain or entering the system. As the most simple approach for hydrodynamic dispersion an isotropic dispersion of $D = 10^{-5}$ m^2/s was chosen.

RESULTS OF THE SIMULATION

The calculated flow field is quite complicated. Owing to the different heights, at which the borehole penetrated the fractures, the water is more attracted to point I than to point II. For this reason a discontinuity in the velocity field across the fracture intersection can be observed. Figure 2 shows the velocity field in a section containing the upper two thirds of fractures I and II. The scale of the velocity vectors is chosen to be different so as to indicate the pathlines of the flow field. (The shape of the fractures is distorted in these figures because the plot software used only allows projections on the x-z plane or the y-z plane, respectively.)

Water at the centre of the initial concentration cone is accelerated from $8.7 \, 10^{-5}$ m/s to $6.6 \, 10^{-4}$ m/s at point II and $1.6 \, 10^{-3}$ m/s at point I. The corresponding grid Peclet numbers are 11, 66 and 235, respectively. Figure 3 shows the results of the transport simulation as a chronological sequence. The plots indicate lines of equal concentration every 64,000 s, in fracture I on the lefthand side and in fracture II on the righthand side.

In the first instance, the concentration cloud moves along curved stream-lines to the fracture intersection and is distorted accordingly (figure 3, t_0 and t_1). After passing the intersection, a slight bending of the contour lines occurs in fracture II due to the discontinuities in the flow field (figure 3, t_2). The mixing process then lowers the concentrations of the subsequently divided plume in the ratio of the upstream water flows (figure 3, t_3 and t_4).

Certain oscillations occur upstream of the intersection (figure 3, t_2 and t_3). These are caused by the steep concentration gradient created by the passing plume at the intersection. When the cloud is transported beyond this line, the oscillations disappear again (figure 3, t_4 and t_5). (The same affect applies analogously to the sinks.)

CONCLUSIONS

The improved advection behaviour is sufficiently well demonstrated for a fracture geometry. Despite the fact that the grid peclet numbers lie in the range of 11 to 235 and that the plume consists of a cone with quite a small base area, the transport process is simulated well by the numerical scheme.

ACKNOWLEDGEMENTS

This paper is based upon work commisioned by the Bundesanstalt für Geowissen-schaften und Rohstoffe BGR.

REFERENCES

[1] Gärtner, S. Zur diskreten Approximation kontinuumsmechanischer Bilanz-gleichungen, Bericht Nr. 24/1987, Institut für Strömungsmechanik und Elektronisches Rechnen im Bauwesen, Universität Hannover, 1987

[2] Leismann, H. M. Berechnung von Ausbreitungsvorgängen im Grundwasser mit der Methode der finiten Elemente, Ph.D. thesis, Institut für Hydro-mechanik, Universität Karlsruhe, 1987

[3] Helmig, R., Kröhn, K.-P., Shao, H., Wollrath, J., ROCKFLOW, Theorie und Benutzeranleitung, Institut für Strömungsmechanik und Elektronisches Rechnen im Bauwesen, Universität Hannover, 1986-1989

[4] Kröhn, K.-P., Wollrath, J., Zielke, W. Modelling Transport in Fractured Rock by the Finite Element Method, in Contaminant Transport in Ground-water (Ed. Kobus, H.E. and Kinzelbach, W.), pp. 291 to 297, Proceedings of the International Symposium on Contaminant Transport in Groundwater, Stuttgart, West-Germany, 1989. A.A. Balkema/Rotterdam/Brookfield, 1989

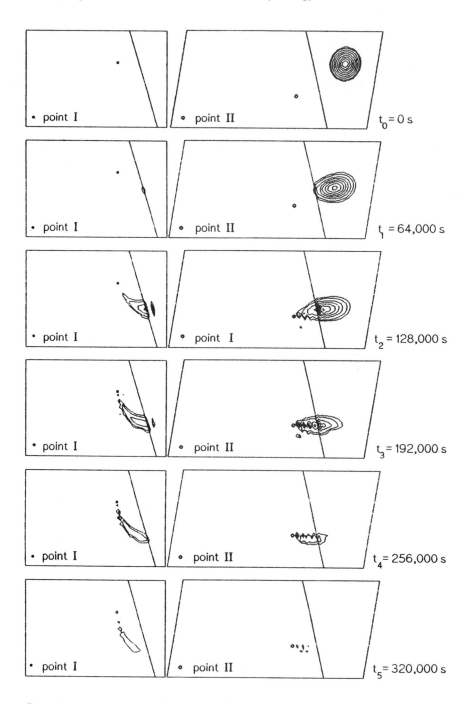

Figure 3   Contour lines of 5%, 10%, 20%, 30%, ... , 80% concentration at
different times of the simulation

SECTION 5 - GROUNDWATER TRANSPORT CONTAMINATION PROBLEMS

Simulation and Analytical Tools for Modeling Ground Water Quality

W. Woldt(*), I. Bogardi(*), W.E. Kelly(*), A. Bardossy(**)

(*) Department of Civil Engineering, University of Nebraska, Lincoln, Nebraska, 68588-0531, USA
(**) Institute for Hydrology and Water Resources, University of Karlsruhe, Karlsruhe, West Germany, D7500

ABSTRACT

A methodology is developed to explore the effects of two sources of uncertainty in modeling groundwater contamination processes. The first source of uncertainty is related to the initial conditions describing the suspected plume location. The second source of uncertainty concerns variability of aquifer parameters that influence contaminant movement. The relative contribution of these two sources to the overall uncertainty is investigated. It is shown with a case study example that the imprecision in initial conditions may lead to higher level of uncertainty in the transport modeling process than the uncertainty caused by imprecise aquifer parameter identification. The methodology can also be used as a screening tool to decide on the need of further remedial action planning.

INTRODUCTION

The purpose of this paper is to develop and test a groundwater model that may be used to quantify the need for additional remedial action planning in an existing groundwater plume. Application of groundwater models includes control of conjunctive surface groundwater systems, prevention of saltwater intrusion, aquifer dewatering, and restoration of contaminated aquifers. There has recently been a steady increase in research that incorporates uncertain elements into management models [1].

In general there have been two main approaches to incorporate random fields into groundwater models. The first is to include random parameters as coefficients in the partial differential equations that describe the system to be modeled (Gelhar [2], Bakr et al. [3], and Graham and McLaughlin [4]). Then the stochastic differential equations are solved by using the theory of spectral analysis to analyze perturbed forms of the equations.

A second approach is the linking of Monte Carlo simulation techniques with various groundwater modeling methods. This technique employs the repeated process of: 1) generation of a random field, 2) conditioning of the field to observed data, and 3) using the conditioned data as input to a groundwater model. Each time the process is repeated a new realization of aquifer parameters are generated and used in the groundwater model.

One of the principle advantages of using simulation combined with modeling is the ability to consider relatively complex groundwater systems. An application of this methodology by Gorelick [5] examined the influence of spatially correlated transmissivity fields upon aquifer restoration strategies. In a related work, Wagner and Gorelick [6] incorporate aquifer parameter uncertainty into a management model for optimal design of a remediation scheme. In this application, first-order first- and second-moment analysis of the parameters are used to transfer their uncertainty to the management model. Another paper by Wagner and Gorelick [7] describes the use of conditional simulation in which numerous log-hydraulic conductivity fields are generated by the turning bands method. Simulated fields are then used as input to a groundwater model that has been linked to a nonlinear optimization model for identification of reliable remediation strategies.

In summary, much of the prior efforts have been directed toward stochastic modeling of aquifer parameters such as hydraulic conductivity and interstitial velocity. In the case of an existing contamination event, two general approaches may be used to predict the effects. One approach attempts to recreate the historical events and model the system, using stochastic aquifer parameters, from the beginning of the failure. The other uses available data to formulate a deterministic interpretation of the existing plume. This interpretation is then used as initial conditions in a model that considers aquifer parameters as realizations of a stochastic process. In both cases the initial conditions are assumed to be known precisely and uncertainty is introduced from the aquifer parameters.

In addition, most of the efforts related to groundwater pollution management have addressed the problem of planning and managing restoration activities. This seems to imply that the decision has been made to design a remediation scheme, yet there is very little research in the literature to guide the decision related to the need for a restoration program. Restoration planning efforts have employed rather complex numerical models, that according to Chu et al. [8] require an extensive amount of good quality data that may not be available. Another consequence of using complex numerical models is the extensive use of 2-dimensional models to predict solute transport. However, as pointed out by Dagan [1], many problems of interest such as solute transport occur at the local scale where flow and/or transport processes are of a 3-dimensional nature.

METHODOLOGY

A methodology will be presented to assess the need for remedial action planning within a 3-dimensional, stochastic, analytical groundwater modeling framework. To accomplish this objective, 3-dimensional conditional simulation using the method of turning bands will be used to generate several possible realizations of contamination plume position and intensity. A unidimensional simulation of log-hydraulic conductivity will be used to generate variable aquifer parameters. Then results of both simulations will be used as input to a 3-dimensional analytical solute transport model to predict contamination levels at specified locations and times. A schematic of the primary components is included as Figure 1.

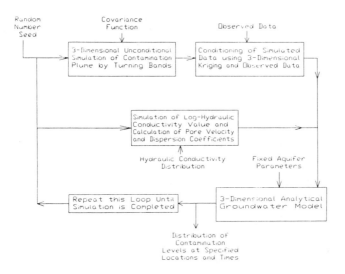

Figure 1. Schematic of Simulation Algorithm

Simulation of the Contamination Plume

There are several multidimensional simulation techniques described in the literature including matrix models such as the nearest neighbor method used by Smith and Freeze [9] and spectral techniques described by Mejia and Rodriguez-Iturbe [10]. Here we used the turning bands simulation method which takes a different approach such that the basic concept is to transform a multidimensional problem into the sum of a series of equivalent unidimensional simulations [11].

Simulation of Aquifer Parameters

Aquifer parameters that are simulated for input to the 3-dimensional analytical model include interstitial (pore) velocity, and dispersion coefficients in the longitudinal, transverse, and vertical directions. Since the analytical model assumes uniform aquifer parameters throughout the modeled region, only one value for each of the simulated parameters needs to be calculated for each simulation run.

Simulated values of interstitial velocity can be computed by first simulating a hydraulic conductivity K for the area of interest. Assuming fixed values for local hydraulic gradient, dh/dx, and local effective porosity P, interstitial velocity can be computed as follows:

$$V^* = \frac{K \; dh/dx}{P} \tag{1}$$

The term *local* as used in this paper is defined according to the local scale analysis described by Dagan [1].

Values for dispersion coefficients in the longitudinal *Dx*, transverse *Dy*, and vertical *Dz* directions may be computed from the simulated interstitial velocity and representative dispersivity values, *ax, ay, az,* for each of the spatial dimensions.

$$D_x = V^* a_x \ , \qquad D_y = V^* a_y \ , \qquad D_z = V^* a_z \qquad (2)$$

A single realization of hydraulic conductivity can be simulated using standard distribution generation techniques [12]. As additional data on the spatial distribution of aquifer parameters is obtained, it would be possible to model a system using simulated input to distributed parameter numerical models instead of a lumped parameter analytical model.

Three Dimensional Groundwater Model
A 3-dimensional analytical groundwater model is used to predict the contamination level at specified locations and future times. The selected locations and times may correspond to a region of interest, a municipal well, or a regulatory compliance surface. A modified version of an existing computer code by Wagner et al. [13] capable of solving the 3-dimensional analytical solute transport equation has been adapted for this paper.

The analytical solution is constrained by the following assumptions: 1.Completely saturated flow regime.   2. Aquifer properties constant and uniform.   3. Groundwater flow is horizontal, continuous and uniform.   4. Point contaminant source. 5. Aquifer is infinite in extent.   6. Source located at origin of coordinate system.   7. Mass flow rate of source is constant. 8. Initial concentration of contaminant is zero.   While these assumptions appear to substantially limit the applicability of an analytical approach to modeling actual contamination events, many of the limitations can be eliminated by using the principle of superposition.

. The differential equation describing constituent mass concentration in a porous medium is a linear partial differential equation.   Walton [14] and Haitjema [15] describe the use of superposition with analytical models, which can be utilized to overcome many of the limitations and solve contamination problems of considerable complexity.   The computer program developed by Wagner et al. [13] incorporates image sources to eliminate the assumption of infinite aquifer extent in the vertical direction. Consequently, a saturated thickness must be specified, and the aquifer is assumed to extend to infinity in the horizontal dimension.   The limitation of a single source located at the origin may be circumvented by using superposition in space, and systematically moving the origin to each of the multiple source locations.   The assumption of a constant source mass flow rate may be avoided by using the principle of superposition in time.   In this case a variable source rate of contamination is approximated by the sum of a finite number of constant source rates distributed in time.

Application of the superposition principle in both space and time results in the ability to model an aquifer restricted only by limitations 1 through 4 noted previously.   Of these limitations, the condition that a contamination source be a point type source can be alleviated by computing the total mass of contaminant introduced by the source, and placing this mass at the source centroid. The other remaining limitations are saturated flow, constant and uniform aquifer properties, and horizontal, continuous, uniform groundwater flow.

Since the model requires constant and uniform aquifer properties, a single value of interstitial velocity and a dispersion coefficient for each of the three orthogonal axis directions are generated for each model run. Different realizations of the simulated aquifer properties are generated for each simulation run.

The ability to model numerous, variable rate sources is utilized to consider uncertainty in initial conditions related to contamination plume location and intensity. Concentration levels that have been generated at specified grid coordinates by the plume simulation program are input to the model as a short term pulse when the model is initiated.

The complete process of contaminant plume simulation, aquifer parameter simulation, and analytical modeling is then repeated numerous times to generate several possible realizations of contamination levels at specified locations and times. The distribution of contaminant levels may then be characterized and the probability of exceeding a threshold level can be determined.

ILLUSTRATIVE EXAMPLE

The proposed methodology has been tested by application to a groundwater contamination event at a low-level radioactive waste processing facility near Wood River Junction, Rhode Island. The site has been investigated by the U.S.G.S. and documented in Open File Report 84-725 by Ryan et al. [16].

A plan view of the study area is presented in Figure 2 with the source being an industrial processing site and direction of groundwater movement indicated on the figure.

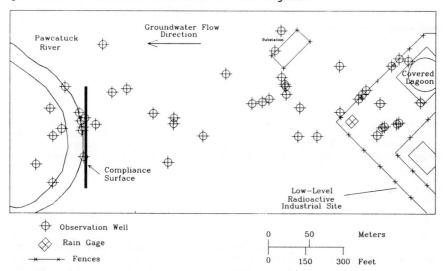

Figure 2. Plan View of Study Area

Contamination at the site is the result of 14 years of waste disposal from the industrial facility during the period 1966 to 1980. Liquid wastes containing both radionuclides and other chemical solutes have percolated from uncovered ponds and trenches located at the facility into an unconfined highly permeable sand and gravel aquifer. The resultant plume extends about 2300 feet from the source to the Pawcatuck River. Processing at the plant has been terminated and material from the bottom of the ponds and trenches has been removed and shipped off site for proper disposal.

An objective of the case study is to assess the effects of uncertainty in both aquifer parameters and initial conditions describing the plume location. The effect of uncertainty is analyzed by modeling the same system with different stochastic inputs. The first simulation considers aquifer parameters as stochastic with constant initial conditions. The second simulation uses constant aquifer parameters with stochastic initial conditions. The final run considers both aquifer parameters and initial conditions as stochastic input.

Results of each run are analyzed by estimating the probability that contamination levels will exceed regulatory limits at a predefined compliance surface shown in Figure 2 with a vertical depth of 40 feet beginning 5 feet below ground surface. The surface is approximated by 15 points determined by a 5 by 3 mesh of observation locations. The model computes contamination levels after 500 days have elapsed from the initiation of each run. The selection of compliance surface and time variables are used for example purposes only, and may be changed to reflect individual site requirements.

Study Site Parameters
The study site has a constant saturated thickness of 80 feet with a constant effective porosity equal to 0.38 and a hydraulic gradient of 0.0053 as reported by Ryan and Kipp [17]. A longitudinal dispersivity value of 120 ft^2/day was taken from previous work by Kipp et al. [18]. Transverse dispersivity was assumed to be one order of magnitude smaller than longitudinal. Vertical dispersivity was assumed to be two orders of magnitude smaller than longitudinal. The contaminant was assumed to be completely conservative with a retardation coefficient equal to 1.0 and a decay constant of zero.

Statistical parameters that characterize the log-hydraulic conductivity field are taken to be μ=4.94 and σ^2=0.10. The mean log-hydraulic conductivity of 4.94 corresponds to a transformed mean of 147.6 ft/day. A variance of 0.10 corresponds to a transformed variance of 2420 ft^2/day^2. These values are representative of the values reported by Ryan and Kipp [17] and Kipp et al. [18]. Table 2 presents a summary of the study site parameters.

Initial Conditions
Stochastic initial conditions were generated using direct observation well data located by plan view in Figure 2. Well clusters of 2 to 6 wells are located at eight of the observation wells. Specific conductance is used as an indicator of the more hazardous Strontium-90 contamination. A relationship between elevated conductivity and Strontium-90 levels has been confirmed by Woldt et al. [19]. Observed conductivity data can be characterized by a lognormal distribution, hence all simulation is performed in log transform space. After completion of the

Table 2. Aquifer Parameters

Saturated thickness	80 feet
Effective porosity	0.38
Retardation coefficient	1.0
First order decay factor	0.0
Average interstitial velocity	f(K)
Hydraulic conductivity (K)	$\mu=147.6$ ft/day $\sigma^2=2420$ ft^2/day^2
Longitudinal dispersivity	120 feet
Transverse dispersivity	12 feet
Vertical dispersivity	1.2 feet

3-dimensional simulation an inverse transform is performed, and the conductivity data is converted to mg/L total dissolved solids using the relationship from Todd [20]: 1 mg/L = 1.56 μmhos.

Structural analysis of the borehole data was used to evaluate covariance functions for conditional simulation of the uncertain plume location as described by Woldt et al. [19]. Simulated contamination values were generated on a grid measuring 100' by 100' in the horizontal-spatial dimension with 10' grid increments in the vertical dimension for a total of 1482 simulated values. A 3-dimensional representation of a simulated plume (Figure 3) has been created using National Center for Atmospheric Research (NCAR) graphics on a Cray supercomputer. Resulting simulated values were used as initial conditions for the groundwater model by considering them as a short term, 0.5 day source.

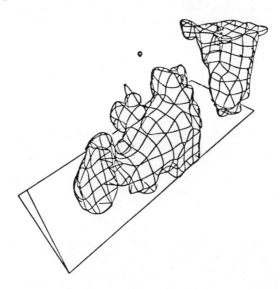

Figure 3. Three Dimensional Iso-surface of Conductivity
Equal to 1500 μmhos

Simulation Runs
The complete geostatistical/groundwater simulation procedure was executed 100 times for each run. Each generation of a random contamination field and solution for contamination levels at the compliance surface (15 locations based on 5x3 mesh) after 500 days required approximately 45 minutes CPU time on a 20 Mhz personal computer based on 80386/80387 architecture.

RESULTS

Examination of the frequency distributions for each run revealed a lognormal type distribution. Mean and variance are tabulated for each simulation in Table 3. The statistics are presented for a single location in the aquifer corresponding to the geometric center of the compliance surface.

Table 3. Statistical Summary of Contamination Levels
at the Center of Compliance Surface (mg/L)

Aquifer	Initial	Mean	Std. Dev.	P(C〉400)
Stochastic	Constant	97.2	62.7	0.0207
Constant	Stochastic	154.5	77.8	0.0109
Stochastic	Stochastic	195.8	186.5	0.1361

Based on these results it appears that imprecision in contamination plume location leads to a greater level of uncertainty in the modeling process than does uncertainty in aquifer parameters. It is interesting to note that the expected value of contamination level has also increased. In addition, uncertainty in both aquifer parameters and initial conditions causes the standard deviation to increase by a factor of 2.97 compared to uncertainty in aquifer parameters alone. The combined uncertainty also produced a mean contamination level greater than either of the single element uncertainty models. These results suggest that available resources should be allocated in a greater proportion toward contamination assessment than aquifer parameter identification.

The analysis can also be used to evaluate the need for further remedial action planning by computing the probability of exceeding regulatory limits established for the contaminant of interest. In this case the EPA drinking water standard for Strontium-90 is 8pCi/L which corresponds to a specific conductance of 630 μmhos and a total dissolved solids value of 400 mg/L. Hence the probability of exceeding 400 mg/L can be used as an indicator that Strontium-90 levels are exceeding the regulatory limit. These probabilities have been tabulated in Table 3 for each simulation run. It is apparent that consideration of two sources of uncertainty produces an increase in the probability that levels will exceed regulatory standards. These results imply that an increased model uncertainty generates results that tend to favor further remedial action planning.

CONCLUSIONS

1. Geostatistical simulation and groundwater modeling can be combined to identify the need for a clean-up of an existing plume.
2. Two major uncertainties of the groundwater contamination process can be simultaneously considered: the inaccuracy of the estimated plume and the uncertainty of aquifer parameters.

3. The methodology is general enough to accommodate various types of groundwater models; a 3-dimensional analytical model with lumped parameter for input was used in the present example.
4. Under case study conditions, the uncertainty caused by imprecise initial conditions is higher than the uncertainty caused by imprecise aquifer parameters.
5. The combined effect of both uncertainties has considerably increased the risk of groundwater contamination at selected compliance surfaces and times.

REFERENCES

1. Dagan, G. Statistical Theory of Groundwater Flow and Transport Pore to Laboratory, Laboratory to Formation, and Formation to Regional Scale, Water Resources Research, Vol.22, No.9, pp. 120S-134S, 1986.

2. Gelhar, L. Stochastic Subsurface Hydrology From Theory to Applications, Water Resources Research, Vol.22, No.9, pp. 135S-145S, 1986.

3. Bakr, A.A., Gelhar, L.W., Gutjahr, A.L., and MacMillan, J.R. Stochastic Analysis of Spatial Variability in Subsurface Flows, 1, Comparison of One- and Three-Dimensional Flows, Water Resources Research, Vol.14, No.2, pp. 263-271, 1978.

4. Graham W. and McLaughlin D. Stochastic Analysis of Nonstationary Subsurface Solute Transport, 1, Unconditional Moments, Water Resources Research, Vol.25, No.2, pp. 215-232, 1989.

5. Gorelick, S.M. Sensitivity Analysis of Optimal Groundwater Contaminant Curves: Spatial Variability and Robust Solutions, in Proceedings, NWWA Conference Solving Groundwater Problems With Models, pp. 133-146, National Water Well Association, Dublin, Ohio, 1987.

6. Wagner, B.J. and Gorelick, S.M. Optimal Groundwater Quality Management Under Parameter Uncertainty, Water Resources Research, Vol.23, No.7, pp. 1162-1174, 1987.

7. Wagner, B.J. and Gorelick, S.M. Reliable Aquifer Remediation in the Presence of Spatially Variable Hydraulic Conductivity: From Data to Design, Water Resources Research, Vol.25, No.10, pp. 2211-2225, 1989.

8. Chu, W., Strecker, E.W., and Lettenmaier, D.P. An evaluation of Data Requirements for Groundwater Contamination Transport Modeling, Water Resources Research, Vol.23, No.3, pp. 408-424, 1987.

9. Smith, L. and Freeze, R.A. Stochastic Analysis of Steady State Groundwater Flow in a Bounded Domain, 2, Analysis of Uncertainty in Prediction, Water Resources Research, Vol.15, No.6, pp. 1543-1559, 1979.

10. Mejia, J. and Rodriguez-Iturbe, I. On the Synthesis of Random Fields from the Spectrum: An Application to the Generation of Hydrologic Spatial Processes, Water Resources Research, Vol.10, No.4, pp. 705-711, 1974.

11. Journel, A.G., and Huijbregts, Gh.J. Mining Geostatistics, Academic Press, Orlando, Florida, 1978.

12. Ang A.H-S. and Tang, W.H. Probability Concepts in Engineering Planning and Design, Volume II: Decision, Risk, and Reliability, John Wiley and Sons, New York, 1984.

13. Wagner, J., Watts, S.A., and Kent, D.C. Plume 3D: Three Dimensional Plumes in Uniform Groundwater Flow, R.S. Kerr Environmental Research Laboratory, U.S.E.P.A. project number EPA/600/2-85/067, 1985.

14. Walton, W.C. Selected Analytical Methods for Well and Aquifer Evaluation, Bulletin 49, Illinois State Water Survey, Urbana, Illinois, 1962.

15. Haitjema, H.M. Modeling Three-Dimensional Flow in Confined Aquifers by Superposition of Both Two- and Three-Dimensional Analytic Functions, Water Resources Research, Vol.21, No.10, pp. 1557-1566, 1985.

16. Ryan, B.J., DeSaulniers, R.M., Bristol, D.A., and Barlow, P.M. Geohydraulic Data for a Low-Level Radioactive Contamination Site, Wood River Junction, Rhode Island, U.S.G.S. Open File Report, 84-725, 1984.

17. Ryan, B.J. and Kipp, K.L. Low-level radioactive contamination from a cold-scrap recovery operation, Wood River Junction, Rhode Island, U.S.G.S. Open File Report, 84-066, 1984.

18. Kipp, K.L., Stollenwerk, K.G., and Grove, D.B. Groundwater Transport of Strontium 90 in a Glacial Outwash Environment, Water Resources Research, Vol.22, No.4, pp. 519-530, 1986.

19. Woldt, W., Bogardi, I., Kelly, W.E. and Bardossy, A. Detection of Groundwater Contamination, Part 1: Use of Direct Borehole Data and Multi-dimensional Geostatistics, Working Paper, Department of Civil Engineering, University of Nebraska - Lincoln, 1990.

20. Todd, D.K. Groundwater Hydrology, John Wiley and Sons, New York, 1980.

Acknowledgements   The research on which this paper is based has been supported by the U.S. Geological Survey, Department of Interior, under award number 14-08-0001-G1133, and by the Nebraska Water Resources Center, University of Nebraska - Lincoln. Special thanks are due to Geza Pesti of the University of Nebraska - Lincoln for his valuable assistance.

Groundwater Quality Model with Applications to Various Aquifers

M. Soliman, U. Maniak, A. El-Mongy, M. Talaat, A. Hassan

Irrigation and Hydraulics Department, Faculty of Engineering, Ain Shams University, Cairo, Eygpt

ABSTRACT

A finite element model was developed in order to solve for both regional ground-water flow and conservative solute transport in porous medium. The model was applied to a 55 sq.km groundwater basin in Ruehen region in the Federal Republic of Germany using a network of 1450 elements and 780 nodes. This model was used in simulating a contaminant plume done through injection. Similarly, the model was applied to a 4750 sq.km portion in the Eastern Nile Delta aquifer in Egypt. The model was applied to this portion of the Delta using a network of 543 elements and 310 nodes with the main objective of simulating the problem of salt water intrusion.

MODEL FORMULATION

The groundwater quality problem can be simulated in terms of two partial differential equations. The first equation describes the groundwater flow and the second describes the solute transport in porous media. The quasi three-dimensional flow equation may be written as:

$$\delta/\delta x \ (T_x \ \delta h/\delta x) + \delta/\delta y \ (T_y \ \delta h/\delta y) + l_i \ (H_i - h) \pm Q = S \ \delta h/\delta t \qquad (1)$$

where h is the hydraulic head, T is transmissivity, S is storativity, $\pm Q$ is a sink or a source discharge per unit area, l_i is leakage factor = K_{vi}/d_i, K_{vi} and d_i are vertical hydraulic conductivity and thickness of aquitard i and t is time.

The initial and boundary conditions may be expressed as

$$h \ (x,y,0) = h_0(x,y) \qquad\qquad\qquad\qquad (2-a)$$

$$h \ (x,y,t) = h_1(x,y) \qquad\qquad on \ B_1 \qquad\qquad (2-b)$$

$$q_n(x,y,t) = q_{n1}(x,y) \qquad\qquad on \ B_2 \qquad\qquad (2-c)$$

where h_0 is the initial head, h_1 is the prescribed head on boundary portion B_1, q_n is the component of the specific discharge normal to boundary portion B_2.

In order to describe the solute transport in porous media, the following form .

of the advective-hydrodynamic dispersion equation is used:

$$\delta/\delta x \, (D_x \, \delta C/\delta x) + \delta/\delta y \, (D_y \, \delta C/\delta y) - q_x \, \delta C/\delta x - q_y \, \delta C/\delta y \pm QC' = \delta C/\delta t \qquad (3)$$

where D_x and D_y are the hydrodynamic dispersion coefficients in x and y directions.

$$D_x = a_L \cdot V + D_{md} \quad , \quad D_y = a_T \cdot V + D_{md}$$

where D_{md} is the coefficient of the molecular diffusion, a_L: longitudinal dispersivity in the flow direction, a_T: transversal dispersivity in the transversal direction, C': concentration of the solute of a source/sink of a strength, Q, which is assumed to be known, q is effective pore water velocity in x direction that equals (V/n_e) and n_e is effective porosity.

The initial and boundary conditions may be written as:

$$C(x,y,0) = C_0(x,y) \qquad (4\text{-}a)$$

$$C(x,y,t) = C_1(x,y) \qquad \text{on } B'_1 \qquad (4\text{-}b)$$

$$D_x \, \delta C/\delta x + D_y \, \delta C/\delta y = q_x \, C^* + q_y \, C^* \qquad \text{on } B'_2 \qquad (4\text{-}c)$$

$$D_x \, \delta C/\delta x + D_y \, \delta C/\delta y - q_x \, C - q_y \, C = q_x \, C^* + q_y \, C^* \qquad \text{on } B'_3 \qquad (4\text{-}d)$$

where C_0 is initial concentration and C^* is a known value.

SEQUENCE OF SOLUTION

Following the Galerkin method, Connor [2] and Zienkiewicz [5], the numerical solution for the both hydraulic head and solute concentration was developed. For the time derivative a finite difference solution was applied. Firstly, the flow equation is solved in order to compute the hydraulic head distribution. The velocity components through elements are computed applying Darcy's law. Afterwards, the solute transport submodel is used to compute the concentration distribution. For more details concerning the computational sequence, one can refer to Soliman [4].

APPLICATIONS

Ruehen Valley quality model:
The model area is about 55 sq.km in Ruehen Valley that lies about 40 km northern East from Braunschweig City, Federal Republic of Germany. Ruehen aquifer is a leaky confined aquifer consisting of two layers. The main layer is the sublayer which is a mixture of coarse to fine sand with a thickness varies between 10 m to 85 m. The top layer is a mixture of silty clay and fine sand with a thickness differs from 5 to 40 m. The bottom of the aquifer is an impervious layer of a very compacted clay formation.

The model area was discretized into 1450 elements and 780 nodes as shown in figure 1. Firstly, the steady state within a period of one month was simulated using a zero storage coefficient and a time interval of one week in order to calibrate the transmissivity and leakage factor. The calibration was carried out using the trial and error method. For each run the computed piezometric levels were compared with the observed values for some selected observation wells as guides until a desirable mean convergence less than 40 cm was reached. Figure 2 illustrates a sample of the calibration results in the form of piezometric level isolines at a certain simulation time.

The calibrated parameters were used in order to simulate the unsteady state conditions within a period of 5 months with a constant time interval of one day with the purpose of determining the storage coefficient. The calibration process was carried out until a desired mean convergence less than 60 cm was reached at the end of simulation time.

Using the same network and the results of the flow model in terms of the velocity components through elements, the quality model was constructed and calibrated. The model was used to simulate the propagation of a contaminant plume done by injection at a certain point in the model area. The rate of injection is 10 kg/hour and the total injection time is 2 hours. Figure 3 shows the situation after three weeks in a concentration isolines form.

East Nile Delta quality model:
The model area, that is about 4750 sq.km, lies in the eastern region of the Nile Delta in Egypt. The Delta-aquifer is composed generally of unconsolidated sand and gravel with occasional clay lenses. The aquifer thickness varies between 200 m up to 700 m. The top boundary of these formations is a clay cap aquitard with a thickness differs from 0.0 m up to a value not more than 15 m. The aquifer lies on an aquiclude composed of thick layers of impermeable clays.

The model area was discretized into 543 elements and 310 nodes as shown in figure 4. The steady state was simulated using a prescribed criteria $\pm$ 1.0 m depending on the reliability and availability of the collected data. The result of the last run is presented in figure 5. With the aim of determining the model storativity, a period of about 8 months was simulated with a time interval of 10 days.

Using the same grid of the groundwater flow model and the available chemical data, the solute transport model was constructed and calibrated to determine the model dispersivity. For each run the computed concentration values were compared with the observed values at some points for which data are available. The calibrated values of the model longitudinal dispersivity, (northwards), and lateral dispersivity (eastwards), amount to 100 and 10 km respectively. Figure 6 shows the simulated situation.

CONCLUSIONS AND RECOMMENDATIONS

A developed finite element model has proved to be reliable in simulating both groundwater flow and solute transport problems. The applications of the model to field problems show that the model parameters are strongly dependent of the network's geometry, the boundary conditions and the reliability of field data. It is also concluded that the model dispersivity is many hundred times of magnitude of both theoretical and laboratory values.

Concerning the calibrated model parameters, it is recommended that they should be regarded as apparent values and must be independently checked using more reliable data. It is also strongly recommended that the developed quality model of the East Nile-Delta aquifer should be considered as a first step of the final simulation of the whole aquifer which could be performed when more reliable and enough data were available.

REFERENCES

1. Bear J. Dynamics of fluids in porous media, American Elsevier, New York, 1972.

2. Connor J.J. and Brebbia C.A. Finite Element Techniques For Fluid Flow, Newnes-Butterworths, London/Boston, 1976.

3. Segol G. and Pinder G.F. Transient simulation of saltwater intrusion in south-eastern Florida, Water Res. Res., Vol. 12 (1), pp. 65–70, 1976.

4. Soliman M.M., Maniak U., El-Mongy A., Talaat A.M. and Hassan A. Management of Wadis in Arid Regions with Applications to Quality Models, Ph.D. Thesis, Faculty of Eng., Ain Shams Univ., Cairo, Egypt, 1989.

5. Zienkiewicz O.C. The Finite Element Method in Engineering Science, McGraw-Hill, London, 1971.

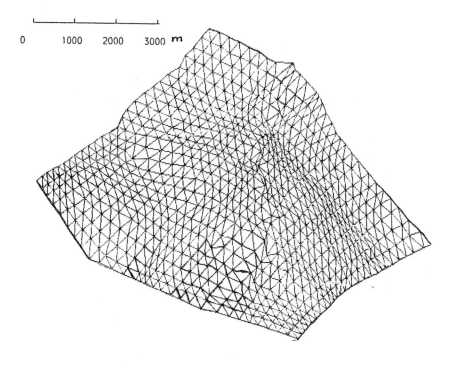

Figure 1. (Ruehen Model) Element network.

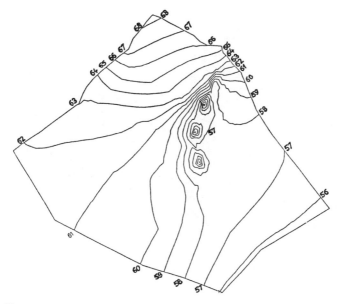

Figure 2. (Ruehen Model Results) Piezometric level in m. + M.S.L.

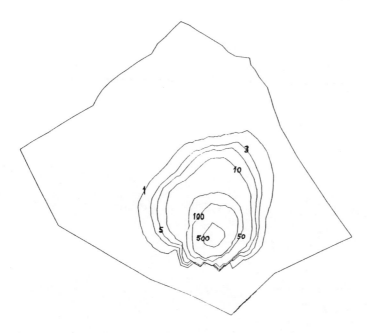

Figure 3. (Ruehen Model Results) Concentration in gm/cu. m. after
21 days injection.

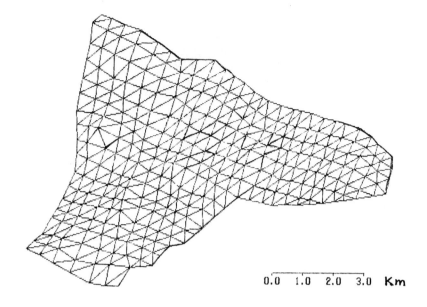

Figure 4. (Delta Model) Element network.

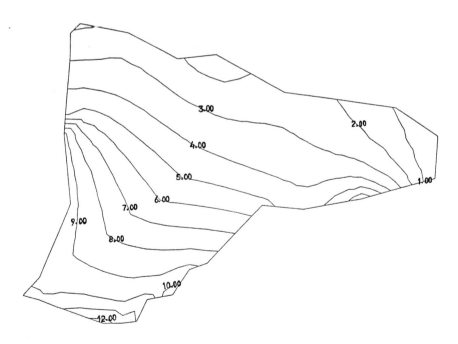

Figure 5. (Delta Model Results) Piezometric level in m. + M.S.L.

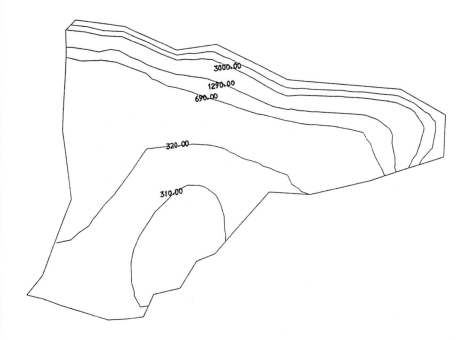

Figure 6. (Delta Model Results) Concentration in ppm.

Galerkin's Finite Element Matrices for Transport Phenomena in Porous Media

M.Ferraresi, G Gottardi

Facoltà di Ingegneria, Università di Bologna,I - 40136 Bologna, Italy

ABSTRACT

Galerkin's finite element technique is applied to contaminant dispersion, adsorption and biological decay in porous media. In two–dimensional configurations, when linear triangular elements are chosen and capacity terms are expressed in diagonal matrix form by using a lumping procedure, the terms of elemental matrices arising from the Galerkin residual method can be written in analytical form of polynomial type, so that spatial integration can be performed analytically instead of numerically.

INTRODUCTION

The phenomena of dispersion, diffusion and convection of pollutants in porous media are governed by two mass conservation equations, both of partial differential type: the first equation, referring to the mass conservation of the mixture of solvent (water) and solute (pollutant), can be written for an elemental volume of aquifer, as (Bear [1]) :

$$-div(\rho\vec{u}) - q = \frac{\partial}{\partial t}(\varphi\rho) \qquad (1)$$

while the second equation, expressing the mass conservation of solute, including adsorption by solid matrix of the porous medium and biological decay of pollutant, reads (Fried [2], Kinzelbach [3]):

$$-div\vec{J} - div(c\vec{u}) - \lambda\varphi c - \lambda\varphi_1 f - q_c = \frac{\partial}{\partial t}(\varphi c + \varphi_1 f) \qquad (2)$$

The adsorption model is supposed to be the linear equilibrium Henry's law:

$$f = \chi c \qquad (3)$$

where λ is the rate of biological decay of pollutant and χ is a repartition constant between the pollutant concentration f adsorbed by the solid matrix and the pollutant concentration c remaining in liquid phase. Concentrations c and f are expressed as mass per volume of mixture and solid respectively. In Equations (1) and (2), velocity $\vec{u}$ follows Darcy's law:

$$\vec{u} = -\frac{[K_p]}{\mu}\left(\vec{grad}\,p + \vec{grad}\,\rho g z\right) \qquad (4)$$

being K_p the permeability tensor of the porous medium, μ the mixture viscosity, p its pressure and z the elevation from a given reference datum, measured vertically.

In Equations (1) and (2) q and q_c are source terms for mixture and solute respectively, given as mass discharge rates per unit volume, (negative for injection); φ is medium porosity, $\varphi_1 = (1 - \varphi)$ and ρ is the density of the mixture. Both density and viscosity are functions of fluid pressure p and pollutant concentration c. In Equation (2) the mass concentration c is usually related to the dispersive mass flux $\vec{J}$ of pollutant per unit surface of porous medium by a Fickian type law as follows:

$$\vec{J} = -\rho[\varphi K^d]\vec{grad}(\frac{c}{\rho})$$ (5)

where K^d is the hydrodinamic dispersion tensor, d_m the molecular diffusion coefficient and K_p the permeability of the porous medium. Equations (1) through (5) completely define the general dispersion problem, provided the geometry of the flow domain has been identified, sources location and intensity have been ascertained, porous medium and fluid physical properties are known, initial and boundary conditions have been determined. The numerical solution of the problem is however cumbersome due to lack of linearity.

ASSUMPTIONS AND RESTRICTIONS OF THE MODEL

The mathematical structure of the general dispersion problem can be considerably simplified and the numerical computation can be consequently greatly reduced if some hypotheses and restrictions are introduced, which in practice can be accepted when they result in negligible effects on the degree of approximation of the solution.

The most relevant hypothesis refers to the pollutant behavior as a *tracer*: pollutant and solvent densities do not differ significantly or, at least, solute concentration does not affect solvent density. Under this assumption, density and viscosity of the mixture become indipendent from solute concentration. Since pressure too, for its usual range of variability in groundwater phenomena, can be regarded as having very slight effects on mixture density and viscosity, these quantities can be considered as constant. Because of these assumptions, Equations (1) and (2) are no longer coupled: they can be linearized and solved separately, with great advantage from the computational point of view. When model applications are restricted to two–dimensional problems, the governing Equations (1), (2), (3) and (5) become:

$$-\rho div\,(h\vec{u}) - hq = \frac{\partial}{\partial t}\,(h\varphi\rho)$$ (6a)

$$-div\,(h\vec{J}) - div\,(hc\vec{u}) - h\lambda\varphi c - h\lambda\varphi_1 f - hq_c = h\frac{\partial}{\partial t}(\varphi c + \varphi_1 f)$$ (6b)

$$f = \chi c$$ (6c)

$$\vec{u} = -[K]\,\vec{grad}\,\psi$$ (6d)

$$\vec{J} = -[\varphi K^d]\,\vec{grad}\,c$$ (6e)

$$\psi = \frac{p}{\rho g} + z$$ (6f)

where all parameters and magnitudes represent now average values along the vertical, being h the aquifer thickness, ψ the piezometric head, and $[K]$ the hydraulic conductivity tensor, linked to the permeability tensor $[K_p]$ by:

$$[K] = \frac{\rho g}{\mu}[K_p]$$ (6g)

$[K_p]$ may be approximately considered dependent on the porous medium only. As far as the dispersion tensor $[K^d]$ is concerned, the assumption is made that at every aquifer location the following expression holds:

$$\varphi K_{ij}^d = \varphi d_m \delta_{ij} + \frac{\alpha_L}{|u|} \begin{bmatrix} u_1{}^2 & u_1 u_2 \\ u_1 u_2 & u_2{}^2 \end{bmatrix} + \frac{\alpha_T}{|u|} \begin{bmatrix} u_2{}^2 & -u_1 u_2 \\ -u_1 u_2 & u_1{}^2 \end{bmatrix} \tag{7}$$

being d_m the molecular diffusion coefficient and δ_{ij} Kronecker's *delta*, u_1 and u_2 the Darcy mixture velocity components along x_1 and x_2 axes respectively, and $|u|$ the velocity modulus.

Let Ω be the two–dimensional flow domain, with boundary Γ. Combining Equations (6a) and (6b) with (6c) and (6d) respectively, the mass conservation equations can be written, by using indicial notation, as follows:

$$\frac{\partial}{\partial x_i}\left(hK_{ij}\frac{\partial\psi}{\partial x_j}\right) - hq = hS\frac{\partial\psi}{\partial t} \tag{8a}$$

$$\text{on } \Omega, \qquad \text{for } i,j = 1,2$$

$$\frac{\partial}{\partial x_i}\left(h\varphi K_{ij}^d\frac{\partial c}{\partial x_j}\right) + \frac{\partial}{\partial x_i}\left(hcK_{ij}\frac{\partial\psi}{\partial x_j}\right) - h\lambda f_r c - hq_c = h\,f_r\,\frac{\partial c}{\partial t} \tag{8b}$$

where $f_r = \varphi + \varphi_1\chi$ is the retardation factor and S is the specific storage. The source or sink term hq represents now the mixture volume of withdrawal or recharge rate per unit area.

Let the boundary Γ be subdivided into five regions Γ_i (i = 1, 5), with $\Gamma = \Gamma_1 \cup \Gamma_2 \cup \Gamma_3 \cup \Gamma_4 \cup \Gamma_5$ In each subregion Γ_i a different boundary condition is specified: the mixture flux is given on Γ_1, the pollutant flux due to dispersion and convection is specified on Γ_2 and Γ_3 respectively, the piezometric head and tracer concentration are known on Γ_4 and Γ_5. Boundary conditions of Neuman type can be expressed as:

$$\left(hK_{ij}\frac{\partial\psi}{\partial x_j}\right)n_i = -h\tilde{q}(x_i,t) \qquad\qquad \text{on } \Gamma_1 \quad (9a)$$

$$\left(h\varphi K_{ij}^d\frac{\partial c}{\partial x_j}\right)n_i = -h\tilde{q}_d(x_i,t) \qquad\qquad \text{on } \Gamma_2 \quad (9b)$$

$$\left(hcK_{ij}\frac{\partial c}{\partial x_j}\right)n_i = -h\tilde{q}_c(x_i,t) \qquad\qquad \text{on } \Gamma_3 \quad (9c)$$

where n_i is the unit vector normal to the boundary, positive outward, while Dirichlet conditions can be written as:

$$\psi(x_i,t) = \tilde{\psi}(x_i,t) \qquad\qquad \text{on } \quad \Gamma_4 \quad (9d)$$

$$c(x_i,t) = \tilde{c}(x_i,t) \qquad\qquad \text{on } \quad \Gamma_5 \quad (9e)$$

In Equations (9) every supersigned symbol in the right hand side represents an assigned quantity. Equations (8) and (9), together with initial conditions, flow domain geometry and porous medium properties, thoroughly and uniquely define the simplified model of tracer dispersion in aquifers.

FINITE ELEMENT FORMULATION

The simplified mathematical formulation derived in the previous paragraph lends itself to satisfactory solution through finite element method. In particular, the flow domain Ω is now subdivided in classical triangular elements Ω^e. Over each Ω^e, functions

$\psi(x_i, t)$, $c(x_i, t)$ and $h(x_i)$ can be approximated by polynomials $\hat{\psi}(x_i, t)$, $\hat{c}(x_i, t)$ and $\hat{h}(x_i)$:

$$\psi(x_i, t) \simeq \hat{\psi}(x_i, t) = \psi_k(t) N_k^e(x_i) \tag{10a}$$

$$c(x_i, t) \simeq \hat{c}(x_i, t) = c_k(t) N_k^e(x_i) \tag{10b}$$

$$h(x_i) \simeq \hat{h}(x_i) = h_k N_k^e(x_i) \qquad for \quad k = n, p, q \tag{10c}$$

where $\psi_k(t)$, $c_k(t)$ and h_k are nodal coefficients and N_k^e the interpolation functions associated with the three nodes n, p, q of element Ω^e.

Let $L_1(\hat{h}, \hat{\psi})$ and $L_2(\hat{h}, \hat{\psi}, \hat{c})$ be the residuals deriving from substitution of Equations (10) in (8a) and (8b) respectively. According to Galerkin's method, the orthogonality condition between residuals and interpolation functions is imposed on Ω^e, yielding for $k = n, p, q$:

$$\int_\Omega^e L_1(\hat{h}, \hat{\psi}) N_k^e(x_i) \, d\Omega = 0 \tag{11a}$$

$$\int_\Omega^e L_2(\hat{h}, \hat{\psi}, \hat{c}) N_k^e(x_i) \, d\Omega = 0 \tag{11b}$$

At each time t, Equations (11) are equivalent to a system of six ordinary differential equations in the nodal unknowns $\psi_k(t)$ and $c_k(t)$, for each linear triangular element. This equivalent system is obtained by substituting in Equations (11) the residual expressions L_1 and L_2 by using the linear approximations (10) and applying the Gauss-Green lemma, thus eliminating second order derivatives. Natural boundary conditions are so automatically introduced. The *element* system can be written in matrix form as:

$$A_{nm}^e \Psi_m + B_{nm}^e \dot{\Psi}_m = Q_n^e \tag{12a}$$

$$T_{nm}^e C_m + R_{nm}^e \dot{C}_m = G_n^e \quad for \qquad n, m = n, p, q \tag{12b}$$

where Ψ_m and C_m are the (3×1) vectors of the piezometric head and tracer concentration at element nodes, $\dot{\Psi}_m$ and $\dot{C}_m$ the (3×1) vectors of their corresponding time derivatives, A_{nm}^e and T_{nm}^e *stiffness* matrices in flow and dispersion equation respectively, B_{nm}^e and R_{nm}^e the corresponding (3×3) *capacity* matrices; Q_n^e and G_n^e are the (3×1) forcing vectors representing sources or sinks. For $m = n, p, q$ (nodes of the triangular element Ω^e), the interpolation functions are

$$N_m^e(x_1, x_2) = (a_m + b_m x_1 + c_m x_2)/2\Delta \tag{13a}$$

where Δ is the element area and coefficients a_m, b_m and c_m depend on the node coordinates only:

$$
\begin{array}{lll}
a_n = x_1^p x_2^q - x_1^q x_2^p & b_n = x_2^p - x_2^q & c_n = x_1^q - x_1^p \\
a_p = x_1^q x_2^n - x_1^n x_2^q & b_p = x_2^q - x_2^n & c_p = x_1^n - x_1^q \\
a_q = x_1^n x_2^p - x_1^p x_2^n & b_q = x_2^n - x_2^p & c_q = x_1^p - x_1^n
\end{array} \tag{13b}
$$

When areal coordinates are used as interpolation functions, the computation of spatial integrals can be easily performed by the following formulas (Connor and Brebbia [4]):

$$\int_{\Omega^e} (N_n^e)^i (N_p^e)^j (N_q^e)^k \, d\Omega = \frac{i! \, j! \, k!}{(i + j + k + 2)!} 2\Delta \tag{13c}$$

$$\int_\Gamma^e (N_n^e)^i (N_p^e)^j \, d\Gamma = \frac{i! \, j!}{(i + j + 1)!} L_{np} \tag{13d}$$

where L_{np} is the length of the triangle side whose corner nodes are n and p.

For $n, m = n, p, q$ the structure of the element matrices is given as follows:

$$A_{nm}^e = \int_{\Omega^e} \hat{h} K_{ij} \frac{\partial N_n^e}{\partial x_i} \frac{\partial N_m^e}{\partial x_j} d\Omega = \frac{\bar{h}}{4\triangle}(K_{11}b_m b_n + K_{12}c_m b_n + K_{21}b_m c_n + K_{22}c_m c_n) \quad (14a)$$

$$B_{nm}^e = \int_{\Omega^e} \hat{h} S N_n^e d\Omega = \frac{\bar{S}\triangle}{12}(2h_n + h_p + h_q) \quad for \qquad n = m$$

$$(14b)$$

$$B_{nm}^e = 0 \qquad\qquad\qquad\qquad\qquad for \qquad n \neq m$$

$$Q_n^e = -\int_{\Omega^e} \hat{h} q N_n^e d\Omega - \int_{\Gamma^e} \hat{h}\tilde{q} N_n^e d\Gamma = -\frac{\triangle}{12}q(2h_n + h_p + h_q) - \frac{L_{pq}}{6}(2h_n + h_p)\tilde{q} \quad (14c)$$

$$T_{1nm}^e = \int_{\Omega^e} \hat{h}\varphi K_{ij}^d \frac{\partial N_n^e}{\partial x_i} \frac{\partial N_m^e}{\partial x_j} d\Omega + \int_{\Omega^e} \hat{h} K_{ij} N_m^e \frac{\partial N_n^e}{\partial x_i} \frac{\partial \hat{\psi}}{\partial x_j} d\Omega \qquad (15a)$$

$$= \frac{\bar{h}\bar{\varphi}}{4\triangle}(K_{11}^d b_m b_n + K_{12}^d c_m b_n + K_{21}^d b_m c_n + K_{22}^d c_m c_n) +$$

$$\frac{\bar{h}}{16\triangle}[(\psi_n b_n + \psi_p b_p + \psi_q b_q)(K_{11}b_n + K_{21}c_n) + (\psi_n c_n + \psi_p c_p + \psi_q c_q)(K_{12}b_n + K_{22}c_n)]$$

$$T_{2nm}^e = \int_{\Omega^e} \hat{h}\bar{\lambda} f_r N_n^e N_m^e d\Omega = \frac{\lambda \bar{f}_r \triangle}{60}(6h_n + 2h_p + 2h_q) \, for \quad n = m$$

$$(15b)$$

$$T_{2nm}^e = \frac{\lambda \bar{f}_r \triangle}{60}(2h_n + 2h_m + h_{k \neq n \neq m}) \qquad\qquad for \quad n \neq m$$

$$T_{nm}^e = T_{1nm}^e + T_{2nm}^e \qquad (15c)$$

$$R_{nm}^e = \int_{\Omega^e} \hat{h} f_r N_n^e d\Omega = \frac{\bar{f}_r \triangle}{12}(2h_n + h_p + h_q) \, for \quad n = m$$

$$(15d)$$

$$R_{nm}^e = 0 \qquad\qquad\qquad\qquad for \quad n \neq m$$

$$G_n^e = -\int_{\Gamma_3^e} \hat{h}\tilde{q}_c N_n^e d\Gamma - \int_{\Gamma_2^e} \tilde{q}_d N_n^e d\Gamma - \int_{\Omega^e} \hat{h} q_c N_n^e d\Omega$$

$$(15e)$$

$$= -\frac{L_{pq}}{6}\tilde{q}_c(2h_n + h_p) - \frac{L_{pq}}{6}\tilde{q}_d(2h_n + h_q) - \frac{\triangle}{12}q_c(2h_n + h_p + h_q)$$

Quantities S and ϕ are assumed constant inside each element, while supersigned terms refer to average values over the three nodes of Ω^e. In order to write elementary vectors and matrices, by means of formulas (14) and (15), indexed terms with n, p and q are meant to cyclically turn. Matrix A_{nm}^e is symmetrical as can be easily recognized by considering that $K_{12} = K_{21}$ in the absolute permeability tensor. On the contrary, matrix T_{nm}^e has not the same property, being the result of two contributions: the dispersion one is symmetrical, being $K_{12}^d = K_{21}^d$, but the convective contribution is not.

Capacity matrices B_{nm}^e and R_{nm}^e are diagonal, due to the *lumping* procedure (Gottardi and Mesini [5],[6]), by which time derivatives are substituted at each node by weighted averages, defined as:

$$\frac{\partial \Psi_m}{\partial t} \int_{\Omega^e} \hat{h} S N_n^e d\Omega = \int_{\Omega^e} \hat{h} S \frac{\partial \hat{\psi}}{\partial t} N_n^e d\Omega \qquad \frac{\partial C_m}{\partial t} \int_{\Omega^e} \hat{h}\varphi N_n^e d\Omega = \int_{\Omega^e} \hat{h}\varphi \frac{\partial \hat{c}}{\partial t} N_n^e d\Omega$$

Should this procedure be not applied, B_{nm}^e and R_{nm}^e would be symmetrical:

$$B_{nm}^e = \int_{\Omega^e} \hat{h} S N_n^e N_m^e d\Omega \qquad R_{nm}^e = \int_{\Omega^e} \hat{h}\varphi N_n^e N_m^e d\Omega$$

Vectors Q_n^e and G_n^e account for forcing actions (sources or sink fluxes) on fluid and tracer respectively, both at internal and boundary nodes.

When the elements Ω^e are assembled to represent the whole flow domain Ω, equations analogous with (12) can be written:

$$A_{nm}\Psi_m + B_{nm}\dot{\Psi}_m = Q_n \tag{16a}$$

$$T_{nm}C_m + R_{nm}\dot{C}_m = G_n \quad n, m = 1, 2, 3, ...N \tag{16b}$$

where global matrices and vectors are obtained by adding corresponding terms of the elemental matrices.

As far as time integration of ordinary differential Equations (16) is concerned, Crank–Nicolson inconditionally convergent method is applied. If $\Delta t^k = t^{k+1} - t^k$ is the time step, the following linear algebraic system can be derived from Equations (16):

$$\left((1-\theta)A_{nm} + \frac{B_{nm}}{\Delta t^k}\right)\Psi_m^{k+1} = -\left(\theta A_{nm} - \frac{B_{nm}}{\Delta t^k}\right)\Psi_m^k + \theta Q_n^k + (1-\theta)Q_n^{k+1} \tag{17a}$$

$$\left((1-\theta)T_{nm}^{k+1} + \frac{R_{nm}}{\Delta t^k}\right)C_m^{k+1} = -\left(\theta T_{nm}^k - \frac{R_{nm}}{\Delta t^k}\right)C_m^k + \theta G_n^k + (1-\theta)G_n^{k+1} \tag{17b}$$

CONCLUSIONS

In order to ascertain the degree of reliability of the formulation, a computer code was implemented and applied to some elementary problems, whose solutions are analytically known. The numerical results for one– and two–dimensional test problems compare favourably with analytical ones, provided the classical criteria for stability and control of numerical dispersion are satisfied.

However, it must be noted that space discretization still poses the most delicate problems, since the proposed formulation relates to the improvement of solution accuracy within each element (with respect to numerical integration) and only marginally does it affect the representation over the whole domain.

REFERENCES

[1] Bear, J. Dynamic of Fluids in Porous Media, American Elsevier, 1972.

[2] Fried J.J. Groundwater Pollution, Elsevier 1975.

[3] Kinzelbach,W. Groundwater Modelling, Elsevier Science Publ., 1986.

[4] Connor, C.A.and Brebbia, C.A. Finite Element Technique for Fluid Flow, Newnes–Butterworths, 1976.

[5] Gottardi G.and Mesini E. A Compressible Two–Phase Immiscible Flow Model by Galerkin's Method, Proc. Seventeeth Annual Pittsburg Conference on Modeling and Simulation, April 24–25, 1986, vol. 17, pp. 1201–1208.

[6] Gottardi G.and Mesini E. A Two–Phase Finite Element Program for Displacement Simulation Processes in Porous Media, Computer and Geosciences, vol. 12, n. 5, pp. 667–695, 1986.

Determining the Relationship Between Groundwater Remediation Cost and Effectiveness

D.P. Ahlfeld

Environmental Research Institute/Dept. of Civil Engineering, University of Connecticut, Storrs, CT 06269, USA

1. Introduction

In recent years, strong public outcry has prompted the creation of various governmental regulatory mechanisms for the remediation of groundwater at uncontrolled hazardous waste sites. These regulatory mechanisms both at the state and federal level have often taken the form of requirements on the water quality standards that must be satisfied in a groundwater remediation effort and the time required to achieve these standards. Often the achievement of these regulatory requirements is difficult, expensive, and time consuming. At the same time the enormity of the problem and the number of uncontrolled waste sites that impact groundwater has made it very difficult to effectively respond to public demand for rapid and complete remediation.

Ultimately the goal of government regulation is to maximize the protection of public health. Clearly this is best achieved by distributing the limited financial resources available as efficiently as possible to remediate groundwater contamination sites. Determining the best possible allocation is, as yet, an unsolved problem. A significant component of this problem, however, is determining the relationship between the engineering costs of remediation and the regulatory requirements for remediation. Understanding these relationships can provide insight into the most cost effective ways to reduce the threat to public health and ultimately provide a means of comparing different sites and the tradeoffs between cost and risk reduction at each site.

In this paper we develop a methodology for addressing this question which utilizes simulation of groundwater physics and optimization methods for determining best remediation strategies. We focus on the most common remedial technology in current practice, pump and treatment methods. In this approach, contaminated water is pumped from the ground and treated. The treated water may then be used, returned to the groundwater system or otherwise disposed of.

2. Background

A detailed review of the use of optimization and simulation in groundwater management has been presented by Gorelick [1983] who describes problems both in groundwater quantity and quality management. Representation of the full convective-dispersive equation was proposed for the groundwater remediation problem by Gorelick et al. [1984]. Ahlfeld, et al. [1986a] used a formulation similar to that introduced by Gorelick, et al. [1984] and introduced a new formulation that explicitly requires that concentrations not increase in the area outside the initial plume boundary. Ahlfeld, et al. [1986b] proposed and demonstrated a methodology for examining the tradeoff between the use of physical containment of contaminated groundwater and hydraulic containment and removal. Wagner and Gorelick [1987] posed a new formulation which incorporates parameter uncertainty. Ahlfeld, et al. [1988a] studied two optimization formulations: a new formulation for minimization of the mass of contaminant in an aquifer and the formulation for specifying a concentration standard described above. [Ahlfeld, et al. 1988b] applied these approaches to the Woburn aquifer.

3. Contaminant Transport Simulation Model

In this section we present the system of partial differential equations which will be used to represent the behavior of contaminants in the ground. The optimization techniques which will be described in the following sections are not dependent on this particular choice of simulation model. We use the well known two-dimensional, coupled model for areal flow and transport of a non-decaying, adsorbing, desorbing solute in a groundwater aquifer described by Bredehoeft and Pinder [1973] and consisting of three partial differential equations:

$$\nabla \cdot bK \cdot \nabla h + \sum_{k=1}^{n_p} q_k \delta(x_k, y_k) = 0 \tag{1}$$

$$\mathbf{v} = -K \cdot \nabla h, \tag{2}$$

$$\nabla \cdot b \, \theta D \cdot \nabla c - b \mathbf{v} \cdot \nabla c - \sum_{k=1}^{n_p} q_k (c - c_{0k}) \delta(x_k, y_k) = \frac{\partial c b R}{\partial t} \tag{3}$$

where b = saturated thickness of the aquifer (l), K = hydraulic conductivity tensor (l/t), h = vertically averaged hydraulic head (l), n_p = number of pumps, q_k = pump rate for pump located at point (x_k, y_k) (l^3/t), $\delta(x_k, y_k)$ = dirac delta function evaluated at point (x_k, y_k) $(1/l^2)$, $\mathbf{v}$ = average Darcy velocity vector (l/t), c = contaminant concentration (m/l^3), c_{0k} = contaminant concentration in pumped fluid at pump point k (m/l^3), R = retardation coefficient defined as $\theta + \rho_s K_D$ (unitless), D = hydrodynamic dispersion tensor

The aquifer parameters b, K, D, and the source/sink concentration c_{0k} will be considered fixed in the analysis that follows. The variables of (1-3) that we will consider to be controllable in the context of engineering design are q_k and (x_k, y_k). In the context of optimization modeling these variables serve as the decision variables with (x_k, y_k) implicitly constrained to lie only at nodal locations.

4. Optimization Formulations For Aquifer Remediation

The contaminated aquifer remediation design problem solved in this paper involves selecting the location of pumping wells and the magnitude and direction of pumping at those wells. We chose as our cost measure a weighted sum of total pumping. At optimality, the model sets a given pump rate to a positive value indicating injection of clean water at that well, zero if no pump is needed, or a negative value indicating extraction at that well. The objective measures the different costs of pumping and water cleaning at each well for injection or extraction. The fixed costs associated with initial construction for the wells and cleaning systems are assumed small in comparison to the long term operating costs and are neglected. To accommodate total pumping we define positive and negative components of the single pump rate at a well where q_j^+ and q_j^- represent injection and extraction respectively at the same well.

Our constraints state that contamination must be less than or equal to a prespecified value at the end of the planning horizon, T, at observation points within the system.

The optimization formulation is stated as :

$$minimize \sum_{j \in J} \alpha_j^+ q_j^+ + \alpha_j^- q_j^- \tag{4}$$

$$such \ that \quad c_{i,T}(\mathbf{q}) \le c_i^* \quad i \ \varepsilon \ I \tag{5}$$

$$q_j = q_j^+ - q_j^- \tag{6}$$

$$0 \le q_j^+ \le qup_j \tag{7}$$

$$0 \le q_j^- \le qlow_j \quad j \ \varepsilon \ J \tag{8}$$

where

J = set of nodes to be considered as potential pump locations,

$\mathbf{q}$ = vector of pump rates located at nodes defined by J with each element defined as the difference of the corresponding positive and negative components of the pump rate,

q_j = the jth element of $\mathbf{q}$,

q_j^+ = the positive (injection) component of q_j,

q_j^- = the negative (extraction) component of q_j,

α_j^+ = unit cost of injection pumping at node j,

α_j^- = unit cost of extraction pumping at node j,

I = set of nodes at which system concentration behavior will be observed,

$c_{i,T}(\mathbf{q})$ = the concentration at node i at the last time period of the simulation as a function of pump rates. This functional relationship is provided by a groundwater transport simulation model,

c_i^* = the specified maximum concentration bound at node i,

$qlow_j$ = lower bound on magnitude of pumping at location j,

qup_j = upper bound on magnitude of pumping at location j.

Equation (5) forces the concentration at the observation points (contents of set I) to be less than or equal to the specified level (c_i^*). Equations (7) and (8) place bounds on the allowable magnitudes of pumping. The design objective (4) minimizes the total pumping required. To evaluate (5) given $\mathbf{q}$, we must solve a groundwater simulation model.

5. Policy Analysis Using the Simulation/Optimization Methodology

The formulation described above and those similar to it have been solved for various hypothetical aquifers [Gorelick et al., 1984, Ahlfeld et al., 1986a, Wagner and Gorelick, 1987] and for real aquifers [Ahlfeld et al., 1988b]. These various research results have demonstrated the utility of formulations of this form for solving remediation design problems given certain regulatory requirements. A new use for the combined simulation and optimization approach, described in the previous section, is to provide the mechanism for determining the relationship between optimal remedial cost and the value of various remediation requirements or parameters involved in determining the optimal cost.

Determining this relationship will consist of repeated solution of the optimization problem described above with different values of the remediation parameters. The result of each subsequent optimization provides a value of the relationship between the remediation cost and the remediation parameter. Generating many of these values will produce a curve that provides the relationship that we seek. We stress at this juncture that the use of the combined simulation and optimiztion procedure is essential to determining the relationship which we seek. Comparing results from repeated simulations is inadequate as some simulated remedial strategies will have greater levels of achievement then others. It is only by using a procedure that will guarantee that each remedial strategy compared will be identical except for the single remediation parameter that is modified, that an unbiased comparison can be made. This procedure will be demonstrated on an aquifer that has the subject of intense study by the author the Woburn aquifer in eastern Massachusetts, USA.

The Woburn aquifer was modeled using a vertically averaged two-dimensional finite element code. The present day condition of the plume was generated by simulating the movement of the plume based on assumed dates and locations of introduction of the contaminant into the groundwater. For a detailed description of the development of the model see

Ahlfeld et al. [1988b].

5.1. Level of Remediation vs Cost of Remediation

As a first example of the ability of the combined simulation and optimization procedure to generate the relationship between remedial cost and remedial parameters we consider the remedial parameter which describes the level of remediation. This level can be expressed as the residual concentration which can be allowed to remain at the conclusion of the remedial effort. While the residual concentration is often set at the drinking water standard for the particular compound, there are various reasons why this value may not be a good choice. For example, a remediation may be the first stage of longer term effort, or relaxation of this requirement may be justified if the water is not to be directly used for drinking so that natural dilution and contaminant degradation may play a role. Finally, the uncertainty associated with measurement of concentration levels may cause doubts to arise as to the necessity of precise compliance with a drinking water standard. To develop this relationship, the regulatory requirement on concentration (c^*) was varied over a range of 34 to 80 ppb at intervals of 2 ppb. A plot of the optimal remedial cost vs the allowed residual concentration is presented as Figure 1. The vertical scale is a normalized cost. Note that as the residual concentration becomes large the costs approach a constant. Whereas when the residual concentration is small, the remedial cost becomes large. Below a residual concentration of 34 no solution can be found. This implies, given the other remediation parameters, that the physical characteristics of the Woburn aquifer prohibit remediation to the desired level.

5.2. Time to Remediation vs Cost of Remediation

One of the remediation parameters in the example above was the time by which residual concentration must be achieved. This parameter (T in Equation 5) is a regulatory requirement which may be flexibly applied. The relationship between this remediation parameter and the remediation cost was generated in a manner similar to that described above, by varying the time by which the concentration requirement must be satisfied and reoptimizing the cost. The results of this analysis are presented in Figure 2 where the horizontal axis is the time to achieve the concentration standard and the vertical axis is a normalized remediation cost. This figure contains three curves. The curve labeled achievement cost represents the cost to achieve the desired standard as a function of the time by which the standard must be achieved. If we consider the remediation effort to have a fixed total time length then we must account for a period of time after the achievement of the standard and before the end of the complete remediation effort. We refer to this time period as the maintenance phase. There is a cost represented by this maintenance phase which declines with longer time to achievement, as longer achievement time implies shorter maintenance time. For example, if the maintenance phase lasts for 10 years then the maintenance cost is the annual maintenance cost (assumed constant in this analysis) times the number of years during which the maintenance is required. Thus a maintenance phase of 5 years duration will have half the cost of a maintenance phase of 10 years duration. The sum of the two costs is presented in the curve labeled total optimal cost. This final curve exhibits interesting behavior. At the left end of the curve, decreasing the time required for remediation sharply increases the cost. For the particular concentration standard imposed, no strategy can be found to satisfy the standard in the time frame of less than 1.5 years. As the time increases, the cost of remediation increases. However, after 2.5 years the remedial cost begins to climb again. This suprising result suggests that waiting too long to clean a site may increase the remedial cost and there may be, for a particular site, an optimal time frame over which the achievement of residual concentration should be required.

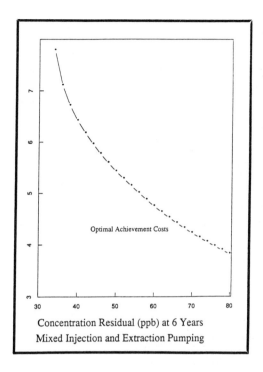

Figure 1 : **Concentration Residual vs Cost of Remediation**

6. Conclusions

A methodology has been presented for determining the relationship between remedial cost and remediation regulatory requirements or design parameters. The methodology is based on a combined simulation/optimization procedure and repeated use of this procedure under varying regulatory requirements. An example has been presented to demonstrate the utility of the methodology on a complex aquifer system and to exhibit the behaviors of the relationship between cost and regulatory requirements. It must be stressed that the results presented here are specific to the Woburn aquifer. Therefore it is unwise to draw general conclusions about the relationship between remediation cost and regulatory requirements. However, these results do suggest that the methodology can provide a powerful tool for determining this relationship at least on a site-by-site basis.

References

Ahlfeld, D.P., Mulvey, J.M., and Pinder, G.F., "Designing Optimal Strategies for Contaminated Groundwater Remediation," *Advances in Water Resources*, vol. 9 , no. 2, June, 1986a.

Ahlfeld, D.P., Pinder, G.F., and Mulvey, J.M., "Combining Physical Containment with Optimal Withdrawal for Contaminated Groundwater Remediation," *Proc. of the VI Intl. Conf. on Finite Elements in Water Resources*, Lisboa, Portugal, June, 1986b.

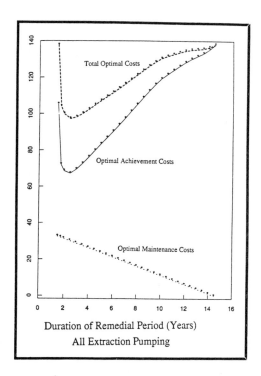

Figure 2 : **Time to Remediation vs Cost of Remediation**

Ahlfeld, D. P., Mulvey, J. M., Pinder, G. F., and , E. F. Wood , "Contaminated Groundwater Remediation Design Using Simulation, Optimization, and Sensitivity Theory 1. Model Development   ," *Water Resources Research*, vol. 24(3), March 1988a.

Ahlfeld, D. P., Pinder, G. F., and Mulvey, J. M., "Contaminated Groundwater Remediation Design Using Simulation, Optimization, and Sensitivity Theory 2. Analysis of a Field Site ," *Water Resources Research*, vol. 24(3), March 1988b.

Bredehoeft, J. D. and Pinder, G. F., "Mass Transport in Flowing Groundwater," *Water Resources Research*, vol. 9, no. 1, February, 1973.

Gorelick, S. M., "A Review of Distributed Parameter Groundwater Management Modeling Methods," *Water Resources Research*, vol. 19, no. 2, 1983.

Gorelick, S. M., Voss, C. I., Gill, P. E., Murray, W., Saunders, M. A., and Wright, M. H., "Aquifer Reclamation Design: The Use of Contaminant Transport Simulation Combined with Nonlinear Programming," *Water Resources Research*, vol. 20, no. 4, 1984.

Wagner, B. J. and Gorelick, S. M., "Optimal Groundwater Quality Management Under Parameter Uncertainty," *Water Resources Research*, vol. 23(7), pp. 1162-1174, July 1987.

Three-Dimensional Modeling of the Subsurface Transport of Heavy Metals Released from a Sludge Disposal

F. De Smedt

Laboratory of Hydrology, Vrije Universiteit Brussel, Pleinlaan 2, 1050 Brussels, Belgium

ABSTRACT

A 3-dimensional model for simulation of groundwater flow and pollutant transport was developed based upon isoparametric hexahedral finite elements. The model considers steady state groundwater flow and transient solute transport including non-linear adsorption. The model was applied to the release of zinc from a sludge disposal, which was excavated in one of the docks of the Antwerp harbor. Results indicate that after 10,000 years the zinc concentrations in the surrounding groundwater layers have not significantly increased, due to the slow water movement and the adsorptive capacity of the ground layers.

INTRODUCTION

During the construction of a tunnel under the river Scheldt at Antwerp, about 330,000 m³ of river sludge had to be removed. This sludge is severely polluted especially by heavy metals. It was decided to dump this material in a pit excavated in one of the harbor docks on the left bank of the river (Fig.1). A detailed study was made in order to assess the environmental impact of this operation, including a hydrogeological study of the possible groundwater pollution.

The dock is located at about 1 km from the river Scheldt. It is 500 m wide and 3500 m long. A constant water level is maintained at +2.5 m. The area between the dock and the river consists largely of raised terrains, with surface levels between +5 m and +3 m. Polders are present to north of the dock, where the water table is artificially maintained at a level of +2 m. The water level in the river fluctuates between 0 m and 5 m, with an average of 2.5 m. The subsoil consist of raised material on an original polder soil and quaternary sediments, which are semi-impervious. Underneath is a 30 m thick permeable tertiary sand layer resting on an impervious clay formation. The sludge disposal was excavated 8 m deep from the bottom of the dock at a level of -18 m, such that the dumped sludge is in contact with the groundwater in the sand layer.

Measurements of the hydrogeological parameters of the different ground layers were performed, as well as chemical analyses of the sludge and the different soil and water types.

THEORY

The groundwater flow equation is given by

$$\nabla(K\nabla h) + R = 0 \tag{1}$$

with h : hydraulic potential [L]
    K : hydraulic conductivity [LT-1]
    R : sink or source term [T-1]
    ∇ : del operator

The solution of this equation is obtained by means of the finite element method, using isoparametric linear hexahedral finite elements. Application of the Galerkin procedure yields the following set of algebraic equations

$$GH = B \tag{2}$$

where H is the vector of nodal hydraulic potential values, G is the conductance matrix and B is the vector containing boundary conditions and other actions of the outside world on the system. The solution of equation (2) is achieved by matrix inversion, for which the conjugate gradient technique was used with preconditioning by the diagonal method.

The pollutant transport equation is

$$n\frac{\partial c}{\partial t} + \rho\frac{\partial s}{\partial t} = \nabla(nD\nabla c) - q\nabla c + R(c_0-c) \tag{3}$$

with c : pollutant concentration $[ML^{-3}]$
    s : adsorption of pollutant $[MM^{-1}]$
    n : porosity
    ρ : soil bulk density $[ML^{-3}]$
    D : dispersion tensor $[L^2T^{-1}]$
    q : groundwater flux given by Darcy's law $[LT^{-1}]$
    c_0 : source pollutant concentration (equal to zero in case of a sink) $[MM^{-1}]$
    t : time [T]

The pollutant adsorption amount is related to the concentration by means of a Langmuir adsorption isotherm

$$s = \frac{kc}{1 + kc/s_m} \tag{4}$$

where k is an adsorption parameter and s_m the maximum adsorption possible, both depending upon the soil type.

An alternative formulation of the adsorption process is possible with a kinetic rate equation

$$\frac{ds}{dt} = \lambda[kc(1-s/s_m)-s] \tag{5}$$

where λ is a kinetic rate coefficient. The same type of finite element approach was used here. The Galerkin method combined with a mass lumping technique for the terms with time derivatives yields the following set of ordinary differential equations

$$M_1 \frac{dC}{dt} + M_2 \frac{dS}{dt} + EC = B'$$ (6)

where C and S are respectively the nodal concentrations and adsorptions, M_1 and M_2 respectively the diagonalized water mass and soil mass matrices, E is the transport matrix, and B' the vector containing the actions of the outside world. This set of differential equations together with the adsorption equations was solved in time by the Euler-Heun predictor-corrector method, such that no matrix inversion was needed. Time intervals were controlled by restricting the maximum variation in nodal concentrations per time step.

The computer model was written in FORTRAN77. A pre-processor was developed for finite element mesh generation and a post-processor for graphical representation of the results.

RESULTS

First, the regional groundwater flow was calculated in the area between the dock and the river. The model was run on a personal computer, such that the amount of elements that could be used was restricted. A total of 1235 elements with 1680 nodes was defined. The different hydrogeological conditions were specified in this mesh. Figure 1 shows the predicted groundwater potentials in the sand layer. It was found that these were nearly uniform with depth. Figure 2 shows a cross-section perpendicular to the dock. It can be concluded that the groundwater flow south of the dock is directed towards the dock, because of the natural recharge and the draining action of the dock. However, north of the dock the groundwater flow is directed from the dock to the polders, because of the artificial drainage system that is maintained in this area. Hence, the main direction of the possible groundwater pollution migration will be to the polder area in the north. The groundwater velocities in this area are small, i.e. in the order of 20 m per year.

For the simulation of the zinc migration a smaller area was selected around the sludge disposal, as shown in Fig. 1. For this region 1309 elements and 1728 nodes were used. The local groundwater flow was recalculated with boundary conditions taken from the regional simulation. The main result of this simulation was that the groundwater underneath the dock proved to be nearly immobile, as the water transfers between the dock and the sand layer are concentrated along the edges of the dock. Next, the transport of zinc from the sludge disposal was simulated.

Realistic values were specified for the molecular diffusion and the dispersion coefficients and for the adsorption parameters. Figure 3 shows a cross-sectional view perpendicular to the dock of the zinc concentrations and adsorptions after 10,000 years. From these it can be clearly seen that while the zinc adsorption in the sludge is about 450 ppm, the resulting concentrations and adsorptions in the surrounding ground layers remain very limited after 10,000 years. The reasons for this are the slow release of zinc from the sludge, the slow groundwater movement, and the adsorption capacity of the different ground layers. Hence, it was concluded that the sludge disposal will have no important impact on the environment and can be carried out as planned.

CONCLUSIONS

A three dimensional simulation of groundwater movement is possible by using hexahedral finite elements and a conjugate gradient matrix inversion technique with diagonal scaling. Transport of solutes including non-linear adsorption can be simulated by using a mass

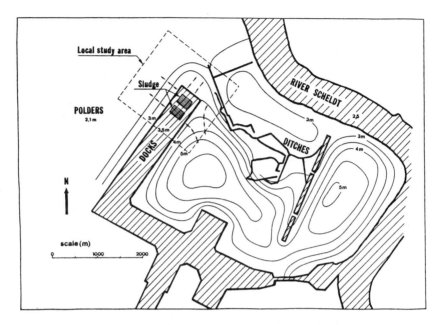

Figure 1. Situation plan and calculated groundwater potentials.

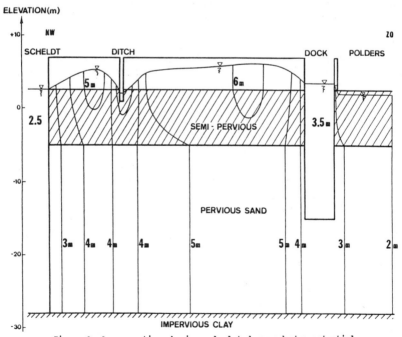

Figure 2. Cross-section showing calculated groundwater potentials.

lumped finite element technique and an explicit predictor-corrector approach for the time dependence. This model can run on PC when nodes and elements are restricted to about 2000 each.

The transport of heavy metals released from a sludge disposal, excavated in a dock of the Antwerp harbor is mainly controlled by adsorption and diffusion process, while advection with the groundwater is slow. It was shown that after 10,000 years the migration of these pollutants remained very limited.

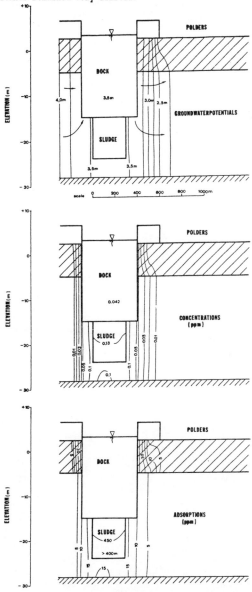

Figure 3. Cross-sections showing potentials, concentrations and adsorptions.

A 1-D Finite Element Characteristics Code for the Transport of Radionuclides in Porous Media

I.A. Arregui, C. Conde, F.J. Elorza, M.J. Miguel

Dpto. de Matemática Aplicada y Métodos Informáticos, ETSI Minas, Universidad Politécnica de Madrid, Rios Rosas, 21, 28003, Madrid, Spain

1. INTRODUCTION.

One of the most probable and dangerous technical problems in an underground radioactive waste disposal is the dissolution and transport of radioactive substances by the water around the site. Recently, a lot of work has been done on the development of radionuclide transport codes and their verification and validation [10]. But, normally, these models regard entities like radionuclide release rates from the waste as merely being input data. A new computer code (TRIMEP-1D) designed to model radionuclide transport in 1-D saturated porous media is presented in this paper. It combines a near-field model, giving radionuclide release rates from the repository, with a transport model, considering conservative and non-conservative physico-chemical mechanisms, including efficient numerical methods for advection dominate cases.

2. PHYSICAL DESCRIPTION OF THE SOURCE TERM.

The transport by underground water of radionuclides stored in underground repositories from the waste packages to the geosphere can be divided in two phases. During the first one (whose length depends on a lixiviation parameter, $\dot{L}_o$ [12]) no transport of the radioactive mass will take place; there will be just a process of corrosion of the waste packages and dissolution of the radionuclides contained.   Nevertheless, this dissolution is not always complete, so that an intermediate phase or storage arises. In the second phase, the radionuclides will be removed from the intermediate phase by secondary dissolution in the water flow, which finally will carry them to the geosphere.   All this is represented in the following figure:

The equations governing these processes can be expressed by the following [8]:

* in the waste packages:

$$\frac{du^{\ell}}{dt} = \lambda_{\ell-1}\, u^{\ell-1} - \lambda_{\ell}\, u^{\ell} - \dot{F}_{\ell} \qquad (1)$$

* in the intermediate phase:

$$\frac{dw^{\ell}}{dt} = \lambda_{\ell-1}\, w^{\ell-1} + \dot{F}_{\ell} - \lambda_{\ell}\, w^{\ell} - \dot{S}_{\ell} \qquad (2)$$

$$(\ell = 1, 2, .., \text{nr. nuclides})$$

where u^{ℓ} and w^{ℓ} are the concentrations of nuclide "ℓ" in the matrix and in the intermediate phase respectively, λ_{ℓ} is the radioactive decay constant, $\dot{F}_\ell$ is the "ℓ" nuclide flux arising from the waste package and $\dot{S}_\ell$ is the flux that groundwater around the storage will be able to carry. This flux depends mainly on the solubility of isotopes. If we call $C(K)$ the solubility of an element (K), the solubility of an isotope "ℓ" of that element is defined:

$$c_\ell = \left[\theta \frac{w^\ell}{\sum\limits_i w^i} + (1-\theta) \frac{\dot{F}_\ell}{\sum\limits_i \dot{F}_i} \right] C(K) \qquad (3)$$

$0 \leq \theta \leq 1$
$\sum$ = sum over all
i elements of K

3. NUMERICAL MODELLING OF THE SOURCE TERM.

Two numerical methods to solve the system (1-2) have been applied here. Gear's method [7] leads to excesively large computation times so we have developed a two-step retrograde method [5], designed for stiff systems, with time step control based in the mass balance.

The integration of equation

$$\frac{du}{dt} = f(u,t) \qquad (4)$$

by this method will lead to the following scheme:

$$u^{n+1} = \Delta t_n \, \beta_2(n) \, f(u^{n+1}, t^{n+1}) + \alpha_{12}(n) \, u^n + \alpha_{22}(n) \, u^{n-1} \qquad (5)$$

where:

$$\alpha_{12}(n) = \frac{(\Delta t_n + \Delta t_{n-1})^2}{(\Delta t_{n-1})^2 + 2 \, \Delta t_n \, \Delta t_{n-1}}$$

$$\alpha_{22}(n) = \frac{- \Delta t_n^2}{(\Delta t_{n-1})^2 + 2 \, \Delta t_n \, \Delta t_{n-1}} \qquad (6)$$

$$\beta_2(n) = \frac{\Delta t_n + \Delta t_{n-1}}{2\,\Delta t_n + \Delta t_{n-1}}$$

with $\Delta t_n = t^{n+1} - t^n$.

For the mass balance, we have evaluated the difference "ε" between the nuclide mass that there would be in the waste package (if lixiviation would not take place, u_a) and the sum of quantities of nuclide in the matrix, the intermediate phase and the total escape to geosphere:

$$\varepsilon_\ell(t^n) = u_a^\ell(t^n) - \left[u^\ell(t^n) + w^\ell(t^n) + e_\ell(t^n) \right] \qquad (7)$$

where $e_\ell(t)$ is solution of the equation:

$$\frac{de_\ell}{dt} + \lambda_\ell\, e_\ell = \lambda_{\ell-1}\, e_{\ell-1}(t) + \dot{S}_\ell(t)$$

$$e_\ell(t=0) = 0. \qquad (8)$$

For this last evaluation, a numerical integration using the non-equidistant abscissae method [6] has been developed, with satisfactory results. A longer description of this approximation can be found in [2]

4. PHYSICAL DESCRIPTION OF THE TRANSPORT MECHANISM.

The evolution of a substance through a porous medium suffers two different sorts of physico-chemical mechanisms: conservative and non-conservative. Here, the mechanisms we are going to consider are, among the conservative ones, advection, molecular diffusion and mechanical dispersion, and, among the non-conservative ones, adsorption, radioactive decay and filiation, first order chemical reactions and dissolution. Considering all these phenomena in one spatial dimension, the partial derivative equations system that governs them is [1]:

$$R_\ell \frac{\partial u^\ell}{\partial t} + \frac{\partial}{\partial x}(v u^\ell) - \frac{\partial}{\partial x}(k \frac{\partial u^\ell}{\partial x}) + R_\ell \lambda_\ell u^\ell - R_{\ell-1} \lambda_{\ell-1} u^{\ell-1} +$$

$$+ \left[\sum_{\kappa \neq \ell} r_d^{\ell \rightarrow \kappa} \right] u^\ell - \sum_{\kappa \neq \ell} r_i^{\ell \rightarrow \kappa} u^\kappa = 0.$$

(9)

$$(\ell = 1, 2, .., \text{nr. nucl})$$

where u^ℓ is the concentration of nuclide "ℓ", v is the pore velocity, k is the dispersion coefficient, R is the retention factor, λ is the decay constant and $r_d^{\ell \rightarrow \kappa}$ and $r_i^{\ell \rightarrow \kappa}$ are the direct and inverse chemical rate constants between elements "ℓ" and "κ". These equations will be solved subject to initial and boundary conditions defined by the source term model, and their physical parameters can be space and time dependent.

5. NUMERICAL MODELLING OF TRANSPORT EQUATION.

The developed code solves the system (9) by a combination of characteristics and finite element methods [11]. The first one gives a good approximation of the advective term, while the second one treats the diffusive part of the equations.

Time integration is performed by an Euler scheme, and the arising algebric equations system is solved by an L-U method, adapted for tridiagonal systems.

The equation system can be expressed in a matrix way as follows:

$$([M] + \Delta t (1-\theta)[K^\ell]) \{u^{\ell(n+1)}\} =$$

$$= ([M] - \Delta t \theta [K^\ell]) \{\bar{u}^{\ell(n+1)}\} + \{f^{\ell(n)}\}$$

(10)

where $[M]$ and $[K]$ are the mass and stiffness matrices, that at every element are expressed by:

$$M = \begin{bmatrix} \frac{h}{3} & \frac{h}{6} \\ \frac{h}{6} & \frac{h}{3} \end{bmatrix} \qquad K^\ell = \begin{bmatrix} \frac{k}{h} + \lambda\frac{h}{3} & -\frac{k}{h} + \lambda\frac{h}{6} \\ -\frac{k}{h} + \lambda\frac{h}{6} & \frac{k}{h} + \lambda\frac{h}{3} \end{bmatrix}$$

(11)

$\{f^{\ell}\}$ is the source vector, $\{\overline{u}^{-\ell}\}$ is the vector of concentrations in points $\overline{x}$ of the characteristic curve:

$$\overline{x}^{n+1} = x^n + \int_{t^n}^{t^{n+1}} v(x,t) \, dt \tag{12}$$

and "θ" is a parameter that allow us to solve the system in an implicit or explicit way. This approximation has been intensively verified by comparison with the numerical results of the codes TROUGH-1D [9] and COLUMN2 [4], with standard upstream boundary conditions [3].

6. EXAMPLE.

To verify the developed code, we have solved a problem composed of a complicated source term coupled to an 1-D geosphere medium. The source term was proposed by Hartley [8] and it consists in 24 radioactive species, grouped in 4 chains and 6 fission products, embedded in a stabilising matrix. The simulation time is one million years. Release rates to the geosphere for the elements of the first chain are shown in Fig. I.

Transport of these elements in the geosphere has been simulated. Concentration versus space graphicae are shown in Fig. II – VI for different times. Physical parameters adopted are:

		λ	R	u_o
$\dot{L}_o$ = 3.65E-4 kg/m^{2}y	Cm-245	8.15E-5	5000	3.4E-3
ρ = 2700 kg/m^3	Am-241	1.60E-3	4000	2.7E-1
q = 4.2 m^3/y	Np-237	3.24E-7	700	3.6E+0
v = 1.E-3 m/y	U-233	4.30E-6	300	1.0E-3
k = 100 m^2/y	Th-229	9.40E-5	20000	2.0E-6

7. FINAL CONSIDERATIONS.

A new 1-D radionuclide transport model including near field modelling has been developed with satisfactory numerical and physical results.   Nevertheless, an effort is needed in two senses: first, to couple this model with heat conduction and rock mechanics models for the near field and, second, to develop a two dimensional transport code including non linear properties (adsorption,...)

8. REFERENCES.

[1]   Arregui, I.A.   (1989).   Resolución de problemas de transporte de radionúclidos mediante una combinación de los métodos de curvas características y elementos finitos. E.T.S.I. Minas. Madrid.

[2]   Arregui, I.A., Conde, C. and Elorza, F.J. (1990). A numerical modelization of the source term for the safety analysis of an underground radioactive waste disposal.   III International Symposium in Numerical Analysis, Madrid.

[3]   Arregui, I.A., Conde, C., Elorza, F.J. and Miguel, M.J. (1989). Comparación de diversos métodos de resolución de la ecuación de migración de radionúclidos.   XI CEDYA / I Congreso de Matemática Aplicada, Málaga.

[4]   Bo, P., Carlsen, L. and Nielsen, O.J.   (1985).   COLUMN2, a computer program for simulating migration.   RISØ National Laboratory, Roskilde, Denmark.

[5]   Crouzeix, M. and Mignot, A.L. (1984). Analyse numérique des équations différentielles.   Masson.

[6]   Davis, Ph.J. and Rabinowitz, Ph. (1984). Methods of Numerical integration. Academic Press, Inc.

[7]   Gear, C.W. (1971). Numerical initial value problems in ordinary differential equations. Prentice-Hall.

[8]   Hartley, R.W. (1985). Release of radionuclides to the geosphere from a repository for high-level waste. Mathematical modelling, results. Polydinamics Ltd.

[9]   Hopkirk, R.J. (1987). Transport of radioactive outflows in underground hidrology. Polydinamics Ltd., Zurich.

[10] INTRACOIN (1984).   International Nuclide Transport Code Intercomparison Study.

[11] Johnson, C.   (1987).   Numerical solution of partial differential equations by the finite element method. Cambridge University Press.

[12] Pigford, T.H. and Chambré, P.L.   (1986).   Reliable predictions of waste performance in a geologic repository. H. C. Burkholder (editor).

9. FIGURES.

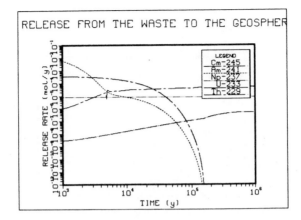

Figure I.

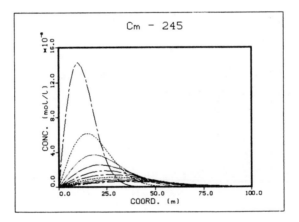

Figure II.

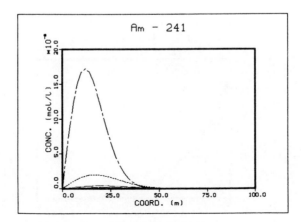

Figure III.

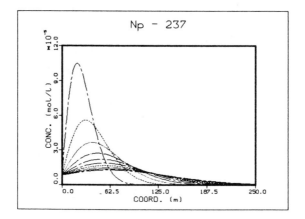

Figure IV.

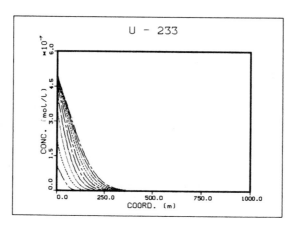

Figure V.

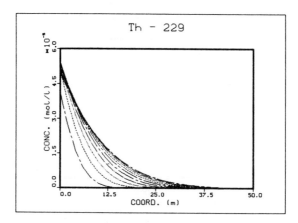

Figure VI.

A New Method for In-Situ Remediation of Volatile Contaminants in Groundwater - Numerical Simulation of the Flow Regime

B. Herrling, W. Buermann

Institute of Hydromechanics, University of Karlsruhe, D-7500 Karlsruhe, West Germany

ABSTRACT

A new hydraulic system for in-situ remediation of volatile contaminants (e.g. chlorinated hydrocarbons) in groundwater is presented. The contaminated groundwater is stripped by air in a below atmospheric pressure field in a special filtered well. The vertical well discharge initiates a circulation flow in the region surrounding the well. Using simplifying assumptions the complex three-dimensional velocity field and the capture zone of a well or well field are computed.

INTRODUCTION

The contamination of groundwater by volatile substances, particularly by volatile chlorinated hydrocarbons, is a significant problem in all industrial countries. As an alternative to conventional hydraulic remediation measures (pumping, off-site cleaning and reinfiltration of the groundwater), a new method for in-situ remediation of groundwater is currently being used at several locations in Germany (Herrling and Buermann [1]). The contaminated groundwater is stripped by air in a below atmospheric pressure field in a so called "underpressure-vaporizer-well" (German: Unterdruck-Verdampfer-Brunnen, abbreviation: UVB), see acknowledgement. The used air, charged with volatile contaminants is cleaned employing activated carbon.

METHOD OF IN-SITU REMEDIATION OF VOLATILE CONTAMINANTS

Using the UVB-method a special well with two filter sections is employed, one at the aquifer bottom and one at the groundwater surface (see figure 1). The borehole reach between the two filter sections should be made impermeable. One well should be used to remediate only one aquifer (phreatic or confined) and should not connect separate aquifers.

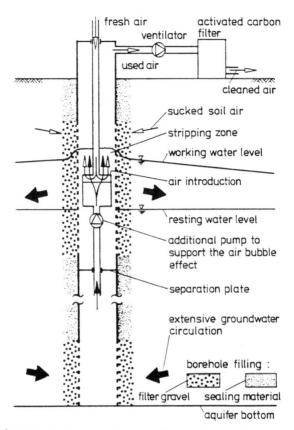

Fig.1   Underpressure-vaporizer-well, called UVB

The upper, closed part of the well is provided with below atmospheric pressure. Fresh air is sucked into the well top, possibly along with soil air from the surrounding unsaturated soil. The fresh air is introduced below the water surface and is mixed with the water by special flow directors. At the same time the rising air bubbles produce a pump effect which moves the water up and causes a suction effect at the well bottom. In recent wells a separating plate and an additional pump (figure 1) are used to support the pump effect of the air bubbles.

On one hand the cleaning effect of the well is based on the reduced pressure which supports the escape of volatile contaminants out of the water, and, as a result of the air intermixing, is also based on the considerable surface area of the air bubbles and on the concentration gradient. In this sense the permanent vibration, caused by the bursting air bubbles, is beneficial to the escaping process of the contaminants. This vibration is transmitted as a compression wave into sediment and fluid and influences possibly the mobility of contaminants even outside the well.

On the other hand the upwards streaming, stripped groundwater leaves the well shaft through the upper filter in the
reach of the groundwater surface, which is lifted in a phreatic
aquifer by the explained pump processes and by the below atmospheric pressure, and returns in an extensive circulation to
the well bottom. In this way the groundwater surrounding the
well is also remediated. The expansion of the groundwater circulation is positively influenced by the anisotropy existing in
each natural aquifer with higher horizontal hydraulic conductivities than vertical ones. The artificial groundwater circulation determines the sphere of influence of a well and is
overlapped with the natural groundwater flow.

OPEN QUESTIONS FOR NUMERICAL INVESTIGATIONS

The influence zone of an UVB in a region where the natural
groundwater flow can be neglected and, of particular importance, the capture zones of an UVB in a natural groundwater
flow field are so far unknown quantities. Of particular significance is the second case, in which a complex three-dimensional flow field results. The contamination plume is captured by
an UVB, similar to a conventional pump well, only within a
definite influx width. This width depends on the natural
groundwater flow, the quantity of discharge through the well,
the geometrical conditions at the position of investigation
(thickness of the aquifer, length of the filter sections of the
well) and on the geological conditions (horizontal and vertical
hydraulic conductivities). The width of the capture zone varies
over depth.

As a contamination plume might be wider than the capture
zone of an UVB of normal size, several UVBs may have to be
used. The local arrangement of a well field along with the
quantity of discharge through each well have to be determined
for each remediation site. For the long term, dimensioning
diagrams of those well fields have to be established, each
diagram dependent on the above mentioned parameters and based
on the dimension analysis. For all of this, many numerical
simulations of the complex three-dimensional flow field have to
be performed. To get quick answers to each special case
suitable numerical computer methods have to be set up.

NUMERICAL COMPUTATIONS

Objectives of the numerical computations
Objectives of the numerical computations are as follows:
- Computation of the complex three-dimensional flow field of a
  single UVB or of an UVB-field in a quick and, concerning the
  computer time needed, inexpensive way. Minimal effort should
  be put into changes of the discretization and addition of new
  data to realize each case study.
- Computation and plotting of the capture zone of the lower
  filter section of an UVB or an UVB-field for achieving a
  dimensioning of a well or well field.

Simplifications and assumptions of the numerical computation
The resulting flow field differs from the natural groundwater flow field only in a limited scope around the UVBs. This is well-founded because sinks and sources are located at the bottom and top of the same aquifer, each at places with the same horizontal coordinates. Then the model area can be limited to the sum of the areas of influence of all the UVBs, each area of influence computed without natural groundwater flow. The flow field downstream of the UVBs is of reduced interest.

For this limited three-dimensional aquifer region the following simplifications and assumptions are used to achieve the above explained objectives of the computation:
- The aquifer thickness is constant.
- Only confined aquifer conditions are considered in the cal-culation,even if the natural aquifer is phreatic.
- Concerning the hydraulic conductivities the aquifer structure is homogenious in its horizontal dimensions. Horizontal lay-ers, each with different conductivities, can be used. The hy-draulic conductivities may be anisotropic but each horizontal layer may have only one vertical and one horizontal conducti-vity.
- The local below atmospheric pressure field near the wells is neglected.
- The computations are made for steady state conditions.
- For estimating the capture zone, only convective transport is considered.

Realization of the numerical computation
The three-dimensional flow field in the above defined, limited aquifer region is obtained by superposition of a horizontal uniform flow field, computed in a vertical cross section and representing the natural groundwater flow, and of radial symmetric flow fields for each UVB. The superposition of the different flow fields with their own discretization is achieved by interpolating and adding the different flow vectors at the various nodes of a simple rectangular grid with variable grid distances, that are independently chosen for each Cartesian coordinate. The rectangular grid can be quickly and simply set up and allows for some refinements near the wells and its fil-ter sections. Using cylindrical coordinates the radial symmetric flow problem of each UVB can be formulated by

$$\frac{\partial}{\partial r} (2 \pi r k_r \frac{\partial h}{\partial r}) + \frac{\partial}{\partial z} (2 \pi r k_z \frac{\partial h}{\partial z}) = 0 \qquad (1)$$

where h is the piezometric head and k_r and k_z are the aniso-tropic hydraulic conductivity coefficients, see figure 2. If rk_r and rk_z are used as modified conductivities, the calcu-lation can simply be handled as in vertical plane computations, see e.g. Zienkiewics [2]. Boundary conditions are

$$h - \bar{h} = 0 \qquad \text{and} \qquad v_i n_i + \bar{v} = 0 \qquad (2)$$

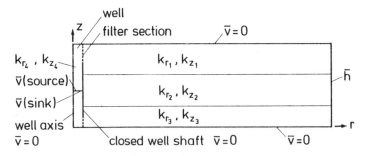

Fig. 2.: Radial symmetric flow domain and boundary conditions

where the bar denotes the prescribed values, v_i the velocity vector and n_i the unit vector normal to the boundary. Within the well shaft (see figure 2) the discretization is horizontally cut which means that two Neumann boundary conditions are prescribed, each with the same discharge quantity but opposite sign. At the right model boundary, hydrostatic conditions are prescribed. This boundary must be so far away from the well that the inflow and outflow is neglectible. The numerical computation has been performed using a Galerkin finite element method with linear shape functions and triangular discretization which allows for simple mesh refinements. The velocity vectors, constant in each element, are interpolated for each global node by weighting with the reciprocal distance beween element centre and node.

   The computation of the surface of the complex, three-dimensional zone which is captured by the lower filter section of an UVB or by those of an UVB-field is carried out using the particle-tracking method. Starting at the upstream boundary, streamlines are calculated for the entire flow domain. This calculation is systematically coupled with a search procedure to find those streamlines that a) end in the lower filter section of an UVB and; b) have a neighbouring streamline that leaves the flow domain at the downstream boundary without being captured by an UVB. These streamlines represent the surface of the capture zone and are the data base for perspective or other plots.

   For the computation of each streamline, the path of a particle is integrated in the three-dimensioanl velocity field. This can be realized using an explicit Euler method with small integration steps. The more sophisticated Runge-Kutta method of fourth order is more accurate and allows for greater integration steps.

SOME FIRST NUMERICAL RESULTS

To demonstrate the complex flow field near an individual UVB the streamlines are shown in a vertical cross section in the

(a)

(b)

(c)

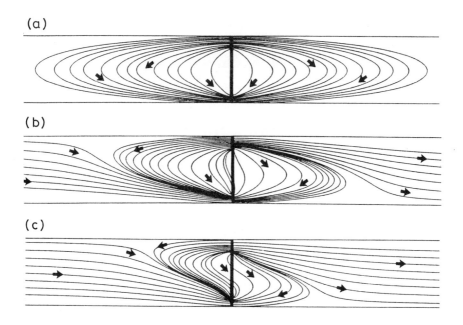

Fig. 3   Streamline for natural velocities (a) 0.0 m/d;
         (b) 0.3 m/d; (c) 1.0 m/d

direction of the natural groundwater flow (symmetry plane of
the flow problem). In figure 3 three case studies are demonstrated with Darcy velocities of the natural groundwater flow
of 0.0 m/d, 0.3 m/d and 1 m/d. All other parameters remain
constant: the discharge through the well shaft is 20.16 m³/h,
the thickness of the aquifer is 10 m, the anisotropic conductivities are 0.001 m/s (horizontal) and 0.0001 m/s (vertical)
and the lengths of the filter sections are 1.2 m and 2.1 m (at
the top).

ACKNOWLEDGEMENT

The authors greatfully acknowledge B. Bernhard (Industrie-
Engineering-GmbH (IEG mbH), D 7410 Reutlingen, F.R. Germany)
for the valuable discussions, the cooperation and support in
studying the UVB-method.

REFERENCES

[1] Herrling, B. and Buermann, W. UVB-Verfahren - Grundprinzip
    und Messungen, in: Untergrundsanierung mittels Bodenluftab-
    saugung und In-Situ-Strippen, Bock, P. et al., Schriften-
    reihe. Angew. Geologie, Uni. Karlsruhe, No.8, in press,
    1990.
[2] Zienkiewicz, O.C. The Finite Element Method in Engineering,
    McGraw-Hill, London, 1971.

Characteristic Methods for Modeling Nonlinear Adsorption in Contaminant Transport

C.N. Dawson(*), M.F. Wheeler(**)

(*) Department of Mathematics, University of Chicago, Illinois, USA

(**) Department of Mathematics, University of Houston, Texas, USA and *Department of Mathematical Sciences, Rice University, Houston, Texas, USA*

Abstract

Numerical methods based on characteristic tracing for simulating contaminant transport with nonlinear adsorption are discussed. Under various assumptions on functional forms, these models can pose severe numerical difficulties. Numerical results in one space dimension for the Freundlich isotherm with $0 < p \leq 1$ are presented.

Introduction

The quality of groundwater at numerous locations in the United States and other countries has been adversely effected by the introduction of a variety of contaminants into the subsurface. In many areas, dangerous compounds such as BTX (benzene, toluene, xylene) and gasoline have escaped from underground storage tanks and other facilities and invaded local aquifers. The dangers this type of contamination poses to the drinking water of the United States have been detailed in two recent reports by the Environmental Protection Agency [1] and the Department of Energy [2].

As noted in these reports, several physical and chemical factors influence the levels and movement of substrates in an aquifer. These include the presence of organisms capable of biodegrading certain substrates, the particular mix of substrates (primary and cometabolic) and other nutrients (minerals, etc.), and environmental factors, such as soil pH, temperature, hydraulic conductivity of the medium, and adsorption.

Under normal soil conditions, the effects of biodegradation, hydraulic conductivity, and adsorption are particularly important to the control and removal of substrates. Natural biodegradation is capable of removing a substantial amount of substrate, in both aerobic and anaerobic conditions. Studies show that in areas where both aerobic and anaerobic biodegradation is occurring, aerobic biodegradation dominates [3, 1]. Thus, while anaerobic biodegradation can take place once oxygen supplies are depleted, biodegradation can be enhanced by the introduction of dissolved oxygen into oxygen-depleted areas. This process, called *in-situ* biorestoration, has been proven effective in numerous field and laboratory studies [4]. The rate and path of substrate movement are effected by soil heterogeneity and adsorption. The adsorption process helps control contaminant spreading, as it slows the overall movement of contaminants.

Given estimates of biodegradation and adsorption rates, and the spatial variability of hydraulic conductivity, numerical simulation can be an effective tool for predicting the movement and biodegradation of contaminants, and for evaluating the relative importance of the chemical and physical parameters. Over the last several years, the authors have developed and tested a numerical algorithm for modeling two-dimensional flow with variable, linear adsorption, variable hydraulic conductivity, tensor diffusion/dispersion, and injection and production wells for the injection of dissolved oxygen and the removal of substrate. This algorithm represents an extension of earlier work of the second author and various collaborators, see for example, [5, 6, 7]. In this procedure, a finite element modified method of characteristics (MMOC) is used to approximate advection and diffusion/dispersion. Darcy velocity and pressure are calculated by a higher-order mixed finite element method [8]. Wells are modeled as point sources and sinks, and nonlinear terms modeling biodegradation are handled by time-splitting. For more details on this algorithm, and for results of numerical simulations of contaminant transport with biodegradation, see [9, 10, 11].

Other characteristic approaches which are very useful for these types of problems are the so-called higher-order Godunov schemes, see for example, [12, 13]. These methods can be combined with mixed finite element procedures to obtain a robust algorithm for problems with nonlinear advection and diffusion [14, 15]. In the case of linear advection, with certain choices of parameters, the Godunov method is equivalent to a finite difference MMOC.

In a recent paper [9], the authors and C. Y. Chiang completed a fairly extensive numerical study on the effects of spatially variable, instantaneous, linear adsorption, and its interaction with variable hydraulic conductivity and biodegradation in two-dimensional contaminant transport. Our goal in this paper is to discuss numerical methods for the extension of these models to the case of nonlinear adsorption, and to present some preliminary one-dimensional numerical results. Once these effects are understood in one dimension, both physically and numerically, a multi-dimensional simulator will be developed.

Adsorption models

Let c denote the concentration of substrate in solution, and let $A = A(c)$ describe the adsorption of substrate. In one space dimension, conservation of mass gives

$$\phi c_t + \rho A_t + q c_x - D c_{xx} = 0, \quad x \in \mathbf{R}, \quad t > 0, \tag{1}$$

where ϕ is porosity, ρ (g/cm^3) is the bulk mass density, q (cm/sec) is the Darcy velocity, and D/ϕ (cm^2/sec) is the sum of the molecular diffusion and mechanical dispersion coefficients.

Dividing the adsorption sites into two classes, one where the adsorption reaction is in equilibrium (the reaction kinetics are fast compared to the rate of transport), and one where reactions are not in equilibrium ("slow" adsorption), gives the following relationships:

$$A = \lambda_1 \Psi(c) + \lambda_2 S, \tag{2}$$
$$S_t = F(c, S), \tag{3}$$

where $\lambda_i \geq 0$, $\lambda_1 + \lambda_2 = 1$, S represents substrate adsorbed at non-equilibrium sites, and F is the overall adsorption rates for these sites.

Various functional forms for Ψ and F can be found in the literature. Under the assumption of instantaneous, linear adsorption, $\lambda_2 = 0$, and $\Psi(c) = K_d c$, where K_d (cm^3/g) is the distribution coefficient [16]. A typical form for F is [17]

$$F(c, S) = k_a \Phi_1(c) - k_s S, \tag{4}$$

where k_a and $k_s > 0$, and $\Phi_1 : [0, \infty) \to [0, \infty)$, with $\Phi_1(0) = 0$, $\Phi_1(c) > 0$ for $c > 0$, and Φ_1 is monotone non-decreasing.

The functions $\Psi(c)$ and $\Phi_1(c)$ are adsorption isotherms. Two specific isotherms which have been proposed are the Langmuir isotherm,

$$f(c) = \frac{K_1 c}{1 + K_2 c}, \quad K_1, K_2 > 0, \tag{5}$$

and the Freundlich isotherm,

$$f(c) = K_d c^p, \quad 0 < p \le 1. \tag{6}$$

Note that in the last example, $f(c)$ is not Lipschitz continuous at $c = 0$ for $0 < p < 1$.

For further discussion of these models, and the mathematical ramifications, see [18, 17].

Numerical Methods for nonlinear adsorption

First, consider the case $\lambda_2 = 0$ in (2), and $\Psi(c)$ is either the Freundlich or Langmuir isotherm, which we will denote by $f(c)$. Then, (1) becomes

$$\phi c_t + \rho f(c)_t + q c_x - D c_{xx} = 0. \tag{7}$$

If $f'(c)$ is bounded for $c \ge 0$, as in the case of the Langmuir isotherm, we rewrite (7) as

$$\gamma(c) c_t + q c_x - D c_{xx} = 0. \tag{8}$$

where $\gamma(c) = \phi + \rho f'(c)$. Expressing $\gamma(c) c_t + q c_x$ as a directional derivative:

$$\gamma c_t + q c_x \equiv \sqrt{\gamma^2 + q^2} c_\tau \equiv \theta c_\tau, \tag{9}$$

we obtain

$$\theta c_\tau - D c_{xx} = 0. \tag{10}$$

In this case, one can easily apply the finite difference or finite element MMOC to (10), as described and analyzed by Douglas and Russell [19].

In the case of the Freundlich isotherm, $f(c) = K_d c^p$, with $0 < p < 1$, the above formulation is not well-defined when $c = 0$, since $f'(0)$ is infinite. In order to understand the behavior of the solution for this case, and for the sake of simplicity, we have implemented the following scheme. Let h and $\Delta t > 0$, $x_j = jh$, for j an integer, and $t^n = n\Delta t$, $n = 0, 1, \ldots$. For a function $f(x,t)$, let $f_j^n = f(x_j, t^n)$. Setting $\mu(c) = c + R_s c^p$, where $R_s = \rho K_d / \phi$, (7) can be expressed in "conservation form"

$$\mu_t + \bar{q}\eta(\mu)_x - \bar{D}\eta(\mu)_{xx} = 0, \tag{11}$$

where $\eta(\mu)$ is the inverse map $\eta(\mu(c)) = c$, $\bar{q} = q/\phi$, and $\bar{D} = D/\phi$. Applying the Godunov scheme to $\mu_t + \bar{q}\eta(\mu)_x$ and including diffusion implicitly, we obtain

$$\frac{\mu_j^{n+1} - \mu_j^n}{\Delta t} + \bar{q}\frac{\eta(\mu_j^n) - \eta(\mu_{j-1}^n)}{h} - \bar{D}(\eta(\mu_j^{n+1})_x)_{\bar{x}} = 0, \tag{12}$$

where $(\eta(\mu_j^{n+1})_x)_{\bar{x}}$ represents central differencing in space.

Since the scheme (12) is explicit, a CFL constraint must be satisfied; that is, $\bar{q}\Delta t/h \leq 1$. In the case $\bar{q}\Delta t/h = 1$, (12) is equivalent to the following MMOC-based scheme. Combining $\phi c_t + q c_x$ into a directional derivative θc_τ, and approximating c_τ and $(c^p)_t$ by backward differences in time, we obtain the following finite difference MMOC approximation to c_j^n:

$$\frac{C_j^{n+1} - \hat{C}_j^n}{\Delta t} + R_s\frac{(C_j^{n+1})^p - (\hat{C}_j^n)^p}{\Delta t} - \bar{D}((C_j^{n+1})_x)_{\bar{x}} = 0, \tag{13}$$

where

$$\hat{C}_j^n = \hat{C}^n(x_j - \bar{q}\Delta t).$$

If $\bar{q}\Delta t/h = 1$, then $\hat{C}_j^n = C_{j-1}^n$. Thus, (13) and (12) are formally equivalent.

It is well-known that the schemes given in (12) and (13) are of low order and excessively diffusive for advection-dominated flow, at least on course meshes. In general, we do not recommend their use. We propose them here as a first approximation to understanding the behavior of solutions in one space dimension. More robust and accurate numerical procedures employing the finite element MMOC and higher-order Godunov techniques are currently under study, and will be discussed in a later paper.

In the more general case, $\lambda_1, \lambda_2 \neq 0$, the system of equations which must be solved is:

$$\phi c_t + \rho(\lambda_1 \Psi(c)_t + \lambda_2 S_t) + q c_x - D c_{xx} = 0, \quad x \in \mathbf{R}, \quad t > 0, \tag{14}$$

$$S_t = F(c, S). \tag{15}$$

If $\Psi'(c)$ is bounded for all $c \geq 0$, set $\gamma(c) = \phi + \rho\lambda_1 \Psi'(c)$ and express $\gamma(c)c_t + q c_x$ as a directional derivative as in (10). Approximating this derivative by backward differences we obtain the following formal expression:

$$\phi \frac{c^{n+1}(x) - \hat{c}^n(x)}{\Delta t} + \rho\lambda_2 \frac{S^{n+1}(x) - S^n(x)}{\Delta t} - D c_{xx}^{n+1}(x) \approx 0, \tag{16}$$

$$S^{n+1}(x) - S^n(x) = \int_{t^n}^{t^{n+1}} F(c(x,t), S(x,t))dt, \tag{17}$$

where $\hat{c}(x, t^n) = c^n(x - q\Delta t/\gamma(c^n(x)))$. Discretizing (16)-(17) would appear to give a nonlinear system of equations; however, an explicit procedure (for example, second order Runge-Kutta) could be used to integrate (17), thus updating S, and the result inserted into (16), where c would be updated. As stated above, numerical methods for the case where $\Psi'(c)$ is unbounded as $c \to 0$ are currently being studied.

Numerical results

In Figure 1, we present numerical results at $t = 150$ *days* for the "equilibrium sorption" case

$$c_t + R_s(c^p)_t + \bar{q} c_x - \bar{D} c_{xx} = 0, \tag{18}$$

where $0 < p \leq 1$. The method outlined in (11) was employed on the domain $0 \leq x \leq 260$ *feet*. We assumed $c(x, 0) \equiv 0$, $c(0, t) = 1$ for $t > 0$, and considered t sufficiently small such that $c(260, t) \equiv 0$. We enforced the CFL constraint $\bar{q}\Delta t/h = 1$, with $\bar{q} = .0007 \; cm/sec$ and $h = .65 \; feet$, and set $\phi = .3$, $\bar{D} = .033 \; cm^2/sec$, and $R_s = 1.7$. We considered $p = 1, .5$, and $.25$.

The results indicate that reducing p results in increased retardation of the substrate front. Moreover, reducing p results in much steeper fronts, as the effective diffusion level is decreased.

One-dimensional numerical results for "slow sorption," with the Freundlich isotherm and $p = .5$, can be found in Knabner [20].

Acknowledgments: The first author acknowledges the support of the National Science Foundation, Grant No. DMS-8807257. The second author acknowledges the support of the Department of Energy, Grant No. DE-FG05-88ER-25060. The authors would also like to thank Hans van Duijn for helpful discussions on nonlinear adsorption.

References

[1] Thomas, J. M., M. D. Lee, P. B. Bedient, R. C. Borden, L. W. Canter, and C. H. Ward, *Leaking underground storage tanks: remediation with emphasis on* in situ *biorestoration*, Environmental Protection Agency, 600/2-87,008, January, 1987.

[2] United States Department of Energy, *Site-directed subsurface environmental initiative, five year summary and plan for fundamental research in subsoils and in groundwater, FY1989-FY1993*, DOE/ER 034411, Office of Energy Research, April 1988.

[3] Kindred, J. S. and M. A. Celia, *Contaminant transport and biodegradation 2. Conceptual model and test simulations*, Water Resour. Res. 25, pp. 1149-1159, 1989.

[4] Raymond, R. L., V. W. Jamison, and J. D. Hudson, *Final report on beneficial stimulation of bacterial activity in ground water containing petroleum products*, Committee on Environmental Affairs, American Petroleum Institute, Washington, D. C., 1975.

[5] Russell, T. F., *Time-stepping along characteristics with incomplete iter-ation for a Galerkin approximation of miscible displacement in porous media*, SIAM J. Numer. Anal. 22, pp. 970-1013, 1985.

[6] Russell, T. F., M. F. Wheeler, and C. Y. Chiang, *Large-scale simulation of miscible displacement*, Proceedings of SEG/SIAM/SPE Conference on Mathematical and Computational Methods in Seismic Exploration and Reservoir Modeling, W. E. Fitzgibbon, ed., Society for Industrial and Applied Mathematics, Philadelphia, pp. 85-107, 1986.

[7] Chiang, C. Y., M. F. Wheeler, and P. B. Bedient, *A modified method of characteristics technique and a mixed finite element method for simula-tion of groundwater solute transport*, Water Resour. Res. 25, pp. 1541-1549, 1989.

[8] Douglas, J. Jr., R. E. Ewing, and M. F. Wheeler, *Approximation of the pressure by a mixed finite element method in the simulation of miscible displacement*, RAIRO Anal. 17, pp. 17-33, 1983.

[9] Chiang, C. Y., C. N. Dawson, and M. F. Wheeler, *Modeling of* in-situ *biorestoration of organic compounds in groundwater*, to appear.

[10] Dawson, C. N., M. F. Wheeler and P. B. Bedient, *Numerical modeling of subsurface contaminant transport with biodegradation kinetics*, Pro-ceedings of National Water Wells Meeting, Houston, Texas, 1987, pp. 329-344.

[11] Wheeler, M. F. and C. N. Dawson, *An operator-splitting method for advection-diffusion-reaction problems*, MAFELAP Proceedings VI, J. A. Whiteman, ed., Academic Press, pp. 463-482, 1988.

[12] Bell, J. B., C. N. Dawson, and G. R. Shubin, *An unsplit higher order Godunov scheme for scalar conservation laws in two dimensions*, Jour. Comput. Physics 74, pp. 1-24, 1988.

[13] Dawson, C. N., *Error estimates for Godunov mixed methods for nonlin-ear parabolic equations*, Ph. D. Thesis, Rice University, 1988.

[14] Dawson, C. N., *Godunov-mixed methods for immiscible displacement*, to appear.

[15] Dawson, C. N., W. A. Kinton, and M. F. Wheeler, *Time-splitting for advection-dominated parabolic problems in one space variable*, Communications in Applied Numerical Methods 4, pp. 413-423, 1988.

[16] Freeze, R. A. and J. A. Cherry, Groundwater, Prentice-Hall, Englewood Cliffs, New Jersey, 1979.

[17] van Duijn, C. J. and P. Knabner, *Solute transport in porous media with equilibrium and non-equilibrium multiple-site adsorption: Travelling waves*, Institut für Mathematik, Universität Augsburg, Report No. 122, 1989.

[18] van Duijn, C. J. and P. Knabner, *Solute transport through porous media with slow adsorption*, Institut für Mathematik, Universität Augsburg, Report No. 54, 1988.

[19] Douglas, J. Jr. and T. F. Russell, *Numerical methods for convection-dominated diffusion problems based on combining the method of characteristics with finite element or finite difference procedures*, SIAM J. Numer. Anal. 19, pp. 871-885, 1982.

[20] Knabner, P., *Mathematische modelle für den transport gelöster stoffe in sorbierenden porösen medien*, Institut für Mathematik, Universität Augsburg, Report No. 121, 1989.

Sorption Studies

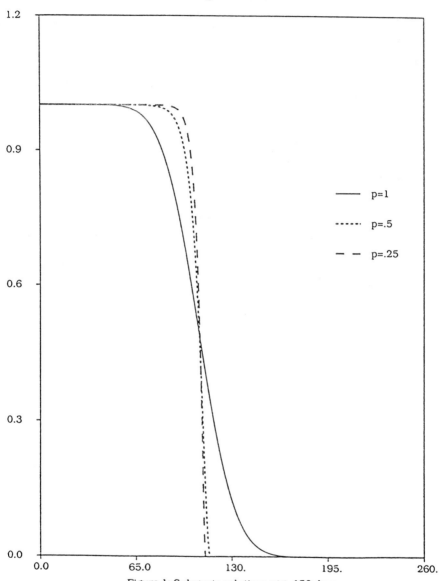

Figure 1: Substrate solutions at t=150 days

Natural Convection in Geological Porous Structures: A Numerical Study of the 2D/3D Stability Problem

D. Bernard, P. Menegazzi

LEPT-ENSAM (URA CNRS 873), esplanade des arts et métiers, 33405 Talence, Cédex, France

INTRODUCTION

Natural convection is a very common phenomenon in geological porous structures. Indeed, due to the existence of the geothermal heat flux and various heterogeneities at different scales, the stability conditions for the saturating fluid are seldom verified. The effects of the induced fluid flow can be:
　　　** direct; modifying the conductive temperature field. This effect is appreciable only when the convective cells are of very large extension (basin scale).
　　　** indirect; coupling the existence of a fluid flow in a non uniform temperature field with some temperature dependent phenomena (hydrocarbon dissolution, geochemical reaction, precipitation/dissolution,...). Those effects may be important even when the direct thermal effects are negligible.

In most cases, numerical simulations of heat and mass transport in geological structures are performed in two dimensions assuming that the variations along the third one are negligible. As far as conduction and forced convection are the only ways of transfer that are considered, 2D calculations are representative of the transport phenomena. When free convection is taken into account it may be untrue. Indeed it is well known that strictly 3D flow can be induced by free convection within porous structures that are a priori reducible to 2D.

In this paper we present a program developed to study this 2D/3D stability problem for geological porous structures. The 2D steady state solution is perturbated adding perturbations periodic in the third direction. After presenting the equations verified by the perturbations, we give their matricial form derived using a classical FE formulation. After some transformations an eigenvalue problem is obtained. The position of the eigenvalue spectrum relative to the real axis of the complex plan gives us the criterion for the stability of the 2D solution to the 3D perturbations.

To validate the method we apply it to classical configurations for which analytical solutions are available. Then we
examine some simplified geological configurations for which no
results exist.

2D NUMERICAL MODEL

DBCONV is a numerical model developed by the authors to simulate heat and mass transfers by natural convection in complex
porous media as, for instance, geological structures. This
program has been presented elsewhere Bernard [1], Menegazzi
[2] so we only recall here its main features.

Natural convection within rigid porous media is classicaly
described by the following set of equations, Combarnous and
Bernard [3] (presented here under dimensionless form):

$$-\nabla \cdot (\underset{\approx}{k} \cdot \nabla p) = Ra^{*} \nabla \cdot (T \underset{\approx}{k} \cdot \underset{\sim}{e}) \qquad (1)$$

$$R^{*} \frac{\partial T}{\partial t} - \nabla \cdot (\underset{\approx}{\lambda}^{*} \cdot \nabla T) + \underset{\sim}{V} \cdot \nabla T = f_{\Omega} \qquad (2)$$

$$\underset{\sim}{V} = -\underset{\approx}{k} \cdot (\nabla p + Ra^{*} T \underset{\sim}{e}) \qquad (3)$$

This set of equations is composed respectively of the
fluid mass conservation equation, of the heat transport
equation and of Darcy's law.

The domain Ω within which natural convection is to be
studied, is a priori reducible to 2D. Two regions are distinc-
ted: Ω_1 the impervious region and Ω_2 the porous region. Into Ω_2
the complete set of equations (1, 2, 3) must be solved, into Ω_1
solely equation (2) has to be considered.

We use the finite element method to solve this problem.
Two triangular grids are used (one for T and one for p) and
linear interpolation is used for temperature and a quadratic
one for pressure. Time integration is performed using a multi-
step scheme.

Several validations has been performed and DBCONV has been
used to study real simplified cases or general academic configurations, Bernard [1].

It is well known that in some cases natural convection may
induce 3D flows within porous structures a priori reducible to
2D, Beck [4], Bories [5]. In order to be able to study this
type of stability problem (the 2D and the 3D temperature, pressure and velocity fields are solutions of the set of equations
but the 2D fields are unstable to small perturbations. The 3D
fields are stable) for complex porous structures as the geological ones, the numerical program ST2D3D has been developed.

2D-3D STABILITY PROBLEM

Consider E0 a 2D steady state solution of equations (1,2,3) given by DBCONV. E0 is perturbated adding to (p0,T0,V0) the small perturbations (Π,θ,ω). The perturbated pressure, temperature and velocity fields verify the same set of equations. Using the properties of the basic state E0, eliminating the velocity perturbation ω and linearizing the obtained equations yield to the following equations for the pressure and temperature perturbations (Π,θ):

$$-\nabla.(\underset{\approx}{k}.\nabla\Pi) = \text{Ra}^* \nabla.(\theta \underset{\approx}{k}.\underset{\sim}{e}) \tag{4}$$

$$R^* \frac{\partial\theta}{\partial t} - \nabla.(\underset{\approx}{\lambda}^*.\nabla\theta) + \underset{\sim}{V}_0.\nabla\theta - (\underset{\approx}{k}.(\nabla\Pi + \text{Ra}^*\theta \underset{\sim}{e})).\nabla T_0 = 0 \tag{5}$$

Solutions of the following form are looked for:

$$\Pi(x,y,z,t) = \Pi(x,z) \ e^{\sigma t - i\alpha y} \tag{6}$$

$$\theta(x,y,z,t) = \theta(x,z) \ e^{\sigma t - i\alpha y} \tag{7}$$

where α is the wave number in the y direction (direction in which geometry and transport properties are assumed to be constant in order to satisfy the assumption of 2D reducibility) and σ the perturbation coefficient of amplitude.

The 2D part of the perturbations satisfy the following equations:

$$\nabla.(\underset{\approx}{k}.\nabla\Pi) + \text{Ra}^* \nabla.(\theta \underset{\approx}{k}.\underset{\sim}{e}) - \alpha^2 k_{yy} \Pi = 0 \tag{8}$$

$$R^*\sigma\theta - \nabla.(\underset{\approx}{\lambda}^*.\nabla\theta) + \underset{\sim}{V}_0.\nabla\theta - (\underset{\approx}{k}.(\nabla\Pi+\text{Ra}^*\theta \underset{\sim}{e})).\nabla T_0 + \alpha^2 \Lambda_{yy} \theta = 0 \tag{9}$$

As classically with the FEM, a weak form of equations (8,9) is writen and using the same 2D grids and the same interpolations as in DBCONV, the following matricial system is obtained for the values of both perturbations on each nodes of the grids (noted $\{\Pi\}$ and $\{\theta\}$ here):

$$\left[A_{11}\right]\{\Pi\} + \left[A_{12}\right]\{\theta\} + \alpha^2 \left[B_{11}\right]\{\Pi\} = 0 \tag{10}$$

$$\left[A_{21}\right]\{\Pi\}+\left[A_{22}\right]\{\theta\}+\alpha^2 \left[B_{22}\right]\{\theta\} = \sigma \left[M\right]\{\theta\} \tag{11}$$

The matrices A, B and M depend:
  1) on the geometry, on the transport properties in the plane (x,z), on the boundary conditions, on the interpolation functions and on the basic state E0. All those informations came from the program DBCONV.

  2) on the transport properties in the direction y.

All the subroutines computing those matrices have been generated by REDUCE ,Hearn [6], a symbolic computing routine which avoided us almost all the errors that migh be expected when manipulating algebraic expressions like the ones appearing in the matrices A, B and M (some of the local matrices are not square matrices because two different interpolations are mixed). For more details refer to Menegazzi [2].

Combining equations (10) and (11) yields to the following eigen value problem:

$$\left[\; D \;\right] \{\theta\} = \sigma \{\theta\} \tag{12}$$

Remembering that σ is the perturbation amplitude coefficient, the stability criteria is given by the sign of the real part of the greater eigen value of D (noted RPGEV):

if RPGEV is < 0 the basic state E0 is stable,
if RPGEV is > 0 the basic state E0 is unstable.

ST2D3D performed all the calculations briefly presented above and can give precise informations for 2D stability problems (the wave number α is equal to 0) or for 3D stability problems occuring in configurations a priori reducible to 2D.

ONE EXAMPLE OF VALIDATION

Consider the configuration presented on Fig.1. For this kind of homogeneous horizontal porous boxes limited by impervious isothermal planes and impervious adiabatic walls, analytical calculations are possible, Bories [5]. The stability criteria is rather simple: the basic state E0 (it is the purely conductive temperature field in this case) is stable relatively to a perturbation if the filtration Rayleigh number Ra* is lower than a marginal stability Rayleigh number Rs*. If Ra* > Rs*, E0 is unstable. Rs* is a function of α, the perturbation wave number in the y direction and of $2i\pi$, the perturbation wave number in the x direction (i=0,1,2). The critical Rayleigh number is the minimum of Rs*.

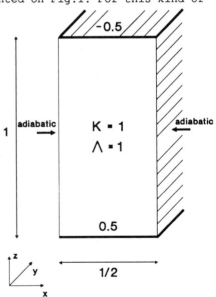

Fig.1 Configuration used to validate the 2D-3D stability program.

On Fig. 2 are presented two theoretical curves:

** $C_0(\alpha)$ = Rs*$(\alpha, i=0)$

** $C_1(\alpha)$ = Rs*$(\alpha, i=1)$

and the numerical results ob
tained for two different grids.

The agreement between
theoretical and numerical results is good in this example
as well as in the other tests
that have been performed,
Menegazzi [2]. This allows to
conclude that the theoretical
approach and the numerical
treatment are correct.

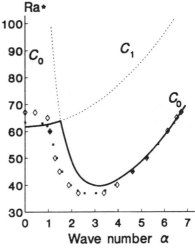

Fig.2   Theoretical   and
numerical results.

A SIMPLIFIED GEOLOGICAL EXAMPLE

This simplified example is shown on Fig.3. Numerical results
given by DBCONV (velocity field on Fig.4) have been published a
few years ago, Bernard [7] and the assumption of 2D flow was
questioned. The program ST2D3D was developed for this purpose
and applying it to this example gives the curve presented on
Fig.5. For α between 2 10^{-2} and 2.23 the RPGEV is positive
meaning that the calculated 2D flow is unstable. A 3D calculation is needed to have precise data on the stable flow.

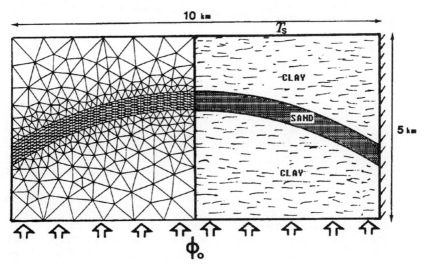

Fig.3 Simplified geological example: FE temperature grid, media, boundary conditions and dimensions.

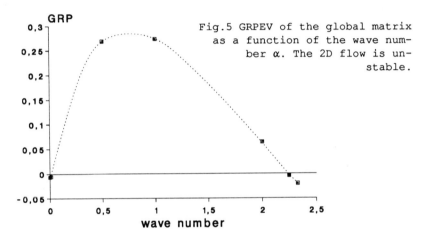

Fig. 4 Fluid
flow pattern in the
porous region. The imper-
vious region is not represented.

Fig.5 GRPEV of the global matrix
as a function of the wave number α. The 2D flow is un-
stable.

1. Bernard,D. A FE model of free convection in geological porous structures, in Num. Methods for Trans. and Hydrol.
Proc. (Ed. Celia, M.A.), pp. 301-306, Proc. of the VII Int.
Conf. Comp. Meth. Water Res., Cambridge, USA, Computational
Mechanics Pub., 1988.
2. Menegazzi, P., Convection dans les structures géologiques
poreuses: étude numérique 2D et stabilité des écoulements,
Thèse de doctorat, Un. Bordeaux I, 1989.
3. Combarnous,M. and Bernard,D., Modeling of free convection in
porous media: from academic cases to real configurations, in
Proc. 25th Nat. Heat Transf. Conf. ASME, Vol.1, pp.735-745,
Houston, USA, 1988.
4. Beck,J.L., Convection in a box of porous material saturated
with fluid, Phys. Fluids, Vol. 15(8), pp.1377-1383, 1972.
5. Bories,S., Natural convection in porous media, NATO Adv.
Study Inst. on Fund. of Tansp. Phen. in Porous Med.,
Delaware, USA, July 1985.
6. Hearn,A.C., REDUCE User's manual, The Rand Corp., Santa
Monica, CA, USA, 1985.
7. Bernard,D., Convection naturelle dans les structures géologiques poreuses: 2 exemples numériques, Bull. Minéralogie,
Vol.111, pp. 601-611, 1988.

Methodological Study for the Evaluation of Vulnerability of Hydrogeological Basin

C. Masciopinto(*), S. Troisi(**), M. Vurro(*)

(*) CNR, Istituto di Ricera Sulle Acque, Reparto Sper. di Bari, via F. De Blasio, 5, 70123 Bari, Italy (**) Universita' Della Calabria, Dipartmento di Difesa del Suolo, Rende(CS), Italy

ABSTRACT

The groundwater quality worsening mainly due to in-progress and forecasting unavoidable human activities calls for domestic strategies (master plan, etc.). The current methodologies for hydrogeological basins vulnerability evaluation are based on the production and consumption of some substances. Even though they are valid for selecting and planning aquifers, they are too qualitative. Such methodologies, in fact, do not taken into account hydrodynamic and hydrodispersive phenomena. This paper proposes a methodological approach based on basin vulnerability for characterizing different aquifer areas. The methodology uses a mathematical model resolving hydrodynamic dispersion equation with finite triangular elements tecnique on a bidimensional horizontal domain. Furthermore, as for the parameteres used in the mathematical model, the methlogy evaluates the influence of their variability on areas magnitude. This methodology together with domestic master plan should improve the effectiveness of local environment clean-up programs.

INTRODUCTION

The problems concerning the acquifer pollution are very actual since these water resources represent a large tank for all the uses. The worsening quality causes a limitation of the drawings, mainly for the potable utilization. The CEE directive no. 75/440 [5], which has been taken into account by Italian Ministry of Health, has regulated the water acceptability qualifications for the potable utilization. Moreover, the actual Italian normative law about the water protection from the pollution is the outcome of a series of studies which have brought about an evolution of the acceptability principles of the water quality indexes: the first Italian law about this subject was the Merli Law (no. 319/76) [1] that also made the writing of reclamation plans.

A fundamental aspect about the formulation of the protection plans of the water resources is the evaluation of the water body pollution causes. For these evaluations, the knowledge of the potential pollutant charges and their distribution in space and time, the location of the areas whether of communication among surface water and groundwater bodies or high density of the anthropic activities are indispensable for the writing of the vulnerability maps. These maps give information about the vulnerability degree at the pollution of the areas under examination [6].

This paper proposes a methodological approach for the evaluation of the real pollution of an acquifer, by using hydrological, geological data and data concerning some anthropic activities present in the basin in question.

IMPROVED METHODOLOGY

The proposed methodology has been developed by considering a well defined hydrogeologic unity. The considered acquifer is located inside an alluvial cone of porous material (sand and gravel) with a good vertical permeability which allows infiltration and accumulation of water into the interstices. The supplying of the alluvial plan itself occurs nearly exclusively thanks to the river. It has been supposed an area of about 90 km² with a river length of about 20 km. The medium thickness is 24 m and the storage volume is about 400 millions of cubic metres. It have been supposed drawings for potable, industrial and agricultural utilization distributed into a part of the domain for a total of 40 millions of cubic metres every year (fig. 1). The range of the hydraulic conductivity in longitudinal direction is from 0.00257 to 0.000064 m/s, whereas the transversal direction is from 0.00130 to 0.000032 m/s; the effective porosity is between 0.2 and 0.1. The longitudinal dispersivity is between 39 and 10 m and the transversal one is between 6 and 2 m.

The quality parameter, considered as reference point to the application of the proposed methodology, are the nitrates. This parameter has been chosen because of its high solubility because it is often present in acquifers below areas subject to agricultural practice and zootechnic activities. Moreover, this parameter is very considerable in order to establish the possible potable utilization of the water resource, as established from the above-said normative

Fig. 1 The aquifer and its different zones: A pumping zone, B recharge zone (nitrate concentration 8 ppm), C recharge zone (nitrate concentration 30-50 ppm)

laws. Particularly, the reference values have been 25 and 50 ppm, deduced from the DPR 515/82 to realize the CEE directive, and 5 and 50 ppm, deduced from the DPCM of the 8th february 1985 [4]. From the existing reclamation plans and potential pollution study [2], it can be deduced the entity of the pollutant charges let into the water bodies and their distribution in space and time. About the considered aquifer the immission area has been subdivided into two parts according to the nitrate concentration present in the river (fig. 1).

The study of the delimitation of the areas of real pollution of the considered acquifer has been realized by using the IASO mathematical model [8,10] in pc version [9]. This model is based on the balance mass of the pollutant, supposed that it does not interact with the solid matrix, and in absence of abiothic chemical-physical phenomena (sorption, ionic-exchange, decay or production) or biothic chemical-physical phenomena (biochemical degradation, etc.) the following equation of diffusion and transport [3,7] is obtained:

$$\frac{\delta}{\delta x}n\left(Dxx\frac{\delta c}{\delta x}\right) + \frac{\delta}{\delta y}n\left(Dyy\frac{\delta c}{\delta y}\right) - \frac{\delta}{\delta x}(n\,U_x\,c) - \frac{\delta}{\delta y}(n\,U_y\,c) + Qi = \frac{\delta}{\delta t}(c\,n) \qquad (1)$$

where c is the concentration in weight of the pollutant substance, U_x and U_y are the velocity components and Qi is the "source" representing the pollutant immission. The coefficient of of hydrodynamic dispersion is expressed:

$$D_L(x,y,t) = \alpha_L(x,y)\quad U(x,y,t) \; + \; Do$$
$$\qquad\qquad\qquad\qquad\qquad\qquad\qquad (2)$$
$$D_T(x,y,t) = \alpha_T(x,y)\quad U(x,y,t) \; + \; Do$$

where α_L and α_T are parameters which depend only on the solid matrix and are called longitudinal dispersivity and transversal dispersivity; Do is the assumed coefficient of molecular diffusion, equal to 10e-08 m²/s. The IASO/pc procedure allows solution of problems which are connected with the evaluation of the quantity and quality of the water available in an acquifer system and with the water flow. The use of mathematical models giving solutions of the equation of hydrodynamic dispersion, by means of numerical technique of finite elements. The pollutant behaves like a tracer, that is it does not influence the velocity field in a remarkable way, its concentration variation can be calculated through the use of hydrodynamic dispersion equation known velocities. The finite elements which are used are triangularly-shaped with a linear shape function. The integration in time is carried out by using the implicit Crank-Nicolson method.

At first the water table has been calculated by considering the boundary conditions, recharge from the river and pumpings. The figure 2 shows the piezometric head of the aquifer. The obtained velocities are lower than 10e-04 m/s. Afterwards the concentrations of the nitrates in the groundwater have been calculated, starting from continuous and constant immissions in all the recharge area. The so obtained concentrations in each grid node have been reported in an evalutation scale, according to the limits established by the cited normative law. So four areas delimited by the following levels of nitrate concentration are obtained:
- zone 1: inferior to 5 ppm, optimal use for potable purposes;
- zone 2: between 5 and 25 ppm, use allowed for potable purposes with

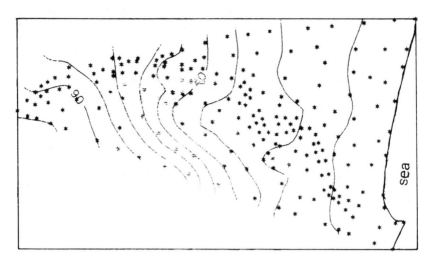

Fig. 2 Piezometric head of the acquifer

checks;
- zone 3: between 25 and 50 ppm, use allowed for potable purposes, but unadvisable;
- zone 4: beyond 50 ppm, use not allowed for potable purposes.

Therefore these areas represent the real pollution of the considered acquifer, according to the hypotheses of previously defined potential pollution and water flow. At last the proposed methodology can complete using a graphic representation of the areas through an interpolation of the IASO outcomes with the Kriging method.

APPLICATIONS

A first application of the mathematical model has been carried out for a sufficently long period of time in order to realize steady state conditions of the nitrate concentrations of the single grid nodes. Neumann conditions are along the cone sides and Dirichlet conditions along the coast. The unsteady state for the adopted parameters is about 15 days. An analysis of the influence of some parameters on the delimitation of zones has been carried out. Remarkable variations of the longitudinal and transversal dispersivity, also of 50%, do not cause significant changes of the range of the zones of real pollution. Variations of the transmissivity influence the duration of the unsteady state of the concentrations of the single points.

Significant changes of the pollution areas are caused by the boundary conditions; so the methodology application to real cases asks for a calibration with experimental analyses of the parameter concentrations. It being understood that the movement and immission conditions of pollutant substance, a second application has been carried out by considering different boundary conditions of the nitrate concentrations. The figure 3 shows the zones of real pollution supposing to assume the concentrations along the cost equal to 10 ppm; otherwise the figure 4 shows the different pollution zones obtained supposing

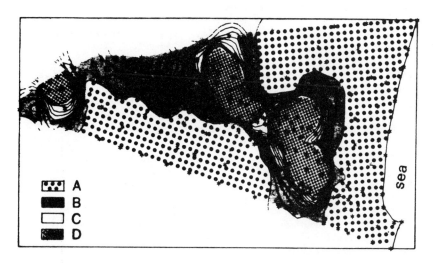

Fig. 3 Aquifer real pollution zones: A zone 1, B zone 2, C zone 3, D zone 4

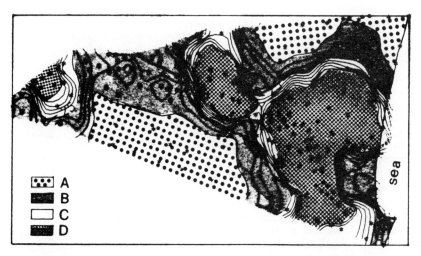

Fig. 4 Aquifer real pollution zone supposing further four points: A zone 1, B zone 2, C zone 3, D zone 4

that the concentration in further four points of the grid is known (variable between 15 and 62 ppm).

CONCLUSIONS

The suggested methodology, in this first working-out phase, gives a mean useful to locate polluted areas of an acquifer, which has undergone different modalities of immission of pollutant substances.

A different influence of the parameters in the equation of the hydrodynamic dispersion has been pointed out. Indeed, in the first phase of the approximation, it seems that the velocity variations of the water in the aquifer influence more than the dispersivity variations for the delimitation of the polluted zones.

The knowledge of this influence is fundamental to plan the operations in order to clean up a polluted acquifer. Indeed these operations can last to a limited time and cannot allow to reach steady state for the concentrations of the grid nodes. The division of an acquifer into zones allows a correct check of the water supplies area for different uses of groundwater resources.

REFERENCES

1. AA.VV. Criteri e limiti per il controllo dell'inquinamento delle acque. Dieci anni di esperienze, Quaderni IRSA, 75, Roma, 1987.
2. Barbiero, G., Cicioni, G. and Spaziani, F.M., Un sistema informativo per la valutazione dell'inquinamento potenziale, Quaderni IRSA, 74, Roma, 1987.
3. Bear, J. Hydraulics of groundwater, Mc Graw-Hill, New York, 1979.
4. D.P.C.M. 8 febbraio 1985 Caratteristiche di qualità delle acque destinate al consumo umano.
5. D.P.R. 515/82 Attuazione della direttiva (CEE) n. 75/440 concernenete la qualità delle acque superficiali destinate alla produzione di acqua potabile.
6. Francani, V. and Civita, M. (Ed.). Proposta di normativa per l'istituzione delle fasce di rispetto delle opere di captazione di acque sotterranee, CNR Gruppo Nazionale Difesa Catastrofi Idrogeologiche, pubb. 75, Milano, 1988.
7. Fried, J.J. Groundwater pollution, Elsevier Sciences Pubbl. Company, New York, 1975.
8. Gambolati, G., Troisi, S. and Volpi, G. Un modello di inquinamento agli elementi finiti in sistemi filtranti, Idrotecnica, 6, pp. 305-316, 1982.
9. Masciopinto, C., Palmisano, V. and Vurro, M. Una procedura di calcolo per lo studio della propagazione di un inquinante in falda: IASO/pc, Quaderni IRSA 90, Roma, 1990.
10. Troisi, S., Volpi, G. and Vurro, M. Studio di problemi d'inquinamento in un acquifero costiero pugliese con modelli agli elementi finiti, Acqua e Aria, 4, pagg. 341-348, 1985.

Application of Quasi-Newton Methods to Non-linear Groundwater Flow Problems

J.D. Porter, C.P. Jackson

AEA Technology, Harwell Laboratory, Oxfordshire, England

ABSTRACT

This paper describes the application of a number of Quasi-Newton methods to a test case of coupled groundwater flow and salt transport. The calculations were carried out using the Harwell finite-element code for groundwater flow and mass transport, NAMMU, on the Harwell CRAY 2. A number of iterative methods were employed and the effect of line searches was also investigated. The Broyden method was the most successful, and it was found to be robust and to give significant savings in computation time compared with the Newton-Raphson method. Comments are also made on the effect of using CRAY Assembler versions of core routines.

INTRODUCTION

The flow of constant-density groundwater in a fully saturated domain is described by a linear equation which can generally be solved numerically within an acceptable expenditure of computing time (although the cost of three-dimensional calculations may be significant). However, when part of the domain becomes unsaturated, or when the water contains a solute that can strongly affect its density, then the movement of groundwater and solute is described by a system of coupled non-linear equations. Such problems generally prove to be very difficult and expensive to solve (see eg [1,8]). It is therefore worthwhile to evaluate alternative solution techniques and this paper describes the application of a number of Quasi-Newton methods to a test case of saline groundwater flow.

QUASI-NEWTON METHODS

Quasi-Newton or Modified-Newton methods are widely applied to non-linear problems in fields such as optimisation and structural engineering [2-7]. These methods seek to reduce the costs associated with a full Newton-Raphson calculation. For two- and three-dimensional problems the most expensive part of each Newton-Raphson iteration is the solution of a matrix equation involving the Jacobian. Quasi-Newton methods reduce the cost of each iteration after the first by approximating the inverse Jacobian matrix by a low rank update of the original inverse, and exploiting the fact that once the LU decomposition of the initial Jacobian has been carried out it is cheap to solve subsequent systems with the same matrix. The price paid for using these approximations is generally a deterioration from the quadratic convergence of the Newton-Raphson method to superlinear convergence, leading to more iterations. To ensure that the approximation to the Jacobian is reasonable, Quasi-Newton methods are required to satisfy the condition:

$$\delta_n = H_{n+1} r_n \, , \tag{1}$$

where

$$r_n = f_{n+1} - f_n \, , \tag{2}$$

f_n is the vector of residuals, δ_n is the update of the solution vector and H_n is the approximation to the inverse of the Jacobian at iteration n.

There are many possible Quasi-Newton methods, corresponding to different approximations to the Jacobian matrix. In the following sections some of these methods are outlined, and their application to a suitable test case is discussed to ascertain which is the most useful for non-linear groundwater flow systems.

Broyden method
Perhaps the simplest way of updating the Jacobian matrix, or its inverse, H, is to add a rank 1 matrix, giving an update of the form

$$H_{n+1} = H_n + auv^T , \tag{3}$$

where u, v are vectors and a is a scalar. a, u and v can be determined by the Quasi-Newton condition (1) and an orthogonality condition suggested by Broyden (see [2, 3, 4]). Together these conditions give the Broyden update formula:

$$H_{n+1} = \left\{ I + \frac{(\delta_n - H_n r_n) \delta_n^T}{\delta_n^T H_n r_n} \right\} H_n , \tag{4}$$

where I is the identity matrix.

BFGS method
Perhaps surprisingly, the rank one update was not the first to be discovered. The earliest formula to be suggested was one of the family of rank two formulae, which have the general form

$$H_{n+1} = H_n + auu^T + bvv^T , \tag{5}$$

where a and b are scalars and u and v are vectors. Various rank two formulae can be obtained, depending on the choices made for the vectors u and v. The rank two update that has been found to work best in practice is the Broyden, Fletcher, Goldfarb and Shanno (hence BFGS) method [2,4]. For H, the inverse of the Jacobian, this can be written as:

$$H_{n+1} = H_n + \left\{ I + \frac{r_n^T H_n r_n}{\delta_n^T r_n} \right\} \frac{\delta_n \delta_n^T}{\delta_n^T r_n} - \left\{ \frac{\delta_n r_n^T H_n + H_n r_n \delta_n^T}{\delta_n^T r_n} \right\} , \qquad (6)$$

in the notation of the previous subsection.

Line searches

It is often reported that the performance of an iterative method is greatly improved by the use of line searches [2-7]. The basic idea of a line search is to find a parameter γ_n, such that at each iteration

$$x_{n+1} = x_n + \gamma_n \delta_n \qquad (7)$$

is a better estimate of the true solution than that obtained with $\gamma_n=1$, and thus leads to faster convergence. γ_n is usually determined from the minimisation of a suitable quantity (e.g. the sum of squares of the nodal residuals).

The Broyden and BFGS methods with and without line searches were implemented in the Harwell finite-element groundwater flow and transport code NAMMU [9] as an alternative to the standard Newton-Raphson solver. The methods were tested on several small algebraic problems and then applied to the more realistic test case of non-linear groundwater flow described below.

TEST CASE : DENSITY-DEPENDENT GROUNDWATER FLOW

This test case was chosen to be representative of situations of potential interest but less demanding in terms of computer time than, say, the HYDROCOIN Phase 1 example of saline groundwater flow [8]. The physical problem considered was a convection cell driven by density differences. The equations which must be solved in this case are:

$$\frac{\partial}{\partial t} (\rho \varphi) + \nabla.(\rho \varphi \underline{V}) = 0 , \qquad (8)$$

$$\frac{\partial}{\partial t} (\rho \varphi C) + \nabla.(\rho \varphi \underline{V} C) = \nabla.(\rho \varphi \underline{\underline{D}}.\nabla C) , \qquad (9)$$

$$\underline{q} = \varphi \underline{V} = \frac{-k}{\mu} (\nabla p - \rho \underline{g}) , \qquad (10)$$

where $\underline{V}$ is the fluid velocity, C is the mass fraction of the concentration of salt in solution, $\underline{\underline{D}}$ is the hydrodynamic dispersion tensor, $\underline{q}$ is the specific discharge, p is the fluid pressure, $\underline{g}$ is the acceleration due to gravity,

$$\left(\underline{\underline{D}} \right)_{ij} = D_p \delta_{ij} + (a_L - a_T) V_i V_j / |\underline{V}| + a_T |\underline{V}| \delta_{ij} , \qquad (11)$$

and

$$\rho = (C/\rho_b + (1-C)/\rho_0)^{-1} . \qquad (12)$$

D_p is the pore-water diffusion coefficient, (which is the molecular diffusion modified by a tortuosity factor), and other symbols are explained in Table 1.

Details of the geometry and boundary conditions are shown in Figure 1 and the values of the physical parameters are given in Table 1.

TABLE 1 : Parameters of Test Case

k	Permeability of Rock	:	$10^{-12}m^2$
φ	Porosity of Rock	:	0.2
μ	Viscosity of Water	:	10^{-3} kg m^{-1}s^{-1}
ρ_o	Density of Fresh Water	:	1000 kg m^{-3}
ρ_b	Density of Salt Water	:	1200 kg m^{-3} (fully saturated)
a_L	Longitudinal Dispersion Length	:	20m
a_T	Transverse Dispersion Length	:	2m
D_p	Pore-Water Diffusion Coefficient	:	$10^{-6}m^2s^{-1}$
	Length of Cell	:	400m
	Height of Cell	:	100m

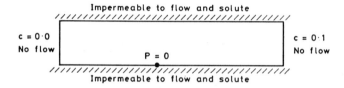

FIGURE 1. TEST CASE

The equations were solved using the NAMMU finite-element code [9]. NAMMU provides a large number of options, allowing the user to select the equations to be solved, method to be adopted and the element type and basis functions to be used. In this case finite-element versions of equations (8), (9) and (10) were solved for pressure, salt concentration and the two velocity components. A Galerkin finite-element method was used with quadrilateral elements and with quadratic basis functions for all variables.

The test case was solved on four different mesh sizes (5x5, 10x10, 20x20 and 30x30 elements) using the following methods to deal with the non-linearity:

a)   Newton-Raphson iteration (standard method for NAMMU)
b)   BFGS method without line searches
c)   BFGS method with line searches
d)   Broyden method without line searches
e)   Brodyen method with line searches.

The Newton-Raphson method successfully converged to a solution on all mesh sizes and these solutions and the times required to obtain them were then used as a yardstick against which the performance of the other methods could be judged. The solution obtained for the concentration on the finest mesh with the Newton-Raphson method is shown in Figure 2. (The vertical direction has been scaled so that the contours may be more clearly seen.)

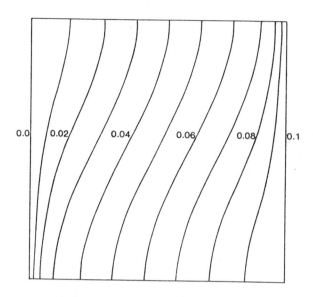

FIGURE 2. CONCENTRATION CONTOURS

Results: BFGS Method

When applied to the saline convection-cell test case this method was not successful. When the BFGS method was used after one Newton-Raphson iteration it failed to converge on all grid sizes. In the field of optimisation it is common to use the identity matrix as the initial estimate of the inverse of the Jacobian. This was also tried for the test case, but without success.

Contrary to expectation, the use of line searches with the BFGS method did not improve matters but actually resulted in a more rapid failure of the method.

It is clear from these results that the BFGS method is completely unsuitable for our density-dependent groundwater flow problem.

Results: Broyden Method

The Broyden method with line searches failed to converge for the saline convection-cell test case. However, the Broyden method without line searches successfully converged on all mesh sizes and the graphical output obtained was generally indistinguishable from the corresponding case using Newton-Raphson iteration. Very small differences (a few per cent at most) were visible in the numerical values at the nodes. For all practical purposes these small differences are negligible. Table 2 compares the number of iterations and computing time required by the two methods to converge to the same tolerance.

It was noticeable that the convergence of the Broyden method was not monotonic. Although the Broyden method always required more iterations than the Newton-Raphson method, the Broyden method uses at most the same and generally rather less computing time to solve a given problem because one iteration of the Broyden method is considerably less expensive than one Newton-Raphson iteration. Note that the saving obtained with the Broyden method increases with the size of the mesh.

TABLE 2 : Results for Test Case

Grid Size (Elements) (Number of Unknowns)	Number of Iterations (i)		Time in Solver (ii)		Ratio of Solver Times
	(NR)	(BY)	(NR)	(BY)	(BY/NR)
5x5 (484)	5	1+20	4.45	3.74	0.84
10x10 (1764)	5	1+15	36.9	19.9	0.54
20x20 (6724)	4	1+14	479	161	0.34
30x30 (14884)	4	1+13	2250	607	0.27

Notes: (i)   NR means Newton-Raphson iteration
BY means 1 initial Newton-Raphson iteration followed by Broyden iterations (hence 1+20 etc)

(ii)   Solver times in seconds

CONCLUSIONS

The following conclusions concerning the application of Quasi-Newton methods to non-linear groundwater flow problems may be drawn from this study:

1)   the Broyden method gives a useful reduction in computing time and appears to be robust;

2)   the saving obtained with the Broyden method increases with the problem size;

3)   the BFGS method is unsuccessful for the non-linear groundwater flow problems considered here;

4)   line-search techniques do not improve the performance of either of the Quasi-Newton methods investigated here.

Finally, this work indicates that useful savings in computing time can be obtained through the use of alternative solvers and suggests that it would be beneficial to investigate other approaches.

ACKNOWLEDGEMENTS

The funding for this research provided by the United Kingdom Department of the Environment, as part of its radioactive waste management research programme, is gratefully acknowledged. The results will be used in the formulation of Government policy but at this stage they do not necessarily represent policy.

Funding for this work from the Commission of the European Communities is also gratefully acknowledged.

REFERENCES

1.   Nuclear Energy Agency/Swedish Nuclear Power Inspectorate. The International HYDROCOIN Project, Background and Results, OECD Paris, 1987.

2.   Fletcher, R. Practical Methods of Optimisation. (2nd Ed.) John Wiley and Sons 1987.

3.   Broyden, C.G. Mathematics of computation, Vol. 21, pp.368-381, 1967.

4.   Dennis, J.E. Jr. and More, J.J. Siam Review Vol. 19, pp.46-89, 1977.

5.   Brodie, K.W., Gourlay, A.R. and Greenstadt, J.J. Inst. Maths. Applics. Vol. 11, pp.73-82, 1973.

6.   Taylor, C., Hinton, E. and Owen, D. (Eds) Numerical Methods for Non-Linear Problems (vol. 2), Pineridge Press, Swansea, 1984.

7.   Crisfield, M.A. Variable Step Lengths for Non Linear Structural Analysis, Transport and Road Research Laboratory Report 1049, 1982.

8.   Herbert, A.W., Jackson, C.P. and Lever, D.A. Wat. Resour. Res. Vol.   24, pp.1781-1795, 1988.

9.   Atkinson, R., Herbert, A.W., Jackson, C.P. and Robinson, P.C. and Williams, M.G. NAMMU User Guide (Release 4), UKAEA Report AERE-R12354, 1986. (Draft)

APPENDIX : THE EFFECT OF USING CRAY ASSEMBLER ROUTINES

NAMMU is used by a number of customers on a variety of computers. This paper is therefore mainly concerned with describing techniques which can benefit all users employing the standard Fortran version of NAMMU. However, given the interest at this conference in the solution of water resources problems on supercomputers, this Appendix gives an indication of the effect of replacing core solver routines with versions in CRAY Assembler. The routines in question are part of the frontal solver and perform either one or two steps of the Gaussian elimination on the frontal matrix. As one might expect the reduction in CPU time obtained with the use of these routines is dramatic. The CPU time for the 20x20 test case described above (solved by the Newton-Raphson method) is reduced from 479s to 34s, (on the Harwell CRAY2). This is actually less than the time required by the Broyden method (82s) when the same Assembler routines are used. Unfortunately an Assembler version of the Broyden solver was not available, so a direct comparison of the two solvers when coded in Assembler was not possible.

A 3-D Finite Element Code for Modelling Salt Intrusion in Aquifers

L. Brusa(*), G. Gentile(*), L. Nigro(*), D. Mezzani(**), R. Rangogni(**)

(*) CISE Tecnologie Innovative SpA, PO Box 12081 - 20134 Milan, Italy (**) ENEL - Hydraulic and Structual Research Center, Via Ornato No.90-14-20162 Milan, Italy

ABSTRACT

The paper presents the 3-D finite element code SIQUAF for the simulation of water flow and salt transport in saturated/unsaturated porous media. The code is based on an hydraulic head/concentration formulation. A Fractional Step method is used for the solution of the advective/convective equation. The program is designed for the parallel computer ALLIANT FX/80.

INTRODUCTION

Pollution of fresh water aquifers is becoming more and more important as far as design, construction and management of industrial plants are concerned. In particular, for thermal electric plants near the sea it is quite important to study the consequences of salt intrusion into the aquifers, to avoid the degradation of fresh water quality.

To analyze such complex phenomena the 3-D non linear finite element code SIQUAF has been developed. The computer code treats non-steady problems and computes both the flow field and the salt concentration (or other non-reactive pollutants) in the saturated and unsaturated zones. The code can model real complex situations such as anisotropy of soil, presence of wells, trenches ecc.

The classical Galerkin variational formulation has been used both for the flow and the transport equations. For the flow field the formulation allows to handle symmetric positive definite matrices also in presence of convective terms. The transport of salt concentration is computed applying a fractional step method [1] for time discretization of the governing equation, so as to separate the convective part (non symmetric) from the diffusive part (symmetric).

A second order accurate explicit Taylor-Galerkin method [2] is employed to solve the convective step and an implicit time integration is performed to solve the differential equations of the diffusive step.

The paper describes the mathematical model and the numerical schemes used in SIQUAF code. Numerical examples are not presented here, being the validation of the code in progress.

MATHEMATICAL MODEL

The proposed model is based on two equilibrium equations governing the behaviour of the hydraulic head and of the dissolved salt concentration [4],[5]. The first equation is diffusive and describes the flow of the fluid in the porous medium:

$$\frac{\partial(\rho\theta)}{\partial t} + \vec{\nabla}\cdot(\rho\vec{q}) = S; \quad \vec{q} = -K\left[\vec{\nabla}h + (h-z)\frac{\vec{\nabla}\rho}{\rho}\right] \qquad (1)$$

where $\vec{q}$ is the Darcy velocity vector; h is the hydraulic head; $\rho = \rho(C)$ is the fluid density depending on the salt concentration C; z is the vertical coordinate; $\theta = \theta(z-h)$ is the water content; $K = K(\theta)$ is the hydraulic conductivity depending on θ; S is the volumetric flow rate of sources (or sinks) per unit volume of the porous medium.

The following initial and boundary conditions are associated to eq. (1):

$$h(t_0) = h_0; \quad h|_{\Gamma_1} = \bar{h}; \quad q_n|_{\Gamma_2} = \bar{q}_n; \quad \Gamma = \Gamma_1 \cup \Gamma_2$$

with q_n the Darcy velocity normal to the boundary Γ_2.

The convective-diffusive equation which describes the transport of the dissolved salt is:

$$\frac{\partial(\rho\theta C)}{\partial t} + \vec{\nabla}\cdot[\rho C\vec{q} - \theta D\vec{\nabla}(\rho C)] = 0 \qquad (2)$$

where D is the diffusion tensor.
Taking into account eq.(1), eq.(2) can be written as follows:

$$\rho\theta\frac{\partial C}{\partial t} + \rho\vec{q}\cdot\vec{\nabla}C - \vec{\nabla}\cdot[\theta D\vec{\nabla}\cdot(\rho C)] = -CS \qquad (3)$$

The initial and boundary conditions associated to eq.(3) are:

$$C(t_0) = C_0; \quad C|_{\Sigma_1} = \bar{C}; \quad D\vec{\nabla}(\rho C)\cdot\vec{n}|_{\Sigma_2} = \bar{f}$$

where $\Gamma = \Sigma_1 \cup \Sigma_2$ and $\vec{n}$ is the outward normal on Σ_2

DISCRETIZATION OF THE DIFFUSION EQUATION FOR THE HYDRAULIC HEAD

Galerkin finite element technique applied to eq.(1) produces a discretized problem with non symmetric matrices. To avoid this drawback, the following change of variable has been done:

$$\Phi = \rho(h-z) \tag{4}$$

With this assumption, eq.(1) takes the form:

$$\frac{\partial(\rho\theta)}{\partial t} - \vec{\nabla}\cdot(K\vec{\nabla}\Phi + K\rho\vec{\nabla}z) = S \tag{5}$$

Eq.(5) is firstly discretized in time and subsequently in space by means of finite element method. For time discretization the α-method [6] is used:

$$\rho^{n+1}\theta^{n+1} = \rho^n\theta^n + \Delta t_n\vec{\nabla}\cdot(K^*\vec{\nabla}\Phi^* + K^*\rho^*\vec{\nabla}z) + \Delta t_n S^* \tag{6}$$

where $\Delta t_n = t_{n+1} - t_n$, indexes n and n+1 refer to variables computed at time t_n and t_{n+1} respectively and:

$$\Phi^* = \alpha\Phi^{n+1} + (1-\alpha)\Phi^n; \quad 0 < \alpha \le 1$$

A similar notation holds for ρ^*, k^*, S^*. In the computation of ρ^*, ρ^{n+1} is approximated by:

$$\rho^{n+1} = \rho^n + \frac{\Delta t_n}{\Delta t_{n-1}}(\rho^n - \rho^{n-1}) \tag{7}$$

so that

$$\rho^* = \rho^n + \alpha\frac{\Delta t_n}{\Delta t_{n-1}}(\rho^n - \rho^{n-1}) \tag{8}$$

By defining:

$$\gamma = \frac{d\theta}{d(h-z)} \tag{9}$$

one has:

$$\theta^{n+1} = \theta^n + \gamma^*\left(\frac{\Phi^{n+1}}{\rho^{n+1}} - \frac{\Phi^n}{\rho^n}\right) \tag{10}$$

and eq.(6) takes the form:

$$\gamma^*\Phi^{n+1} - \gamma^*\frac{\rho^{n+1}}{\rho^n}\Phi^n + (\rho^{n+1} - \rho^n)\theta^n +$$

$$-\Delta t_n\vec{\nabla}\cdot[\alpha K^*\vec{\nabla}\Phi^{n+1} + (1-\alpha)K^*\vec{\nabla}\Phi^n + K^*\rho^*\vec{\nabla}z] - \Delta t_n S^* = 0 \tag{11}$$

The finite element discretization of eq.(11) is straightforward and leads to the following system of non linear algebraic equations:

$$[B^* + \alpha\Delta t A^*]\Phi^{n+1} = [G^* - (1-\alpha)\Delta t A^*]\Phi^n + F^* \tag{12}$$

where:

$$B^* = \sum_e \int_{\Omega^{(e)}} \gamma^* N^{(e)} N^{T(e)} d\Omega^{(e)} \qquad A^* = \sum_e \int_{\Omega^{(e)}} \vec{\nabla} N^{(e)} \cdot K^* \vec{\nabla} N^{T(e)} d\Omega^{(e)}$$

$$G^* = \sum_e \int_{\Omega^{(e)}} \frac{\rho^{n+1}}{\rho^n} \gamma^* N^{(e)} N^{T(e)} d\Omega^{(e)}$$

$$F^* = -\sum_e \int_{\Omega^{(e)}} (\rho^{n+1} - \rho^n) \theta^n N^{(e)} + K^* \rho^* \Delta t_n \frac{\partial N^{(e)}}{\partial z} - \Delta t N^{(e)} S^* d\Omega$$

$$- \Delta t_n \sum_{e'} \int_{\Gamma_2^{(e')}} N^{(e')} \rho^* \overline{\vec{q}_n} d\Gamma^{(e')}$$

In previous equations $N^{(e)}$ is the vector of the test functions in the e-th element.

DISCRETIZATION OF THE CONVECTIVE-DIFFUSIVE EQUATION OF SALT CONCENTRATION

The Fractional Step Method [1] is used to solve eq.(3). According to this method, problem (3), with the associated initial and boundary conditions, is splitted in a convective and in a diffusive problem for $t_n \leq t \leq t_{n+1}$.

The concentration C^{n+1} is obtained as follows:

i) compute $\tilde{C}^{n+1}$ by solving the problem:

$$\rho\theta\frac{\partial\tilde{C}}{\partial t} + \rho\vec{q} \cdot \nabla\tilde{C} = 0; \quad \tilde{C}(t_n) = C^n; \quad \tilde{C}|_{\tilde{\Sigma}_1} = \overline{C} \qquad (13)$$

where $\tilde{\Sigma}_1$ is the portion of Σ_1 with inward concentration flux.

ii) Solve the diffusive problem:

$$\rho\theta\frac{\partial C}{\partial t} - \vec{\nabla} \cdot (\theta D\vec{\nabla}(\rho C)) = -CS \qquad (14)$$

$$C(t_n) = \tilde{C}^{n+1}; \quad C|_{\Sigma_1} = \overline{C}; \quad D\vec{\nabla}(\rho C) \cdot \vec{n}|_{\Sigma_2} = \overline{f}$$

Problem (13) is solved by means of second order Taylor-Galerkin method [2]. According to this scheme $\tilde{C}^{n+1}$ is approximated by the Taylor expansion:

$$\tilde{C}^{n+1} = \tilde{C}^n + \Delta t \frac{\partial \tilde{C}}{\partial t}\Big|_n + \frac{\Delta t^2}{2}\frac{\partial^2\tilde{C}}{\partial t^2}\Big|_n \qquad (15)$$

By using eq.(13) to evaluate $\frac{\partial c}{\partial t}\Big|_n$ and $\frac{\partial^2 c}{\partial t^2}\Big|_n$ one has:

$$\rho^n \theta^n (\tilde{C}^{n+1} - C^n) = -\Delta t_n \left[\rho^n \vec{q}^n + \frac{\Delta t_n}{2} \frac{\partial (\rho \vec{q})}{\partial t} \Big|_n - \Delta t_n \frac{\vec{q}^n}{\theta^n} \frac{\partial (\rho \theta)}{\partial t} \Big|_n \right] \cdot \vec{\nabla} C^n$$

$$+ \frac{\Delta t_n^2}{2} \vec{\nabla} \cdot \left[\frac{\rho^n \vec{q}^n}{\theta^n} (\vec{q}^n \cdot \vec{\nabla} C^n) \right] \tag{16}$$

The application of the finite element method to eq.(16) gives:

$$M^n (\tilde{C}^{n+1} - C^n) = R^n \tag{17}$$

where:

$$M^n = \sum_e \int_{\Omega^{(e)}} \rho^n \theta^n N^{(e)} N^{T(e)} d\Omega^{(e)}$$

$$R^n = \sum_e \int_{\Omega^{(e)}} N^{(e)} \left(-\frac{\rho^n \vec{q}^n}{2} - \frac{\rho^{n-1} \vec{q}^{n-1} - \vec{q}^n \rho^n}{2\Delta t_{n-1}} - \Delta t_n \frac{\vec{q}^n}{\theta^n} \frac{\rho^n \theta^n - \rho^{n-1} \theta^{n-1}}{\Delta t_{n-1}} \right) \cdot \vec{\nabla} C^n d\Omega +$$

$$- \frac{\Delta t_n^2}{2} \sum_e \int_{\Omega^{(e)}} (\vec{\nabla} N^{(e)} \cdot \rho^n \vec{q}^n) \left(\frac{\vec{q}^n}{\theta^n} \cdot \vec{\nabla} C^n \right) d\Omega^{(e)} \tag{18}$$

$$+ \frac{\Delta t_n^2}{2} \sum_{e'} \int_{\Sigma - \Sigma_1} (N^{(e')} \rho^n \vec{q}^n \cdot \vec{n}) \left(\frac{\vec{q}^n}{\theta^n} \cdot \vec{\nabla} C^n \right) d\Sigma^{(e')}$$

The spatial discretization of eq.(14) is performed by means of finite element method assuming that ρ, θ, D are evaluated at $t = t_n$ and are elementwise constant. The problem is therefore reduced to the solution of the following system of differential equations:

$$M^n \frac{dC}{dt} + K^n C = T^n; \quad C(t_n) = \tilde{C}^{n+1} \tag{19}$$

where M^n defined as in (18) and:

$$K^n = \sum_e \int_{\Omega^{(e)}} [\vec{\nabla} N^{(e)} \cdot \theta^n D^n \rho^n \vec{\nabla} N^{(e)T} + S N^{(e)} N^{(e)T}] d\Omega^{(e)}$$

$$T^n = \sum_e \int_{\Sigma_2^{(e)}} N^{(e)} f d\Sigma_2^{(e)}$$

Eq.(19) is solved by means of the α-method:

$$[M^n + \alpha \Delta t_n K^n] C^{n+1} = [M^n - (1-\alpha) \Delta t K^n] \tilde{C}^{n+1}$$

$$+ \Delta t [\alpha T^{n+1} + (1-\alpha) T^n] \tag{20}$$

COMPUTATIONAL SCHEME AND NUMERICAL METHODS

The approximations performed in the definition of the discretized mathematical model decouple the hydraulic head equation from the concentration equation. This has been done to reduce the computing time

which otherwise may be prohibitive for large scale real problems.
The computational scheme is therefore as follows:

i) compute the hydraulic head at $t = t_{n+1}$ by using the density at $t = t_{n+1}$ obtained through an extrapolation from its values at $t = t_n$ and $t = t_{n-1}$. Iterations are performed in this step to take into account non-linearities depending only on the hydraulic head;

ii) compute the concentration at $t = t_{n+1}$ by using the Darcy velocity, the density and the water content evaluated at $t = t_n$.

The nonlinear equations (12) are solved by a fixed point iteration scheme [6], in which matrices A^*, B^* and G^* can be updated at each iteration or can be considered constant within each time step.

The linear equations solver used in the SIQUAF code is based on the frontal method and is designed to exploit the architectural features of shared memory parallel computers.

REFERENCES

1. Marchuk, G.I. Methods of Numerical Mathematics, Springer Verlag, 1982.
2. Donea, J. A Taylor-Galerkin Method for Convective Transport Problems, Int. J. Num. Meth. Eng. 20, pp. 101-119, 1984.
3. Brusa, L. and Riccio, F. Substructure Technique for Parallel Solution of Linear Systems in Finite Element Analyses, in Parallel computing: Methods, Algorithms and Applications, Ed. D.J. Evans, C. Sutti; Adam Hilyer, 1988.
4. Bonneton, M., Jardin P. and Usseglio Polatera, J.M. Code DEDALE-2D: Dossier de Validation, Report LHF n.C0053, Grenoble, May 89.
5. Bear J., Hydraulic of Groundwaters, Mc Graw-Hill, 1979.
6. Bon, E., Nigro, L. and Sampietri ,C. Numerical Solution of Thermal Processes with State or Phase Change: the Computer Code ATEN-2D, Num. Meth. for Non Linear Problems, Ed. C. Taylor et al., Pineridge Press, 1986.

A Eulerian-Lagrangian Finite Element Model for Coupled Groundwater Transport

G.Gambolati, G. Galeati(*), S.P. Neuman(**)

Dipartmento di Metodi e Modelli Matematici per le Scienze Applicate, Univesità degli Studi di Padova, Italy

INTRODUCTION

Modeling of saltwater intrusion into coastal aquifers requires a coupled formulation between groundwater flow and solute concentration c in the subsurface. For the concentration commonly encountered in seas and oceans (between 25 and 40 Kg/m^3) the dependance of the seepage velocity on c is weak and, as a major result, partially uncoupled numerical solutions may be developed which allow for a much more efficient simulation of long-term transient responses.

In recent years the numerical modeling of seawater intrusion in subsurface systems has attracted a growing interest both in 2-D and 3-D frameworks (see, for example, Frind[5,6]; Voss[13]; Huyakorn et al.[9]; Diersch[4]). The finite element technique for both flow and transport is most commonly employed. Coupling is solved by a quasi-Newton type method wherein the potential and the convection-dispersion problems are separately solved and linked through iterations.

In the present communication we propose a new model for

Permanent address: (*) ENEL-CRIS, Serv. Idrol., Venezia, Italy

(**) Department of Hydrology and Water Resources, University of Arizona, Tucson, Arizona, USA

the coupled and partially uncoupled analysis of groundwater flow and salt dispersion and convection based on a Eulerian-Lagrangian approach for the contamination problem combined with a traditional finite element method for the flow problem. Unlike the schemes developed by Pinder and Cooper[11] and Sanford and Konikow[12], the present approach implements a "single step reverse tracking" technique similar to the one introduced by Neuman[10] and Baptista et al.[1]. The overall model is implemented in a FORTRAN code called CUNEO.

Three features of CUNEO should be particularly attractive: a) the option for partial decoupling which makes the model particularly suitable and efficient for long term transient simulations; b) the symmetric positive definite nature of the resulting algebraic equations which are solved by the cost-effective and accurate accelerated conjugate gradient method (Gambolati and Perdon[8]); c) the low degree of artificial dispersion due to the integration of the convective component along the characteristic lines. Moreover, the model is conceptually suited for a parallel architecture where the fate of selected groups of particles can be traced independently and simultaneously by the available concurrent processors.

In the sequel, after a short description of the numerical model, an application of CUNEO to the prediction of the salt toe encroachment is shown with an emphasis on the reliability of partially uncoupled solutions vs the more expensive coupled ones.

NUMERICAL SOLUTION BY A MIXED EULERIAN-LANGRAGIAN FINITE ELEMENT APPROACH

The governing equations for density dependent transport problems, such as the one describing saltwater intrusion into coastal aquifers, are:

$$\frac{\partial}{\partial x_i}\left[K_{ij}\left(\frac{\partial h}{\partial x_j} + \varepsilon c \eta_j\right)\right] = S_s \frac{\partial h}{\partial t} + n \frac{\rho_o}{\rho}\varepsilon \frac{\partial c}{\partial t} - q \qquad (1)$$

$$i = 1,2$$

$$\frac{\partial}{\partial x_i}\left(n\,D_{ij}\frac{\partial c}{\partial x_j}\right) - \frac{\partial}{\partial x_i}(v_i c) = n\frac{\partial c}{\partial t} - qc^* \qquad (2)$$

where h is the equivalent freshwater head (see Frind[5] and

Huyakorn et al.[9]), c is the relative salt concentration, K_{ij} and D_{ij} are the hydraulic conductivity and dispersion tensors, respectively, S_s is the specific elastic storage coefficient, n is the medium porosity, v_i is the Darcy velocity, ρ_o and ρ are the fresh and the contaminated water density, respectively, q is the volumetric flow rate per unit volum, c* is the relative solute concentration in the injected (withdrawn) fluid and:

$$\varepsilon = (\rho_{max} - \rho_o)/\rho_o$$
$$\eta_1 = 0 \qquad \eta_2 = 1$$

The dispersion tensor D_{ij} is related to the Darcy velocity through the longitudinal and transversal dispersivities α_L and α_T (Bear[2]). Boundary conditions for eqs. (1) and (2) may be of type 1, 2, and 3; for a thorough discussion, see Galeati and Gambolati[7].

Note that eqs. (1) and (2) are coupled through c and v_i. The solution procedure can be summarized as follows:

1) eq. (1) is integrated over the 2-D domain by the Galerkin technique using finite elements
2) the nodal concentration values which appear in the body force and boundary flux vector are intially related to the old time level t
3) integration in time is performed with the fully implicit Crank-Nicolson scheme
4) Darcy velocies are determined over each element by differentiating the linear representation of the potential solution and transferred to each node by Yeh's[14] scheme
5) eq. (2) is solved by a Eulerian-Lagrangian formulation similar to the one proposed by Neuman[10]
6) the advective component is handled by the so called "single step reverse particle tracking"; a modified Euler scheme (Costantinides[3]) and a linear interpolation scheme are used to locate the virtual position of the grid nodes at time t and to estimate the associated concentration values, respectively
7) the residual dispersive part is integrated by traditional finite elements on the same fixed grid used for the flow problem
8) integration in time of the total derivate is carried out by the fully implicit backward difference scheme

9)   if a fully coupled model is implemented, eqs. (1) and (2) at the time level t+ Δt are repeatedly solved until a convergence criterium is met. At each iteration the body force vector of eq. (1) and the flow field, i.e. the virtual location of the grid nodes at time t, are updated

10) if a partially coupled model is adopted, only the first iteration is performed, i.e. the flow field is computed using the concentration values at time t and the transport problem is also solved only once during each time step using Darcy velocities based on the initial potential prediction at time t+ Δt.

NUMERICAL RESULTS

The performance of the partially uncoupled model vs the fully coupled one has been analyzed on a real example of saltwater intrusion into an aquifer of Southern Italy. The drawdown is caused by pumpage needed to dewater a coastal area where a large thermo-electrical plant is to be built. The simulation is carried out in a vertical cross section orthogonal to the coastline. Figure 1 shows the subsurface system being simulated along with the parameter values and the boundary conditions. On the seaward boundary a zero diffusive flux of contaminant is prescribed where outflow occurs. Where inflow occurs a concentration equal to the sea value (25 Kg/m^3) is imposed.

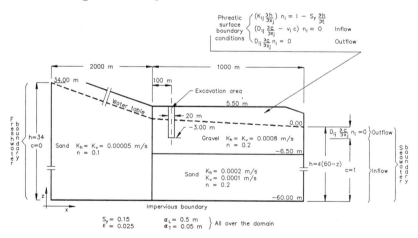

Figure 1. Picture of the simulated area

The upper part of Figure 2 shows the results from a

fully coupled model which performs quasi–Newton type iterations until convergence is achieved. Figure 2a) gives the steady state salt distribution whithin the aquifer before the start of pumping. The (relative) concentration 60 days after startup is shown in Figure 2b). A comparison with

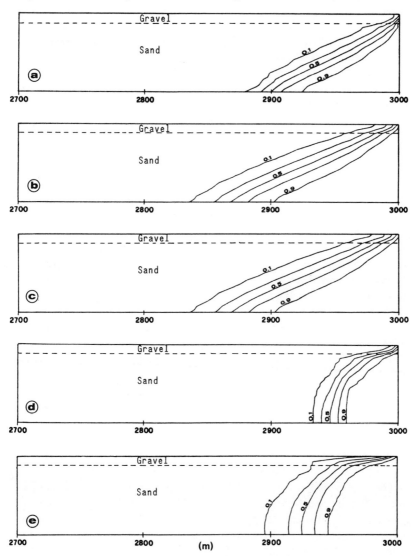

Figure 2. Numerical salt concentration as providede by the model for the sample problem. Fully coupled model, t=0 (a) and t=60 days (b); partially coupled model, t=60 days (c); fully uncoupled model, t=0 (d) and t=60 days (e).

the outcome from a partially uncoupled model is provided in
Figure 2c). Note that the differences are so small that they
cannot be appreciated on the scale of the figure. Figures
2d) and 2e) yield the corresponding results for a totally
uncoupled simulation, including the initial steady state
configuration. It is worth observing that coupling between
flow and transport of contaminant is important to provide
reliable predictions of the salt convection and dispersion.
At the same time Figure 2 shows that in transient
simulations the activation of the quasi-Newton scheme within
each time step is not needed and the partially uncoupled
model may be used as well with a large saving of
computational burden.

CONCLUSION

A numerical model for density dependent transport problems
based on a finite element formulation for the flow problem
and on a Eulerian-Lagrangian approach for the salt intrusion
has been developed and implemented in a FORTRAN code called
CUNEO. The code allows for both fully coupled and partially
coupled transient simulations. CUNEO has been successfully
tested against available 2-D numerical solutions of sample
problems and turns out to be computationally efficient as
the simultaneous algebraic equations to be solved for the
potential and the contamination problems are symmetric
positive definite. Thus a solver relying on preconditioned
conjugate gradients has been implemented into the model.
Numerical experiments also show that partial decoupling of
the transient numerical equations does not introduce
appreciable errors in the predicted concentration values.
This adds to the efficiency of the model since the iterative
quasi-Newton type cycle may be avoided with a large saving
of computer resources.

Acknowledgment. This work has been in part supported by the
Italian CNR, Gruppo Nazionale per la Difesa delle Catastrofi
Idrogeologiche, Linea di Ricerca no. 4, and Progetto
Finalizzato "Sistemi Informatici e Calcolo Parallelo".

REFERENCES

1. Baptista A.M., Adams E.E. and Stolzenback K.D., The
    combined use of the finite element method and the method
    of characteristics, Proc. 5th Int. Conf. on Finite

Elements in Water Resour., Springer-Verlag, 1984.

2. Bear J., Hydraulic of Groundwater, Mc Graw Hill, N.Y., 1979.

3. Costantinides A., Applied Numerical Methods with Personal Computers, McGraw Hill, 1987.

4. Diersch H.J., Finite element modelling of recirculating density driven saltwater intrusion processes in groundwater, Adv. Water Resour., 11, 25-43, 1988.

5. Frind E.O., Simulation of long-term transient density-dependent transport in groundwater, Adv. Water Resour., 5, 73-88, 1982a.

6. Frind E.O., Seawater intrusion in continuous coastal aquifer-aquitard systems, Adv. Water Resour., 5, 89-97, 1982b.

7. Galeati G. and Gambolati G., On boundary conditions and point sources in the finite element integration of the dispersion-convection equation, Water Resour. Res., 25, 847-856, 1989.

8. Gambolati G. and Perdon A.M., The conjugate gradients in subsurface flow and land subsidence modelling, "Fundamentals of Transport Phenomena in Porous Media", Bear J. e Corapcioglu M.Y. Eds., Nato ASI Series, Martinus Nijoff Publ., 1984.

9. Huyakorn P.S., Andersen P.F., Mercer J.W. and White Jr. H.O., Saltwater intrusion in aquifers: development and testing of a three-dimensional finite element model, Water Resour. Res., 23, 293-312, 1987.

10. Neuman S.P., Adaptive Eulerian-Lagrangian finite element method for advection-dispersion, Int. J. Num. Methods Engng., 20, 321-337, 1984.

11. Pinder G.F. and Cooper H.H., A numerical technique for calculating the transient position of the saltwater front, Water Resour. Res., 6, 875-882, 1970.

12. Sanford W.E. and Konikow L.F., A two constituent solute transport model for groundwater having variable density, U.S. Geological Survey, Water Resour. Invest. Report 85-4279, 1985.

13. Voss C.I., SUTRA: A finite-element simulation model for saturated-unsaturated, fluid-density-dependent groundwater flow with energy tranport of chemically-reactive simple-species solute transport, U.S. Geological Survey, Water Resour. Investigation 84-4369, 1984.

14. Yeh G.T., On the computation of Darcian velocity and mass balance in the finite element modeling of groundwater flow, Water Resour. Res., 17, 1529-1534, 1981.

An Application of the Mixed Hybrid Finite Element Approximation in a Three - Dimensional Model for Groundwater Flow and Quality Modelling

R. Mose(*,**), Ph. Ackerer(**), G. Chavent(***)

(*) Ecole Nationale des Ingénieurs des Travaux Ruraux et des Techniques Sanitaires, 1 quai Koch, F - 67070 Strasbourg Cédex, France (**) Institut de Mécanique des Fluides, Université Louis Pasteur - URA CNRS 854, 2 rue Boussingault, F - 67083 Strasbourg Cédex (***) Institut National de Recherche en Informatique et en Automatique, Domaine de Voluceau - Rocquencourt, F - 78153 Le Chesnay Cédex

ABSTRACT

A three dimensional groundwater solute transport model, MARCHAL, is developed based on the mixed hybrid finite element (MHFE) method for the groundwater flow equation and the "random walk" method for the contaminant transport equation.

The stochastic particle tracking method is well adapted for contaminant transport modelling because, in application to three dimensional simulations, where discretisation requirements of standard methods usually cannot be satisfied, the "random walk "is the only method which has no numerical dispersion. Almost all groundwater flow codes first calculate the pressure P and then the velocity by a numerical differentiation of the numerically determined scalar field P which is often a source of unacceptable errors. On the other hand, stochastic particle tracking methods need the derivatives of the velocities : for this purpose, classical finite differences or finite element methods are not accurate enough.

The basic idea of the MHFE procedure is to approximate both the pressure P and the velocity simultaneously. This procedure is based on the Raviart-Thomas space's theory (Raviart [1]), which is at the center of any mixed approximation (Meissner [2], Chang [3]).

THE RAVIART THOMAS SPACE ON ONE ELEMENT E

We consider the parallelepiped $E=[0,l_x] \times [0,l_y] \times [0,l_z]$ with faces A_i ($i=1,...$ 6). The lowest order Raviart Thomas space over E is a finite dimensional subspace $\mathbf{X_E}$ of $H(\text{div}, E) = \{\mathbf{q} \in (L^2(E))^2 / \nabla \mathbf{q} \in L^2(E)\}$, where $L^2(E)$ is the set of all functions on E whose space has finite integral, having the following properties :

- $\forall \mathbf{q} \in \mathbf{X_E}$, $\nabla \mathbf{q}$ is constant over E

- $\forall i=1,...6$, $\mathbf{q}.\gamma_E$ is constant over the face A_i, γ_E being the unit exterior normal vector of the face A_i

- any vector field $\mathbf{q} \in \mathbf{X_E}$ is perfectly determined by the knowledge of its flux Q_i through the faces A_i ($i=1,...$ 6).

We use then as basis functions for X_E the vector fields w_j defined by :

$$\int_{A_i} w_j \cdot \gamma E = \delta_{ij}$$

(1)

With this basis we can write any vector $q \in X_E$ as :

$$q = \sum_{j=1}^{6} Q_j w_j$$

(2)

where the six degrees of freedom Q_j are the flux of q through all six faces of E. We shall notate the integral of the scalar product of the basis functions

$$A_{ij} = \int_E w_i w_j$$

(3)

THE MIXED APPROXIMATION

We are going to investigate a way of approximating a pressure field P over a domain Ω of $\mathbf{R}^3$ and the associated velocity field $q = -K.\nabla P$ (when Ω is covered by a grid made of parallelepipeds E) where $\mathbf{K}$ is the permeability tensor. Thereafter we shall assume that the chosen axis are parallel to the principal directions of anisotropy. So that $\mathbf{K}$ becomes :

$$\mathbf{K} = \begin{bmatrix} k_{xx} & 0 & 0 \\ 0 & k_{yy} & 0 \\ 0 & 0 & k_{zz} \end{bmatrix}$$

We shall present in this paragraph the mixed approximation, which consists in approximating simultaneously the pressure P and the velocity field q

On each element E, we approximate P and q by :

$P_E \in \mathbf{R}$ approximation of the mean of P on E

$TP_{Ei} \in \mathbf{R}$ approximation of the mean of P on Ai (i=1,...,6)

$q_E \in X_E$ approximation of $q = -K.\nabla P$ on E

where X_E is the Raviart Thomas space on E described before.

As we have seen that q_E was perfectly known once its fluxes through the six faces were known, the approximation of P and q on E is perfectly determined when one knowns the thirteen degrees of freedom :

- P_E
- TP_{Ei} i=1,...,6
- Q_{Ei} i=1,...,6 (components of q_E)

These thirteen numbers cannot be chosen arbitrarily, as the quantities P and q are related by :

$$q = -K.\nabla P$$

(4)

In order to find the consistency relation which we shall impose to the thirteen unknowns on E, we write (4) in a variational form over the element E. By multiplying scalarly by a test function s_E, we obtain esily the consistency equation for the approximate quantities q_E, P_E (over E) and TP_E (over the aperture of E, ∂E):

$$\int_K \left(K_E^{-1} q_E \right) \cdot s_E = \int_E P_E \nabla s_E - \sum_{j=1}^{6} \int_{Aj} TP_{Ej} \, s_E \gamma_E \qquad \forall s_E \in X_E$$

(5)

Taking for test function s_E successively the basis functions w_j of X_E and using that :

$$\int_E \nabla w_i = 1 \text{ and } q_E = \sum_{j=1}^{6} Q_{Ej} w_j \tag{6}$$

we obtain with the A_{ij} notation (defined before):

$$Q_{E1}\frac{A_{1i}}{k_{xx}} + Q_{E2}\frac{A_{2i}}{k_{xx}} + Q_{E3}\frac{A_{3i}}{k_{yy}} + Q_{E4}\frac{A_{4i}}{k_{yy}} + Q_{E5}\frac{A_{5i}}{k_{zz}} + Q_{E6}\frac{A_{6i}}{k_{zz}} = P_E - TP_{Ei} \quad \forall i=1,\dots,6 \tag{7}$$

We define then :

$$B_E = \begin{bmatrix} \dfrac{l_x}{k_{xx}l_yl_z}\begin{bmatrix} 2 & -1 \\ -1 & 2 \end{bmatrix} & 0 & 0 \\[2em] 0 & \dfrac{l_y}{k_{yy}l_xl_z}\begin{bmatrix} 2 & -1 \\ -1 & 2 \end{bmatrix} & 0 \\[2em] 0 & 0 & \dfrac{l_z}{k_{zz}l_xl_y}\begin{bmatrix} 2 & -1 \\ -1 & 2 \end{bmatrix} \end{bmatrix}$$

We can write now the consistency equation (using a matrix notation) :

$$B_E Q_E = P_E DIV_E^T - TP_E \tag{8}$$

where DIV_E is the elementary divergence matrix ($DIV_E = [1\ 1\ 1\ 1]$).

Let us now write down all the relations to be satisfied by P_E, TP_E, q_E :

- on each element E, P_E, TP_E and Q_E must satisfy the consistency equation (8)

- we must have the continuity of pressures, that means :

$$TP_{E,A} = TP_{E',A} \quad \text{for each internal face A} \tag{9}$$

- we must have the continuity of fluxes, that means :

$$Q_{E,A} = Q_{E',A} \quad \text{for each internal face A} \tag{10}$$

Summarizing, we see that P_E and q_E yield a consistent approximation of P and $q = -K.\nabla P$ as soon as (8), (9), (10) are satisfied.

THE FINITE ELEMENT APPROXIMATION FOR THE CONTINUOUS PROBLEM

The three dimensional movement of groundwater of constant density through porous earth material may be described by the partial differential equation :

$$c\, \partial P/\partial t - \nabla (K.\nabla P) = f \tag{11}$$

where   - c is the specific storage of the porous material (L^{-1})
           - P is the pressure (L)
           - K is the permeability tensor ($L.T^{-1}$)
           - f is the source or sink of water (T^{-1})

On each element we have the equation (11). By using the same variational formulation over each element E, we obtain using the matrix notation :

$$|E|c_E \frac{P_E^n - P_E^{n-1}}{\Delta t} + DIV_E \, Q_E^{n(\theta)} = F_E^{n(\theta)} \quad \forall E, \forall n$$

(12)

where
- F_E = approximation of $\int_E f$
- c_E = approximation of c over E
- $\theta \in [0,1]$ is the parameter of the θmethod used for the time discretization

$$(F_E^{n(\theta)} = (1-\theta) \, F_E^{n-1} + \theta \, F_E^n)$$

- $|E|$ is the volume of the element E.

THE MIXED HYBRID FINITE ELEMENT APPROXIMATION

We choose here as mains unknowns for the solution of the equations (8),(9),(10) and (12) TP_A. A being the faces of the elements E (one pressure value per face) : this choice makes the mixed hybrid approximation.
Of course equation (9) will be automatically satisfied as soon as $TP_{E\,A} = TP_A$
$\forall A, \forall E \supset A$.
For sake of simplicity, we shall suppose throughout the following paragraph that the pressure boundary condition will be $TP_A = 0$. We denote by A^{MH} the set of faces except faces with imposed pressure.
We can now eliminate the auxiliary unknowns P_E and Q_{EA} and we present in this paper just the elliptic case.

THE ELLIPTIC CASE

For the elliptic problem, $c_K = 0$. Then we choose $\theta = 1$ and we obtain an equation for the TP_A unknowns which can be rewrite in a matricial form :

$$(M^{MH} - N^{MH}) \, TP = F^{MH} - G^{MH}$$

(13)

where
- M^{MH} is the mixed hybrid mass matrix

$$M_{A,A'}^{MH} = \sum_{E \supset A \, and \, A'} B_{EAA'}^{-1} \quad \forall A, A' \in A^{MH}$$

(14)

- N^{MH} is the mixed hybrid rigidity matrix

$$N_{A,A'}^{MH} = \sum_{E \supset A \, and \, A'} \lambda_{EAA'} \quad \forall A, A' \in A^{MH}$$

(15)

$$\text{if} \, \lambda_{EAA'} = \frac{\alpha_{EA}\alpha_{EA'}}{\alpha_E} \quad \text{with} \, \alpha_{EA} = \sum_{A' \subset \partial E} B_{EAA'}^{-1} \, \text{and} \, \alpha_E = \sum_{A \subset \partial E} \alpha_{EA}$$

(16)

- F^{MH} = source vector

$$F_A^{MH} = \sum_{E \supset A} \frac{\alpha_{EA}\alpha_{EA}}{\alpha_E} F_E$$

(17)

- G^{MH} = vector of "flux imposed"
$G_A^{MH} = Q_{eA}$ if A is a face with imposed flux

$= 0$ if not

Once TP_A are known, we obtain P_E and Q_{EA} by the local equations.
Remember that we can calculate the velocity at any point of the modelled
zone by :

$$q_E = \sum_{A \supset E} Q_{EA} w_A \quad \text{where } w_A \text{ are the basis functions of } X_E$$

And the normal component of the velocity is <u>continuous</u> from one element
to another.

COMPARISON BETWEEN THE MIXED HYBRID AND THE FINITE ELEMENT APPROXIMATION FOR THE VELOCITY FIELD

We compare now the accuracy of the velocity field calculated by the mixed
hybrid and the classical finite element approximation (e.g. Pinder and Gray
[4]). We use for that an analytical solution (e.g. Bear and Verruijt [5]), which
refers to a system of well in an homogeneous aquifer of infinite extent, with
a uniform flow at infinity (our dataset refers to a region of 29750x29750 m
with a uniform flow at infinity of 10^{-6} m/s in the x direction and a single
well in the point x=y=14750 m pumping 0,1 m^3/s). The network is made of
irregular elements (each square used for the mixed approximation being
replaced by two triangles for the finite element method) with a refined area
around the pumping well. Our grid system is chosen so that at the boundaries
we have the same velocity field with the two approximations and the
analytical solution.

We calculate the velocity field given by the finite element method (more
precisely the value of the velocity magnitude $|q| = (q_x^2 + q_y^2)^{0.5}$) in the
refined area calculated at each centre of mass of the triangular elements. We
calculate at the same points the velocities given by the analytical solution
(figure 1) and the mixed hybrid approximation.

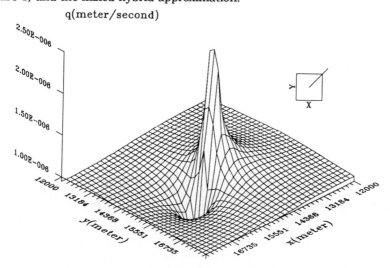

fig.1 Velocity field (analytical solution)

We compare thereafter the relative deviation between the analytical
solution and the FE method (figure 2) and the relative deviation between the
analytical solution and the MH approximation (figure 3).

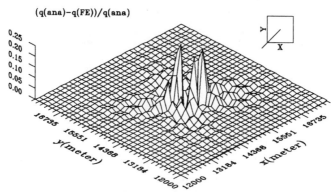

fig.2 Relative deviation between analytical and FE solution

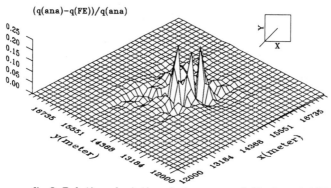

fig.3 Relative deviation between analytical and MH solution

The "interpolation" made in the mixte elements gives as good results as values given at the centre of mass of the finite elements. But the FE method stops there while the mixed hybrid method gives the velocity everywhere else with the continuity of the normal component of the velocity vector from one element to another.

CONCLUSION

This example shows that the mixed hybrid approximation is a powerful numerical tool ; but we don't have to forget that we pay this power with a larger number of unknowns : practically we think that this method is particularly well adapted to a transport model based on a particle tracking method where the continuity of the normal component of the velocity is essential. However it is possible to use the velocity field given by the mixed hybrid approximation for transport model based on finite differences or finite elements method.

REFERENCES

[1] Raviart P.A., and Thomas J.M., A mixed finite method for the second order elliptic problems, in Mathematical Aspects of the Finite Element

Method, Rome 1975, Lecture Notes in Mathematics, Springer-Verlag, New York, 1977.

[2] Meissner U., A mixed finite element model for use in potencial flow problem, Int. J. Numer. Methods Eng., 6, pp.467-473, 1973.

[3] Chiang C.Y., Wheeler M.F. and Bedient P.B., A modified method of characteristics technique and mixed finite elements method for simulation of groundwater solute transport. Water Ressources Research,   Vol. 25 (7), pp.1541-1549, 1989.

[4] Bear J. and Verruijt A. , Modelling groundwater flow and pollution. Reidel Publishing Company, Dordrecht, 414 pp.,1987.

[5] Pinder G.F. and Gray W.G., Finite element simulation in surface and subsurface hydrology. Academic Press, New-York, 295 pp., 1977.

A Practical Application and Evaluation of a Computer-Aided Multi-Objective Decision Algorithm used in the Selection of the Best Decision Algorithm used in the Selection of the Best Solution to a Problem of Salinity Intrusion in Ghana

F.D. Nerquaye-Tetteh, N.B. Ayibotele

Water Resources Research Institute, (Council for Scientific and Industrial Research), P O Box M32, Accra, Ghana

ABSTRACT

An application of a dimensionless multi-objective outranking algorithm enabled a classic evaluation of several alternative solutions to a problem of salinity intrusion in the tidal estuary of the Pra river in Ghana which sometimes affects an important water supply system at Daboasi during the low flow season. This algorithm iteratively employs principles of set theory in a composite analysis of performance matrices and preference graphs under control vectors, threshold stipulations and kernel stability tests. The convergence, capacity, adaptability, limitations and possible improvements and extensions of this algorithm are discussed.

INTRODUCTION

Multi-objective analyses have the capacity and global appeal to provide strategic, systematic and usually convincing basis for decision making in the various facets of the development process. In engineering intervention in particular, they form an exclusive and unsuperseded concept. For example, the multiplicity and complexity of objectives, constraints, foci of interest and decision makers typical of issues commonly arising in the sustainable management and utilization of water resources provide a real task for the analyst to which multi-objective analyses have often proved indispensable.

This has therefore led to a diversity of multi-objective formulations at various levels of sophistication. There are for example the classical generalized optimization methods applied to multiple objectives with n-dimensional vectors of decision variables subject to constraints of the feasible region. Though in such cases there is an unorthodox and intuitive meaning to the concept of optimality due to the necessary tradeoffs between

the objectives to obtain the best solution, Kindler [1]. There
are also composite factorial analyses using various methods
ranging from pure Set Theory to Statistical Analysis which
provide possible approaches to multi-objective analysis.
Chatfield [2], Makarov [3] and Topcheyev [4] have demonstrated
the principles in this group. Finally, other approaches are
also suggested following a conceptual interpretation of Game
Theory with contenders as decision makers using mixed strategies
for evaluating objectives subject to chance constraints. The
basic principles of this category have also been described by
Thomas [5], Taha [6] and others.

The application of such concepts and methodologies are
varied and extensive as can be found in models for:
- the selection of the best action to a problem eg. the
  outranking decision algorithm to be discussed.
- facility siting and location analysis eg. the Regional
  Energy Facility Location Model (RELM) of Church et al
  [7].
- multiple capacity expansion models in capital project
  financing ie. following Loucks et al [8].
- scheduling of projects with multiple objectives of
  achievement ie. following Smith et al [9].
- multi-lattice structure control models eg. as used in
  Binary Automatic Control systems, Emelyanov [10].
All of which, in their extreme form, pose multiple considerations for the analyst.

This paper therefore evaluates one such application in the
selection of the best action to solve an undesirable problem of
seawater intrusion in the Pra river in southwestern Ghana. This
intrusion problem is a nuisance in the low flow season of
February to April whenever it penetrates the river up to an
important water supply intake located 20km inland at Daboasi and
sometimes beyond.   Whenever this occurs, the problem is
transferred to the main consumers in Sekondi-Takoradi (with a
population of 178,256, by the 1984 census) as illustrated by the
conceptual model in Figure 1.   This leads to consequent water
shortage due to plant shutdown, associated risk of consumers
resorting to other water sources of doubtful quality or
consumers having to contend with saline water supply if put
through the distribution system. This situation is aggravated
by the fact that the Daboasi Waterworks is the larger of 2
Plants to the twin cities. It has a direct run-of-the river
abstraction subtending a catchment of 22,730km^2 at Daboasi and
has a present treatment capacity of 6mgd though capable of
expanding to 18mgd. Comparatively the other plant at Inchaban
depends on a 380 million gallon reservoir located on the small
Anunkwari river with a 64km^2 catchment area and now has a
treatment capacity of 4mgd after recent construction works.

The Ghana Water and Sewerage Corporation (GWSC) therefore
commissioned the Water Resources Research Institute (WRRI) under

POINT	ALTERNATIVE SOLUTIONS	RANK	NOTES
P1	E — Breakwater	—	Pi ≈ ith possible point of intervention
P2	E — Channel modification	—	(i ≈ 1,....6)
	X1 — Weir construction	1	where, P1 = at the source
	X2 — Multipurpose reservoir	2	P2 = within the river
P3	X3 — Resited intake	3	P3 = within the plants
	E — Desalination	—	P4 = within the interlinks of the Daboasi
	E — No Action	—	and Inchaban plants
	X4 — Controlled water intake	7	P5 = within the Sekondi — Takoradi
			supply area
P4	X5 — Joint plant management	8	P6 = Among water supply systems in that
			part of the country
P5	X6 — Potable water importation	6	
	X7 — Conservation policy schemes	5	
	E — Redundant distribution storage	—	Rank = the rank of the alternative after analysis
P6	X8 — Intertie of supply systems	4	E = alternative solution eliminated after
			preliminary screening
			Xj = jth alternative considered for analysis
			(j = 1,......8)

Table 1 : Alternative solutions proposed for the Daboasi seawater intrusion problem.

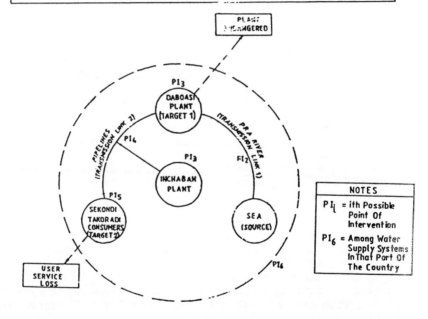

Fig. 1. A Conceptual Model Of The Problem Of Sea Water Intrusion At Daboasi.

a detailed terms of reference to study the frequency and persistence of the problem, identify alternative remedial measures and recommnend the best.   This has been accomplished and the full study report has been presented by Ayibotele & Nerquaye-Tetteh [11].

The crux of this presentation is therefore focused on the aspect of this study which deals with a simultaneous analysis to suggest the best solution among a set of 8 under 11 criteria of technical, economic, social and environmental interest using a dimensionless multi-objective outranking algorithm from set theory. This algorithm is evaluated regarding its convergence, capacity, adaptability, limitations, possible improvements and extensions.

THE STRUCTURE AND APPLICATION OF THE ALGORITHM

This algorithm, taken after Makarov [3], is of a dimensionless structure with applied principles from set and graph theory. It is illustratively discussed here by the Daboasi case.

Definition of Basic Input Structure
The algorithm essentially used a group of basic input sets, matrices and vectors in subsequent iterations which were defined as follows:
- Set $X = \{X_j, j = 1,.....,n\}$ of n compared solutions as in Table 1.
- Set $I = \{I_i, i = 1,.....,m\}$ of m evaluating criteria as detailed in Table 2 and so selected to best evaluate the given set X. For this problem, they were evolved from technical, economic, social and environmental considerations.
- Set $K = \{K_i, i = 1,.....,m\}$ which correspondingly specified ranges of evaluation of elements of set I.

Two control vectors were also defined as follows:
- k^{th} iteration weights vector $\underline{W}^{(k)} = (W_1{}^k \ W_2{}^k \ ....W_m{}^k)$ with elements indicating the relative importance the decision maker accords corresponding criteria of evaluation at the kth iteration.
- k^{th} maximal score difference vector $\underline{S}^{(k)} = (S_1{}^k \ S_2{}^k \ ....S_m{}^k)$ with each element $S_i{}^k$ corresponding to a criterion and calculated as the range of the corresponding scale $K_i{}^k$.

Cross entries for each alternative Xj under each criterion Ii, described as the Jij performance score, were then assigned to indicate the worth of an alternative under a criterion according to its scale Ki. A collection of these cross entries then yielded the matrix $[PSM]^{(k)}$ known as the performance scores matrix for the kth iteration.

Hence simulating the good judgement of a decision maker,

the following sets, vectors and matrix were completed as illustrated in Table 2 for the initial assignments to the Daboasi problem when k=0.

CRITERION EMPHASIS	SETS \ X I \	INITIAL PERFORMANCE SCORES MATRIX [PSM]$^{(0)}$								K	$\underline{S}^{(0)}$	$\underline{W}^{(0)}$
		X1	X2	X3	X4	X5	X6	X7	X8			
Simplicity	I1	4	2	6	6	3	5	6	2	1-10	9	6
Reliability	I2	8	10	6	3	5	5	4	10	1-10	9	9
Ease of Phasing	I3	3	2	4	6	6	7	10	10	1-10	9	7
Low Investment Cost	I4	3	1	5	6	4	6	8	3	1-10	9	10
Low Operating Cost	I5	10	9	4	6	3	2	7	1	1-10	9	10
More Benefits	I6	9	10	7	5	6	5	6	8	1-10	9	10
Permanency	I7	8	10	6	2	5	2	5	8	1-10	9	8
Readiness	I8	6	4	6	7	5	6	6	4	1-10	9	10
Env. Enhancement	I9	8	9	9	5	7	7	8	9	1-10	9	7
Less Negative Impacts	I10	6	6	9	8	5	8	8	7	1-10	9	7
Aesthetic Appeal	I11	9	10	6	8	7	8	8	5	1-10	9	4

NOTES: Set X   - set of alternative solutions {Xj,j=1,...,n}.
        Set I   - set of evaluating criteria   {Ii,i=1,...,m}.
        Set K   - set of evaluating scales   {Ki,i=1,...,m}.
        $\underline{S}^{(0)}$ - initial maximal score difference vector.
        $\underline{W}^{(0)}$ - initial weights vector.
        [PSM]$^{(0)}$ - initial performance scores matrix.

Table 2 : Initial performance scores matrix, scales and vectors used to assess alternatives.

Iterative Reduction Analysis

Once the assignments were made, an iteration was performed as described in the following 3-phased procedure to establish single and composite pairwise preference relations:

Step 1   Preference graphs such as those shown in Figure 2 were drawn for each criterion using the scores such that, an arc was defined only when the performance scores dominated in the direction of comparison ie. Jij(x) > Jij(x') for alternative x compared to x' and preferred to it. When the scores were equal, the arc was not drawn.

Step2 From these graphs, a composite development of simultaneous pairwise comparisons were made over all possible pairs of alternatives and criteria using 2 indices known respectively as concordance and discordance indices subject to threshold stipulations to obtain a single composite graph Gd(X,U) as shown in Figure 3.

Where the concordance index between any 2 compared alternatives x and x' was defined as follows:

$$c(x,x') = \frac{1}{W} \sum_{Ii \in C(x,x')} Wi \qquad (1)$$

and

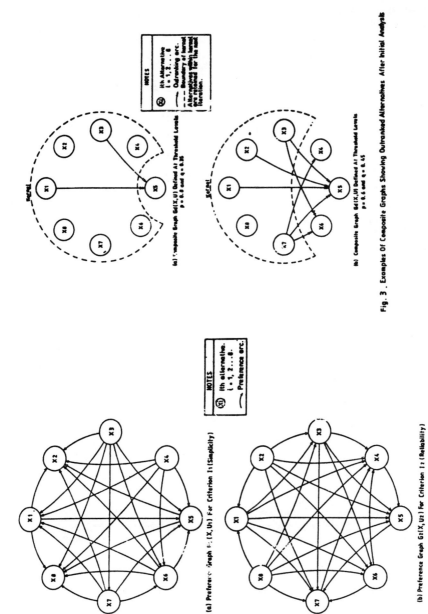

(a) Composite Graph Gd(X,U) Defined At Threshold Levels
p = 0.6 and q = 0.35

(b) Composite Graph Gd(X,U) Defined At Threshold Levels
p = 0.6 and q = 0.45

Fig. 3 . Examples Of Composite Graphs Showing Outranked Alternatives After Initial Analysis

(a) Preference Graph G₁(X, U₁) For Criterion 1 (Simplicity)

(b) Preference Graph G₂(X, U₂) For Criterion 1 2 (Reliability)

Fig. 2 . Examples Of Preference Graphs For Two Criteria In Initial Analysis.

$$C(x,x') = \{Ii:(x,x') \in Ui\} \tag{2}$$

in which,

(x,x') – preference arc between compared pair of alternatives,

$c(x,x')$ – concordance index between x and x',

$C(x,x')$ – class of all criteria under which $Jij(x) > Jij(x')$,

$\in$ – 'belong to',

:   – 'such that',

Ii   – i^{th} criterion,

Ui   – set of feasible arcs (x,x'),

Wi   – weight for criterion Ii in class $C(x,x')$,

W   – sum of all weights, $\sum\limits_{i=1}^{11} Wi$.

and similarly the discordance index was defined as:

$$d(x,x') = \begin{cases} 0 & \text{, if set } D(x,x') \text{ is empty.} \\[1em] \dfrac{1}{Si^{max}} \; \max\limits_{Ii \in D(x,x')} |Jij(x')-Jij(x)| \\[1em] & \text{, if set } D(x,x') \text{ is not empty.} \end{cases} \tag{3}$$

and

$$D(x,x') = \{Ii:(x,x') \notin Ui\} \tag{4}$$

in which,

$d(x,x')$ – discord index between alternatives x and x',

$D(x,x')$ – class of criteria not in $C(x,x')$,

$\notin$ – 'does not belong to',

Si^{max} – maximal scale interval (difference between the extreme scores of the scales taken over all criteria in $D(x,x')$),

$|Jij(x')-Jij(x)|$ – modulus of the score difference.

Essentially the concordance index gave the degree of satisfaction in preferring x to x' when simultaneously compared over all the criteria in which x was preferred to x'. Whilst the discordance index gave the degree of dissatisfaction in choosing option x over x' over all criteria in which it was not preferred to x'.

Thus the computed indices for all possible pairs formed the matrices shown in Table (3).

COMPARED OPTIONS	INITIAL CONCORDANCE MATRIX [C]⁽ᵒ⁾								COMPARED OPTIONS	INITIAL DISCORDANCE MATRIX [D]⁽ᵒ⁾							
x\ x'	X1	X2	X3	X4	X5	X6	X7	X8	x\ x'	X1	X2	X3	X4	X5	X6	X7	X8
X1	-	0.489	0.466	0.545	0.807	0.545	0.545	0.455	X1	-	0.222	0.333	0.333	0.333	0.444	0.778	0.778
X2	0.432	-	0.466	0.545	0.625	0.545	0.545	0.364	X2	0.222	-	0.444	0.556	0.444	0.556	0.889	0.889
X3	0.420	0.455	-	0.466	0.534	0.648	0.467	0.534	X3	0.667	0.556	-	0.222	0.222	0.333	0.667	0.667
X4	0.455	0.455	0.466	-	0.534	0.295	0.114	0.534	X4	0.667	0.889	0.444	-	0.333	0.222	0.444	0.778
X5	0.193	0.261	0.125	0.386	-	0.318	0.102	0.455	X5	0.778	0.667	0.444	0.333	-	0.333	0.444	0.556
X6	0.341	0.455	0.239	0.261	0.500	-	0.102	0.534	X6	0.889	0.889	0.444	0.444	0.333	-	0.556	0.667
X7	0.341	0.455	0.352	0.693	0.693	0.659	-	0.534	X7	0.444	0.667	0.222	0.111	0.111	0.111	-	0.667
X8	0.341	0.273	0.398	0.466	0.545	0.466	0.386	-	X8	1.000	0.889	0.444	0.556	0.222	0.333	0.667	-

(a) Initial Concordance Matrix [C]⁽ᵒ⁾   (b) Initial Discordance Matrix [D]⁽ᵒ⁾

Table 3 : Initial Concordance and Discordance Matrices.

Then the distinct dominance between any alternative pairs was tested by the following stipulation:

$$c(x,x') \geq p \text{ and } d(x,x') \leq q \tag{5}$$

in which,

p, q - threshold levels chosen closer to 1 and 0 respectively.

Whenever it was satisfied, it was said that alternative x 'outranked' x'. A single composite dominance graph Gd(X,U) was then drawn to define all the outranked or complete dominance states among the alternatives at the specified threshold level. Figure (3) shows the composite graph for the initial iteration and illustrates the outranking principle for:

p=0.80 and q=0.35 :

X1, X3 >> X5

in which,
>>   - 'outrank(s)'

p=0.60 and q=0.45 :

X1, X2, X3, X7 >> X5
X7 >> X4
X3, X7 >> X6

in that order.

Thus by varying the threshold levels tolerably, all the weakest alternatives were unambiguously eliminated at that iteration.

Step 3   The retained alternatives were then constituted into a reduced set R which was tested for external and internal stability as follows:

- External stability;

$$\forall x' \in X\text{-}R \quad \diagdown \quad x \in R : \{x,x'\} \in Upq \qquad (6)$$

in which,

R   -   retained set of alternatives.

Upq   -   dominance arcs set in the composite graph $Gd(X,U)$.

$\forall$ -   'for every'

$\diagdown$ -   'there is'

All other variables and symbols were similarly defined as before.

- Internal stability;

$$\forall x',x'' \in R : \{x',x''\} \notin Upq \qquad (7)$$

in which,

x', x" = pair of retained alternatives.

These stability tests respectively confirmed that for every outranked alternative there was at least a retained alternative in set R which outranked it and that none of the retained alternatives was outranked by any other retained alternative at this stage of the analysis. When the tests proved positive, the retained set R formed a kernel which was ready for the next iteration.

The iterations were thus repeated for each retained set R according to these steps, imposing successive convergence till a single best solution was obtained in the kernel. At each iteraion a revision of the source matrices, scales and vectors were made to impose finer and a more stringent constraints for effective evaluation. It was also necessary to confirm eliminated solutions by scanning the assigned performance scores again to note how they were really precipitated from the kernel. In all 3 iterations were made , the details of which can be found in Ayibotele & Nerquaye-Tetteh [11]. The resulting ranking of the alternatives is as shown in Table 1.

Appraisal   On this multi-objective basis therefore, the best alternative was identified as the weir construction near the Daboasi intake to check the salinity intrusion. However a study of the performance matrices in the final iterations, when the comparison became keener, showed that this alternative exhibited weaknesses in the environmental criteria. This was due to the possible ecological imbalance by its construction, the need to contend with a pondage in that area and the possibility of tides occasionally overtopping it from the downstream and rendering the pondage saline etc. The experience of the McConnell weir on the Belfast River Lagan which resulted in anaerobic conditions as related by Wilson [12] is worth noting in this respect. Hence certain design conditions are really necessary to minimise or even eliminate these weaknesses eg. the critical choice of weir height and the use of blind weir with bypasses hydraulically and structurally designed to keep salinity out of the pondage whenever the intrusion comes up the downstream face as well as maintaining a reasonable backwater profile.

EVALUATION AND EXTENSION OF ALGORITHM

Following the description of the structure and application of the algorithm, it can now be evaluated for its efficiency in convergence, capacity and adaptability as well as any limitations in structure or application. Finally, possible improvements and extensions of the algorithm are also suggested.

Efficiency
The efficiency of the algorithm is evaluated according to the following considerations:

Convergence   The convergence in the algorithm is evaluated in due regard of its ability to identify the best solution, the correct ranking of alternatives and minimise the number of iterations till convergence.

For the best solution, it was found that the algorithm is very efficient since that solution should have been cumulatively robust to all the composite comparisons with all other retained alternatives and over all iterations under strictly weighted criteria, threshold stipulations and stability tests.

In ranking, it was found that an alternative was eliminated at any iteration once there was at least one other alternative which outranked it. However this was rather limiting because assuming that the dominating alternative had been temporarily isolated from the analysis, the eliminated alternative might have performed well against some other alternatives. Yet under the present structure, it would have already been ousted. This therefore illustrates a weakness in the efficieny of this present algorithm for precise ranking.

Lastly, the number of iterations for this 11x8 sized problem has been impressive especially because alternatives could be eliminated in batches at each iteration and not necessarily singularly which would have required at least 8 iterations. In some iterations, 3 alternatives could even be eliminated at a stretch by varying the threshold stipulations tolerably.

<u>Capacity</u>   The dimensionless structure of the algorithm is very accomodating since it allows for the inclusion of varied and often non-commensurate or intangible evaluating criteria.

The size of matrix and vectorial arguments is not limiting when solving large problems. Especially the existence of the algorithm also as a Computer Aided Design (CAD) Software really facilitates analysis since the task of the analyst is highly reduced to focusing on the critical choice of assignments rather than a maze of computations.

<u>Adaptability</u> The dimensionless application again renders it adaptable to a wide range of decision problems involving choices or assessments under multiple objectives and constraints.

Its versatility in handling enlarged input sets also makes it adaptable to a good range of problems in different fields of development.

In application, it can even be adaptable to desk top calculations for a considerably sized problem though it is a CAD algorithm. Infact, for the 11x8 sized matrices of the Daboasi case, calculations were conveniently done by an ordinary calculator since a CAD software was lacking and there was not enough time to design and test a program.

<u>Decision Making</u>   Its efficiency in facilitating the decision making process is unmatched and worth commending. It provides satisfaction in substantive decision making in a simple yet quantitative way. The simultaneous provision of a ranking among compared alternatives together with the series of performance characteristics at each iteration, also facilitates posterior evaluation of the ranking obtained.

On-line communication with decision makers for choice of inputs can even be allowed for, by interactive analysis to get actual simulations of their decisions.

Limitations

<u>Structural</u>   The discordance index is highly influenced by the worse deviation under a single criterion to which an alternative may be very weak compared to another. This actually is one of the causes of the weakness in the convergence efficiency for ranking.

<u>Influence of poor or misguided technical judgement</u> The assignments to the algorithm require some substantial skill by the analyst to simulate an 'informed decision maker' since the algorithm is quite sensitive to poor or misguided judgements which reflect in the choice of assignments and threshold stipulations.

Uncertainty in exact simulation of the decision maker's
judgement   There is some uncertainty in simulating best assignments to the model inputs by different would be decision makers.
There is also the difficulty in having final decision makers
actually do this sort of analysis themselves or have it done
under their jurisdiction especially when they are not very
informed.

Possible Improvements and Extensions of Algorithm

Improvements   The problem in the convergence efficiency of ranking could be resolved by using repeatedly the robust feature of
the algorithm at obtaining the best solution. Thus, once a best
solution is obtained for the complete set of alternatives, that
alternative can be isolated and the remaining alternatives taken
through the whole analysis again to yield the second best and so
on. In which case the CAD version will definitely be necessary.

Conversely, when an alternative is outranked during an
iteration and therefore has to be eliminated, the dominating
alternative is rather isolated and the analysis continued till a
single weakest solution is obtained in the kernel which is then
isolated from further analysis.   Simultaneously, as a useful
variation, the number of times it was outranked can also be
noted to show how weak it really is. The reduced set can then
be similarly analysed till the complete inverse ranking is
obtained.

The structural limitation in the computation of the
discordance index can also be modified by the use of a weighted
sum of maximal deviations over the class of criteria $D(x,x')$ for
which the arc set (x,x') is not defined.

The influence of poor or misguided technical judgement and
uncertainty in the exact simulation of the decision maker's
judgement can be addressed at the primary level by employing a
skilled and informed analyst to preferably use a CAD software
for thorough analysis. At the secondary level, principles of
Parametric Sensitivity Analysis as outlined by Vajda [13] can be
used to organise the investigation of variants. The large
influence of parameters and their sensitivity to a range of
variation can be assessed to confirm the robustness of decisions
and choices. By these, checks can be put in the algorithm to
control the input of infeasible assignments which could lead to
contradictions. Then at the tertiary level, actual simulations
can be done using real decision makers to provide assignments to
the algorithm through interviews or survey questionaires which
can then be investigated using principles of Parametric
Sensitivity Analysis and Game Theory on decision makers with
different strategies and requirements.

Extensions Structurally, the algorithm can therefore be extended in sophistication by the suggested improvements. A plot of

the path of convergence and its sensitivity to parameters can be obtained and controlled.

In application, there are also several possibilities for extension. For example, the transposition of the source matrices to compare criterion impacts across a set of projects of varied importance can help identify worse impacts in project planning. Its possible application in several fields of study portrays the immense potential.

CONCLUSIONS AND RECOMMENDATIONS

It can now be agreed that multi-objective analyses have their exclusive place in the decision making process as illustrated by the practical application of this algorithm to the salinity intrusion problem at the Daboasi Waterworks in Ghana.

The presentation and evaluation of this algorithm has primarily identified and focused on its efficiencies, limitations, improvements and extensions. Most of the structural weaknesses highlighted have been suggestively corrected and it is hoped that in further application, more avenues of improvement will be demonstrated. The possible use of principles of Parametric Sensitivity Analysis and Game Theory can lead to even more classic and efficient variants of this algorithm.

The extension of this algorithm cannot be narrowly charted because it has a real potential.

REFERENCES

1.   Kindler, J. On the Multi-Objective Framework of Environmentally Sound and Sustainable Water Resources Management, International Journal of Water Resources Development, Vol 4, No 2, pp. 117-123, 1988.

2.   Chatfield, C. The Design and Analysis of Experiments. Chapters 10 & 11, Statistics for Technology, pp. 224-287, Chapman and Hall, London and New York, 1983.

3.   Makarov, I.M. Multi-Objective Decision Analysis for Large System Computer-Aided Design. Chapter 9, Management and Control in Large Systems, (Ed. Voronov, A.A.), pp. 242-257, Mir Publishers, Moscow, 1986.

4.   Topcheyev, Y.I. Multi-Attribute Feasibility Evaluation of Prospective System Designs. Chapter 10, Management and Control in Large Systems, (Ed. Voronov, A.A.), pp. 258-270, Mir Publishers, Moscow, 1986.

5.   Thomas, L.C. Games, Theory and Applications, Ellis Horwood Limited, Chichester, 1984.

6. Taha, H.A. Decision Theory and Games. Chapter 10, Operations Research - An Introduction, pp. 322-359 Collier-Macmillan Publishers, London and New York, 1976.

7. Church, R., Cohon, J. and ReVelle, C. Energy Facility Siting by multi-objective location analysis, in Modeling and Simulation. Volume 9 Part 1: Energy and Power System Modeling - Ecological and Biomedical Modeling (Ed. Vogt, W.G. and Mickle, M.H.), pp. 1-10, Proceedings of the Ninth Annual Pittsburgh Conference, Pittsburgh, USA, 1978. Instrument Society of America, Pittsburgh, Pennsylvania, 1978.

8. Loucks, D.P., Stedinger, J.R. and Haith, D.A. Water Resouces Planning Under Uncertainty. Chapter 3, Water Resources Systems Planning and Analysis, pp. 94-191, Prentice-Hall Inc., New Jersey, 1981.

9. Smith, A.A., Hinton, E. and Lewis, R.W. Critical Path Analysis. Chapter 8, Civil Engineering Systems Analysis and Design, pp. 276-308, John Wiley and Sons, Chichester, New York, 1983.

10. Emelyanov, S.V. Binary Automatic Control Systems, Mir Publishers, Moscow, 1987.

11. Ayibotele, N.B. and Nerquaye-Tetteh, F.D. Studies into the Problem of Saline Intrusion at the Sekondi-Takoradi Water Supply Intake Works on the Pra River at Daboasi, Draft Final Report, Water Resources Research Institute, (Council for Scientific and Industrial Research), Accra, 1989.

12. Wilson, J., Upgrading the Lagan, World Water, Vol 10 No 5., pp. S14-15, 1987.

13. Vajda, S. Algorithms. Chapter 2, Linear Programming Algorithms and Application, pp. 16-49, Chapman and Hall, London and New York, 1981

SECTION 6 - CHEMICAL REACTION PROBLEMS IN POROUS MEDIA

Numerical Simulation of the Transport of Adsorbing Solutes in Heterogeneous Aquifers

A.J. Valocchi

Department of Civil Engineering, University of Illinois at Urbana - Champaign

INTRODUCTION

Development of scientifically sound management strategies for protecting groundwater resources must be based upon a firm understanding of contaminant behavior in natural subsurface environments. Realistic pollution scenarios share two important characteristics. First, surface-chemical reactions such as adsorption are important due to the enormous solid-water interfacial area of natural porous media. Second, groundwater aquifers exhibit significant small-scale, three-dimensional spatial variability in permeability.

Adsorption reactions during porous media flow can be described using either an equilibrium or kinetic approach. Most studies to date have assumed that the kinetic processes are extremely fast and hence the adsorption reactions are governed by conditions of local chemical equilibrium; this is often denoted as the "local equilibrium assumption" (LEA). The LEA results in significant conceptual as well as mathematical simplification. There have been several recent theoretical [14] and laboratory [4] investigations which have focused upon the effect of nonequilibrium adsorption processes on solute transport. These studies show that kinetic adsorption effects cause increased spreading and failing of contaminant plumes and breakthrough curves. This behavior was exploited by Valocchi [14] in the development of quantitative criteria to determine conditions for which the LEA is valid for homogeneous soils. The criteria are based upon quantification of the influence of kinetics relative to the other mechanisms that cause plume spreading (e.g., molecular diffusion, mechanical dispersion, time variation of the input concentration).

For nonreactive solutes, unique field-scale effects are well-known to be caused by heterogeneities in the aquifer permeability [5,7]. Spatial variability manifests itself upon contaminant plume behavior in several ways; for example, field-scale dispersion coefficients are much larger than their laboratory counterparts and they demonstrate a "scale effect" whereby their value increases with distance from the pollutant source. These phenomena are often referred to collectively as "macrodispersion". For the case of adsorbing solutes, both macrodispersive and kinetic processes will contribute to the overall spreading of the solute plume, and therefore LEA validity will depend upon the heterogeneous structure of the aquifer. We have been utilizing a numerical modeling approach to examine the issue of LEA validity for transport in randomly heterogeneous aquifers. This paper reviews the computational techniques we have developed and reports some recent results.

DESCRIPTION OF NUMERICAL MODELS

First, we give a brief overview of the steps involved in the numerical modeling experiments. The hypothetical problem domain is illustrated schematically in Figure 1 for the two-dimensional case. The aquifer is discretized into blocks and the boundary conditions correspond to mean one-dimensional steady flow from left to right. The numerical experiments consist of the following steps:

(1) Assign hydraulic conductivity (K) values to each grid block. This is accomplished by generating a sample realization of lognormally distributed, stationary random field with isotropic exponential spatial covariance. The generation is performed using the computer code TURN developed by Tompson et al. [13].

(2) Determine the steady state velocity field by solving the groundwater flow equation for the hydraulic conductivity field realization generated in step (1). To accomplish this, we developed a highly efficient finite difference - conjugate gradient technique; this technique is described in greater detail by Meyer et al. [11] and is discussed further below.

(3) Simulate the transport of a theoretically adsorbing contaminant instantaneously input at the source location shown in Figure 1. To accomplish this, we developed a special particle tracking technique which is described in greater detail by Valocchi and Quinodoz [16] and is discussed further below.

(4) Analyze and interpret the results of the transport simulation. We have focused upon spatial moment analysis, that is, the temporal behavior of the lower-order spatial moments of the traveling plume.

(5) Repeat steps (3) and (4) for a range of adsorption reaction rates.

(6) Repeat steps (1) through (5) for varying degrees of aquifer

heterogeneity (e.g., for **varying values** of the standard deviation of the ℓn-K field).

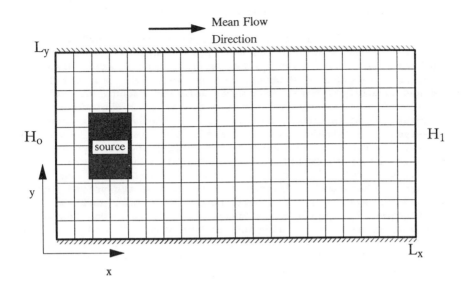

Figure 1. Schematic illustration of problem domain used in the numerical experiments (two-dimensional case).

Groundwater Flow Model

The governing equation for steady-state groundwater flow through a three-dimensional, heterogeneous aquifer is

$$\nabla \cdot K(x,y,z)\ \nabla h = 0 \tag{1}$$

where h is the piezometric head and K(x,y,z) is the hydraulic conductivity. We discretize equation (1) using a standard seven-point, block-centered finite technique equivalent to that implemented in the USGS MODFLOW model [9]. The hydraulic conductivity is constant within a grid block but varies from block to block; block conductivity values are assigned by the random field generation code. The boundary conditions correspond to one-dimensional mean flow (see Figure 1); constant head (Dirichlet) conditions apply on the two opposite faces in the direction of flow, and no flow (Neumann) conditions apply on the other faces.

Extremely fine spatial resolution is necessary to describe accurately the spatial variability, and one-million grid blocks are easily needed in three-dimensions. The need for a large number of grid points has been discussed by Tompson et al. [13] and Ababou et al. [1]. Such large problems are only computationally feasible on advanced scientific supercomputers having special high-speed architecture. We have developed and implemented our flow code on a vector machine (Cray 2) and a vector-parallel machine (Alliant FX/8); these machines are operated at the University of Illinois under the auspices of the National Center for Supercomputing Applications (NCSA).

The discretized form of the groundwater flow equation (1) is a large, sparse, symmetric and positive definite system of linear algebraic equations, Ax=b. The matrix A has a seven-diagonal structure, as shown by McDonald and Harbaugh [9]. For the large systems considered here, iterative solution techniques must be utilized. A powerful and popular iterative technique is the conjugate gradient (CG) method [3]. Standard CG, however, converges too slowly to be practically useful. Favorable convergence rates can be achieved by multiplying the system by a so-called preconditioning matrix C. That is, the system $CAx = Cb$ is solved, and this is called the preconditioned conjugate gradient method (PCG). Favorable properties of the precondioning matrix, C, are: (1) it should approximate A^{-1}; and (2) the calculation of C should be computationally "easy". One choice is $C = D^{-1}$, where D is a diagonal matrix with $D_{ii} = A_{ii}$. This preconditioner is easy to implement, and we have implemented it by diagonally scaling the original system in the manner described by Meyer et al [11]; we refer to this method as DSCG.

The most popular choice for C^{-1} is the incomplete Cholesky factorization of the matrix A [10]. This method is referred to as ICCG. ICCG has been demonstrated to be highly effective for solution of the groundwater flow problem on traditional serial computers [6,8]. However, as discussed by Meyer et al. [11], the ICCG algorithm includes recursive operations which prevent its vectorization. In fact, for several test problems run on the Cray 2, ICCG took more CPU time than DSCG. In our model, we have implemented polynomial preconditioning (PPCG), which does not entail any recursive operations and thus is well suited to vector and vector-parallel machines. In this method the preconditioning matrix C is taken to be a polynomial in the matrix A, i.e., $C = P_m(A)$, where m is the degree of the polynomial. We use for P_m the Chebyshev polynomials. Further details are given by Meyer et al. [11] and Ashby [2].

We have tested our code on a series of three-dimensional test problems that are described more fully by Meyer et al. (1989). The problem domain is a cube of length $L = 25\lambda$ (λ is the correlation length of the $\ln$-K field). Table 1 shows the results for the largest problem with 980,000 grid blocks which correspond to 4 grid points per correlation length, and for $\sigma_y = 2.3$ (σ_y is the standard deviation of the $\ln$-K field). In Table 1, m is the degree of the preconditioning polynomial, and m = 0 corresponds to DSCG; also $\kappa(CA)$ is an estimate of the condition number of the preconditioned matrix. The timings in Table 1 are for the Cray 2 and are approximately 15 times faster than those reported by Meyer et al. [11] which are for the Alliant FX/8. Our results

Table 1.   CG Times on the Cray 2 for the 3D
Test Problem with N = 980,000 and σ_y = 2.3

m	Iterations	Seconds	DSCG/PPCG	κ(CA)
0	1127	350.73	1.00	45,093
2	384	217.40	1.61	5,019
4	236	195.26	1.80	1,808
6	170	185.93	1.89	924
8	152	207.19	1.69	559
10	131	213.38	1.64	372

illustrate the importance of choosing the proper iterative method
and preconditioner.   To emphasize this point, we compare our
results with those of Ababou et al. [1] who used the strongly
implicit procedure (SIP) to solve a problem similar to that in
Table 1.   The PPCG time is more than fifteen times faster than
SIP.   The reason for PPCG's superiority is this: it is better
suited to the vector architecture of the Cray and Alliant than is
SIP, which is highly recursive.

Following solution of the groundwater flow equation (1) for
h, the pore-water velocity is calculated from Darcy's Law as

$$\mathbf{v} = - \frac{1}{n} K(x,y,z) \; \nabla h \qquad (2)$$

where $\mathbf{v}$ is the pore-water velocity vector and n is the porosity,
which we assume is constant.   The block-centered difference
approximation to (1) gives h at the center of each grid cell.
Therefore, the first-order derivatives of h appearing in (2) are
approximated by a centered finite difference about the midpoint of
a face separating two adjacent cells.   In order to conserve mass,
K(x,y,z) in (2) is taken to be the harmonic mean of the hydraulic
conductivity values in the two adjacent cells.

Contaminant Transport Model
The   groundwater   flow   model   yields   the   three-dimensional
distribution of hydraulic head and pore-water velocity.   The
velocity field is then input into a solute transport model which
includes the processes of advection and adsorption.   The governing
solute transport equation is:

$$\frac{\partial c}{\partial t} + \frac{\partial s}{\partial t} + \nabla \cdot \mathbf{v}c = 0 \qquad (3)$$

where c is the dissolved phase concentration, s is the adsorbed
phase concentration, and $\mathbf{v}$ is the pore-water velocity vector
defined by (2).   Although local scale hydrodynamic dispersion is
neglected in (3), macrodispersive mixing will result from the
heterogeneous velocity field.   In addition to the transport
equation (3), an adsorption rate law is also required.   We use the
first-order, linear reversible rate law given by

$$\frac{\partial c}{\partial t} = k_f \ c - k_r \ s = k_r \ (K_d \ c - s) \qquad (4)$$

where k_f and k_r are the forward and reverse adsorption rate coefficients, respectively, and the equilibrium distribution coefficient, K_d, can be expressed as $K_d = k_f/k_r$.

Because of the fine spatial resolution required to capture accurately the effects of small-scale heterogeneity, conventional finite difference or finite element approximations to (3) and (4) are computationally infeasible for three-dimensional problems. Therefore, we developed a highly efficient random walk-particle tracking technique based upon a stochastic analog to the reaction equation (4). The continuous solute concentration field is approximated by a large number of particles, each of which can exist in either the dissolved or adsorbed phase. A stochastic model is used to compute phase transition and the fraction of an arbitrary time step spent by a particle in the dissolved (i.e., mobile) phase. At each time step, a particle is advected only for that portion of time spent in the dissolved phase. Although the technique cannot be vectorized easily, it is highly concurrent and thus well suited to parallel computers. To date, the model is restricted to two space dimensions and spatially uniform reaction parameters. Further details are given by Valocchi and Quinodoz [16].

EXAMPLE RESULTS FOR A TWO-DIMENSIONAL PROBLEM

Comparison of Nonreactive Results with Stochastic Theory
For the case of nonreactive solute transport in randomly heterogeneous porous media, there are some analytical stochastic results reported by Gelhar and Axness [7] and Dagan [5]. If we assume ergodicity, we can compare our numerical results, which consist of spatial averages of a single realization, to the analytical results, which are in terms of ensemble averages. The two-dimensional problem domain considered here is similar to that shown in Figure 1 with $\Delta x = \Delta y = 1$, $L_x = 1000$, and $L_y = 500$.

The first item to check is the random K field generated by the program TURN. In this example, we use the following input statistics for TURN: $\lambda/\Delta X = 5$, K_G (geometric mean K) = 1, and $\sigma_y^2 = 0.5$. The sample output statistics were $K_G = 0.9997$ and $\sigma_y^2 = 0.4881$; also the sample and theoretical values of λ and the spatial covariance function agreed very closely. In agreement with the results of other investigators [1,13], we have found the numerical results to be sensitive to both the number of grid points per ℓn-K correlation length and the number of ℓn-K correlation lengths contained within the domain.

Since it is the velocity field heterogeneity that gives rise to macrodispersion, it is crucial to check the statistics of the velocity field calculated by the groundwater flow model. In our example problem, the mean value of v_x should theoretically be equal to $K_G(H_0-H_1)/(Ln)$, which equals 5×10^{-3} for the parameters we have used. The sample mean of the computed v_x field was 4.94 $\times 10^{-3}$, in close agreement with the theory. The sample mean of the computed v_y field was -7.6×10^{-6} whereas the theoretical value should be zero. Theoretical values for the velocity field variance and spatial covariance function have recently been given by Rubin [12]. The theoretical variance of v_x is $(U^2\sigma_y^2)(3/8)$ and v_y is $(U^2\sigma_y^2)(1/8)$, where U is the mean longitudinal velocity (i.e., the mean of v_x). The sample variance of the computed v_x field is 4.48×10^{-6}, whereas the theoretical value equals 4.47 $\times 10^{-6}$ if the sample statistics of U and σ_y^2 are used, or 4.69 $\times 10^{-6}$ if the theoretical values of U and σ_y^2 are used. The sample variance of the computed v_y field is 1.69×10^{-6} which is somewhat larger than the theoretical value.

Finally, we check the results of the particle tracking model. We compute the mean and variance of the particle positions in the longitudinal (x) direction and compare these values with the theoretical results of Dagan [5]. We have always found that the time rate of change of the mean particle position equals the mean longitudinal velocity (U), in agreement with theory. In contrast to the case for the mean particle position, we have found our results for the particle position variance to be highly sensitive to the size of the input zone. Figure 2 shows a plot of dimensionless variance, defined as the longitudinal variance of the particle position divided by $\sigma_y^2\lambda^2$, versus the dimensionless time, T*, defined as Ut/λ (i.e., the mean particle position at time t divided by the ℓn-K correlation length). The solid line is the theory of Dagan; the computed results correspond to a reactangular input zone from $y = 50$ to 450 and $x = 100$ to 200 (input A), and $x = 300$ to 400 (input B). One thousand particles were used, but the results were essentially the same for ten thousand. The fluctuation of the computed results about the theory appears reasonable since one would expect any single realization to differ from the ensemble.

Kinetically Adsorbing Solutes

The particle tracking experiments shown in Figure 2 were repeated for the case of kinetically adsorbing solutes. Although no detailed results will be presented here, we have found that kinetics increases the spatial variance of the contaminant plume relative to the case of equilibrium adsorption. For large travel times (T* > 50) and relatively fast reaction rate ($k_r t > 10$), the impacts of adsorption kinetics and spatial heterogeneity upon the plume variance appear additive. In this case the impact of kinetics upon the longitudinal variance of the plume, denoted as VAR_x^K, follows the theory presented by Valocchi [15] for the simplified case of stratified aquifers with spatially uniform reaction parameters. That is,

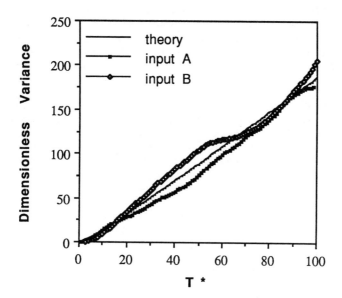

Figure 2.   Comparison between theoretical and numerical
results for the dimensionless longitudinal
variance of particle positions.

$$VAR_x^K = 2 \frac{U^2}{R^3} \frac{K_d}{k_r} t \qquad (5)$$

where $R = 1 + K_d$ is the retardation factor.   Therefore,   the
overall longitudinal plume variance will be

$$VAR_x = VAR_x^K + VAR_x^H \qquad (6)$$

where VAR_x^H is the variance due to the spatial heterogeneity of the
K field.   Dagan [5] has shown (given the many assumptions inherent
in his stochastic theory) that $VAR_x^H = 2\sigma_y^2 \lambda(U/R) t$.   Thus equation
(6) provides a basis for quantifying the importance of adsorption
kinetics relative to spatial heterogeneity.   For the case of early
travel time (which may be more important in practice), we have
found the impacts of kinetics and heterogeneity to be nonadditive
and to be interrelated in a complex way.

CONCLUSIONS

Large-scale numerical simulation models of reactive solute
transport provide an important research tool for investigating
fundamental aspects of field-scale transport processes.   The

problem studied here focuses upon the importance of kinetic adsorption reactions in a heterogeneous flow field. Successful development of large simulation models must take advantage of computational methods particularly suited for supercomputer architectures.

ACKNOWLEDGMENTS

This paper is based upon research supported in part by the U.S. Geological Survey under grant 14-08-0001-G1299 and through the State Water Resources Research Institute program. This work utilized the Cray 2 system at the National Center for Supercomputing Applications at the University of Illinois at Urbana-Champaign. I would like to thank M. Holst and P. Meyer for performing the computations reported in Table 1. I am also grateful to H. Quinodoz who provided the results shown in Figure 2.

REFERENCES

1. Ababou, R., Gelhar, L.W. and McLaughlin, D. Three-Dimensional Flow in Random Porous Media, Tech. Rept. No. 318, Parsons Laboratory, MIT, Cambridge, MA 02139, 1988.

2. Ashby, S.F. Polynomial Preconditioning for Conjugate Gradient Methods, Dep. Comput. Sci. Rep. UIUCDCS-R-87-1355, Univ. of Illinois at Urbana-Champaign, 1987.

3. Ashby, S.F., Manteuffel, T.A. and Saylor, P.E. A Taxonomy for Conjugate Gradient Methods, Dep. Comput. Sci. Rep. UIUCDS-R-88-1414, Univ. of Illinois at Urbana-Champaign, 1988.

4. Bouchard, D.C., Wood, A.L., Campbell, M.L., Nkedi-Kizza, P. and Rao, P.S.C. Sorption Nonequilibrium during Solute Transport, J. Contam. Hydrol., 2, 209-223, 1988.

5. Dagan, G. Theory of Solute Transport by Groundwater, Ann. Rev. Fluid Mech., 19, 183-215, 1987.

6. Gambolati, G. and Perdon, A. The Conjugate Gradients in Flow and Land Subsidence Modeling, Fundamentals of Transport Phenomena in Porous Media, (Eds. Bear, J. and Corapcioglu, Y.), pp. 953-984, Martinus Nijoff, Dordrecht, Netherlands, 1984.

7. Gelhar, L.W. and Axness, C. Three-Dimensional Stochastic Analysis of Macrodispersion in Aquifers, Water Resour. Res., 19(1), 161-180, 1983.

8. Kuiper, L.K. A Comparison of Iterative Methods as Applied to the Solution of the Nonlinear Three-Dimensional Groundwater Flow Equation, SIAM J. Sci Stat. Comput., 8(4), 521-528, 1987.

9. McDonald, M.G. and Harbaugh, A.W. A Modular Three-Dimensional Finite Difference Ground Water Flow Model, USGS, 1984.

10. Meijerink, J.A. and van der Vorst, H.A. An Iterative Solution Method for Linear Systems of Which the Coefficient Matrix is a Symmetric M-Matrix, Math. Comput., 21(137), 148-162, 1977.

11. Meyer, P.D., Valocchi, A.J., Ashby, S.F. and Saylor, P.E. A Numerical Investigation of the Conjugate Gradient Method as Applied to Three-Dimensional Groundwater Flow Problems in Randomly Heterogeneous Porous Media, Water Resour. Res., 25(6), 1440-1446, 1989.

12. Rubin, Y. Stochastic Modeling of Macrodispersion in Heterogeneous Porous Media, Water Resour. Res., 26(1), 133-142, 1990.

13. Tompson, A.F.B. Ababou, R. and Gelhar, L.W. Implementation of the Three-Dimensional Turning Band Random Field Generator, Water Resour. Res., 25(10), 2227-2244, 1989.

14. Valocchi, A.J. Validity of the Local Equilibrium Assumption for Modeling Sorbing Solute Transport through Homogeneous Soils, Water Resour. Res., 21(6), 808-820, 1985.

15. Valocchi, A.J. Theoretical Analysis of Deviations from Local Equilibrium during Sorbing Solute Transport, through Idealized Stratified Aquifers, J. Contam. Hydrol., 2, 191-207, 1988.

16. Valocchi, A.J. and Quinodoz, H.A.M. Application of the Random Walk Method to Simulate the Transport of Kinetically Adsorbing Solutes, in Groundwater Contamination (Ed. Abriola, L.M.), pp. 35-42, IAHS Publ. No. 185, 1989.

A Eulerian-Lagrangian Localized Adjoint Method for Reactive Transport in Groundwater

M.A. Celia, S. Zisman

Water Resources Program, Dept. of Civil Engineering and Operations Research ,Princeton University, Princeton, NJ 08544 USA

ABSTRACT

The Eulerian-Lagrangian Localized Adjoint Method (ELLAM) has recently been developed for advection-diffusion transport equations. The method is a generalization of traditional Eulerian-Lagrangian Methods (ELM's) in that it subsumes many ELM formulations. It also has the important property of demonstrable mass conservation, which is due to its systematic treatment of boundary conditions. The ELLAM approximation may be extended to the case of reactive transport. All advantages of the original ELLAM formulation are maintained.

INTRODUCTION

Reactive transport in groundwater systems is an increasingly important problem in water resources. It is now widely recognized that both chemical and biological processes can significantly affect the subsurface transport of many contaminants. Efficient and robust numerical simulators are required to effectively analyze reactive transport problems.

It is well known that numerical solutions for transport are often inadequate for problems of practical interest. These include the important case of advection-dominated transport. To treat this case, many approximations have been developed. As discussed in some detail by Celia et al. [9], two broad classes of approximations may be identified. The first, referred to as the class of Optimal Spatial Methods (OSM's), contains approximations rooted in the idea of using a semi-discrete approximation based on an optimal approximation to the spatial derivatives alone. Examples of these approximations include the original finite difference approximation of Allen and Southwell [1], the quadratic Petrov-Galerkin (P-G) method of Christie et al. [11], the streamline upwind P-G method of Hughes and coworkers ([4],[18]), the exponential P-G method of Hemker [14], and the method of Optimal Test Functions (OTF) presented by Celia, Herrera, and coworkers ([6],[7],[8]). The essential equivalence of these methods was shown by Bouloutas and Celia [3]. Performance of these methods is characterized by large time truncation errors and Courant number limitations.

A second broad class of approximations may be identified as Characteristic Methods (CM's). The approximations in this class have as their common element a Lagrangian treatment of the advection process. This is based on recognition of the potential importance of the secondary characteristics [10] in the transport equation. Again there are many examples from the literature, including Eulerian-Lagrangian Methods ([2],[19]), Method of Characteristics [20], Modified Method of Characteristics ([13],[21]), and Operator Splitting Methods ([12],[22]). These methods usually provide significant reduction in time truncation error as well as alleviation of the Courant number restrictions. However, one problem that has been persistent is an essentially ad-hoc treatment of boundary conditions, and none of these methods appears to possess the conservative property.

The Optimal Test Function approach [7] is based on special definition of test functions, based on solutions to the homogeneous adjoint operator equation. Such a procedure is referred to as a Localized Adjoint Method (LAM). Because the LAM approach is general [16], it is not restricted to only the spatial dimension, and can therefore be extended beyond the OTF semi-discrete approximation. In particular, by judicious definition of space-time test functions, still based on homogeneous adjoint solutions, the LAM formulation may combine with the CM philosophy to yield the Eulerian-Lagrangian LAM, or ELLAM. This has been done by Celia and coworkers [9], for the case of nonreactive transport. In that reference, the treatment of boundary conditions was illustrated in detail, and the ELLAM approximation was shown to possess the conservative property. This paper will develop the ELLAM approximation for the more general case of reactive transport. Example solutions will be presented to indicate the behavior of the method and to illustrate the importance of proper treatment of boundary conditions.

THE ELLAM APPROXIMATION FOR REACTIVE TRANSPORT

The general philosophy of the LAM approach is based on the Algebraic Theory of Numerical Methods presented by Herrera ([15],[16],[17]). The general methodology will not be presented here. It has been developed for ordinary differential equations by Herrera, Celia, and coworkers ([5],[17]). Application to multi-dimensional steady state equations may be found in [6], while application in the context of semi-discretization of transient partial differential equations (which leads to the OTF approximations) may be found in ([7],[8]). The development of ELLAM for nonreactive transport equations is given in [9]. This presentation will focus only on the specific example of reactive transport.

In this regard, consider the following model reactive transport equation,

$$\mathscr{L}u \equiv \frac{\partial u}{\partial t} + V\frac{\partial u}{\partial x} - D\frac{\partial^2 u}{\partial x^2} + Ku = f(x,t), \quad \begin{array}{l} x \in \Omega_x = [0,\ell] \\ t \in \Omega_t = [0,T] \end{array} \tag{1}$$

subject to any combination of boundary conditions (first, second, or third type) as well as suitable initial conditions. The adjoint operator associated with the operator $\mathscr{L}$ is

$$\mathscr{L}^* w = -\frac{\partial w}{\partial t} - V\frac{\partial w}{\partial x} - D\frac{\partial^2 w}{\partial x^2} + Kw \tag{2}$$

The LAM approach is initiated by writing the weak form of equation (1). Let $w(x,t)$ be a test function (whose definition will be determined as part of the solution procedure, as dictated by LAM). Then the weak form of equation (1) is

$$\int_0^T \int_0^\ell (\mathscr{L}u-f) \, w(x,t) \, dx \, dt = 0 \tag{3}$$

Let the space-time domain $\Omega \equiv \Omega_x \times \Omega_t$ be discretized into E space-time subintervals, or elements, Ω_e ($e=1,2,\ldots,E$). Presuming sufficient continuity (in this case $u \in \mathbb{C}^1[\Omega_x] \times \mathbb{C}^0[\Omega_t]$ and $w \in \mathbb{C}^{-1}[\Omega]$), the integral of equation (3) may be written as a simple sum of elemental integrals, viz.,

$$\sum_{e=1}^E \int_{\Omega_e} (\mathscr{L}u-f) \, w(x,t) \, dx \, dt = 0 \tag{4}$$

The key to LAM methods is repeated application of multi-dimensional integration by parts (Green's Theorem) until all derivatives are transferred from u to w. This produces the adjoint operator acting on w. Each element integral in equation (4) is therefore replaced by a boundary integral (evaluated around the boundary of the element) and an interior element integral that has as part of its integrand the adjoint operator acting on w. The philosophy of LAM is to choose $w(x,t)$ such that it satisfies the homogeneous adjoint equation locally, that is, within each element. This is the criterion for defining the test functions w; it serves to eliminate interior element integrals, leaving only boundary evaluations.

For ordinary differential equations, the solution space of the homogeneous adjoint equation ($\mathcal{L}^* w = 0$) is finite-dimensional. Therefore test functions may be chosen that span the solution space. This leads to exact solutions [17]. However, in the case of partial differential equations, the solution space is usually infinite-dimensional, so that only a subset of the solution space may be spanned. This precludes exact solutions, and also necessitates care in choosing which subspace to span. As illustrated in [9], there is a natural choice for the space-time test functions (and the associated space-time elements) that yields a general Characteristic Method (CM) approximation, which is precisely the ELLAM. For the case of non-reactive transport, the test function definition is [9],

$$
(w_{NR})_i^{n+1}(x,t) = \begin{cases} \dfrac{x-x_{i-1}}{\Delta x} + V \dfrac{t^{n+1}-t}{\Delta x}, & (x,t) \in \Omega_i^1 \\[2ex] \dfrac{x_{i+1}-x}{\Delta x} - V \dfrac{t^{n+1}-t}{\Delta x}, & (x,t) \in \Omega_i^2 \\[2ex] 0, & \text{all other } (x,t) \end{cases} \tag{5}
$$

where NR denotes nonreactive. The elements Ω_i^1 and Ω_i^2 are illustrated in Figure 1, as is a typical test function. Subscript i denotes spatial location, superscript n denotes time step number, and constant node spacing (Δx) is assumed in writing equation (5). The test function $w_i^{n+1} \in \mathbb{C}^{-1}[\Omega_t] \times \mathbb{C}^0[\Omega_x]$ is nonzero over only one time step (from n to n+1) with discontinuities in time aligned along the lines (x, t^n) and (x, t^{n+1}). It has discontinuities in spatial derivative that coincide with the secondary (hyperbolic) characteristics of the equation that pass through nodes x_{i-1}, x_i, x_{i+1} at time t^{n+1}. The function is defined on a fixed grid, analogous to backward-tracking ELM's. Notice that this is but one of an infinite number of choices of test functions that satisfy the homogeneous adjoint equation. However it is the one that most closely corresponds to the CM approximation that we seek.

Incorporation of the reaction term can again be accomplished in a number of ways. For this presentation we have chosen to incorporate the effect of the reaction with the advection terms. That is to say, the reaction will force the test function to change its value along a characteristic (see Figure 2), instead of remaining constant as in the nonreactive case (see Figure 1). The functional form used for the test function in the reactive case is therefore given by a direct modification of the test functions defined above, namely

$$(w_R)_i^{n+1}(x,t) = [(w_{NR})_i^{n+1}(x,t)]\{\exp[K(t-t^n)]\}. \tag{6}$$

The element definitions, grid points, etc. remain as defined in Figure 1. A typical reactive test function is shown in Figure 2.

Given the space-time element definition and the location of discontinuities in the test functions, a general ELLAM approximating equation may be written for all test functions that are zero at all points between t^n and t^{n+1} at the spatial boundaries, in this case x=0 and x=ℓ. For such functions, application of the LAM procedure (integration by parts, etc.), produces the following equation

$$\int_{x_{i-1}}^{x_{i+1}} u(x,t^{n+1})w_i^{n+1}(x,t^{n+1})dx - \int_{x_{i-1}^*}^{x_{i+1}^*} u(x,t^n)w_i^{n+1}(x,t^n)dx$$

$$- D \left\{ \int_{t^n}^{t^{n+1}} u(x_\ell^i(t),t)\left[\frac{\partial w_i^{n+1}}{\partial x}\right]_{x_\ell^i(t)} dt + \int_{t^n}^{t^{n+1}} u(x_c^i(t),t) \right.$$

$$\left. \cdot \left[\frac{\partial w_i^{n+1}}{\partial x}\right]_{x_c^i(t)} dt + \int_{t^n}^{t^{n+1}} u(x_r^i(t),t)\left[\frac{\partial w_i^{n+1}}{\partial x}\right]_{x_r^i(t)} dt \right\}$$

$$= \int_\Omega f \, w_i^{n+1} \, dx \, dt \tag{7}$$

where [·] denotes a jump operator, and the space-time lines x_ℓ, x_c, and x_r correspond to characteristic lines, as shown in Figure 1. Detailed evaluation of the integrals in equation (7) for the nonreactive test functions of equation (5) is given in [9]. For the reactive transport case, evaluation of the integrals using the definition of w and a piecewise linear spatial interpolator for u yields

$$[\frac{e^\kappa}{6} - \rho\gamma\theta] U_{i-1}^{n+1} + [\frac{2e^\kappa}{3} + 2\rho\gamma\theta] U_i^{n+1} + [\frac{e^\kappa}{6} - \rho\gamma\theta] U_{i+1}^{n+1}$$

$$= (\alpha_1) U_{i-Nc-2}^n + (\alpha_2) U_{i-Nc-1}^n + (\alpha_3) U_{i-Nc}^n + (\alpha_4) U_{i-Nc+1} \tag{8}$$

where $\kappa=K\Delta t$, $\rho=D(\Delta t)/(\Delta x)^2$, $\gamma=(e^\kappa-1)/\kappa$, θ is the usual time weighting parameter ($0\leq\theta\leq1$), Nc is the truncated integer value of the Courant number Cu, Cu=V(Δt)/(Δx), and f is assumed to be zero. The coefficients α_i are scalars that arise from evaluations in equation (7) at time level n. Detailed expressions for these coefficients may be found in Zisman [23].

BOUNDARY CONDITIONS

The ELLAM approach inherently accommodates any boundary conditions into the approximating equations. This is accomplished by systematic application of the LAM integrations, coupled with careful evaluation of the boundary integrals when the domain boundary is encountered. Such integrals are important when a test function has nonzero value at a boundary of the domain. An example is illustrated in Figure 3, where the test function w_1^{n+1} is shown for a case of $1 < Cu < 2$. As shown in detail in [9], this case leads to the following ELLAM equation

$$
\int_{x_0}^{x_2} u(x,t^{n+1}) w_1^{n+1}(x,t^{n+1}) dx - \left[\int_{x_0}^{x_2^*} u(x,t^n) w_1^{n+1}(x,t^n) dx \right.
$$

$$
+ V \int_{t^n}^{t^{n+1}} u(0,t) w_1^{n+1}(0,t) dt \left. \right] - D \left\{ \int_{t_1^*}^{t^{n+1}} u(0,t) \frac{\partial w_1^{n+1}}{\partial x}(0,t) \, dt \right.
$$

$$
+ \int_{t_1^*}^{t^{n+1}} u(x_c^1(t),t) \left[\frac{\partial w_1^{n+1}}{\partial x} \right]_{x_c^i(t)} dt + \int_{t^n}^{t^{n+1}} u(x_r^1(t),t)
$$

$$
\cdot \left[\frac{\partial w_1^{n+1}}{\partial x} \right]_{x_r^i(t)} dt \left. \right\} + D \int_{t^n}^{t^{n+1}} \frac{\partial u}{\partial x}(0,t) w_1^{n+1}(0,t) dt
$$

$$
+ D \int_{t^n}^{t_1^*} u(0,t) \frac{\partial w_1^{n+1}}{\partial x}(0,t) \, dt \; = \int_{\Omega} f \, w_1^{n+1} \, dx \, dt \tag{9}
$$

The importance of equation (9) is that boundary evaluations of both the function and its spatial derivative are present in the equation. This means that any type of boundary condition may be easily accommodated. But it also means that both the function and its derivative should be solved for at the spatial boundaries. When the boundary integrals in all equations are treated as in equation (9), the ELLAM approximation can be shown to possess the conservative property. Without this complete treatment of boundary fluxes, mass conservation can in general not be achieved. Additional details of boundary condition implementation, for both inflow and outflow boundaries, may be found in [9]. The treatment for reactive transport is identical to that for the nonreactive equation, with differences arising from the definition of w.

APPLICATION

To demonstrate the ELLAM formulation for reactive transport as well as the importance of proper implementation of boundary conditions, two standard model equations are solved. The first is a nonreactive example that is meant to illustrate the influence of proper boundary condition treatment. The example is a standard one-dimensional Gaussian hill transport problem over domain $[0, \ell]$, except that the initial and boundary conditions are assigned to correspond to an initial Gaussian hill centered at x= -0.25 (with σ_0=0.033). The analytical solution is used to define the initial and boundary conditions, with the imposed condition at x=0 being a first-type condition. Relevant parameters are: V=5.5, Δx=0.02, Δt=0.01, θ=0.5, D=.00001, resulting in Grid Peclet and Courant numbers of Pe=1100 and Cu=2.75. In Figure 4 are solutions that employ both the ELLAM boundary condition formulation and a simplified implementation that ignores the diffusive flux component of the approximating equations. The solutions show dramatic differences, with the ELLAM solution clearly superior.

The second example solves the reactive case with zero initial condition and boundary condition u(0,t)=1. The following parameters are used: V=1.66, Δx=0.02, Δt=0.1, θ=0.5, Cu=8.33, Pe=20. Again the two numerical solutions correspond to ELLAM boundary implementation versus simple boundary treatment. The differences seen in Figure 5 are again dramatic.

In both examples, linear interpolation at the feet of the characteristics was used. Overall, the ELLAM approach to Eulerian-Lagrangian approximations is limited, like other ELM's, by problems of solution interpolation at the characteristic feet (see [2]). It also faces the same challenges as other ELM's when dealing with variable velocity fields. However, ELLAM provides expressions (for example, equations (7) and (9)) that inherently yield general Characteristic Method approximations with proper incorporation of boundary conditions. This leads naturally to ELM's that possess the conservative property, and allows any combination of boundary conditions to be easily incorporated into the numerical approximation. We are currently extending ELLAM to multiple dimensions and to problems with variable coefficients.

ACKNOWLEDGEMENTS

This work was supported in part by the U.S. Environmental Protection Agency under Agreement CR-814946 and by the National Science Foundation under Grant 8657419-CES. Although the research described in this article has been funded in part by the U.S.E.P.A., it has not been subjected to Agency review and therefore does not necessarily reflect the views of the Agency and no official endorsement should be inferred.

REFERENCES

[1] Allen, D. and R. Southwell, "Relaxation methods applied to determining the motion in two dimensions of a fluid past a fixed cylinder," *Quart. J. Mech. Appl. Math.*, 8, 129-145, 1955.

[2] Baptista, A.M., "Solution of advection-dominated transport by Eulerian-Lagrangian methods using the backward methods of characteristics," PhD Thesis, Dept. Civil Eng., MIT, 1987.

[3] Bouloutas, E.T. and M.A. Celia, "An analysis of a class of Petrov-Galerkin and Optimal Test Functions methods," *Proc. Seventh Int. Conf. Computational Methods in Water Resources* (Celia, et al., eds.), 15-20, 1988.

[4] Brooks, A. and T.J.R. Hughes, "Streamline upwind Petrov-Galerkin formulations for convection dominated flows with particular emphasis on the incompressible Navier-Stokes equations," *CMAME*, 32, 199-259, 1982.

[5] Celia, M.A. and I. Herrera, "Solution of general ordinary differential equations by a unified theory approach," *Numerical Methods for PDE's*, 3(2), 117-129, 1987.

[6] Celia, M.A., I. Herrera, and E.T. Bouloutas, "Adjoint Petrov-Galerkin methods for multi-dimensional flow problems," *Finite Element Analysis in Fluids*, Chung and Karr, eds., UAH Press, 965-970, 1989.

[7] Celia, M.A., I. Herrera, E. Bouloutas, and J.S. Kindred, "A new numerical approach for the advective-diffusive transport equation," *Numerical Methods for PDE's*, 5, 203-226, 1989.

[8] Celia, M.A., J.S. Kindred, and I. Herrera, "Contaminant transport and biodegradation: I. A numerical model for reactive transport in porous media," *Water Resources Research*, 25, 1141-1148, 1989.

[9] Celia, M.A., T.F. Russell, I. Herrera and R.E. Ewing, "An Eulerian-Lagrangian localized adjoint method for the advection-diffusion equation," to appear, *Advances in Water Resources*, 1990.

[10] Celia, M.A. and W.G. Gray, *Fundamental Concepts for Applied Numerical Simulation*, to be published by Prentice-Hall, 1990.

[11] Christie, I., D.F. Griffiths, and A.R. Mitchell, "Finite element methods for second order differential equations with significant first derivatives," *Int. J. Num. Meth. Engrg.*, 10, 1389-1396, 1976.

[12] Espedal, M.S. and R.E. Ewing, "Characteristic Petrov-Galerkin subdomain methods for two-phase immiscible flow," *CMAME*, 64, 113-135, 1987.

[13] Ewing, R.E., T.F. Russell, and M.F. Wheeler, "Convergence analysis of an approximation of miscible displacement in porous media by mixed finite elements and a modified method of characteristics," *CMAME*, 47, 73-92, 1984.

[14] Hemker, P.W., "A numerical study of stiff two-point boundary value problems," PhD Thesis, Mathematisch Centrum, Amsterdam, 1977.

[15] Herrera, I., "Unified approach to numerical methods, Part I: Green's formula for operators in discontinuous fields," *Num. Meth. for PDE's*, 1(1), 25-44, 1985.

[16] Herrera, I., "Unified approach to numerical methods, Part II: Finite elements, boundary elements, and their coupling," *Num. Meth. for PDE's*, 1(3), 159-186, 1985.

[17] Herrera, I., L. Chargoy, and G. Alduncin, "Unified approach to numerical methods, Part III: Finite differences and ordinary differential equations," *Num. Meth. for PDE's*, 1(4), 241-258, 1985.

[18] Hughes, T.J.R. and A. Brooks, "A theoretical framework for Petrov-Galerkin methods with discontinuous weighting functions: Applications to the streamline-upwind procedure," in *Finite Elem. in Fluids*, Vol. 4, Gallagher, et al., eds., 47-65, 1982.

[19] Neuman, S.P., "Adaptive Eulerian-Lagrangian finite element method for advection- dispersion," *IJNME*, 20, 321-337, 1984.

[20] Pinder, G.F. and H.H. Cooper, "A numerical technique for calculating the transient position of the saltwater front," *Water Resources Research*, 6(3), 875-882, 1970.

[21] Russell, T.F., "Time stepping along characteristics with incomplete iteration for a Galerkin approximation of miscible displacement in porous media," *SIAM J. Num. Anal.*, 22, 970-1013, 1985.

[22] Wheeler, M.F. and C.N. Dawson, "An operator-splitting method for advection-diffusion- reaction problems," *MAFELAP Proc. VI*, Whiteman, ed., Academic Press, 463-482, 1988.

[23] Zisman, S., "Simulation of contaminant transport in groundwater systems using Eulerian-Lagrangian localized adjoint methods," MS Thesis, Dept. Civil Eng., MIT, 1989.

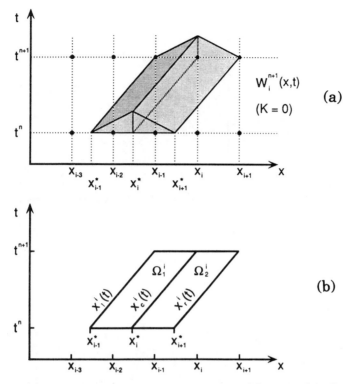

FIGURE 1 – (a) Test function for nonreactive case, (b) space-time elements.

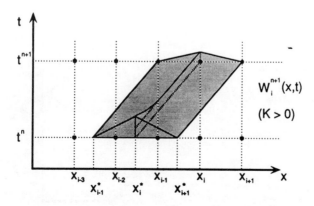

FIGURE 2 – Test function for the reactive case.

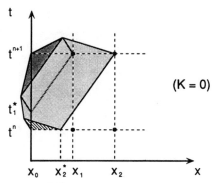

(K = 0)

FIGURE 3 - Test function $w_1^{n+1}(x,t)$, showing boundary intersection.

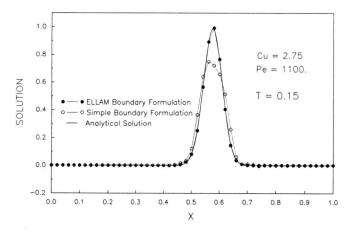

FIGURE 4 - Gauss hill example for the nonreactive case.

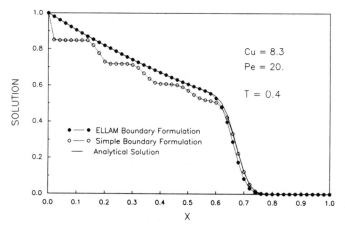

FIGURE 5 - Reactive case using standard step-front problem.

Particle-Grid Methods for Reacting Flows in Porous Media: Application to Fisher's Equation

A.F.B. Tompson(*), D.E. Dougherty(**)
(*) Earth Sciences Department, Lawrence Livermore National Laboratory, Livermore, CA 94550 USA (**) Department of Civil Engineering, University of California, Irvine, CA 92717 USA

Introduction

Recent investigations have shown the particle tracking method to be a viable and efficient method for solving equations describing nonreactive mass transport in porous media (Tompson and Dougherty [1]). The approach is particularly useful for solving multidimensional, densely gridded problems (Tompson and Gelhar [2]), is readily implemented on parallel and vector architecture computers (Baptista and Turner [3]; Dougherty and Tompson [4]), and may be extended for use in multicomponent or nonconservative problems (Dougherty et al. [5]). In the discussion below, we briefly outline a specific framework in which the method can be applied to reactive problems and demonstrate its use to solve a particular diffusion-reaction system described by the so-called Fisher's Equation.

Simplified Problem Description

A reactive transport problem in a porous medium may generally be described by a set of mass balance equations of the form

$$\frac{\partial(\epsilon^\alpha c_i^\alpha)}{\partial t} + \nabla \cdot (\epsilon^\alpha \mathbf{J}_i^\alpha) - \epsilon^\alpha r_i^\alpha = 0, \tag{1}$$

where c_i^α and $\mathbf{J}_i^\alpha$ are the average mass density and mass flux of a particular chemical constituent i in phase α, ϵ^α is the phase void fraction, and r_i^α is the rate of mass production or loss due to phase change or chemical reaction. In this paper, we focus on the solution of (1) for a single component in a liquid (groundwater) phase. In this case (deleting super and subscripts for clarity), the flux $\mathbf{J}$ is defined in terms of a relative groundwater velocity $\mathbf{v}$ by

$$\mathbf{J} = \mathbf{v}c - \mathbf{D} \cdot \nabla c, \tag{2}$$

and r will be introduced later. The dispersion tensor $\mathbf{D}$ is defined by

$$\mathbf{D} \approx (\alpha_T |\mathbf{v}| + \mathcal{D})\mathbf{I} + (\alpha_L - \alpha_T)\frac{\mathbf{v}\mathbf{v}}{|\mathbf{v}|}. \tag{3}$$

where α_L and α_T are the longitudinal and transverse dispersivities and $\mathcal{D}$ is an effective molecular diffusivity.

Operator Splitting Approach

The numerical solution of the resulting system is formulated in terms of a time splitting approach. The change of concentration at a point may be found by integrating the mass balance equation over a time interval Δt of interest:

$$\Delta c = -\int_t^{t+\Delta t} \nabla \cdot (\mathbf{v}c)dt + \int_t^{t+\Delta t} \nabla \cdot (\mathbf{D} \cdot \nabla c)dt + \int_t^{t+\Delta t} rdt. \tag{4}$$

Equation (4) may also be expressed as

$$\Delta c = \Delta c_A + \Delta c_D + \Delta c_R, \tag{5}$$

in terms of individual concentration changes resulting from advective, diffusive, and reactive processes, respectively. The idea behind this approach is that different methods may used to evaluate the integrals or approximate the spatial operators appearing within them (*e.g.*, Wheeler and Dawson [6]). Within (4), for example, one might tailor the solution process by using a spatial approximation appropriate for hyperbolic operators in the first integral, one suitable for parabolic operators in the second (*e.g.*, Douglas and Russell [7]), and a time integration scheme fitted to the reaction time scales in the third.

Here, we are concerned with solving (4) within the framework of a generalized particle technique. A random walk approach will be used to determine the advective and diffusive concentration changes, $\Delta c_{AD} = \Delta c_A + \Delta c_D$, and one or more ODE schemes will be used to evaluate the reactive changes Δc_R.

Particle Formulation

In the particle formulation, the spatial distribution of the constituent mass is represented as a finite system of N particles. Each particle will be characterized by a position $\mathbf{X}_p(t)$ within the computational region Ω_c, a mass $m_p(t)$, and, perhaps, a phase pointer indicating whether the particle is "active" in the liquid system. The number of active particles will be denoted by N_p. At any time, the set of particle attributes will define the state of the system. A simulation involves changing the particle attributes or their active numbers over small time intervals according to the flux or reactive forces embodied in (4).

Treatment of Advection-Diffusion Processes

We use a two-step process to advance a chemical transport system from one time level to the next. In *Step I*, the particle locations will be changed using a random walk particle method (RWPM):

$$\mathbf{X}_p \leftarrow \mathbf{X}_p + \mathbf{A}(\mathbf{X}_p)\Delta t + \mathbf{B}(\mathbf{X}_p) \cdot \mathbf{Z}_p\sqrt{\Delta t}, \tag{6}$$

where $\mathbf{A}$ is a "coherent" forcing vector, $\mathbf{B}$ is a deterministic, square scaling matrix, and $\mathbf{Z_p}$ is a vector of three independent random numbers with mean zero and unit variance. In the limit as $N_p \rightarrow \infty$, the particle mass density that evolves from repeated application of (6) will satisfy the nonreactive form of (1) if $\mathbf{A}$ and $\mathbf{B}$ are chosen to be

$$\mathbf{A} \equiv \mathbf{v} + \nabla \cdot \mathbf{D} \tag{7}$$

$$\mathbf{B} \cdot \mathbf{B}^T \equiv 2\mathbf{D} \tag{8}$$

(Kinzelbach [8]; Tompson and Dougherty [1]).

The particle mass density $\hat{c}$ may be defined at any time from the collection of active particle locations and masses:

$$\hat{c}(\mathbf{x}, t) = \sum_{p \in N_p} m_p(t)\delta(\mathbf{x} - \mathbf{X}_p(t)). \tag{9}$$

Because this representation is discontinuous, a smoothed approximation

$$\tilde{c}(\mathbf{x}, t) = \int_{\Omega_c} \hat{c}(\mathbf{x}', t)\varsigma(\mathbf{x} - \mathbf{x}')d\mathbf{x}' = \sum_{p \in N_p} m_p(t)\varsigma(\mathbf{x} - \mathbf{X}_p(t)) \tag{10}$$

is used in its place. The function $\varsigma(\mathbf{x})$ should be ideally be chosen such that $\int_V \varsigma(\mathbf{x}) d\mathbf{x} = 1$ is satisfied and (10) is unaffected by coordinate rotations. The simplest type of smoothing approximation involves the selection of a small, symmetric region $\mathcal{V}$ around $\mathbf{x}$ and dividing the mass found within $\mathcal{V}$ by its volume V. This amounts to the choice of $\varsigma = 1/V$ within $\mathcal{V}$ and zero elsewhere. The relative smoothness of this representation can be described by an error $e = c - \tilde{c}$, and will depend on the support and form of $\varsigma(\mathbf{x})$ (Raviart [9]) as well as the particle "resolution", loosely defined here as the number of particles used to represent a unit mass in a unit volume.

If we consider a set of N_g "grid" points at which $\tilde{c}^\alpha$ is to be evaluated then (10) may be alternatively written as

$$\tilde{\mathbf{c}} \leftarrow \mathcal{F} \cdot \mathbf{m}, \tag{11}$$

where $\tilde{\mathbf{c}}$ is a vector of N_g concentrations, $\mathbf{m}_p$ is a vector of N_p particle masses, and $\mathcal{F}$ is an $N_g \times N_p$ weighting matrix. Conversely, one may envision an inverse relationship of the form

$$\mathbf{m} \leftarrow \mathcal{B} \cdot \tilde{\mathbf{c}}, \tag{12}$$

in which a group of concentration values is mapped into a set of masses at known particle locations. Although the forward interpolator (11) is relatively easy to construct, it is generally difficult to develop a backward interpolator (12) that is consistent with it in the sense that $\mathcal{F} \cdot \mathcal{B} = \mathbf{I}$, because the backward system will usually be underconstrained. One system in which this will work involves breaking up the region Ω_c into a set of N_g nonoverlapping volumes $\mathcal{V}_g$ centered around a regular grid of points and using the simple uniform weighting functions described above.

Initial conditions are implemented by mapping specified concentrations onto a set of particles within Ω_c in a manner consistent with (12). Nonzero Dirichlet boundary conditions may be specified along certain boundary "layers" or regions in the same way, although one must ensure the integrity of these conditions over a time step Δt by adjusting layer thickness or size of Δt. Nonzero flux conditions may be specified by introducing fixed amounts of particle mass past a boundary at an appropriate rate. Zero-flux conditions may be specified by bouncing particles off boundaries. Zero-valued Dirichlet conditions cannot be implemented in the traditional sense; one may either enforce a no-flux boundary, or let the boundary be open or "absorbing" in which particles freely pass through (never to be seen again).

Treatment of Reaction Processes

In *Step II* of this algorithm, the abundence of constituent mass is changed to represent the contributions of the reactions over Δt. In general, concentration changes must be computed and used to modify the number of particles N_p or their masses. This approach involves (i) mapping the particle masses and locations that result after *Step I* onto a grid of concentrations $\tilde{\mathbf{c}}$ using (11), (ii) determining a set of concentration changes $\Delta \tilde{\mathbf{c}}_R$ by integrating the rate term in (1), and (iii) mapping these into changes of particle mass or number using (12). In this sense, we call this a particle-grid solution method. If, on the other hand, the reaction processes described by r can be expressed on a mass basis, as in the case of radioactive decay, then the particle masses may be directly modified. In other cases, N_p may be directly modified using a statistical approach (Valocchi and Quinodez [10]) provided an appropriate transitional probability for the reaction being considered can be identified.

There are two important concerns at this stage. If concentration estimates made after *Step I* are used to predict concentration changes, then the estimation error e will be propagated through the reaction calculation, and a certain error feedback loop may be established. The magnitude of this error will depend on the type of reaction being considered, the type function ς used to evaluate concentrations, and the particle resolution. The other issue involves the specific way in which the reaction integral in (4) is evaluated. The smaller the time scale of the reaction process relative to Δt, the stiffer the integral becomes, and more accurate ODE integration schemes are needed, such as a fourth order Runge Kutta or a more specialized method (Oran and Boris [11]). Other possibilities include using a smaller Δt or a mass balance equation modified for the case of equilibrium reactions.

Solution of Fisher's Equation in One-Dimension

Fisher's equation in one-dimension is given by

$$\frac{\partial c}{\partial t} - \frac{\partial^2 c}{\partial x^2} = c(1 - c), \tag{13}$$

a diffusion-reaction system often used to describe population dynamics. If c is considered positive, solution is bounded $0 \leq c \leq 1$ and propagates in the direction of negative concentration gradients. For the initial condition

$$c(x) = (1 + e^{x/6})^{-2}, \tag{14}$$

the analytical solution is a steady traveling wave $c(x - st)$ moving in the positive x direction at a speed of $s = 5/\sqrt{6}$ (Reitz [12]). Because the analytical solution is sensitive to perturbations in the initial and far field conditions, it is considered to be a good problem to test numerical solution routines.

A particle model was applied to this problem on a prismatic domain lying in $-20 < x < 40$ with a cross sectional area $A = 0.25$ (only one-dimensional movements were allowed). The initial condition (14) was used to specify initial particle distributions within 120 nonoverlapping volumes $\mathcal{V}_g$ centered at 0.5 unit intervals along the x axis. Mass was assigned uniformly to the particles within a cell according to the centroid concentrations $c(x_g, 0)$; in other words, a piecewise box weighting function ς was used in (11). The initial particle distribution was based upon using an initial "resolution" of 600 particles per unit mass and unit volume throughout the entire domain. This meant that all particles initially had a mass of 0.0002. The initial concentration was therefore used to to compute the number of particles assigned to each cell. These were located randomly and uniformly within each $\mathcal{V}_g$.

The leftmost cell was used to enforce a boundary condition of $c(-19.75, t) = 1$. A time step of $\Delta t = 0.1$ was used to advance the simulation. Because this is an advection-free problem, particles were moved by the random displacement only in *Step I*. The average *rms* displacement of a particle is $\sqrt{2 \cdot 1 \cdot \Delta t} = .44$, almost one cell's length. The leftmost edge of the domain was therefore treated as a reflecting boundary to keep what would otherwise by a large number of particles from diffusing upstream, out of the the domain. This helped to maintain the integrity of the boundary condition over Δt.

Following the displacement step, concentrations were estimated for use in *Step II* at the cell centers x_g using (11) and a nonuniform *chapeau* weighting function in (10). The contribution of a particle's mass to the concentration at x_g is weighted by a factor of $\varsigma = (1 - |x_g - X_p|)/\Delta x$ for $x_g - \Delta x < X_p < x_g + \Delta x$, and 0 otherwise. This yields a smoother approximation than that obtained by using the box function, and hence a smaller feedback error e. These intermediate concentrations were used to evaluate $\Delta c_R(x_g)$ using Euler integration. Since this system seeks to maintain $0 \leq c \leq 1$, the largest change c could make using this scheme and time step is 0.025, so no time integration troubles were expected. Fractional concentration changes Δc_R at points x_g were mapped into fractional changes of the particle masses located within the volumes $\mathcal{V}_g$. Note that the forward and backward methods of interpolation (11-12) are different and, hence, inconsistent. Following the reaction step, the boundary condition in the leftmost cell was reestablished for the next time step.

Figure 1 depicts the traveling wave nature of the particle solution (in terms of estimated concentrations) and analytical solution at $t = 0$ and 5. As mass diffuses to the right, the solution becomes nonzero, and the reaction adds mass in the regions where

$0 < c < 1$. In terms of the particle solution, this means that rightmost particles are always growing in mass, while those in the left are not. The average partical mass or particle resolution at the front in figure 1 was over 7 times those used in the initial configuration. Thus, the smoothness of the concentration estimates near the front will degrade with time and the feedback errors e would be expected to increase. Because of this, the problem was run again where the particle mass was redistributed on a cell-wise basis before the start start of every n_r-th time step. This lowered the average particle mass and increased the particle resolution. Beacuse new particles were redistributed uniformly in in individual cells, this procedure will result in a small diffusive error proportional to $\Delta x^2/n_r\Delta t$. Figure 2 shows the computed wave speed of the particle solution using $n_r = 2$ out to $t = 20$. The wave speed, defined by $\int_\infty^\infty c(1-c)dx$, compares favorably with those computed from several finite difference solutions of Reitz [1981]. The solutions were obtained through different combinations of explicit and implicit treatments of the diffusive and reaction terms. The best one (curve D) resulted from the use of a diffusion coefficient modified to account for truncation errors.

These solutions are encouraging because they demonstrate the use of the particle grid method to solve a problem with a nonlinear reaction term. The process of developing the model solutions has shown that the particle grid method may involve unfamiliar kinds of approximation errors. Future directions of research will address these issues, alternative types of reaction treatments, as well as implementation of these two-step models on parallel computing platforms.

References

1 Tompson, A. F. B., and D. E. Dougherty, On the Use of Particle Tracking Methods for Solute Transport in Porous Media, in *Computational Methods in Water Resources, vol 2., Numerical Methods for Transport and Hydrologic Processes*, M. Celia, L. Ferrand, C. Brebbia, W. Gray, and G. Pinder, eds., Elsevier, 1988.

2 Tompson, A. F. B.and L. W. Gelhar, Numerical Simulation of Solute Transport in Randomly Heterogeneous Porous Media, in review, *Water Resources Research*, 1990

3 Baptista, A. M. and P. J. Turner, The Impact of Parallel Processors on the Numerical Solution of the Transport Equation, *EOS, Transactions of the American Geophysical Union*, 70:43:1083, 1989

4 Dougherty, D. E. and A. F. B. Tompson, Implementing the Particle Method on the iPSC/2, this volume, Proceedings Computational Methods in Water Resources, Venice, Italy, June 9–13, 1990.

5 Dougherty, D. E., A. Bagtzoglou, and A. F. B. Tompson, Particle Methods for Reactive Transport: 1. Review and Consistent Formulation, *EOS, Transactions of the American Geophysical Union*, 70:43:1078, 1989

6 Wheeler, M. F., and Dawson, C. N., An Operator-Splitting Method for Advection-Diffusion-Reaction Problems, MAFELAP Proceedings VI (Ed. Whiteman, J. A.), Academic Press, pp. 463–482, 1988.

7 Douglas, J. Jr. and T. F. Russell, "Numerical Methods for Convection Dominated Diffusion Problems basedon Combiningf the Method of Characteristics with Finite Element or Finite Difference Procedures" *SIAM J. Numr. Anal.* 871-885, 1982

8 Kinzelbach, W., "The Random Walk Method in Pollutant Transport Simulation", *Groundwater Flow and Quality Modelling*, E. Custodio, A. Gurgui, and J. P. Lobo Ferreira, eds., D. Riedel, Dordrecht, The Netherlands, 1987

9 Raviart, P. A., "Particle Numerical Models in Fluid Dynamics", in *Numerical Methods for Fluid Dynamics II*, (K. W. Morton and M. J. Baines, eds.), 1986.

10 Valocchi, A. J., and A. M. Quinodez, "Application of the random walk method to simulate the transport of kinetically adsorbing solutes", in *Groundwater Contanmination*, proceedings of the symposium held at the Third IAHS Scientific Assembly, Baltimore, MD, L. M. Abriola, editor, IAHS publication no. 185, 1989

11 Oran, E. S. and J. P. Boris, *Numerical Simulation of Reactive Flow*, Elsevier, New York, 1987

12 Reitz, R. D., A Study of Numerical Methods for Reaction-Diffusion Equations, *SIAM J. Sci. Stat. Comput.*, 2:1:95–106, 1981

Acknowledgements: Portions of this work were conducted under the auspices of the U. S. Department of Energy by Lawrence Livermore National Laboratory under contract W-7405-Eng-48 and were supported by the Subsurface Science Program of the Office of Health and Ecological Research of the US DOE. Additional support was provided by the University of California, Irvine through an allocation of computer time on the Intel iPSC/2 and Convex C240.

Keywords: Solute Transport, Reacting Flows, Particle Method

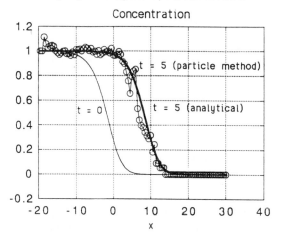

Figure 1: Traveling Wave Simulation

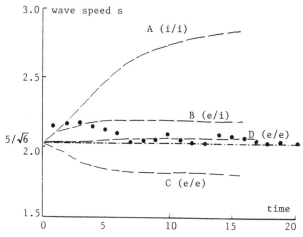

Figure 2: Estimated wave speeds. Dots are from particle solution; lines from Reitz [12]; e and i indicate explicit or implicit treatment of diffusion/reaction term

Simulation of Coupled Geochemical and Transport Processes of an Infiltration Passage Introducing a Vectorized Multicomponent Transport-Reaction Model

M. Vogt(*), B. Herrling(**)

(*) Lahmeyer International GmbH, Lyoner Strasse 22, 6000 Frankfurt/Main 71, West Germany (**) Inst. of Hydromechanics, University of Karlsruhe, Kaiserstrasse 12, 75 Karlsruhe, West Germany

ABSTRACT

A vectorcomputer orientated multicomponent transport-reaction model is presented. Using the described optimizations and vectorizations it is possible to realize even large-scale simulations of coupled chemical and transport processes within a reasonable CPU-time-scale. This model was applied to field data of an infiltration passage. A complete water contents analysis consisting of about 50 chemical species is transported along an infiltration passage of 200 m length. Precipitation/dissolution of several carbonate minerals and mixing of water of different origins is incorporated to simulate complex water quality changes.

INTRODUCTION

Since several years a number of computer models was developed for the simulation of transport processes of several chemical species, which are undergoing coupled geochemical reaction processes. Most of the existing models were just coupled models of existing transport codes and one of the well known geochemical equilibrium models (PHREEQE, WATEQ etc.). Due to the scale of the mentioned processes in terms of multitude of reactive species, dimension of space - for a correct and numerical stable simulation of the transport processes the flow regime has to be discretized very fine - and multitude as wells as complexity of the homogeneous or heterogeneous reaction processes, simulations of field-scale processes are very CPU-time intensive.

INTRODUCTION TO THE MULTICOMPONENT TRANSPORT-REACTION MODEL

The transport part of the combined model (Vogt [1]) solves the transient convection dispersion differential equation in two dimensions, vertical integrated and in a vertical cross section, as well. For the integration in space the finite element method is applied. The integration in time is done by a weighted finite difference method. The weighting factors are choosen according to the Leismann-Scheme (Leismann and Frind [2]), which results in a symmetric coefficient-matrix.

The conception of the chemical part is that one of an 'Open Chemical System' (Vogt [1]) where several different kinds of kinetic and equilibrium reactions

-	ion exchange	-	precipitation/dissolution
-	ad-/desorption	-	complexation
-	acid-base reactions	-	redox-reactions

can be implemented following simple directions.

The solution of kinetic equations is trivial from the numerical point of view. The major restriction in the pratical use of kinetic expressions is the lack of speed coefficients for the reactions which have to be measured in very time consuming experiments in the laboratory.

The solution of equilibrium equations is obtained by using/formulating thermodynamic laws or balance equations
-      mass, charge and electron (operational valences) balance
-      thermodynamic equilibrium equations, the equilibrium coefficients undergo a temperature correction by the Van't Hoff Differential Equation
-      mineral saturation
-      ionic strength
-      activity coefficients (Debeye-Hückel, Güntelberg, Davies etc.).

A transformation of this equations results into a highly nonlinear algebraic equation system, which is solved by a modified Newton Raphson procedure.

Within the presented model the well known 'Two Step Procedure' is applied to couple the solution of the transport equation with that one of the chemical equilibrium part (Figure 1). As shown in figure 1, for a transient computation four CPU-time intensive main loops have to be passed.

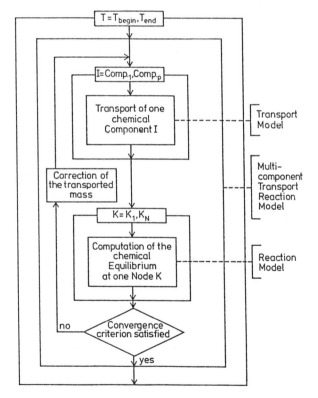

Figure 1:   Scheme of the coupled model using the 'Two Step Pocedure'

OPTIMIZATION OF THE COMPUTER CODE

To make the model applicable even for large scale simulations or for the simulation of field data the model was systematically optimized and vectorized.

The solution of the transport equation can be optimized for all iterations within one time-step under certain assumptions. If one assumes the same material parameters, the dispersion coefficients and porosity, for one finite element, which is always valid, there exists only one symmetric coefficient matrix uniform for all transported components. Using a special way of building in Dirichlet boundary conditions (Vogt [1]) for the solution of the transport equation there has to be realized only one unique decomposition of the coefficient matrix. Then for all iterations and all transported components within one time step only backward substitutions have to be done, which are very fast in terms of CPU-time.

Assuming also a steady state velocity field for the entire transient computation, which many users of such models do, there has to be realized only one unique decomposition of the coefficient matrix. Finally for all time-steps, all iterations and all transported components there have to be performed only backward substitutions.

To optimize the chemical equilibrium solver, that part of the model was vectorized. The vectorization was done by solving the entire equilibrium system parallel for all nodes of the finite element discretization (Vogt [1]).

To verify the efficiency of the presented optimizations and vectorizations, benchmark tests were performed on the vector computers CRAY 2 and CYBER 205 and on a SIEMENS mainframe, which is no vector computer (Figure 2). For this benchmark test a model with 535 nodes, 950 elements, 16 equations within the equilibrium system, 6 transported components and 625 time-steps was used. The CPU-time on the CRAY 2 decreased by the factor 27, on the CYBER 205 by the factor 43 and on the non vector computer by 4.

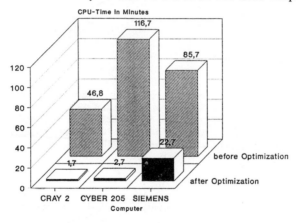

Figure 2:   Benchmark test

SIMULATIONS WITHIN AN INFILTRATION PASSAGE

Within a main research program of the German Research Foundation (DFG), 'Polutants in Groundwater', this model was applied to field data of an infiltration field, 'Waterwork Hengsen' (Figure 3), which is part of the water supply system of the city of Dortmund, FRG. Within this area there has been serious problems with Iron (Fe) and Maganese (Mn) precipitations.

The simulations, which will be discussed within this paper, are the very first attempts to model the complex coupled transport and chemical processes within this infiltration passage. Therefore first of all the main interest was to simulate the general trend of the chemical quality changes the infiltration water undergoes during the infiltration passage. Later on, after having a good representation of the chemical algorithm, the model will be expanded more suffisticated for the exact simulation of the transport, as well. Hence the multicomponent transport-reaction model was easily reduced to one dimensions.

Storage Reservoir Hengsen

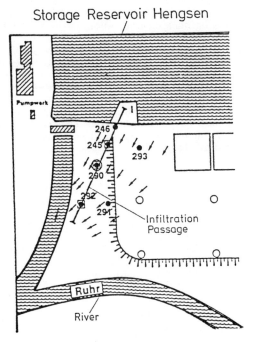

Figure 3:   Infiltration area

The infiltration passage (Figure 3) is about 200 m long, beginning at the observation well no. 246 close behind the storage reservoir Hengsen leading along the observation wells 245 and 290 to no. 292. For the one dimensional investigation a major problem is the change of the flow direction of the infiltrating water. For the first 100 m from well 246 to well 290 it flows along the measurement points. But for the last 100 m to observation well 292 the flow direction changes. This was considered within the model by introducing source terms of the known water quality of observation well no. 291.

To simulate the chemical equilibrium of the water contents analysis at obervation well no. 246, 13 master species, $Na^+, K^+, Fe^{2+}, Fe^{3+}, Mn^{2+}, Ca^{2+}, Mg^{2+}, SO_4^{2-}, Cl^-, CO_3^{2-}, NH_4^-, H_2S$, pH were considered within the transport part. Additional to that master species 41 related species and corresponding reactions

$$Fe^{3+} + H_2O \Longleftrightarrow FeOH^{2+} + H^+$$
$$Fe^{2+} + 3H_2O \Longleftrightarrow Fe(OH)_3^- + 3H^+$$
$$Fe^{3+} + Cl^- \Longleftrightarrow FeCl^{2+}$$
$$Fe^{3+} + 3Cl^- \Longleftrightarrow Fe(Cl)_3$$
$$Ca^{2+} + SO_4^{2-} \Longleftrightarrow CaSO_4$$
$$NH_4^+ \Longleftrightarrow NH_3 + H^+$$
$$H^+ + CO_3^{2-} \Longleftrightarrow HCO_3^-$$
$$Na^+ + HCO_3^- \Longleftrightarrow NaHCO_3$$
$$K^+ + SO_4^{2-} \Longleftrightarrow KSO_4^-$$
$$Mg^{2+} + HCO_3^- \Longleftrightarrow MgHCO_3^+$$
$$Ca^{2+} + OH^- \Longleftrightarrow CaOH^+$$
$$Ca^{2+} + CO_3^{2-} \Longleftrightarrow CaCO_3$$
$$H_2S \Longleftrightarrow H^+ + HS^-$$
$$Fe^{3+} + 2H_2O \Longleftrightarrow Fe(OH)_2^+ + 2H^+$$
$$Fe^{3+} + 4H_2O \Longleftrightarrow Fe(OH)_4^- + 4H^+$$
$$NH_4^+ + SO_4^{2-} \Longleftrightarrow NH_4SO_4^-$$
$$H_2O \Longleftrightarrow H^+ + OH^-$$

$$Fe^{2+} + H_2O \Longleftrightarrow FeOH^+ + H^+$$
$$Fe^{3+} + SO_4^{2-} \Longleftrightarrow FeSO_4^+$$
$$Fe^{3+} + 2Cl^- \Longleftrightarrow Fe(Cl)_2^+$$
$$Fe^{2+} + SO_4^{2-} \Longleftrightarrow FeSO_4$$
$$Mg^{2+} + OH^- \Longleftrightarrow MgOH^+$$
$$HCO_3^- + H^+ \Longleftrightarrow H_2CO_3$$
$$Na^+ + CO_3^{2-} \Longleftrightarrow NaCO_3^-$$
$$Na^+ + SO_4^{2-} \Longleftrightarrow NaSO_4^-$$
$$Mg^{2+} + CO_3^{2-} \Longleftrightarrow MgCO_3$$
$$Mg^{2+} + SO_4^{2-} \Longleftrightarrow MgSO_4$$
$$Ca^{2+} + HCO_3^- \Longleftrightarrow CaHCO_3^+$$
$$H^+ + SO_4^{2-} \Longleftrightarrow HSO_4^-$$
$$HS^- \Longleftrightarrow H^+ + S^{2-}$$
$$Fe^{3+} + 3H_2O \Longleftrightarrow Fe(OH)_3 + 3H^+$$
$$Fe^{2+} + 2H_2O \Longleftrightarrow Fe(OH)_2 + 2H^+$$
$$2H^+ + SO_4^{2-} \Longleftrightarrow H_2SO_4$$
$$Mn^{2+} \Longleftrightarrow Mn^{3+} + e^-$$

$$Mn^{2+} + Cl^- \quad <==> \quad MnCl^+$$
$$Mn^{2+} + 3Cl^- \quad <==> \quad MnCl_3^-$$
$$Mn^{2+} + 3H_2O \quad <==> \quad Mn(OH)_3^- \quad + 3H^+$$
$$Mn^{2+} + HCO_3^- \quad <==> \quad MnHCO_3^+$$

$$Mn^{2+} + 2Cl^- \quad <==> \quad MnCl_2$$
$$Mn^{2+} + H_2O \quad <==> \quad MnOH^+ + H^+$$
$$Mn^{2+} + SO_4^{2-} \quad <==> \quad MnSO_4$$

were implemented into the 'Open Chemical System'. To simulate the quality changes along the passage the possibilty of precipitation/dissolution of another 7 carbonate minerals

$$FeCO_3 \qquad <==> \quad Fe^{2+} + CO_3^{2-} \qquad\qquad \text{Siderite}$$
$$MgCO_3 \qquad <==> \quad Mg^{2+} + CO_3^{2-} \qquad\qquad \text{Magnesite}$$
$$CaMg(CO_3)_2 \quad <==> \quad Ca^{2+} + Mg^{2+} + 2CO_3^{2-} \quad \text{Dolomite}$$
$$CaCO_3 \qquad <==> \quad Ca^{2+} + CO_3^{2-} \qquad\qquad \text{Calcite}$$
$$CaSO_4 \quad + 2H_2O \quad <==> \quad Ca^{2+} + SO_4^{2-} + 2H_2O \quad \text{Gypsum}$$
$$FeOOH \quad + 3H^+ \quad <==> \quad Fe^{3+} + 2H_2O \qquad\qquad \text{Goethite}$$
$$MnCO_3 \qquad <==> \quad Mn^{2+} + CO_3^{2-} \qquad\qquad \text{Rhodocrosite}$$

was implemented. The spacial discretization of the infiltration was represented by 80 one dimensional elements of 2,50 m length.

A steady state flow field was assumed according to a measured mean flow regime. A transient computation was performed. For the initial conditions of the entire infiltration passage the chemical equilibrium situation of the observation well no. 246 was choosen. For the inflow a Dirichlet type boundary condition of the value of the initial condition was assumed for all 13 transported master species. To simulate the change of the water quality on the way between well no. 246 and 245 the possibility of precipitation/dissolution of the mentioned minerals was switched on one after another along the flow path.

Figure 4 shows results obtained after an infinite number of time steps. The graphs for Fe^{2+} and Mn^{2+} increase rapidly by switching on the influence of the minerals Siderite and Rhodocrosite. Mixing in water with the quality of observation well 291 the concentrations decrease slightly. The graph of CO_3^{2-} is mainly influenced by the dissolution of Calcite, Siderite and Rhodocrosite. The pH-value can be assumed as a measurement of the overall combination of all implemented reactions and minerals. It can be seen that the general trend of the analysis can be simulated by the multicomponent transport-reaction model.

SUMMARY AND OUTLOOK

A multicomponent transport-reaction model was introduced which was optimized and vectorized to enable even simulations of field scale. The model was applied to field data to simulate the water quality changes within an infiltration passage. Introducing precipitation/dissolution of several minerals and the mixing of water of different origins it was possible to simulate the general trend of the measured quality changes. In future research the model will be expanded to two dimensions to get a better approach for the transport processes. Also an investigation will be done where kinetic expressions for the precipitation/dissolution of several minerals will be implemented.

ACKNOWLEDGEMENT

The simulations carried out on the field data of the 'Waterwork Hengsen' were financially supported by the Deutsche Forschungsgemeinschaft (DFG), grant no. Th 159/15.

REFERENCES

1. Vogt, M. Ein vektorrechnerorientiertes Verfahren zur Berechnung von großräumigen Multikomponenten Transport-Reaktions Mechanismen im Grundwasserleiter. Dissertation. University of Karlsruhe. VDI-Verlag in press, 1990

2. Leismann, H.M. and Frind, E.O. A symmetric matrix time integration scheme for the efficient solution of advective-dispersion problems. Water Res. Res., 25, 6, 1989

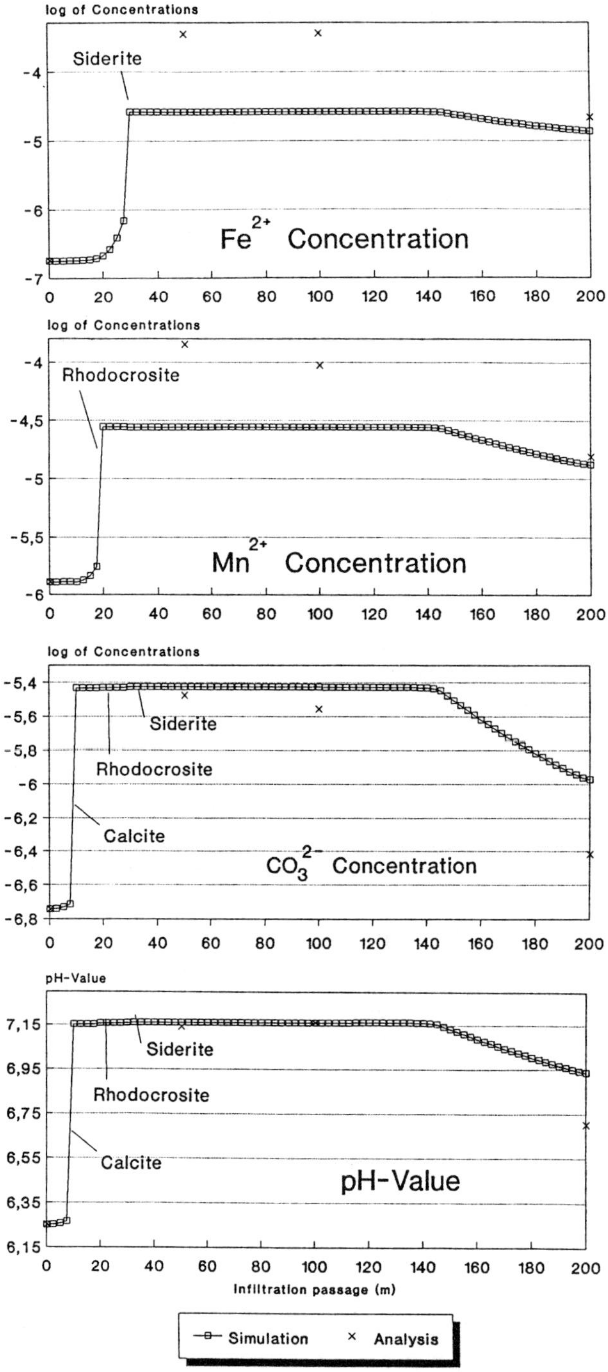

Figure 4:   Graphs for Fe²⁺, Mn²⁺, CO₃²⁻ concentrations and the pH value

Modelling of Pollutant Transport in Groundwater Including Biochemical Transformations

W. Schäfer, W. Kinzelbach

Department of Civil Engineering, Kassel University, D - 3500 Kassel, West Germany

ABSTRACT

A numerical multi-species solute transport model is presented which incorporates two-dimensional advective/dispersive transport in a saturated aquifer and pollutant degradation via heterotrophic bacterial activity. A split operator technique is used in the solution of the coupled physical and biochemical equations. The numerical problems encountered in this method are discussed. Further, the two-step method is compared to a linearized one-step solution scheme. Finally, the model is applied to a field case of in-situ remediation. The numerical model allows to recognize the relative importance of transport processes and processes governing the enhanced microbial activity.

INTRODUCTION

Most organic pollutants in the groundwater environment experience transformation through bacterial action. A correct interpretation of the concentrations of such degradable pollutants in groundwater by means of numerical models must therefore take into account biochemical effects. Simple empirical models lead to non-prognostic data fitting. Prognostic models must comprise the simultaneous modelling of all significantly interacting species involved. This demands coupling of the physical and biochemical processes governing the fate of the organic pollutant in the aquifer. In the model presented here the physical transport processes considered are advection and dispersion while heterotrophic aerobic and denitrifying bacterial growth represent the biochemical processes.

The resulting system of partial and ordinary differential equations can only be solved numerically. In a two-step or split operator approach the solution of the equation system for purely physical transport and the system accounting for biochemical transformations are separated in time and coupled iteratively. The decoupling of actual simultaneous processes simplifies the numerical solution of the related equations and makes it easier to incorporate modifications of the biochemical system, but it may introduce additional numerical errors.

For analytical purposes a coupled mechanistic model can be used to distinguish between transport and chemistry effects on solute concentration. Its prognostic ability renders it a useful tool in the decision making process in the field of groundwater pollution management, for example in the case of designing in-situ remediation measures.

THE BIO-CHEMICAL MODEL

Fig. 1 shows the scheme of the biochemical model describing the interaction of 8 species in three phases: The bacteria are assumed to be attached to the aquifer matrix with dissolved organic carbon (DOC) as the only substrate at their disposal. To oxidize this substrate the optionally anaerobic bacteria use dissolved oxygen (DO) as long as it is available and then switch to nitrate as electron acceptor. While DO and nitrate are supplied to the aquifer by groundwater recharge or injection, for DOC further sources are considered: the additional release of organic material from a source in the aquifer itself (e.g. a little mobile organic pollutant) and the utilizable portion of dead bacteria. Other possibly interacting chemical species are assumed to be available in sufficient quantities in the aquifer in order not to be limiting. Their impact is therefore neglected.

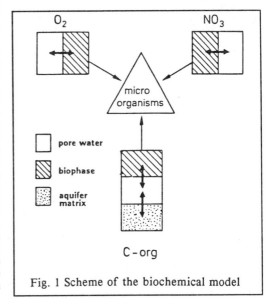

Fig. 1 Scheme of the biochemical model

Bacterial growth is based on Monod-type kinetics and switching between aerobic and anaerobic growth is implemented by a DO dependent weight function:

$$F(O) = 0.5 - \frac{1}{\pi} \arctan((O_2 - O_t)f_{sl})$$

A more detailed description of the biochemical model is given in Kinzelbach and Schäfer[1].

COUPLING OF CHEMISTRY AND TRANSPORT

To calculate the concentrations of chemically interactive species in the aquifer with a numerical model, coupling of terms describing biochemical transfers to those describing physical transfers by advection and diffusion/dispersion is necessary. The system of coupled transport equations is of the form:

$$\frac{\partial(\theta_i c_i)}{\partial t} = -\vec{\nabla}(\vec{v}_i \theta_i c_i) + \vec{\nabla}(D_i \theta_i \vec{\nabla} c_i) + \sum_{j=1}^{n} S_{ij}(c_1, ..., c_n) \quad i = 1, ..., n$$

where v_i and D_i are zero for an immobile species.
The coupling terms S_{ij} are in general non-linear functions of the concentrations.

Several solution strategies exist. In the direct solution method discretization leads to a system of nonlinear algebraic equations which are solved simultaneously (one-step method). In the solution linearization of the coupling terms and iteration may be applied. If the reaction terms are completely expressed in terms of concentrations of the previous iteration step, direct solvers can be applied advantageously as the system matrix remains unchanged during the iteration with the corrected chemical terms appearing on the right-hand side only (e.g. Frind et al.[2]). The solution scheme is of the form:

$$\frac{c_i - c_i(t)}{\Delta t} = L(c_i) + S_i^{old}$$

Here an alternative approach is used, where the system of partial differential equations is solved by a two-step or split operator method (e.g. Noye[3]). In principle, this method views the reactive transport equations as the sum of a pure transport equation of a tracer and a purely chemical equation:

$$\frac{\partial c}{\partial t} + \vec{\nabla}(\vec{v}c) - \vec{\nabla}(D\vec{\nabla}c) = S(c) \qquad \frac{1}{2}\frac{\partial c}{\partial t} + \vec{\nabla}(\vec{v}c) - \vec{\nabla}(D\vec{\nabla}c) = 0$$

$$\frac{1}{2}\frac{\partial c}{\partial t} = S(c)$$

The method can be improved by introducing iteration. An iterative scheme is obtained by using an implicit transport calculation in the first half-step of the iteration with chemical fluxes from the preceding iteration, and an implicit calculation of the chemical fluxes in the second half-step with explicit advective/dispersive fluxes from the preceding transport calculation. For sufficiently small time steps the total concentration changes found in the two iteration half-steps converge, as in both half-steps the total mass balances due to chemistry and transport are calculated. Fig. 2 shows a schematic description of this iteration procedure.

In order to solve the equations for flow and transport, a two dimensional finite difference method weighted both with respect to time and space is used (Herzer[4]). The equation system of biochemical dynamics is solved by means of a Newton-Raphson method.

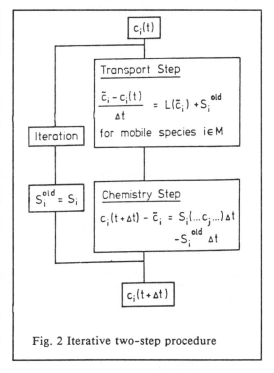

Fig. 2 Iterative two-step procedure

Discretization requirements
The total numerical error of the two-step method is composed of the usual error encountered in conservative tracer transport calculation and an additional error introduced by separating transport and chemistry.

The numerical error of mere transport can be kept small if numerical mixing is kept smaller than physical mixing by diffusion/dispersion. This implies a grid-Peclet number smaller than 2. Further, a sufficient discretization of lateral gradients is required which is of the order of the lateral dispersivity in the vicinity of a point source. Finally the Courant-number should be smaller than 1 (Frind and Pinder[5], Kinzelbach and Frind[6]). An overview of numerical methods to solve the transport equations in groundwater is given by Kinzelbach[7].

While non-adherence to the discretization conditions of physical transport introduces numerical dispersion in upwind-schemes, it shows in the form of oscillations in central difference schemes. The appearance of negative concentrations due to numerical oscillations which may be acceptable in tracer transport calculations is often fatal in the computations in the chemical half-step. Therefore the acceptance of some numerical dispersion is preferable to the occurence of oscillations.

For nonlinear kinetic reactions the time step of the transport scheme must stay small compared to the chemical time scale in order to give accurate results. Larger time steps are possible if implicit solution schemes are used in the chemical step. Here it must be remembered, that the system of nonlinear differential equations describing the biochemical transformations is stiff, the typical time constants of different processes being of different orders of magnitude. Accurate solutions would require a time resolution smaller than the time constant of the fastest process irrespective of whether the process has already died out or not. Economy is possible by applying a standard method for stiff systems of equations, such as the Gear-method (Hindmarsh[8]) which is an implicit method including a control of the time step such that in active regions of the time function the step is decreased while it is increased in regions where the time functions are less steep.

As no analytical solution exists to test complex chemical transport models objectively, it is important to analyze the possible errors. A pragmatic method consists of consecutive refinement of spatial and temporal discretization and comparison of results. More details on the numerical aspects of coupling transport and chemistry by means of the two-step method are given in Herzer and Kinzelbach[9].

Comparison of the linearized one-step method and the two-step method
The coupled transport model was run for a one-dimensional test column in order to compare computational effort for three different solution methods:

/1/ A linearized one-step scheme with explicit chemical terms
/2/ An iterative two-step method with a simple Newton-Raphson method to solve the chemical subsystem
/3/ An iterative two-step method using the Gear-method to calculate chemical transformations

The normalized cpu-time requirement in order to achieve comparable accuracy of results was 1 for method /2/ (fastest method), 1.9 for method /3/ and 2.4 for method /1/. While method /1/ saves time by not having to iterate in the calculation of the chemical fluxes it is less efficient overall due to the much smaller time step needed in the essentially explicit approach. /2/ is faster than /3/ because the resolution in time is not refined during iteration in the implicit chemical half-step. But one has to be aware that due to the stiffness of the system /2/ may lead to an unsatisfactory resolution of the fastest processes involved in biochemical dynamics (usually bacterial growth). It is therefore advisable to have at least one model run using a method such as /3/ to be able to recognize possible inaccuracy of the faster but more inflexible method /2/.

APPLICATION TO IN-SITU REMEDIATION

A field case of in-situ remediation of a shallow aquifer polluted with rather insoluble or strongly adsorbed aliphatic and aromatic hydrocarbons is reported by Battermann and Werner[10]. Hydrocarbons of this type cannot be flushed with water efficiently. Therefore it was attempted to stimulate the activity of the autochtonous heterotrophic microflora to make it use the organic pollutant as carbon and energy source and thus degrade it to CO_2 and water. Due to the pollution this aquifer had excess amounts of electron donors (i.e. the organic pollutant) but it was lacking related electron acceptors. In order to enhance microbial activity, nitrate was supplied to the aquifer as electron acceptor by recharge wells. The possibility to degrade aliphatic and aromatic hydrocarbons via denitrification was examined and proofed for example by Zeyer et al.[11] and Riss and Schweisfurth[12] during laboratory column studies. The configuration of the above-mentioned remediation action is shown in fig. 3.

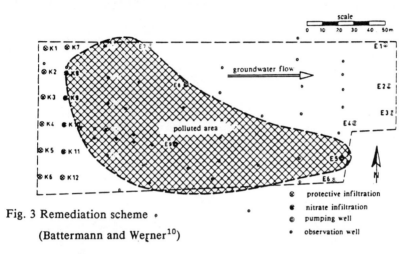

Fig. 3 Remediation scheme

(Battermann and Werner[10])

In order to simulate the behaviour of bacteria, DO, nitrate and the pollutant a two dimensional horizontally plane version of the coupled transport model presented in the preceding sections was used.

The measured breakthrough curves for a non-reactive tracer (sodium chloride) at the three recharge wells E7, E8 and E9 were used to calibrate the hydraulic parameters permeability and effective aquifer thickness, whereas the breakthrough curves for nitrate at these three wells served to evaluate the biochemical parameters. The comparison between these two different break-through curves in one well supplies two kinds of information: the area between the tracer and the nitrate curve is a measure for the amount of pollutant degraded and their differences in shape give clues due to the kinetics of the biochemical transformations including diffusion-limited exchange processes (fig. 4).

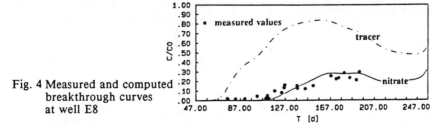

Fig. 4 Measured and computed breakthrough curves at well E8

The different slopes of the two curves indicate that a diffusion process is limiting microbial activity. If the exchange between the different phases in the aquifer were fast, the breakthrough curve of nitrate should be as steep as or even steeper than that of the tracer. In our model, two possible exchange processes exist which may be limited by diffusion:

/1/ Exchange of organic carbon between the aquifer matrix and the mobile pore water.

/2/ Exchange of dissolved species between pore water and the immobile biophase.

If process /1/ was diffusion-limited, the concentration of DOC (i.e the dissolved pollutant) would dramatically drop at the beginning of the remediation action due to microbial consumption and insufficient delivery of DOC. The measured concentrations for DOC did not show an abrupt but a gradual decline and thus process /2/ was considered to be diffusion-limited. In that case the concentration of species in pore water would not indicate any restriction of bacterial growth, while a low and limiting concentration could exist in the immobile biophase.

Fig 5 shows computed isolines for bacterial mass, DOC and nitrate 179 days after the beginning of the remediation. As can be seen, the highest actual concentration of microorganisms exists in the middle of the polluted area where there is also a steep gradient towards the still unaffected part of the spill. From the moving front of bacteria back towards the nitrate injection wells, the bacterial mass gradually declines. There the organic pollutant is already largely degraded, as is visible from the low DOC-concentrations in this region. A typical 2-D effect is the enhanced growth of bacteria at the edges of the spill due to nitrate supply from the outside. The isolines of nitrate concentrations show the steepest gradients (apart from the one between infiltration and protective well) in the same location as those of bacterial mass. This is due to enhanced microbial nitrate consumption at the actual front of the bacteria distribution. Because of missing bacterial activity due to the lack of an electron donor nitrate is transported to far greater downstream distances

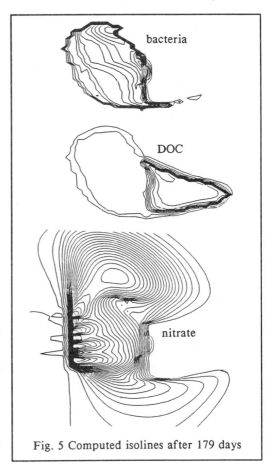

Fig. 5 Computed isolines after 179 days

outside of the polluted area than inside the spill. Here the effect of competitive transport and chemistry phenomena can be shown: Passing by the polluted area nitrate has almost reached the recharge wells on the righthand side of the spill, where a nitrate breakthrough could wrongly be interpreted as a breakdown of microbial activity and achievement of the remediation goal.

LIST OF SYMBOLS

O_2	: concentration of dissolved oxygen	$[M/L^3]$
O_t	: threshold concentration of dissolved oxygen	$[M/L^3]$
f_{s1}	: steepness of the switch function	$[-]$
c_i	: concentration of participant i	$[M/L^3]$
θ_i	: volume fraction of phase containing participant i	$[-]$
$\vec{v}_i$	: vector of average phase velocity of participant i	$[L/T]$
D_i	: tensor of hydrodynamic dispersion of participant i	$[L^2/T]$
$\vec{\nabla}$	: nabla-operator $(\partial/\partial x, \partial/\partial y)$	$[-]$
L	: differential operator of advection/dispersion	$[-]$
$\tilde{c}$	: intermediate concentration	$[M/L^3]$

ACKNOWLEDGEMENT

This research was financially supported by the Deutsche Forschungsgemeinschaft.

REFERENCES

1. Kinzelbach, W. and Schäfer, W. Coupling of chemistry and transport. IAHS Publication No. 188: Groundwater Management: Quantity and Quality, 237-261, 1989.
2. Frind, E. O., Sudicky, E. A. and Molson, J. W. Three-dimensional simulation of organic transport with aerobic biodegradation, IAHS Publication No. 185: Groundwater Contamination, 89-96, 1989.
3. Noye, J. Time-splitting the one-dimensional transport equation, North-Holland Mathematics Studies 145, Numerical Modelling: Application to marine systems, 271-295, Elsevier Science Publishers B. V. (North-Holland), Amsterdam, 1987.
4. Herzer, J. Chemflo-Dokumentation eines Schadstofftransportmodells für mehrere wechselwirkende Komponenten im Grundwasser, Programm-dokumentation Nr. 89/34 (HG118), Institut für Wasserbau, Universität Stuttgart, Stuttgart, F. R. Germany, 1989.
5. Frind, E. O. and Pinder, G. F. The principal direction technique for solution of the advection-dispersion equation, Proceedings of the World Congress of the International Association for Mathematics and Computers in Simulation (IMACS), Concordia University, Montreal, 1982.
6. Kinzelbach, W. and Frind, E. O. Accuracy criteria for advection-dispersion models. Finite Elements in Water Resources, Proceedings of the 6th International Conference, Springer Verlag, 489-504, 1986.
7. Kinzelbach, W. Numerische Methoden zur Modellierung des Transports von Schadstoffen im Grundwasser, Schriftenreihe gwf Wasser Abwasser, 21, R. Oldenbourg Verlag, München Wien, 1987.
8. Hindmarsh, A. C. GEAR: Ordinary differential equation system solver, Lawrence Livermore Laboratory Report UCID-30001, Revision 3, 1974.
9. Herzer, J., and Kinzelbach, W. Coupling of transport and chemical processes in numerical transport models, Geoderma, 44:115-127, 1989.
10. Battermann, G. and Werner, P. Beseitigung einer Untergrundkontamination mit Kohlenwasserstoffen durch mikrobiellen Abbau, gwf-Wasser/Abwasser, 125:366-373, 1984.
11. Zeyer, J., Kuhn, E. P., and Schwarzenbach, R. P. Rapid microbial mineralization of toluene and 1,3-dimethylbenzene in the absence of molecular oxygen. Applied and Environmental Microbiology, 52, 4: 944-947, 1986.
12. Hoos, E and Schweisfurth, R. Untersuchung über die Verteilung von Bakterien von 10 bis 90 Meter unter Bodenoberkante. Vom Wasser 58:103-112, 1982.

The Modeling of Radioactive Carbon Leaching and Migration from Cement Based Materials

J.F. Sykes(*), R.E. Allan(**), K.K. Tsui(**)
(*) Department of Civil Engineering, University of Waterloo, Waterloo, Ontario, Canada (**) Civil Eng. and Arch. Dept., Ontario Hydro, Toronto, Ontario, Canada

ABSTRACT

Heavy water reactors can produce radioactive wastes with high levels of carbon 14. Laboratory studies show that cement-based materials can significantly retard carbon 14 migration through the process of carbonation. In the carbonation process, the solution non-radioactive carbon and carbon 14 species combine with the portlandite in the cement to form calcite. The resulting bound carbon can undergo isotopic exchange with the solution carbon species. A two-dimensional finite element model of the migration, carbonation and exchange process was developed. In the model, the nonlinear mass conservation equations for the two carbon species are fully coupled. The bound concentrations are determined at the gauss points of the elements. The developed carbon-14 leach model is very sensitive to the rate at which carbonation proceeds; rapid carbonation can result in the development of sharp fronts for the solution concentrations.

INTRODUCTION

Heavy water reactors can produce radioactive wastes with high levels of carbon 14. This nuclide requires special consideration for its long term management because of its anionic nature and long half life. Previous studies have indicated that the geosphere surrounding a low- to intermediate-level waste management site would not prevent such nuclides from reaching receptor points at unacceptable levels (Allan,1987 & Russell,1987). For this reason, the engineered barriers of a management facility would have to be designed to provide adequate retention of C-14 nuclides over a long time period.

The retention of carbon by cement-based materials can be significant under certain conditions. This property makes the use of such materials in engineered barriers an attractive option. Preliminary laboratory studies on the use of cement in a management site have been encouraging. Carbonate ions in solution will react with portlandite ($Ca(OH)_2$) in cement and precipitate out as calcite ($CaCO_3$) (Reardon et al, 1987). In the first stage of carbonation, the pore surface areas of the cement are carbonated. This leaves the surfaces coated with a calcite layer. The presence of the coating retards further carbonation of the cement because the carbon must diffuse through the calcite layer in order to utilize any of the additional carbonation capacity of the cement. This slower carbonation via diffusion of carbon through the calcite layer makes up the second stage of carbonation. Relative to the primary carbonation, the second stage of carbonation is negligible.

The carbon required for preliminary carbonation is usually greater than the amount of free carbon that exists in the pore water. This implies that carbonates will be drawn in from the surrounding groundwater to the cement by diffusion and/or advection until the primary carbonation capacity is attained. Under a high flow, some carbonates could be transported through uncarbonated cement. However, high advective flow rates should not exist in a well-designed management facility.

During and after carbonation, C-14 can be transferred into and out of the porewater by isotopic exchange with nonradioactive carbon (C). The exchange process is not yet fully quantified but laboratory work on this subject is ongoing.

WASTE CONCEPTUAL MODEL

The basic management facility system is represented by a cement matrix containing immobilized carbon 14 surrounded by cement backfill materials. The species simulated in the leaching process are C (nonradioactive carbon), C-14 (radioactive carbon) and $\overline{Ca}$ (uncarbonated calcium). The carbons can be in the groundwater or in the cement matrix, while the calcium is bound in the cement. The initial total amount of calcium is assumed to be equivalent to the available primary carbonation capacity of the cement. The initial concentration of C-14 in the groundwater is assumed to be zero. The nonradioactive carbon concentration at the facility boundaries is held constant at the expected ambient groundwater concentration. A strict mass balance of the C-14 is maintained in the model. The initial mass can be depleted by decay and the movement of C-14 out of the system.

SYSTEM EQUATIONS

The two chemical processes that are considered in the numerical solution are carbonation and isotope exchange. The relevant carbonation reactions are:

$$Ca(OH)_2 + H_2CO_3 \leftrightarrow CaCO_3\downarrow + H_2O$$

$$CaO \cdot SiO_2 \cdot H_2O + H_2CO_3 \leftrightarrow CaCO_3\downarrow + \text{silica gel} + H_2O$$

On a mole basis, the above equations can be combined and represented in terms of calcium and nonradioactive carbon by:

$$C + \overline{Ca} \rightarrow \overline{C} . \tag{1}$$

In equation (1), $\overline{Ca}$ represents the moles of calcium hydroxide and hydrous calcium silicate that are available for primary carbonation but have not undergone carbonation, C represents the moles of carbon in the solution carbonic acid and $\overline{C}$ represents the bound carbon in the calcite. Carbon 14 must also be considered in the carbonation process:

$$C_{14} + \overline{Ca}_{14} \rightarrow \overline{C}_{14} . \tag{2}$$

Where C_{14} represents the moles of carbon 14 in solution carbonic acid and $\overline{C}_{14}$ represents the moles of bound carbon 14. Based on the law of mass action, the carbonation reactions represented by equations (1) and (2) can be modeled as:

$$\rho_b \frac{\partial \overline{C}}{\partial t} = k_1 (\rho_b \overline{Ca}) (\theta C) \tag{3}$$

and

$$\rho_b \frac{\partial \overline{C}_{14}}{\partial t} = k_2 (\rho_b \overline{Ca}) (\theta C_{14}) \tag{4}$$

respectively where k_1 and k_2 are the rate constants with the units of $L^3/(T \cdot mole)$ and ρ_b is the bulk density of the concrete. The representative equation for the immobile $\overline{Ca}$ is thus:

$$\rho_b \frac{\partial \overline{Ca}}{\partial t} = -k_1 (\rho_b \overline{Ca}) (\theta C) - k_2 (\rho_b \overline{Ca}) (\theta C_{14}) \tag{5}$$

where θ is the porosity of the concrete. For the bound carbon:

$$\overline{C} + \overline{C}_{14} = \overline{C}_T \tag{6}$$

where $\overline{C}_T$ is the total bound carbon expressed on a mole basis. $\overline{C}_T$ will change over time as a result of the carbonation processes represented by equations (3) and (4). Differentiating equation (6) with respect to time and using equations (3), (4) and (5) gives:

$$\frac{\partial \overline{C}_T}{\partial t} = -\frac{\partial \overline{Ca}}{\partial t} \tag{7}$$

The leaching of $\overline{C}_{14}$ is considered to be a stoichiometric kinetic isotope exchange process expressed as:

$$C_{14} + R \cdot \overline{C} \leftrightarrow C + R \cdot \overline{C}_{14}. \tag{8}$$

where R represents the adsorbent material (ie. concrete). It is assumed that the flow rates in the concrete will not become high enough to cause the dissolution of calcium carbonate ($CaCO_3$). A kinetic exchange model was selected as it permits the pore concentration of the carbon species to be zero; an equilibrium exchange model requires nonzero pore concentrations for the carbon.

The kinetic isotope exchange reaction of equation (8) can be modeled as:

$$\rho_b \frac{\partial \overline{C}}{\partial t} = a_1 \theta C - a_2 \rho_b \overline{C} \tag{9}$$

and

$$\rho_b \frac{\partial \overline{C}_{14}}{\partial t} = a_1 \theta C_{14} - a_2 \rho_b \overline{C}_{14} \tag{10}$$

where the first term on the right hand side of each equation represents the uptake of solution carbon by the cement matrix, and the second term represents the release of bound carbon from the cement matrix to the solution phase. The units of the exchange rate coefficients are $(1/T)$. The use of the same exchange rate coefficients (a_1 and a_2) for both carbon-14 and carbon-12 implies that the equilibrium selectivity coefficient for the reaction of equation (8) is unity. The equilibrium selectivity coefficient can be expressed by:

$$K_{C-14}^c = K_{eq} = \left(\frac{\overline{C}}{C} \right)_{eq} \left(\frac{C_{14}}{\overline{C}_{14}} \right)_{eq} \tag{11}$$

When carbonation and isotope exchange occur simultaneously, reaction equations (3) and (4) can be added to equations (9) and (10) respectively to yield:

$$\rho_b \frac{\partial \overline{C}}{\partial t} = a_1 \theta C - a_2 \rho_b \overline{C} + k_1 (\rho_b \overline{Ca}) (\theta C) \tag{12}$$

and

$$\rho_b \frac{\partial \overline{C}_{14}}{\partial t} = a_1 \theta C_{14} - a_2 \rho_b \overline{C}_{14} + k_2 (\rho_b \overline{Ca}\,) (\theta C_{14})$$ (13)

By substituting equations (12), (13) and (5) into the temporal derivative of (6) with (7), it can be shown that the exchange rate coefficients a_1 and a_2 are related by:

$$a_2 = \frac{a_1 \theta}{\rho_b} \left(\frac{C + C_{14}}{\overline{C} + \overline{C}_{14}} \right)$$ (14)

Substituting this relationship into equations (12) and (13) yields:

$$\rho_b \frac{\partial \overline{C}}{\partial t} = a_1 \theta \frac{\overline{C}_{14}}{\overline{C} + \overline{C}_{14}} C - a_1 \theta \frac{\overline{C}}{\overline{C} + \overline{C}_{14}} C_{14} + k_1 (\rho_b \overline{Ca}\,) (\theta C\,)$$ (15)

and

$$\rho_b \frac{\partial \overline{C}_{14}}{\partial t} = a_1 \theta \frac{\overline{C}}{\overline{C} + \overline{C}_{14}} C_{14} - a_1 \theta \frac{\overline{C}_{14}}{\overline{C} + \overline{C}_{14}} C + k_2 (\rho_b \overline{Ca}\,) (\theta C_{14})$$ (16)

In summary, the carbonation and isotope exchange reactions are modeled by equations (5), (15) and (16).

For carbon, the equation of mass conservation is:

$$\rho_b \frac{\partial \overline{C}}{\partial t} + \theta \frac{\partial C}{\partial t} = \frac{\partial}{\partial x_i} \left[\theta D_{ij} \frac{\partial C}{\partial x_j} \right] - \theta v_i \frac{\partial C}{\partial x_i}$$ (17)

where D_{ij} represents the hydrodynamic dispersion tensor, (which includes the dispersivity and diffusion coefficients), and v_i is the interstitial pore water velocity. The equation of mass conservation for carbon 14 can be similarly stated as:

$$\rho_b \frac{\partial \overline{C}_{14}}{\partial t} + \theta \frac{\partial C_{14}}{\partial t} = \frac{\partial}{\partial x_i} \left[\theta D_{ij} \frac{\partial C_{14}}{\partial x_j} \right] - \theta v_i \frac{\partial C_{14}}{\partial x_i} - \lambda \rho_b \overline{C}_{14} - \lambda \theta C_{14}$$ (18)

where λ is the radioactive decay coefficient for C-14.

Equations (17) and (18) together with equations (5), (15) and (16) are fully coupled and nonlinear. The equations describe the carbonation process, the kinetic exchange of carbon 14 with carbon in concrete and the transport of carbon in the pore water. All five species, bound and solution nonradioactive carbon, bound and solution carbon 14 and the primary carbonation capacity in terms of calcium are included.

In the two-dimensional numerical model GWPC, the preceding equations together with the boundary conditions are solved numerically using the finite element method. Specifically, the Galerkin Method of Weighted Residuals is used with a mixed isoparametric quadrilateral element. The nonlinear Galerkin equations are integrated using a Gauss Legendre quadrature. The solution C and C-14 are nodal variables while the bound phases of C, C-14 and calcium are gauss point quantities.

The nonlinear mass conservation equations (17) and (18) are fully coupled and are solved simultaneously for the nodal variables using a pseudo Newton-Raphson solution scheme with an Aitken accelerator and a Gauss elimination matrix solver. At each iteration, a full Newton-Raphson scheme is used to determine updates for the coupled gauss point bound phase quantities in the discretized forms of equations (5), (15) and (16). A steady state flow equation was used to determine the saturated interstitial pore water velocities .

SIMULATION AND CONCLUSIONS

The numerical properties of the developed two-dimensional model were investigated in a one-dimensional analysis. The 5.0 m spatial domain was discretized into 40 quadrilateral elements of length 0.125 m. Eight elements at either end were used to represent the 1.0 m concrete barrier or buffer surrounding the 3.0 m waste zone. The Darcy velocity through the system was assumed to be 0.002 m/yr. The effective porosity for both the concrete barrier and concrete waste was 0.22. The longitudinal dispersivity was assumed to be 5.0 m; the molecular diffusivity of the soluble carbon species in the concrete was 7.76×10^{-4} m²/yr. The selected parameter values yield a Peclet number of 0.025.

The initial solid phase concentrations of $\overline{Ca}$, $\overline{C}_{14}$ and $\overline{C}$ in the barrier were 2690, 0 and 10 moles/m³ respectively. For the concrete waste zone, the initial solid phase concentrations for the same species were 2160, 108 and 432 moles/m³ respectively. The concentration of soluble carbon in the water surrounding the waste package was assumed to be 5.0 moles/m³. This concentration was imposed at both ends of the spatial domain using a Dirichlet boundary condition; this assumes that there is an infinite supply of carbon in the surrounding groundwater and that the migration of carbon through the concrete is rate limiting.

From a numerical perspective, the most interesting parameter in the developed model is the carbonation rate coefficient. The length of the zone over which carbonation occurs is related to the magnitude of the coefficient; smaller values result in slower carbonation and hence a larger active carbonation zone. Conversely, larger values reduce the width of the active carbonation zone, such that a sharp front can occur. In the upgradient direction, carbonation is complete and the solution carbon concentrations can be at their equilibrium values. In the active carbonation and down-gradient zones, the solution carbon concentrations will approach zero. When the length of the active carbonation zone approaches the length of an element, numerical oscillations in the calculated solution and bound phase concentrations are observed. For the problem investigated in this paper, no oscillations were observed with a carbonation rate coefficient of 0.001 m³/yr/mole. Oscillations in the calculated bound phase concentrations began to occur with a rate coefficient of 0.01 m³/yr/mole and were well developed with a value of 0.03 m³/yr/mole (Figure 1). A value of 0.1 m³/yr/mole resulted in extreme oscillations in both the solution and bound phase concentrations. The calculated solution C-14 concentrations for the cases corresponding to Figure 1 are depicted in Figure 2; there is less evidence of numerical oscillation. The results indicate that the spatial and temporal discretization must be carefully selected in order to minimize oscillations occurring as a result of a rapid reaction rates.

REFERENCES

Allan, R.E. Reactor Waste Modelling 1987 Progress Report, Design & Development Division - Generation Report No. 87379, Ontario Hydro, Toronto, Canada. 1987.

Reardon, E.J., & Abouchar, J.A. Geochemical Aspects of the Carbonation of Grout: 2. Reactions Between Carbonate-Form Exchange Resins and a Grout Slurry, Dept. of Earth Sciences, University of Waterloo, Final Report Submitted to Ontario Hydro. 1987

Russell, S.B. Assessing Low and Intermediate Level Radioactive Waste Disposal - 1987 Progress Report, Design and Development Division - Generation Report No. 87533, Ontario Hydro, Toronto, Canada. 1987

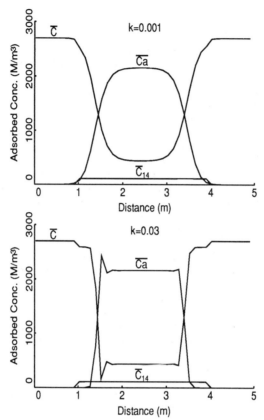

Figure 1: Adsorbed Concentrations at 50,000 Years for Different Carbonation Rate Coefficients (m³/yr•M)

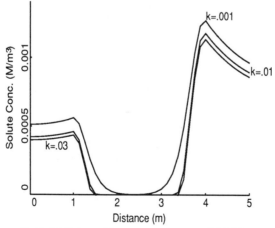

Figure 2: C-14 Solute Concentrations at 50,000 Years for Different Carbonation Rate Coefficients (m³/yr•M)

A Lagrangian-Eulerian Approach to Modeling Multicomponent Reactive Transport

G.T. Yeh, J.P. Gwo

Department of Civil Engineering, The Pennsylvania State University, University Park, PA 16802, USA

ABSTRACT

A Lagrangian-Eulerian (LE) approach is applied to solving reactive chemical transport problems of multi-component systems. The problem of negative concentrations is overcome with the LE approach. Two techniques of treating the Lagrangian step are included: one is to use the retarded velocity and the other is to adopt the pore velocity. The advantages and disadvantages of both techniques are compared and discussed. An attendant advantage of using the LE approach is that the resulting matrix equation is positive definite; thus one can use iteration methods, for example, the conjugate gradient method, to solve the matrix equation with convergency guaranteed.

INTRODUCTION

Concentrations of contaminants undergoing transport may vary over several orders of magnitudes within short distances of their point of release. These variations occur in response to hydrological factors and chemical interactions with the media. In advection-dominated hydrogeochemical transport problems, negative concentrations can occur, especially in sharp front cases, when traditional numerical methods of finite element or finite difference are used to solve the governing equations. Negative concentrations of chemical species are absurd and unacceptable from a chemical viewpoint. Upstream weighting methods can be used to circumvent these problems, but they require the use of very small spatial grid and time-step size to reduce induced numerical dispersion. Lagrangian-Eulerian approaches have been used to overcome the problem of negative concentrations without excessive restriction on the time-step size, Molz, et. al. [1], Neuman [2], Yeh [4]. For a single-component transport with linear isotherm, either the retarded velocity can be used, e.g., Molz et. al. [1] and Yeh [4], in the Lagrangian step, or the pore velocity can be adopted, e.g.,

Neuman [2] and Yeh [4]. For multi-component transport problems in which complexation, sorption, reduction/oxidation, and precipitation/dissolution are considered, the retarded velocity is not known *a priori* and must be obtained along the solution in the nonlinear iterative procedures. Alternatively, the pore velocity can be used in the Lagrangian step at the expense of possibly yielding inaccurate and/or zigzag solutions.

GOVERNING EQUATIONS OF MULTI-COMPONENT TRANSPORT

The general transport equation governing the temporal-spatial distribution of any chemical component in a multi-component system can be derived based on the conservation of mass as given by Yeh and Tripathi [3]

$$\theta \frac{\partial T_j}{\partial t} + \mathbf{V} \cdot \nabla C_j = \nabla \cdot (\theta \mathbf{D} \cdot \nabla C_j), \quad j = 1, 2, .., N \quad (1)$$

where θ is the moisture content, T_j is the total analytical concentration of the j-th component, t is the time, C_j is the total dissolved concentration of the j-th component, $\mathbf{V}$ is Darcy's velocity, $\mathbf{D}$ is the dispersion coefficient tensor, and N is the number of chemical components. Equation (1) is not a complete formulation since it involves more than one independent set of variables, T_j's, and C_j's. The relationship between T_j's and C_j's must be prescribed to complete the formulation. Implicit functional relationships between T_j's and C_j's have been derived based on geochemical equilibria by Yeh and Tripathi [3].

LAGRANGIAN-EULERIAN (LE) APPROACHES

The system of equations in Eq. (1) coupled with appropriate initial and boundary conditions constitute a mathematical statement of physical problems of multi-component chemical transport in subsurface media. Analytical solutions for this general system do not exist, and it would be insurmountably difficult to obtain one because of the high nonlinearity of the system of equations governing chemical equilibria. Numerical algorithms must be devised to solve the problem.

Experience has shown that conventional finite element methods (FEMs) or finite difference methods (FDMs) perform well for dispersion dominant transport problems. For advection dominant transport problems, negative concentrations may result with conventional FEMs or FDMs. To alleviate the problem of negative concentrations, upstream FDMs or FEMS have been used at the expense of introducing numerical dispersion. Mixed LE methods, have recently been employed by Molz et al. [1], Neuman [2], and Yeh [4] to solve subsurface contaminant transport problems. In these mixed LE methods, one adopts a Lagrangian viewpoint when dealing with the advection terms and an Eulerian viewpoint when dealing with all other terms in the transport equations. Thus, the methods combine the simplicity of the fixed Eulerian grid with the computational power of the Lagrangian approach to deal with advection dominant transport. A brief outline of the LE approaches and a comparison with the upstream FDM approaches follows.

For a single-component transport problems without the reaction of precipitation-dissolution, Eq. (1) can be written

$$\theta R_d \frac{\partial C}{\partial t} + \mathbf{V} \cdot \nabla C = \nabla \cdot (\theta \mathbf{D} \cdot \nabla C), \qquad R_d = 1 + \frac{dS}{dC} \qquad (2)$$

where R_d is the retardation factor. Upstream finite difference of Eq. (2) for one-dimensional cases leads to:

$$C_i^{n+1} = C_i^n + \frac{V \Delta t}{\theta R_d \Delta x}(C_{i-1}^n - C_i^n) + \frac{D \Delta t}{R_d \Delta x^2}(C_{i+1}^{n+1} - 2C_i^{n+1} + C_{i-1}^{n+1}) \qquad (3)$$

where C_i^n is the value of C at $t = n\Delta t$ and $x = i\Delta x$, Δt is the time step size, and Δx is the spatial grid size.

On the other hand, in the mixed LE approach, using the "retarded" velocity V^* $(= V/\theta R_d)$, one can write Eq. (2) as:

$$\theta R_d \frac{DC}{Dt} = \nabla \cdot (\theta \mathbf{D} \cdot \nabla C), \qquad \frac{dx^*}{dt} = \frac{V}{\theta R_d} = V^* \qquad (4)$$

where x^* is the location of Lagrangian particle representing the chemical traveling with a velocity of V^*. A finite difference approximation of Eq. (4) yields:

$$C_i^{n+1} = C_i^* + \frac{D \Delta t}{R_d \Delta x^2}(C_{i+1}^{n+1} - 2C_i^{n+1} + C_{i-1}^{n+1}) \qquad (5)$$

where C_i^* is the concentration of a fictitious particle that would arrive at node i at time $(n + 1)$. The accuracy of the mixed LE approach depends on how C_i^* is obtained. If a linear interpolation is used, then C_i^* is given by:

$$C_i^* = C_{i-k}^n + (V^* \Delta t - k \Delta x)(\frac{C_{i-k-1}^n - C_{i-k}^n}{\Delta x}) \qquad (6)$$

where k is an integer representing the number of spatial points the fictitious particle must travel over to reach node i in the time interval Δt with a velocity of V^*. When $k = 0$, substitution of Eq. (6) into Eq. (5) yields an equation identical to Eq. (3). Thus, the upstream difference to deal with the velocity term is a special case of the LE approach of using the linear interpolation to compute the concentration of the fictitious particle. The approach of finding fictitious particles at discrete time intervals so as to make them coincide with the nodes of fixed grids at the end of each time step is referred to as single-step reverse particle tracking by Neuman [2].

It is a well known fact that the upstream difference of the velocity term produces numerical dispersion, and can be concluded that single-step reverse particle tracking of the Lagrangian approach, using linear interpolation, also generates numerical dispersion because it is a generalized approximation of the upstream difference. To diminish numerical dispersion, the grid size must be refined. For the upstream finite difference, the time step size must be correspondingly reduced because stability and convergence often require that the Courant number (= $V^*\Delta t/\Delta x$) be less than one to prevent the sum of the first two terms on the right hand side of Eq. (3) from becoming negative. It is rarely practical to use both fine grid size and small time step to make numerical dispersion negligible in comparison to physical dispersion. On the other hand, in the mixed LE approach, the time step size does not have to be reduced corresponding to the reduction of grid size because C_i^* will not become negative if none of the node concentrations at the previous time is negative. Hence, in the mixed LE approach, only the grid size must be small to provide accurately convergent, stable solutions with the Courant number well in excess of one.

When the "retarded" velocity is known *a priori,* the retarded velocity is used in the Lagrangian approach to obtain C^*. In reactive chemical transport modeling, the retardation is usually not known *a priori* when the reaction is nonlinear. Under such circumstances, the retarded velocity must be obtained along with the nonlinear iterative solution procedure of Eq. (2). Alternatively, the pore velocity can be used to compute C^* as suggested by Neuman [2]. To illustrate how this can be achieved and to demonstrate that it can be reduced, as a special case, to upstream numerical methods and to the case of using retarded velocity, let us consider Eq. (2) written in a different form:

$$\theta\left(\frac{DC}{Dt} + K_d \frac{\partial C}{\partial t}\right) = \nabla\cdot(\theta\mathbf{D}\cdot\nabla C), \qquad \frac{dx}{dt} = \frac{V}{\theta} = V_f, \qquad K_d = \frac{dS}{dC} \qquad (7)$$

where K_d is the distribution coefficient and x is the location of the Lagrangian particles representing the fluid or the dissolved chemical traveling at a velocity of V_f. Finite difference approximation of Eq. (7) leads to:

$$C_i^{n+1} + K_d C_i^{n+1} = C_i^* + K_d C_i^n + \frac{D\Delta t}{\Delta x^2}(C_{i+1}^{n+1} - 2C_i^{n+1} + C_{i-1}^{n+1}) \qquad (8)$$

where C_i^* is the concentration of a fictitious particle that would arrive at node i at time (n + 1), and is given by the following formula:

$$C_i^* = C_{i-k}^n + (V_f \Delta t - k \Delta x)(\frac{C_{i-k-1}^n - C_{i-k}^n}{\Delta x}) \qquad (9)$$

where k is an integer representing the number of spatial points the fictitious particle must travel over to reach node i in the time interval Δt with a velocity of V_f. Clearly, for the special case of k = 0, the substitution of Eq. (9) into Eq. (8) yields an equation identical to Eq. (3). Hence, for the special case, the LE approach of using either the retarded velocity or the pore velocity becomes the upstream difference approximation.

For the case of a linear isotherm when K_d is independent of the concentration C, the LE approach of using the pore velocity offers no advantages over that of using the retarded velocity. However, for the case of a nonlinear isotherm such as Langmuir or Freundlich sorption model, the choice of velocity makes a significant difference. The locations of all fictitious particles (which are equal to the number of spatial grid points) are the same for all iterations in the nonlinear-solution loop if the pore-velocity approach is taken, whereas the locations of the fictitious particles are different for each iteration in the nonlinear-solution loop if the retarded velocity approach is taken. Hence, using the former approach, one only has to compute the locations of all fictitious particles once for the nonlinear loop; using the latter approach, one has to recompute the locations of the fictitious particles as many times as the number of iterations required to reach a convergent solution. However, using the pore velocity, one may generate inaccurate and/or zigzag solutions because of the presence of the term $\partial C/\partial t$.

A simple one-dimensional example is used to illustrate the differences in accuracy between retarded-velocity and pore-velocity based simulations using the computer code developed by Yeh [4]. The initial conditions are: C = 0 at t = 0 for 0 < x < 200 cm. The boundary conditions are: C = 1 at x = 0 and $\partial C/\partial x = 0$ at x = 200 cm. The Darcy velocity is V = 1.0 cm/d, the moisture content is $\theta = 0.2$, the distribution coefficient is $K_d = 0.6$ cm^3/g, and the dispersion coefficient is D = 0.1 cm^2/d. For numerical computations, the region is divided into 40 rectangular elements with element size = 5 x 1 cm. This results in a mesh Peclet number $P_e = 50$. Figure 1 shows the comparison of the simulation using the retarded-velocity approach (Fig. 1a) and the pore-velocity approach (Fig. 1b) with different time-step sizes: $\Delta t = 0.5, 1.6, 4.0, 6.4,$ and 12.8 d, respectively. For these time-step sizes, the mesh Courant number based on the retarded velocity are 0.125, 0.4, 1.0, 1.6, and 3.2 respectively, whereas the mesh Courant numbers based on the pore velocity are 0.5, 1.6, 4.0, 6.4, and 12.8, respectively. It is seen that, based on the retarded velocity, the accuracy of the solution increases with the time-step size, whereas, based on the pore velocity, the accuracy of

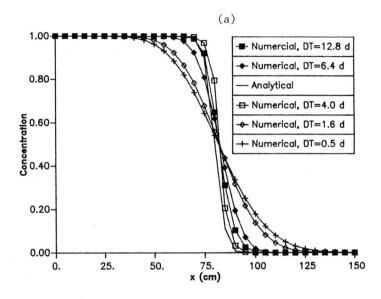

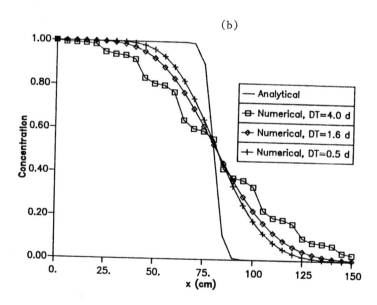

Fig. 1. Comparison of simulations using (a) retarded-velocity
and (b) pore-velocity based approaches.

solution deteriorates with the increasing of time-step size. This deterioration is caused by the presence of the term $\partial C/\partial t$. It is further seen that for the retarded-velocity based approach, the mesh Courant number can be increased to well in excess of 1 to reduce numerical dispersion.   For the pore-velocity based approach, the mesh Courant number must be less than 2 before the zigzag type of solutions develops.

MULTI-COMPONENT TRANSPORT USING THE LE APPROACH

For multi-component hydrogeochemical transport problems, the pore-velocity approach may become advantageous in terms of computational time but at the expense of yielding zigzag solutions when the transport is advection dominant.   The advantage of computational effort with pore-velocity based approach can be demonstrated as follows.   When the retarded velocity is used in the Lagrangian step, Eq. (1) is written:

$$\theta R_{dj}[\frac{DC_j}{Dt}+(\sum_{\substack{i=1\\i\neq j}}^{N} f_{ij}\frac{\partial C_i}{\partial t})] = \nabla\cdot(\theta \mathbf{D}\cdot\nabla C_j) \quad , \quad j = 1, 2, \ .., \ N \quad (10a)$$

$$\frac{dx_j^*}{dt} = \frac{V}{\theta R_{dj}} = V_j^* \quad , \quad j = 1, 2, \ ..., \ N \quad (10b)$$

where

$$f_{ij} = \frac{\partial S_j}{\partial C_i} + \frac{\partial P_j}{\partial C_i} \quad , \quad R_{dj} = 1 + f_{jj} \quad (11)$$

in which R_{dj} and V_j^* are the retardation factor and retarded velocity, respectively, of the j-th chemical component and x_j^* is the location of the Lagrangian particle representing the j-th chemical traveling at a velocity of V_j^*.

Examining Eqs. (10) and (11), we find at least three deficiencies when we use the retarded-velocity appraoch. First, $\partial P_j/\partial C_i$ cannot be determined, either explicitly or implicitly, as a function of C_i because the species concentration p_i is not involved in the mass action law of the precipitation.   Hence, the precipitation process cannot be dealt with properly. Second, since f_{ij} is not zero in general, the total dissolved concentration C_i appears in the j-th equation of Eq. (10), which governs the transport of the j-th component.   This almost makes the simultaneous solution of all N PDEs in Eq. (10a) imperative.   Third, the retarded velocity not only is a function of the primary dependent variables (PDVs) C_j's but is also different for each component.   This requires numerous evaluations of the locations of the fictitious Lagrangian particles for every iteration and for each component in the nonlinear-solution loop.   To overcome these problems, one can use pore velocity in the Lagrangian step but at the expense of possibly yielding inaccurate and/or zigzag solutions.   Both approaches would not produce negative concentrations though.

When the fluid particle velocity is used in the Lagrangian step, Eq. (1) is written in the following form:

$$\theta \left(\frac{DC_j}{Dt} + \frac{\partial S_j}{\partial t} + \frac{\partial P_j}{\partial j} \right) = \nabla \cdot (\theta \mathbf{D} \cdot \nabla C_j) \quad , \qquad j = 1, 2, .., N \qquad (12a)$$

$$\frac{d\mathbf{X}}{dt} = \frac{V}{\theta} = V_f \qquad (12b)$$

where $\mathbf{X}$ is the location of the Lagrangian particle representing the fluid or the dissolved chemicals all traveling at a velocity of V_f. A numerical approximation of Eq. (12) - for example, the backward time difference and either finite element or finite difference spatial discretization - yields

$$\left(\frac{[M]}{\Delta t} + [D] \right) \{ T_j^{n+1} \} = \frac{[M]}{\Delta t} \left(\{ C_j^* \} + \{ S_j^n \} + \{ P_j^n \} \right) +$$

$$[D] \left(\{ S_j^{n+1} \} + \{ P_j^{n+1} \} \right)] \quad , \qquad j = 1, 2, .., N \qquad (13)$$

where $[M]$ is the mass matrix resulting from the terms associated with the time derivatives, $[D]$ is the stiff matrix resulting from the dispersion terms, $\{ T_j^{n+1} \}$ is the column vector representing the total analytical concentration of the j-th component at all spatial nodes at the new time $(n + 1)$, $\{ C_j^* \}$ is the column vector representing the total dissolved concentration of the fictitious particles that would arrive at spatial nodes at time $(n + 1)$ for the j-th chemical, $\{ S_j^n \}$ and $\{ S_j^{n+1} \}$ are the column vectors representing the total sorbed concentration of the j-th chemical at all spatial nodes at the old time n and new time $(n + 1)$, respectively, and $\{ P_j^n \}$ and $\{ P_j^{n+1} \}$ are the column vectors representing the total precipitated concentration of the j-th chemical at all spatial nodes at the old time n and new time $(n + 1)$, respectively. For a single-step reverse particle tracking method, $\{ C_j^* \}$ is, of course, interpolated from $\{ C_j^n \}$.

From Eq. (13), it is seen that we choose the total analytical concentrations of all components as the primary dependent variables (PDV's). This makes it easy to include the precipitation process as most chemical equilibrium models are designed to solve for the species distributions given the total analytical concentrations of all components. Note that only the PDV for one component appears in each of the N equations in Eq. (12). Therefore, it is easy to use an iterative solution of N equations one by one sequentially, whereas in the retarded velocity approach, one normally uses an iterative solution of N equations simultaneously for the reason given previously. Finally, the locations of all the fictitious particles have to be computed only once. These locations are then used for all components and need not be updated during the nonlinear iteration loop. Thus, we can expect that all three deficiencies resulting from using the retarded velocity approach can be eliminated by using the fluid particle velocity approach in the Lagrangian step.

Finally, it is noted that Eqs. (5), (8), and (13) obtained with the LE approach can be written as matrix equations with positive definite matrices. Thus, one can use iteration methods, for example, the conjugate gradient method, to solve these equations with convergency guaranteed.

CONCLUSIONS

Two Lagrangian-Eulerian approaches can be used to solve the advection-dominant reactive hydrogeochemical transport to overcome the problem of negative concentrations: one uses the retarded velocity and the other uses the pore velocity for the computation of Lagrangian concentrations. It is shown that both approaches are reduced to upstream finite difference approximations for the special cases. It is also shown that for single component transport problems with linear isotherm sorption, the approach using the pore velocity is reduced to that of using the retarded velocity when the mesh Courant number is less than 1. To overcome the problem of negative concentrations, spatial grid size must be refined and time step must be correspondingly decreased for upstream numerical schemes but need not be for the Lagrangian-Eulerian approaches. It is demonstrated that excessive numerical dispersion can be overcome by increasing the time-step size using the retarded velocity approach. It is also shown and demonstrated that the deficiencies associated with the method of using the retarded-velocity can be eliminated by using the pore-velocity approach but at the expense of possibly yielding inaccurate and/or zigzag solutions.

REFERENCES

1. Molz, F. J., M. A. Widdowson, and L. D. Benefield.
    Simulation of microbial growth dynamics coupled to nutrient
    and oxygen transport in porous media. Water Resour. Res.
    Vol. 22, 1207-1216, 1986.

2   Neuman, S. P. Adaptive Eulerian-Lagrangian finite element
    method for advection-dispersion, Intern. Jour. Num. Meth.
    Eng. Vol. 20, 321-337.

3. Yeh, G. T. and V. S. Tripathi: HYDROGEOCHEM: A Coupled
    HYDROlogical Transport and GEOCHEMical Equilibrium of Multi-
    Component Systems. ORNL-6371. Oak Ridge National Lab., Oak
    Ridge, Tennessee, 1990.

4. Yeh, G. T. LEWASTE: A Lagrangian-Eulerian Approach to
    Modeling WASTE Transport through Saturated-Unsaturated Media
    PSU Technical Report, The Pennsylvania State University,
    University Park, PA 16802.

A Particle Tracking Method of Kinetically Adsorbing Solutes in Heterogeneous Porous Media

R. Andricevic, E. Foufoula-Georgiou

St. Anthony Falls Hydraulic Laboratory,
Department of Civil and Mineral Engineering,
University of Minnesota, Minneapolis, MN 55414
USA

ABSTRACT

A particle tracking approach is presented for simulating field scale transport of kinetically adsorbing solutes. The chemical kinetic non–equilibrium is modeled as a birth and death process and is coupled with the particle tracking approach by using the first two moments of the particle waiting times spent in the liquid phase. The total fraction of the adsorbed solute at any time and position is then analytically determined from the computed transition probabilities. The use of a birth and death process is especially attractive since it can be easily extended to the case of multi component chemical reactions and exhibits moderate computational requirements.

INTRODUCTION

Pollutant fate and transport in natural aquifers is highly dependent on the solute partitioning between solid and liquid phase. Such a partitioning will be referred to herein as sorption. Sorption reactions are one of the most important processes governing the fate of many organic pollutants whose transport is generally retarded by the presence of sorption mechanism. The local equilibrium assumption (LEA) has been widely used to date to describe fast chemical reactions because of its mathematical simplicity. Several recent investigations (e.g. Jennings and Kirkner [6], Valocchi [12], Bahr and Rubin [2]) reported the conditions for which LEA is applicable. However, some other studies (e.g. Bahr and Rubin [2], Roberts, et. al. [9]) indicated the importance of studying non–equilibrium phenomena in order to properly describe the adsorbing solute transport. The non–equilibrium phenomena have been examined extensively in laboratory experiments focusing on the kinetic non–equilibrium (e.g., the chemical will be adsorbed slowly) and/or physical non–equilibrium which describes the physical resistances encountered by a chemical trying to reach the sorption sites of the porous medium during its movement through the soil.

In order to study coupled phenomena of chemistry and transport at the field scale the numerical technique needs to posses negligible numerical dispersion and be computationally efficient for three dimensional applications. The particle tracking approach meets these requirements (Kinzelbach [7]) and it is especially attractive for problems involving chemical reactions. Apart from being free from numerical dispersion (in the classical sense), the treatment of chemical reactions on the particle level provides more flexibility and allows better insight from the chemical modeling standpoint. This technique typically represents the solute mass as a large collection of particles; each particle is moved with deterministic and random displacements over discrete increments of time. The magnitude of each displacement depends upon the velocity and dispersion field.

The next step involves the description of the considered chemical reaction. This study deals with the first–order reversible chemical kinetic with constant reaction rates. This type of reactions describes slow adsorption–desorption processes or particle exchange between mobile and immobile pore water with given forward and reverse reaction rates.

The key problem in modeling kinetic non–equilibrium is how to determine the fraction of the time step that each particle spends in the liquid phase. The importance of accurately estimating the time spent in liquid phase stems from the fact that only dissolved particles participate in the transport. In other words, the effective time spent in the liquid phase is needed to accurately predict the solute behavior. A full generation of the waiting times in the liquid phase can be a "brute–force" solution which suffers from computational inefficiency.

Under these conditions we propose to use the birth and death process with the corresponding reaction rates to estimate the first two moments of the particle waiting times spent in the liquid phase. The transport model will then use this waiting time to move particles to the new position. The total fraction of the adsorbed solute at any time and position will be analytically determined from the transition probabilities computed following the birth and death–type process.

PARTICLE TRACKING APPROACH

Consider the transport model given by the convection–diffusion equation in two horizontal dimensions which describes the large scale movement of pollutants in groundwater. The transport equation of the conservative solute (Bear [3]) is given by:

$$\frac{\partial c}{\partial t}+u_x\frac{\partial c}{\partial x}+u_y\frac{\partial c}{\partial y} = \frac{\partial}{\partial x}(D_{xx}\frac{\partial c}{\partial x}+D_{xy}\frac{\partial c}{\partial x})+\frac{\partial}{\partial y}(D_{yy}\frac{\partial c}{\partial x}+D_{yx}\frac{\partial c}{\partial x}) \tag{1}$$

where c is the concentration, u_x is the velocity in x direction, and D is the dispersion tensor with, for example, D_{xx} being the dispersion coefficient in the x direction. The random walk method, as a special case of the particle tracking approach, has a full analogy with the Fokker–Plank equation (Tompson et. al. [11]). In particular, the particle distribution can be obtained from a random walk model with step equations of the form (Kinzelbach [7])

$$x(t+\Delta t) = x(t) + \overset{*}{u_x} \Delta t + z_1\sqrt{2\alpha_l u \Delta t}\,\frac{u_x}{u} - z_2\sqrt{2\alpha_t u \Delta t}\,\frac{u_y}{u}$$

$$(2)$$

$$y(t+\Delta t) = y(t) + \overset{*}{u_y} \Delta t + z_1\sqrt{2\alpha_l u \Delta t}\,\frac{u_y}{u} - z_2\sqrt{2\alpha_t u \Delta t}\,\frac{u_x}{u}$$

where

$$\overset{*}{u_x} = u_x + \partial D_{xx}/\partial x + \partial D_{xy}/\partial y$$
$$\overset{*}{u_y} = u_y + \partial D_{yy}/\partial x + \partial D_{yx}/\partial y$$
$$u = \sqrt{u_x^2 + u_y^2}$$

and z_1 and z_2 are random numbers $N(0,1)$, and α_l and α_t are the longitudinal and transversal dispersivities, respectively.

In order to account for kinetic non–equilibrium the transport equation (1) needs to be modified. For illustration purposes, we will present the case of one–dimensional, steady flow in homogeneous porous medium

$$\frac{\partial c}{\partial t} + \frac{\partial s}{\partial t} = D_x \frac{\partial^2 c}{\partial x^2} - u_x \frac{\partial c}{\partial x}$$

$$(3)$$

where s is the adsorbed concentration (mass/fluid volume). In the process of coupling a chemical reaction with the transport model there is a need to evaluate the chemical reaction rate which is now a part of the transport equation (3). In this study the first order reversible linear reaction rate expression will be described as

$$\frac{\partial s}{\partial t} = k_1 c - k_2 s$$

$$(4)$$

where s and c are the concentrations in the solid and liquid phases, respectively; k_1 (time^{-1}) and k_2 (time^{-1}) are the adsorption and desorption rate coefficients, respectively.

Ahlstrom et al. [1] solved the transport problem and then locally applied the discretized form of (4) in space to determine the adsorbed concentration. This approach is evidently hampered with numerical errors unless the time step is very small to justify the decoupled process. Valocchi and Quinodoz [13] presented a way of generating the waiting times from the probability distributions of $\beta(\Delta t)$ (the fraction of time that a particle spends in the liquid phase during a time interval Δt) as given by Keller and Giddings [8]. Again problems arise in case of fast reactions when the necessary number of generated waiting times may be too high.

At this point of the development we propose the use of birth and death type process to model kinetic non–equilibrium adsorption given by (4). The first two moments of the total time spent in the liquid phase during the time step Δt will be derived by simulation. The choice of Δt is usually determined from transport modeling as a function of the velocity field.

Continuous time discrete state Markovian model: the birth and death process

The chemical reaction in (4) can be represented in a stochastic framework as a special case of a birth and death process, specifically a two-state continuous time Markov chain (Valocchi and Quinodoz [13]). Between transitions (phase changes) each particle stays in liquid or solid phase an amount of time $(w_1$ and w_2, respectively) which is exponentially distributed as

$$f_1(w_1) = k_1 exp(-k_1 w_1) \quad ,w_1 \geq 0 \tag{5a}$$

$$f_2(w_2) = k_2 exp(-k_2 w_2) \quad ,w_2 \geq 0 \tag{5b}$$

The waiting times w_1 and w_2 are independent random variables. This   directly implies that the counting process $N(t)$ of transitions from liquid to solid or solid to liquid phase is a renewal process with interarrival times $w=w_1+w_2$. The first two moments of the interarrival time w are easily obtained as

$$E(w) = \frac{k_1+k_2}{k_1 k_2} \tag{6}$$

$$Var(w) = \frac{1}{k_1^2} + \frac{1}{k_2^2} \tag{7}$$

The key problem is to determine the total time each particle stays in the liquid phase during a time step Δt which is usually determined from the particle tracking method as a function of the velocity field.

Let $\Delta t^\star$ denote the total waiting time each particle stays in the liquid phase during a given actual time step Δt. $\Delta t^\star$ can be expressed as

$$\Delta t^\star = \sum_{i=1}^{N(\Delta t)} w_{1,i} \tag{8}$$

where $\{N(\Delta t)\}$ is the random number of transitions from solid to liquid phase during the actual time interval Δt and $w_{1,i}$ is the time spent in the liquid phase during transition i. The first two moments of the counting renewal process $N(\Delta t)$ can be obtained from the mean and variance of the interarrival times w making use of some asymptotic results (Ross [10])

$$\lim_{t \to \infty} E[N(t+\Delta t)-N(t)] = \lim_{t \to \infty} E[N(\Delta t)] \approx \frac{\Delta t}{E(W)} = \frac{k_1 k_2}{k_1+k_2}\Delta t \tag{9}$$

$$\lim_{t \to \infty} Var[N(t+\Delta t)-N(t)] = \lim_{t \to \infty} Var[N(\Delta t)] \approx \frac{Var(W)}{(E(W))^3} \Delta t =$$

$$= \frac{k_1 k_2 (k_1^2 + k_2^2)}{(k_1 + k_2)^3} \Delta t \tag{10}$$

Since the waiting times $w_{1,i}$ in liquid phase between transitions are independent and identically distributed random variables but they are dependent on the counting process of transitions from solid to liquid phase, the mean and variance of $\Delta t^\star$ cannot be easily derived analytically. They can however be derived by simulation as, in fact, can the whole probability distribution function of $\Delta t^\star$.

Under assumption of normality of the random variable $\Delta t^\star$, each particle will be advected and dispersed in the transport model with the time step

$$\Delta t^\star = E(\Delta t^\star) + z \sqrt{Var(\Delta t^\star)} \tag{11}$$

where z is a standard normal random deviate and the mean and variance of $\Delta t^\star$ were estimated by simulation. Note that for very slow reaction rates (for which the average number of transitions $E[N(\Delta t)]$ during Δt might be small) the assumption about normality should be checked and if not appropriate the time step for each particle can be generated from the numerically derived probability distribution function of $\Delta t^\star$.

The next step requires the calculation of the transition probability which will be used to obtain the adsorbed fraction of particles spatially distributed for the next time step. Transition probabilities represent the probability of transition between any combination of two states, given the finite time length. Since the two–state continuous time Markov process used to model kinetic non–equilibrium is stationary, it is only the time length between the initial and final times that is important. The transition probability functions are found by solving a system of Kolmogorov differential equations (Ross [10]) and for the case of finite Δt they read as follows:

$$P_{11}(\Delta t) = \frac{k_2}{k_1 + k_2} + \frac{k_1}{k_1 + k_2} \exp[-(k_1 + k_2)\Delta t] \tag{12}$$

$$P_{22}(\Delta t) = \frac{k_1}{k_1 + k_2} + \frac{k_2}{k_1 + k_2} \exp[-(k_1 + k_2)\Delta t] \tag{13}$$

$$P_{21}(\Delta t) = 1 - P_{22}(\Delta t) = \frac{k_2}{k_1 + k_2} [1 - \exp[-(k_1 + k_2)\Delta t]] \tag{14}$$

$$P_{12}(\Delta t) = 1 - P_{11}(\Delta t) = \frac{k_1}{k_1 + k_2} [1 - \exp[-(k_1 + k_2)\Delta t]] \tag{15}$$

where the first sub–index denotes the initial phase and second sub–index denotes the final phase and all equations are written as functions of the actual time step Δt. Equations (12) – (15) must be coupled with the particle tracking method to provide at each time step the fraction of the particles (transition probability) adsorbed on the solid phase.

The presented formulation of chemical kinetics allows us to explicitly account for total waiting times in each phase as well as to calculate the fraction of particles adsorbed on the solid matrix at any given time instant. Furthermore, the generality of this technique enables the treatment of more than two phases in the transport processes. Some of the available analytical solutions for simplified cases will be used to verify the proposed coupled technique for transport of reactive solutes.

NUMERICAL EXAMPLE

The first case considered represents a continuous injection at the left boundary, homogeneous pore water velocity u=1 without dispersion, and rate coefficients $k_1=k_2=1.0$. This case was chosen based on the availability of an analytical solution (Bolt, 1982). The concentration profile for liquid phase was calculated and analytical solution was compared to the proposed method at T=100 (Figure 1.), showing very good agreement. Simulation with the particle tracking technique used 5,000 particles and required about 3 minutes CPU time on an Apollo DN 10000 workstation.

The second case represents an instantaneous point injection of liquid phase solute with same characteristics as above. The concentration profile corresponding to T=100 is shown on Figure 1. Note that an increased fluctuation for the first case is due to the fact that particles were constantly added to maintain continuous injection, while in the second case a total of 5,000 particles were instantly injected at T=0.

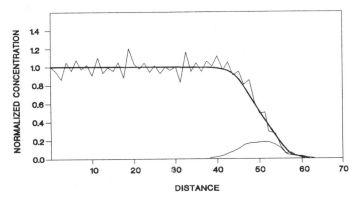

Figure 1. Concentration profiles and analytical solution at T=100.

For this case, the distribution of Δt^* did not show significant deviation from a Normal distribution. Note that because the computation of the whole probability distribution of Δt^* or its first two moments is done off–line, the computational requirements of the method are not increased as compared to the standard random walk model. Here we have only considered a simple case with no dispersion and the effects of variable $\Delta t^{\star}$ are only reflected on the variability of the solution around the theoretical line. Research is underway to extend the proposed method to three dimensional cases and spatially variable reaction rates.

REFERENCES

1. Ahlstrom, S., Foote, H., Arnett, R., Cole, C. and Serne, R. Multi Component Mass Transport Model: Theory and Numerical Implementation, Rept. BNWL 2127, Battelle Pacific Northwest Lab., Richland, 1977.

2. Bahr, J.M. and Rubin, J. Direct Comparison of Kinetic and Local Equilibrium Formulations for Solute Transport Affected by Surface Reactions, Water Resour. Res., Vol 3, pp. 438–452, 1987.

3. Bear, J. Dynamics of Fluids in Porous Media, Elsevier, New York, 1979.

4. Bolt, G.H. Movement of Solutes in Soils: Principles of Adsorption/exchange Chromatography. Chapter 9, Soil Chemistry, B, Physico–Chemical Models, (Ed. Bolt, G.H.), pp. 285–348, Elsevier, New York, 1982.

5. Bouchard, D.C., Wood, A.L., Campbell, M.L., Nkedi–Kizza, P. and Rao, P.S.C. Sorption nonequilibrium during solute transport, J. Contam. Hydrol., Vol. 2, pp. 209–223, 1988.

6. Jennings, A.A. and Kirkner, D.J. Instantaneous Equilibrium Approximation Analysis, J. Hydraul. Eng., Vol 110, pp. 1700–1717, 1984.

7. Kinzelbach, W. The Random Walk Method in Pollutant Transport Simulation, in Groundwater Flow and Quality Modeling (Ed. Custodio E. et. al.), pp. 227–245, D. Reidel, Dordrecht, 1988.

8. Keller, R.A. and Giddings, J.C. Multiple zones and spots in chromatography, J. Chromatog., Vol 3, pp. 205–220, 1960.

9. Roberts, P.V., Goltz, M.N. and Mackay, D.M. A Natural Gradient Experiment on Solute Transport in a Sand Aquifer, 3, Retardation Estimates and Mass Balances for Organic Solutes, Water Resour. Res., Vol 13, pp. 2047–2058, 1986.

10. Ross, S.M. Introduction to Probability Models, Academic Press, San Diego, 1985.

11. Tompson, A.F.B., Vomvoris, E.G. and Gelhar, L.W. Numerical Simulation of Solute Transport in Randomly Heterogeneous Porous Media: Motivation, Model Development, and Application, Rept. R88–05, Ralph M. Parsons Laboratory, M.I.T., 1988.

12. Valocchi, A.J. Validity of the local equilibrium assumption for modeling sorbing solute transport through homogeneous soils, Water Resour. Res., Vol 6, pp. 808–820, 1985.

13. Valocchi, A.J. and Quinodoz, H.A.M. Application of the Random Walk method to Simulate the Transport of Kinetically Adsorbing Solutes, in Groundwater Contamination (Ed. Abriola, L.M.), pp. 35–42, Proceedings of the Third Scientific Assembly of the IAHS, Baltimore, 1989.

An Adaptive Petrov-Galerkin Finite-Element Method for Approximating the Advective-Dispersive-Reactive Equation

F.H. Cornew, C.T. Miller

Department of Environmental Sciences and Engineering, CB# 7400, 105 Rosenau Hall, University of North Carolina, Chapel Hill, NC 27599-7400, USA

INTRODUCTION

The advective-dispersive-reactive (ADR) equation is frequently used to describe contaminant transport in groundwater systems. Accurate approximation of the ADR equation using standard Bubnov-Galerkin methods for realistic parameter values often leads to sharp fronts that require fine spatial and temporal discretization schemes. Considerable attention has been given to this problem in the literature over the last two decades. Three general approaches to solving sharp-front problems have emerged: upstream weighting methods, adaptive mesh methods, and Eulerian-Lagrangian methods. The adaptive Petrov-Galerkin method considered here is a combination of the first two approaches for solving sharp-front problems.

PETROV-GALERKIN METHOD

The general advective-dispersive-reactive equation (ADR) is:

$$\frac{\partial C}{\partial t} = \nabla \cdot (\mathbf{D}_h \cdot \nabla C) - \vec{v} \cdot \nabla C + R + \Gamma(C) \tag{1}$$

where C is a solute concentration; t is time; $\mathbf{D}_h$ is a hydrodynamic dispersion tensor; $\vec{v}$ is a mean pore velocity vector; R represents all solute mass transfer, degradation, and transformation reactions; and $\Gamma(C)$ represents all source or sink functions. A one-dimensional form of the ADR equation is:

$$R_f \frac{\partial C}{\partial t} = D_x \frac{\partial^2 C}{\partial x^2} - v_x \frac{\partial C}{\partial x} - kC \tag{2}$$

where R_f is a retardation coefficient that arises from the assumptions of linear local-equilibrium sorption (Miller and Weber[1]), D_x is a longitudinal hydrodynamic dispersion coefficient, v_x is the mean solute pore velocity in the x direction, and k is a first-order solute degradation rate constant.

The weak form of Equation (2), after applying the method of weighted residuals, Green's formula to the dispersive term, and the Crank-Nicolson approximation to the time derivative is:

$$\int_D \left(\frac{D_x}{2} \frac{\partial W_I}{\partial x} \frac{\partial N_J}{\partial x} C_J^{l+1} + \frac{v_x}{2} W_I \frac{\partial N_J}{\partial x} C_J^{l+1} + \frac{k}{2} W_I N_J C_J^{l+1} + \frac{R_J}{\Delta t} W_I N_J C_J^{l+1} \right) dx$$

$$= \int_D \left(-\frac{D_x}{2} \frac{\partial W_I}{\partial x} \frac{\partial N_J}{\partial x} C_J^l - \frac{v_x}{2} W_I \frac{\partial N_J}{\partial x} C_J^l - \frac{k}{2} W_I N_J C_J^l + \frac{R_J}{\Delta t} W_I N_J C_J^l \right) dx$$

$$+ \frac{1}{2} W_I \frac{\partial \hat{C}^{l+1}}{\partial x} \bigg|_0^L + \frac{1}{2} W_I \frac{\partial \hat{C}^l}{\partial x} \bigg|_0^L$$

$$\tag{3}$$

where W_I are weighting or test functions of an unspecified form, N_J are Lagrange polynomial basis or shape functions of an unspecified degree, I and J are implied summations over the number of nodes in the domain $\mathcal{D}$, for $0 \leq x \leq L$, $\hat{C}$ is a trial solution, $l+1$ is the unknown time level, and l is the known time level.

Lagrange polynomial shape functions are of the form:

$$N_j = \prod_{\substack{n=1 \\ n \neq j}}^{n_d+1} \frac{(x - x_n)}{(x_j - x_n)} \tag{4}$$

where the subscript j is used to specify the shape function for node j in an element of degree n_d.

Westerink and Shea[2] evaluated Petrov-Galerkin methods for linear and quadratic shape functions with accompanying test function that were modified by functions both one and two degrees higher than the shape functions. These methods were termed N+1 and N+2 degree upwinding, respectively. For linear elements, quadratic (5) and cubic (6) modifying functions are used, where the subscript denotes the degree of the modifying function:

$$M_2(\xi) = \frac{3}{4}(\xi + 1)(-\xi + 1) \tag{5}$$

$$M_3(\xi) = \frac{5}{8}\xi(\xi + 1)(\xi - 1) \tag{6}$$

The modifying functions are combined with linear Lagrange polynomial basis functions, $N_{l1}(\xi)$ and $N_{l2}(\xi)$ to produce the weighting functions:

$$W_{l1}(\xi) = N_{l1}(\xi) - \alpha M_2(\xi) - \beta M_3(\xi) \tag{7}$$

$$W_{l2}(\xi) = N_{l2}(\xi) + \alpha M_2(\xi) + \beta M_3(\xi) \tag{8}$$

where α and β are second and third degree upstream weighting coefficients, respectively.

For quadratic elements, the cubic modifying function (6) is used in conjunction with a quartic modifying function:

$$M_4(\xi) = \frac{21}{16}\left(-\xi^4 + \xi^2\right) \tag{9}$$

The resultant forms of the quadratic test functions are:

$$W_{q1}(\xi) = N_{q1}(\xi) - \alpha_c M_3(\xi) - \beta_c M_4(\xi) \tag{10}$$

$$W_{q2}(\xi) = N_{q2}(\xi) + 4\alpha_m M_3(\xi) + 4\beta_m M_4(\xi) \tag{11}$$

$$W_{q3}(\xi) = N_{q3}(\xi) - \alpha_c M_3(\xi) - \beta_c M_4(\xi) \tag{12}$$

where N_q is a quadratic Lagrange polynomial basis function, α_c and β_c are third and fourth degree upwinding coefficients for the corner nodes, and α_m and β_m are third and fourth degree upwinding coefficients for the middle node in a quadratic element.

MESH REFINEMENT METHOD

Three different types of adaptive methods have received attention in the computational fluid mechanics literature: the r-method, the h-method, and the p-method. Effective techniques often use the features of two or more of the three basic methods (Oden[3]). The r-methods relocate a fixed number of nodes and elements within the domain to increase nodal densities in regions of the domain that have large errors. The h-methods refine the mesh to increase nodal density in regions of the domain that have large errors. The p-methods increase the degree of shape functions in regions of the domain with high error, while maintaining a constant discretization pattern.

The adaptive method considered here is a decoupled h-method that has a common feature of a r-method: a constant number of global nodes exists at all times within the domain. Adaptation consists of forming an embedded mesh in regions of high error, partially reducing the resultant system of equations, and substituting the partially-reduced entries from the embedded mesh into the global coefficient matrix and solving. This method has been applied to the one-dimensional, nonreactive form of the advective-dispersive equation using linear shape functions by Yeh.[4]

To illustrate the adaptive embedded-mesh method, consider the case where adaptation of the element bounded by global nodes 3 and 4 in Figure 1(a) is desired. This element is subdivided into eight sub-elements, and a new numbering scheme is adopted for the embedded elements. The coefficient matrix for the embedded elements is shown in Figure 1(b), where nonzero entries are indicated by the shaded boxes. Note that the numbering scheme places the global node equations in the last two rows, which are bounded by heavy lines. Partial Gauss elimination is performed through the first n_i rows, where n_i is the number of embedded nodes. The final two partially-reduced equations are then only a function of global nodes and may therefore be incorporated into the global matrix solution, by substituting the reduced embedded solution coefficients in place of the original global element contribution in the coefficient matrix. The partial reduction is also performed on the right-hand-side (RHS) vector and its last two entries are then similarly inserted in the global RHS vector. Concentrations needed to form the RHS vector for the embedded elements are derived from: (1) known initial conditions at the first time step, (2) embedded solutions from the last time step (if available), or (3) interpolation using the trial solution of the global elements.

ALGORITHM

Adaptive Petrov-Galerkin codes were derived using the methods outlined above for the cases of linear and quadratic shape functions for the general finite element solution given by Equation (3). Both codes can simulate a Gaussian or step source, either of which may be set as an initial condition based upon the appropriate analytical solution (Bear[5]). The analytical solutions are also used to compute six measurements of the solution accuracy: integral error, discrete error, peak depression error, oscillation error, phase shift error, and mass conservation error (Westerink and Shea[2]).

The h-level, or number of embedded elements per adapted element, is unconstrained. The criterion for initiating adaption for any element, m, and the allowable

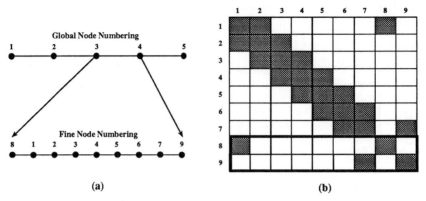

Figure 1. Embedded mesh solution scheme: (a) global and embedded node numbering scheme, (b) embedded region coefficient matrix before elimination.

tolerance level may be selected from:

$$\frac{C^m_{max} - C^m_{min}}{C^m_{max}} > T \tag{13}$$

$$\frac{\Delta C^m}{\Delta x^m} > T \tag{14}$$

$$\Delta C^m > T \tag{15}$$

where C^m_{max} is the maximum concentration for element m, C^m_{min} is the minimum concentration for element m, Δx^m is the length of element m, and T is a tolerance level. Other error criteria are possible, such as schemes that examine the difference between computed solutions as a function of the level of mesh refinement and use this information to guide the refinement pattern. For the case where analytical solutions exist, it would also be possible to control mesh refinement as a function of deviation from the analytical solution. This would provide a mechanism for evaluating the relative cost of competing methods that are required to meet a given error criteria. These investigations have not been performed to date.

All matrices are prefactored by the upper-lower decomposition method to reduce the number of operations performed within the time stepping loop. Refactoring of the global matrices, such as after inserting a subdivided element matrix, is only performed on the rows that require it.

RESULTS

A comparison of the linear and quadratic shape function models was performed for various spatial and temporal discretization patterns for an example where: $v_x = 0.5$, $D_x = 0.0125$, $R_f = 1.0$, $k = 0.0$. The linear shape function results are shown by Figure 2 for the case where the initial conditions of the step source are derived from the analytical solution for $t = 21$, the global mesh Peclet number (P_e) is 200, and the global mesh Courant number (C_r) is 0.8. For the adapted simulations eight elements were embedded in each adapted global element, yielding a $C_r = 0.8$ in the embedded element portion of the domain and a global $C_r = 0.1$. Improved solutions, with oscillations decreasing to about ten percent, were noted for simulations performed for the Bubnov-Galerkin and Petrov-Galerkin cases for a $C_r = 0.1$. The results shown in

Figure 2 demonstrate that sharp front problems cannot be simulated accurately for mesh P_e in the range investigated using either Bubnov-Galerkin or Petrov-Galerkin methods alone. The embedded element adaptive method allows very sharp fronts to be simulated accurately, with the best solution offered by the adaptive Petrov-Galerkin method.

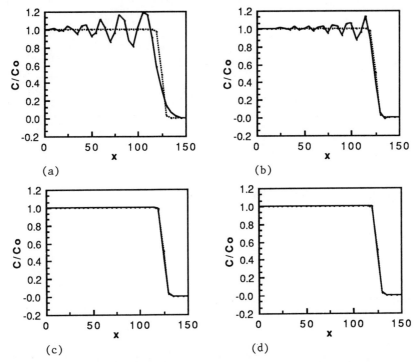

Figure 2. Step source simulation results using linear shape functions: (a) Bubnov-Galerkin, (b) Petrov-Galerkin, (c) adapted Bubnov-Galerkin, (d) adapted Petrov-Galerkin.

Figure 3 shows results for a set of simulations performed using a quadratic shape function model, with other conditions the same as those used for the linear shape function case. The results follow the same trend noted for the linear case; the Bubnov-Galerkin and Petrov-Galerkin methods alone do not provide an accurate representation of the sharp front problem simulated in this case. Adaptive solutions gave good results, while the best solution was obtained using the adaptive Petrov-Galerkin method. Quadratic solutions had slightly less error (integral error and discrete error) than the corresponding linear solutions, but required from 35 to 50 percent more computational effort than the corresponding linear solution for the same discretization scheme.

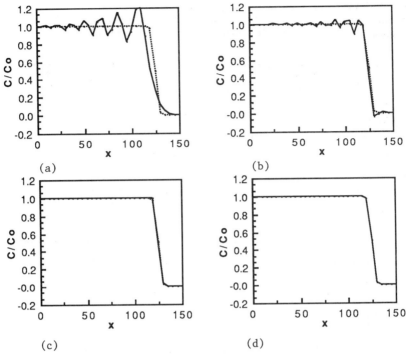

Figure 3. Step source simulation results using quadratic shape functions: (a) Bubnov-Galerkin, (b) Petrov-Galerkin, (c) adapted Bubnov-Galerkin, (d) adapted Petrov-Galerkin.

REFERENCES

1. Miller, C.T. and Weber, W.J., Jr. Modeling the Sorption of Hydrophobic Organic Pollutants by Aquifer Materials II. Column Reactor Systems, Water Research, Vol. 22, pp.465–474, 1988.

2. Westerink, J.J. and Shea, D. Consistent Higher Degree Petrov-Galerkin Methods for the Solution of the Transient Convection-Diffusion Equation, International Journal for Numerical Methods in Engineering, Vol. 28, pp. 1077–1101, 1989.

3. Oden, J.T. Progress in Adaptive Methods in Computational Fluid Dynamics, Chapter 15, Adaptive Methods for Partial Differential Equations (Ed. Flaherty, J.E., Paslow, P.J., Shephard, M.S. and Vasilakis, J.D.), pp. 206–252, Society for Industrial and Applied Mathematics, Philadelphia, 1989.

4. Yeh, G.T. A Zoomable and Adaptable Hidden Fine-Mesh Approach to Solving Advection-Dispersion Equations, in Computational Methods in Water Resources Vol. 2: Numerical Methods for Transport and Hydrologic Processes (Ed. Celia, M.A., Ferrand, L.A., Brebbia, C.A., Gray, W.G. and Pinder, G.F.), pp. 69–74, Proceedings of the 7th Int. Conference on Computational Methods in Water Resources, Cambridge, Mass., U.S.A. 1988. Elsevier, Amsterdam, 1988.

5. Bear J. Hydraulics of Groundwater, McGraw-Hill Inc. New York, New York, 1979.

SECTION 7 - STOCHASTIC PROBLEMS IN GROUNDWATER FLOWS AND TRANSPORT

On Numerical Simulation of Flow Through Heterogeneous Formations

G. Dagan, P. Indelman

Department of Fluid Mechanics and Heat Transfer, Tel Aviv University, Ranat Aviv 69978, Israel

ABSTRACT

The conductivity of heterogeneous natural formations is modeled as a random space function (RSF). Consequently, the continuity equation and Darcy's Law become stochastic partial differential equations. Their solution by Monte Carlo simulations implies a partition of the flow domain in discrete blocks of random properties. The main aim of the study is to develop a methodology to derive the statistical moments of blocks conductivity as function of the given moments of the point values of conductivity and of the block size. The approach is similar to the one leading to effective conductivity, which is indeed a particular case for large block size. The general approach is illustrated for one-dimensional flow, which is solved exactly and by first-order perturbation expansion in the logconductivity variance.

INTRODUCTION

We consider here flow in natural porous formations of large extent, which is governed by the equations of conservation of mass and by Darcy's Law. These equations have to be solved under appropriate boundary and initial conditions in many important applications of groundwater management and pollution control. The necessary input information for any solution, analytical or numerical, is the distribution of the hydraulic properties of the formation (hydraulic conductivity, effective porosity, storativity) throughout the formation. It is an accepted tenet these days that these properties vary in an irregular manner in space, i.e. that natural formations are as a rule *heterogeneous*. The spatial distribution of properties is subjected to uncertainty due its irregular nature and to scarcity of measurements, and this uncertainty is quantified by regarding them as *random*. Stochastic modeling of flow and transport in heterogeneous formations has become an important subject of recent research activity (see, for instance, [2], [3] and [4] for recent reviews).

Except for schematic, idealized, cases, flow and transport problems have to be solved numerically. One of the most powerful technique (e.g. [5],[6]) is that of Monte Carlo simulations. Such simulations comprise three stages: (i) with given probability density functions (p.d.f.) of the values of properties at different points, various realizations of their spatial distributions are generated, (ii) flow and transport is solved in each realization, regarded as a deterministic, conventional, problem and (iii) the outcome of a large number of realizations is employed in order to derive the statistical moments of the dependent variables. Each of these stages poses difficult numerical problems and our concern here is exclusively with stage (i), namely random generation of spatial distributions. In the simplest approach from a conceptual point of view, properties are generated on a dense grid, such as to preserve the statistical structure of the given p.d.f. However, this may lead to a huge number of discrete elements and correspondingly to extremely large matrices, requiring enormous computing capabilities. These can be reduced if a coarser grid, which does not attempt to simulate the pointwise spatial distribution of properties, but provides a sufficiently detailed information for many applications of interest, is adopted. The aim of this study is to provide a methodology to generate the random properties of numerical blocks in Monte Carlo simulations. Such a methodology is a prerequisite to any numerical simulation of flow, which in turn is the starting point of transport simulations. The present study refers to saturated steady flow of water, the only property of concern being the hydraulic conductivity or transmissivity.

STATEMENT OF THE PROBLEM

Let Ω be the flow domain, $\partial\Omega$ its boundary and $x(x,y,z)$ a Cartesian coordinate (Fig. 1a). The water head $H(x)$ and the specific discharge $q(x)$ satisfy the continuity equation and Darcy's Law

$$\nabla . q = 0 \quad ; \quad q = -K(x) \nabla H \qquad (1)$$

where $K(x)$ is the spatially variable conductivity and pumping or recharging wells are regarded as parts of $\partial\Omega$. Boundary conditions are expressed, for instance, by assigning H or q_ν (the normal component of q), on $\partial\Omega$. K is regarded as a given and *stationary* RSF (random space function). In line with existing approaches we operate with $Y = \ell n\ K$, which is assumed to be normal. Consequently, its *unconditional* statistical structure is characterized completely by the expected value $\langle Y \rangle$=const and the two-point covariance $C_Y(x_1,x_2) = C_Y(r)$, where $r=x_1-x_2$. In the simplest representation C_Y is isotropic, i.e. a function of $r=|r|$ and may be characterized in turn by the variance $\sigma_Y^2=C_Y(0)$ and the correlation scale I_Y. These are of course restrictive assumptions, but our aim here is to develop the fundamental concepts and to illustrate them in a simple manner, with further generalizations left for future studies. Hence, with given $\langle Y \rangle$ and C_Y, the mathematical problem for the exact solution is to determine the RSF H and q which satisfy the stochastic partial differential equations (1) in Ω and given boundary conditions on $\partial\Omega$ (for simplicity, the latter are assumed to be deterministic).

In a numerical framework the domain Ω is partitioned in elements ω (Fig. 1b) and Eqs. (1) are discretized, H and q being represented as unknown random sets H, q of their values at the nodes. To simplify the analysis we assume that K or Y are taken as constant in each element ω, and we denote the vectors of their discrete values by K or Y. In each of the realizations of Y, the numerical problem becomes a traditional one, namely solving the discretized version of (1). The vector Y is statistically characterized by its expected value $\langle Y \rangle$ and the variance-covariance matrix C_Y of its values. The problem we are going to address here is: how should $\langle Y \rangle$ and C_Y be selected such as to render the numerical solution H and q a "good approximation" of the exact solution H, q of (1), for given point distribution of Y(x) characterized by $\langle Y \rangle$, C_Y?

As we shall show in the sequel, this is a complicated problem. Quite paradoxically, it has a simple solution in two extreme cases. First, with ℓ the length scale of the elements ω (Fig. 1b), a fine partition with $\ell \ll I_Y$ (say ℓ=0.1 I_Y) is considered. In this case one may assume that by generating K=exp(Y) from the multivariate normal Y with $\langle Y \rangle$=$\langle Y \rangle$, C_Y=C_Y, i.e. $\sigma_Y^2=\sigma_Y^2$ and I_Y=I_Y , will lead to approximations $H \simeq H$ and $q \simeq q$ of increasing accuracy as ℓ is decreased. This was the approach underlying previous simulations [5], [6]. However, as mentioned in Introduction, the number of nodes may be prohibitively large at this limit, since generally I_Y is much smaller than L, the length scale of Ω. Furthermore, in many applications we may not be interested in the details of the solution, but in some gross features of the system response. At the other extreme, we may be interested in determining only $\langle H \rangle$ and $\langle q \rangle$ or equivalently their space averages $\overline{H}$ and $\overline{q}$ over large volumes (areas) relative to I_Y. In this case and under some limitations, the solution is achieved by modeling the formation as homogeneous and of constant *effective* conductivity K_{ef} (or the associated Y_{ef}) and the problem becomes a deterministic one. Most of the theory of heterogeneous materials is concerned with the determination of effective properties (e.g. see [1] for a general approach and [3] for porous formations). A huge body of literature has been devoted to this topic and effective properties have been determined by various approaches in terms of the underlying point-value distribution and for average uniform flow, i.e. for $J = - \nabla \langle H \rangle = $ const and for $L \gg I_Y$. Furthermore, the same properties were assumed to apply, even if the average flow is nonuniform.

Hence, at the one extreme, for $\ell \langle \langle I_Y$, one has $\langle Y \rangle = \langle Y \rangle$, $\sigma_Y^2 = \sigma_Y^2$ and $I_Y = I_Y$, whereas at the other one, for $\ell \rangle \rangle I_Y$, $Y = Y_{ef}$ and $\sigma_Y^2 \simeq 0$, $I_Y \to \infty$. As stated in Introduction we are interested in partitions with ℓ of the same order as I_Y, i.e. a ratio ℓ / I_Y which is small enough to permit one to determine an approximation of the statistical structure of H and q, but large enough in terms of efficient use of computer time and memory. It is clear that for a given ratio ℓ / I_Y, the values of $\langle Y \rangle$, σ_Y^2 and I_Y will lie somewhere between the above extreme limits. Deriving these values is precisely our aim and those familiar with the difficulties involved in determining effective properties may realize that this is not a simple task.

THE GENERAL FRAMEWORK

The first topic we are going to address is how to qualify the solution of the numerical problem $H(x)$, $q(x)$ as a "good approximation" of the exact one H, q. It is clear that by the process of discretization (Fig. 1b) some information is lost, or in spectral terminology, high frequencies of the Fourier transforms of the solutions are filtered out. Let first define the space averages

$$\overline{H}(x) = \frac{1}{\omega} \int_\omega H(x')\, dx' \quad ; \quad \overline{q}(x) = \frac{1}{\omega} \int_\omega q(x')\, dx' \qquad (2)$$

where x is the coordinate of the centroid of ω. Then in line with [7], we require the numerical solution to be an accurate approximation of the space averages (2) of the exact solution over same ω, i.e. $H(x) \approx \overline{H}(x)$ and $q(x) \approx \overline{q}(x)$ *in a statistical sense*. Before proceeding farther, we refer to the function $e(x) = - q(x).\nabla H(x)$, the rate of dissipation of mechanical energy per unit weight of fluid. This function, which can be rewritten by Eq. (1) as $e = K \nabla H.\nabla H = q.q/K$, has been found to be quite useful in deriving the effective conductivity [3]. It has the distinguished properties of being a scalar, positive definite and additive function. Furthermore, integration over an arbitrary domain ω yields by Green's theorem and by Eq. (1), $\overline{e} = (1/\omega) \int_\omega e\, dx = (1/\omega) \int_{\partial\omega} q_\nu H\, dx$. In particular, if $H = -J.x$ on $\partial\omega$, it follows that $\overline{e} = J.\overline{q} = K_\omega(J.J)$, where K_ω is a conductivity tensor defined as the coefficients of proportionality in the linear relationship between the gradient J applied on $\partial\omega$ and the response $\overline{q}$. We adopt the random space function $e(x)$ as the basic one rather than H and q separately, and qualify the numerical solution as a good approximation of the exact one by the requirement $e(x) \simeq \overline{e}(x)$ *in a statistical sense*, $e = - \mathbf{q}.\nabla H$ being   the dissipation in the numerical solution. Let $\langle \bar{e} \rangle$ and $C_{\bar{e}} (\mathbf{x_1}, \mathbf{x_2})$ be the expected value and the covariance of $\bar{e}$, respectively. We formulate now the aim of the present study in a more restricted, but definite, manner, as follows: determine the statistical properties of the discretized blocks $(\langle Y \rangle$ and $\mathbf{C_Y})$ such as to satisfy the relationships

$$\langle e(\mathbf{x}) \rangle = \langle \bar{e}(\mathbf{x}) \rangle \quad , \quad C_e(\mathbf{x_1}, \mathbf{x_2}) = C_{\bar{e}} (\mathbf{x_1}, \mathbf{x_2}) \tag{3}$$

Although Eqs. (3) ensure that the equality $e=\bar{e}$ is satisfied only in a weak sense, the equivalence is a strong one if e is normal.

The problem defined by Eqs. (3) is still a formidable one, because e and the corresponding space average $\bar{e}$ *are not known a-priori* and in general we have first to solve the exact problem in order to be able to apply (3), i.e. a vicious circle is created. Thus, further simplifications are needed if we wish to determine general relationships between the moments of Y and those of Y, independent of the shape of Ω and of the particular boundary conditions. Toward this aim we define an additional length scale λ, characterizing the nonuniformity of the average flow, i.e. $\lambda \sim |\nabla \langle H \rangle| / |\nabla^2 \langle H \rangle|$. We make the additional assumption of slowly varying flow at the heterogeneity scale, i.e. $I_Y \ll \lambda$ and furthermore, we also limit the extent of the discretized blocks by the similar inequality $\ell \ll \lambda$. Similar requirements for the definition of macroscopic variables at the pore-scale [3] tend to indicate that this condition is not too restrictive, although we may expect delicate problems in the case of strong nonuniformity which occur, for instance, in the neighborhood of wells. Discarding for the time being the special problems arising in some neighborhoods of singular points on the boundary, the above inequalities ensure that the average flow is *locally* uniform, i.e. the average head gradient can be regarded as constant over a domain containing a large number of elements ω. With these qualifications we focus our equivalence criteria by considering the heterogeneous formation of stationary logconductivity Y(x), but with boundary condition $H = - \mathbf{J}.\mathbf{x}$ on $\partial\Omega$, with $\mathbf{J}$ a constant vector. The relationships (3) are now applied to the exact solution and to the numerical one for this flow.

If L, the length scale of Ω, is much larger than both I_Y and ℓ, which we assume in any case, we may let L expand to infinity and then e, as well as ∇H and $\mathbf{q}$, become stationary RSF, such that $\langle e \rangle$ does not depend on $\mathbf{x}$, whereas C_e depends only on $\mathbf{r}=\mathbf{x_1}-\mathbf{x_2}$.

For the sake of simplicity, let the statistical structure of the given Y(x) be defined by the three constant parameters $K_G =\exp(\langle Y \rangle)$, σ_Y^2 and I_Y, where K_G is the conductivity geometric mean. Under the stipulated conditions $\langle e \rangle$ is constant and it is equal to $\langle \bar{e} \rangle$ (2), since space and ensemble averaging are commutative. Similarly, the variances σ_e^2 and $\sigma_{\bar{e}}^2$ are constant. Based on the linearity of (1) and on dimensional analysis we may write

$$\langle \bar{e} \rangle = \langle e \rangle = K_G \, J^2 \, a(\sigma_Y^2) \quad ; \quad \sigma_{\bar{e}}^2 = K_G^2 J^4 \, b(\sigma_Y^2, \ell/I_Y)$$
$$I_{\bar{e}} = I_Y \, c(\sigma_Y^2, \ell/I_Y) \quad ; \quad \sigma_e^2 = K_G^2 J^4 \, c(\sigma_Y^2, 0) \tag{4}$$

where a, b and c are dimensionless, unknown functions.

The effective conductivity K_{ef} is equal to the product $K_G \, a(\sigma_Y^2)$ and it has been evaluated in the past for various types of C_Y [3]. If C_Y is anisotropic, b is a tensorial function of σ_Y^2 and additional dimensionless parameters, characterizing the anisotropy of C_Y, show up in (4). The function "a" tends to unity for $\sigma_Y^2 \to 0$, as the formation becomes homogeneous at this limit, and the same is true for b (4). The unknown function $b(\sigma_Y^2, \ell/I_Y)$ is monotonously decreasing, for a fixed σ_Y^2, as ℓ/I_Y increases. This is a general property of space averages [3], and b tends to zero for $\ell/I_Y \to \infty$. A decomposition similar to (4) is valid for $C_{\underset{e}{}}$, which depends, however, on the additional variable r/I_Y.

A representation similar to (4) applies to the solution of the numerical problem, with pertinent a, b and c functions of σ_Y^2, and with K_G, $I_{\underset{e}{}}$ replaced by K_G, I_e, respectively. Once these functions are known, the requirement (3) may be employed in order to derive the moments of Y. This procedure is illustrated in the next Section.

ILLUSTRATION OF APPROACH : ONE-DIMENSIONAL FLOW

To illustrate the approach exposed in the preceding Sections we have solved the problem of one-dimensional flow through a heterogeneous formation (Fig. 2a). This is not a realistic case, since formations are of a two-dimensional structure at the regional scale or a three-dimensional one at the local scale, and we plan to investigate these configurations in the future. The great advantage of the 1D flow is in the simplicity of the solution which leads to conclusions of general interest. The conductivity is a RSF of x, and the flow is driven by a constant gradient J= $\Delta H/L$. We assume that $Y=\ell n K$ is normal and adopt the exponential covariance $C_Y(r) = \sigma_Y^2 \exp(-r/I_Y)$. By the continuity equation (1), the specific discharge q is constant in this case. Furthermore, we assume that $L \gg I_Y$ and ergodicity prevails, i.e. $q=J K_H$ is deterministic, where $K_H = <\exp(-Y)>^{-1} = K_G \exp(-\sigma_Y^2/2)$ is the conductivity harmonic mean. Under these circumstances and by (1) the head gradient and the dissipation have the simple *exact* expressions $dH/dx=-q/K$, $e=q^2/K=$ $J^2 \exp(-\sigma_Y^2 - Y + \langle Y \rangle)$, respectively. Hence, there is no distinction between e and dH/dx in this case, both being proportional to K^{-1}. Still, for the sake of generality, we use e(x) as the basic function. With $\langle e(x) \rangle = J^2 K_H$, the exact expression of the covariance is $C_e(r) = K_G^2 J^4 \{\exp[-Y(x)-Y(x+r)]-\exp(-\sigma_Y^2)\} = K_G^2 J^4 \exp(-\sigma_Y^2)\{\exp[C_Y(r)]-1\}$.

We consider now the numerical solution of 1D flow, by discretizing the flow domain into equal segments of length ℓ (Fig. 2b). Of course, there is no need to solve numerically the problem of average uniform flow, since the exact solution has been achieved. Still, besides the illustrative purpose, even 1D flow problems become complex is the average flow is nonuniform, but slowly varying. Thus the present methodology may serve for Monte Carlo simulations of such flows.

Following the procedure outlined above, we have to determine the covariance of the space average $\bar{e}(x) = (1/\ell) \int_x^{x+\ell} e(x') \, dx'$. $C_{\underset{e}{}}$ could be expressed by an infinite series, which is not reproduced here, and the same is true for the variance

$$\sigma_{\underset{e}{}}^2/(K_G^2 J^4) = b(\sigma_Y^2, \ell') =$$

$$(2/\ell'^2) \sum_{n=1}^{\infty} (\sigma_Y^2)^n \{\ell'/n + [\exp(-n\ell')-1]/n^2\}/n! \quad ; \quad \ell'=\ell/I_Y \qquad (5)$$

This expression simplifies considerably for a first-order perturbation approximation in σ_Y^2, which yields $b = 2\sigma_Y^2 [\exp(-\ell')-1]/\ell'^2$. Both expressions b are represented in Fig. 3 as functions of ℓ' for a few values of σ_Y^2. It is seen that b decreases from its value $b(\sigma_Y^2,0)= 1-\exp(-\sigma_Y^2)$ with increasing ℓ. The small perturbation approximation is valid only for small σ_Y^2, but the 1D flow is an extreme case and better agreement may be expected for 2D or 3D flows [3]. We have also computed the integral scale $c=I_{\bar{e}}/I_Y$ as function of ℓ and σ_Y^2, which is represented in Fig. 4. As expected [3] $I_{\bar{e}}$ increases with the size of the averaging interval and tends asymptotically, for large ℓ, to $\ell/2$. Furthermore, the influence of σ_Y^2 is quite small and the perturbation approximation $c= \ell^2/\{2[\ell^2+\exp(-\ell)-1]\}$ is quite accurate. This results reflects the weak dependence of the auto-correlation coefficient of $\bar{e}$ upon σ_Y^2, i.e. the nonlinearity is mainly in the variance of $\bar{e}$. This results is a good omen for future investigations of 2D and 3D flows, for which the small perturbation approximation is particularly useful.

We turn now to the numerical solution for flow under the same gradient J through a medium made up from blocks of length ℓ set at random and of constant, random, $K=\exp(Y)$ (Fig. 2b). It can be shown that the covariance $C_Y(x)$ is a piecewise linear function, completely determined by the vector $C_Y = C_Y(m\ell)$ $(m=0,1,...)$, where in particular σ_Y^2 is for $m=0$. Assuming that Y is normal and characterized by $K_G=\exp(\langle Y \rangle)$ and C_Y, the dissipation $e=K_G J^2\exp(-Y)$ moments are related to those of Y precisely like the moments of e to those of Y. The covariance C_e cannot reproduce exactly $C_{\bar{e}}$, because of the difference between the shapes of C_Y and C_Y. Hence, the relationship (3) can be satisfied only in an approximate manner, i.e. by requiring $C_e(m\ell)=C_{\bar{e}}(m\ell)$. Once this is done, we obtain the expressions of K_G and C_Y, or the related parameters K_G, σ_Y^2 and I_Y, which determine the statistical structure of the discretized elements of Fig. 1b. These are the main results of our analysis, not reproduced here, but represented graphically in Figs. 5 and 6 as follows.

The ratio K_G/K_G (Fig. 5) is equal to unity for $\ell=0$, as one would expect. Indeed, in this case the numerical solution tends to the exact one. As ℓ increases the ratio decreases and tends, for $\ell\to\infty$ to K_{ef}/K_G, where at present $K_{ef}=K_H= K_G\exp(-\sigma_Y^2/2)$. Again, the small perturbation approximation, also showed in Fig. 5, is of limited accuracy for the higher σ_Y^2. This is a consequence of the one-dimensional nature of the flow, for which K_{ef} is strongly nonlinear in σ_Y^2, but the situation is different for the 2D and 3D flows [3].

The variance σ_Y^2 is represented in Fig. 6, again in its exact form and by the first-order perturbation approximation. As expected, it is equal to σ_Y^2 for $\ell=0$ and drops with increasing ℓ. An interesting finding is that even for relatively large discretized elements of $\ell/I_Y=10$, σ_Y^2 is still quite large and Y has still to be regarded as random. We again expect this drop to be much steeper for 2D and 3D flows.

Last, we have found that the integrals scale I_Y is very close to $I_{\bar{e}}$ and insensitive to the σ_Y^2. In words, the interblock correlation structure can be determined quite accurately by using the first-order approximation of both $I_{\bar{e}}$ (Fig. 4) and C_Y.

CONCLUSIONS AND OUTLINE OF FUTURE INVESTIGATIONS

We have developed a systematic procedure to relate the random properties of discretized blocks of numerical schemes to the continuous and given properties distribution, generalization of [7]. Such a methodology is a prerequisite to numerical modeling of flow through heterogeneous formations by Monte Carlo simulations, if the size of the space elements is not very small compared to the properties correlation scale.

To obtain relatively simple results, we have made several assumptions. Furthermore, we have illustrated the procedure for the simple case of one-dimensional flow. A few directions are envisaged in order to expand the methodology and to generalize it : application to two- and three-dimensional flows, incorporating the effect of average flow unsteadiness and nonuniformity, application to transport processes, generalization to multi-phase flows. These and related topics are considered for future investigations.

REFERENCES

1. Beran, M.J., *Statistical Continuum Theories*, Interscience, New York, 1968

2. Dagan, G., Statistical theory of groundwater flow and transport : pore to laboratory, laboratory to formation and formation to regional scale, Water Resour. Res., 22, 120S-135S, 1986.

3. Dagan, G., *Flow and Transport in Porous Formations*, Springer-Verlag, Berlin-Heidelberg, 1989.

4. Gelhar, L.J., Stochastic subsurface hydrology from theory to applications, Water Resour. Res., 22, 135S-145S, 1986.

5. Freeze, R.A., A stochastic-conceptual analysis of one-dimensional groundwater flow in nonuniform homogeneous media, Water Resour. Res., 11, 725-741, 1975.

6. Smith, L., and R.A. Freeze, Stochastic analysis of steady state groundwater flow in bounded domain, Two-dimensional simulations, Water Resour. Res., 15, 1543-1559, 1979.

7. Rubin, Y. and J.J. Gomez-Hernandez, A stochastic approach to the problem of upscaling of transmissivity in disordered media, 1. Theory and unconditional simulations, Water Resour. Res. (in press).

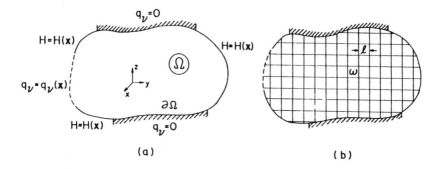

Fig. 1   Sketch of flow domain Ω : (a) exact and (b) partition into blocks ω

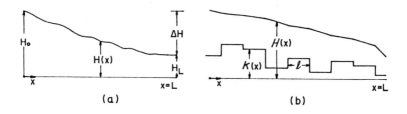

Fig. 2   Sketch of one-dimensional flow: (a) exact and (b) partition into blocks

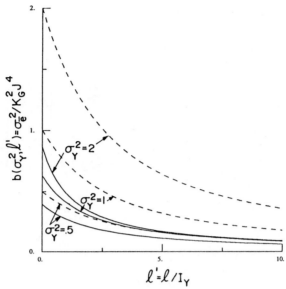

Fig. 3 The dependence of $b(\sigma_Y^2, \ell) = \sigma_{\bar{e}}^2/(K_G^2 J^4)$ (5) upon $\ell' = \ell/I_Y$ for a few values of σ_Y^2 (full line-exact, dashed line-first order perturbation approximation)

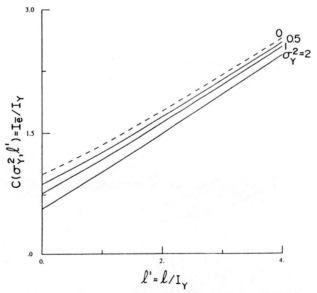

Fig. 4 Same as Fig. 3 for $c(\sigma_Y^2, \ell) = I_{\bar{e}}/I_Y$ (Eq. 5)

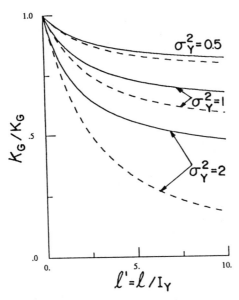

Fig. 5 Same as Fig. 3 for K_G, the blocks conductivity geometric mean

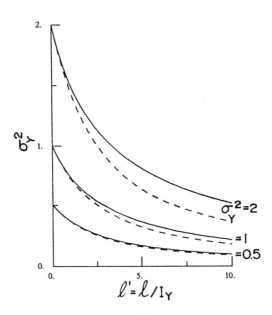

Fig. 6 Same as Fig. 3 for σ_Y^2, the blocks logconductivity variance

Sensitivity Analysis with Parameter Estimation in a Heterogeneous Aquifer with Abrubt Parameter Changes: Theory

S. Sorek, J. Bear

Ben Gurion University of the Negev, Jacob Blaustein Institute for Desert Research, The Water Resources Center, Sede Boqer Campus 84990, Israel and Department of Civil Engineering, Technion IIT, Haifa, Israel

ABSTRACT

Non stochastic concepts are proposed for treating the uncertainty in identifying coefficients appearing in groundwater models, as well as their boundary and initial conditions. As an example, the equation of flow of groundwater in a saturated porous medium is considered. A technique is described for assessing the sensitivity of the estimated initial conditions and coefficients appearing in both the partial differential (balance) equation and in the boundary conditions, to errors in measured piezometric heads. These coefficients may vary in both time and space. Here, the spatial variation is by zones, in each of which the values of the coefficients are constant. A minimum energy criterion, using Hamilton's variational principle is employed for estimating the coefficients of transmissivity, storativity, rate of withdrawal/recharge leakage coefficient, initial and boundary conditions.

SENSITIVITY ANALYSIS

Some of the existing methods for sensitivity analysis are based on the assumption that the governing parameters are continuous in time and space. Thus, when these methods are implemented, (e.g. McElwee and Yukler [5]; McElwee [6]; Cobb et al. [3]; Vemuri et al. [10]; McCuen [4] and Oblow [7]) derivatives of the hydraulic head are taken, for example, with respect to the storativity S, and the transmissivity T, in order to examine their influence on the predicted head distribution.

To stress this point let us consider a solution domain subdivided into zones with different constant values of flow parameters such as transmissivities (T). Hence partial space derivatives of T (i.e. ∇T where ∇ is the gradient operator) are not defined at the zonal boundaries. Since we require partial space derivatives of the hydraulic head (∇h) in order to solve for head distribution, we conclude that the expression

$$\nabla h = \frac{\partial h}{\partial T} \nabla T + \cdots$$

is not valid everywhere in our sub-spatial domain. Instead we resort to a less restrictive continuity demand of the form

$$\frac{\Delta h}{\Delta x_i} = \frac{\Delta h}{\Delta T} \frac{\Delta T}{\Delta x_i} + \cdots$$

where Δ denotes a difference operator and x_i is the spatial coordinate.

Note that the term $(\Delta h / \Delta T)$ is emerging from a different outlook which is not merely a partial derivative approximation (Yeh [11]) used in finite difference or influence coefficient methods.

Groundwater flow in a 2-D leaky confined aquifer is governed by the balance equation

$$\nabla \cdot (\underline{\underline{T}} \cdot \nabla h) = S \frac{\partial h}{\partial t} + Q - \beta(H_s - h) \tag{1}$$

where $h(\underline{x},t)$ = piezometric head, $\underline{\underline{T}} = \underline{\underline{T}}(\underline{x})$-aquifer transmissivity (tensor), $\underline{x}$-position vector, $S(\underline{x})$-aquifer storativity, $\bar{t}$-time, $Q(\underline{x},t)$-sink function (positive for rate of withdrawal, e.g. pumpage, leakage), $\beta(\underline{x},t)$-leakage coefficient and $H_s(\underline{x},t)$-piezometric head in the underlying, or overlying leaky aquifer.

In order to solve (1) for h in a domain R with boundary Γ, we have to specify initial conditions

$$h_{(\underline{x},t)} = h_{0(\underline{x})} \quad ; \quad \underline{x} \in R, \ t=0 \tag{2}$$

and boundary conditions

$$-\underline{\underline{T}} \cdot \nabla h \cdot \underline{n} + \alpha(h - h_\Gamma) = Q_\Gamma \ ; \quad \underline{x} \in \Gamma, \ t \geq 0 \tag{3}$$

where, $\alpha(\underline{x},t)$ is a coefficient that controls the type of condition existing on Γ, and $\underline{n}$ is an outward normal unit vector on Γ. The functions $h_o(\underline{x})$, $h_\Gamma(\underline{x},t)$ and $Q_\Gamma(\underline{x},t)$ are known. Equation (3) expresses a generalized boundary condition. it reduces to a Dirichlet condition when

$$\alpha \to \infty \quad ; \quad h = h_\Gamma \tag{4.1}$$

to a Neumann's condition when

$$\alpha = 0 \ ; \quad -\underline{\underline{T}} \nabla h \cdot \underline{n} = Q_\Gamma \tag{4.2}$$

and to a mixed condition when

$$0 < \alpha < \infty \tag{4.3}$$

Let us assume that the entire domain and its boundaries can be divided into zones with constant and independent values of the relevant coefficients within each of them. Jumps in the values of these coefficients may take place across the boundary between adjacent zones.

Our objective is to determine how sensitive the piezometric head, $h(\underline{x},t)$, at some point $\underline{x}$, is to changes in each of these coefficients at point $\underline{x}'$, taking into account the zonal distribution.

We start from a map $h = h(\underline{x},t)$ obtained by solving the hydrologic model, Eqs. (1)-(4), with prescribed initial values of all $L(=4)$ coefficients C_{ln}^o ($C \equiv \alpha$, β, S, T; $l = 1,2,3,4; n=1,2,...,N$), in all N zones. We now apply a change numbered by k, to a coefficient l in zone n, keeping all other $(LN-1)$ coefficients unchanged, and derive the corresponding water level map. We then express the sensitivity, U_{pn}^l in the form

$$U_{pn}^l = \frac{\Delta h_{pl}^k}{\Delta C_{ln}^k / C_{ln}^o} \tag{5}$$

where

$$\Delta C_{ln}^k = C_{ln}^k - C_{ln}^o \tag{6}$$

$$\Delta h_{pl}^k = h_{pl}^k - h_{pl}^o \tag{7}$$

Here h_{pl}^k is the value attained at point $\underline{x}_p$ by changing the coefficient in zone n to C_{ln}^k, keeping all other coefficients at their previous, $k-1$ values.

The incorporation of the boundary control coefficient $\alpha(\underline{x},t)$ (see Eqs. (3) and (4)) into a systematic sensitivity approach is novel. It enables us to check the influence of the intensity of the boundary condition (i.e. when $0 < \alpha < \infty$) on the head (h) as a function of time and space.

The configuration of the zone boundaries is based on geological and other a-priori known information. However, it is also possible to vary zone boundaries in some systematic way, in order to study the sensitivity of $h(\underline{x},t)$ to such variations.

PARAMETER ESTIMATION

It is customary to estimate aquifer flow parameters by various inverse methods. These methods are often aided by *mathematical techniques*, such as kriging, sto-chastic analysis, gradient oriented schemes and linear interpolation,. Carrera [2] gives an extensive literature survey of these methods.

The novelty of the approach proposed here is that it employs an analogy of a system of mass, spring and damper driven by an impulse force to an aquifer flow model. In view of this analogy, a functional that is based on energy considerations may be constructed. The variation of the functional with respect to model parame-ters is really a search for its extremum, which is described as an optimization pro-cess. This, in turn presents an advantage in being more inherent for the considered phenomena, whereas the use of mathematical distribution functions relies on per-sonal interpretation.

Let us discretize the domain into NE finite elements. Within each element we approximate h by $\hat{h}$.

$$h \cong \hat{h} = \sum_{j=1}^{ne} \xi_j(\underline{x})h_j(t) \tag{8}$$

Here, ne is the number of nodal points prescribed to the element, and $\xi_j(\underline{x})$ is an interpolation function associated with each node that may be defined as:

$$\xi_j(\underline{x}_i) = \delta_{ij} \tag{9}$$

δ_{ij}-being the Kronecker delta function.

Let us assume that the spatial distributions of S and T (as well as other aquifer parameters) are not known. Instead, we have information on observed water levels, $\bar{h}_m(t)$, at M spatial points, $m=1,...,M$, taken at K discrete time levels $k=1,...,K$, during the period $t \in (\tau_o, \tau_f]$, within the domain.

Let us now divide the set of observations into $\overset{o}{M}$ control locations and $\overset{\cdot}{M} = M - \overset{o}{M}$ constraint ones.

In view of (8) we write the evaluated $\hat{h}_{(t)}$ values at the $\bar{m}$ and $\overset{o}{m}$ control and constraint points respectively in the element. We require for each element

$$\eta_{\overline{m}} = [\bar{h}_{\overline{m}} - \sum_{j=1}^{ne} \xi_j(\underline{x}_{\overline{m}})h_j(t)] \leq \varepsilon \ll 1 \;\; ; \;\;\; \overline{m} = 1,2,...,\overline{m}^e \tag{10}$$

$$\eta_m^o = [\bar{h}_m^o - \sum_{j=1}^{ne} \xi_j(\underline{x}_m^o)h_j(t) = 0 \;\; ; \;\;\; \overset{o}{m} = 1,2,...,\overset{o}{m^e} \tag{11}$$

where $\overline{m}^e$ is the number of control points within the element, with $NE_b \times \overline{m}^e = \overline{M}$, and $\overset{o}{m^e}$ is the number of constraint points within the element, with $NE \times \overset{o}{m^e} = \overset{o}{M}$.

We follow ideas from the theory of optimal control regarding dynamic optimization concepts (Pontryagin et al. [8]; Ahmed and Teo [1]). We regard the flow in the aquifer as a process during $(\tau_o, \tau_f]$ that obeys the partial differential equation described in (1). We prescribe an energy functional as our goal function (the performance index in optimal control theory). The inverse problem (the optimization problem) is now defined in terms of finding the head and parameter values (the control vector) that controls the flow process so that the energy goal function will be minimized (i.e. obtain an optimal performance index).

Next, we introduce the energy interpretation of (1). Let us consider a system composed of a mass, a linear spring and a linear damper excited by an impulse. Its force balance equation is

$$m_s \ddot{x} + \bar{k}x + b\dot{x} = F \tag{12}$$

where m_s denotes particle's mass, x denotes the position of the particle, $(\dot{\,})$, $(\ddot{\,})$ denote the first and second time derivatives, $\bar{k}$ denotes the spring's coefficient, b denotes the damper's coefficient, and F denotes the driving force. Let ξ be a spatial coordinate and $G(\xi)$ be a function such that

$$x = G(\xi) \;\; ; \;\; \ddot{\xi} = 0 \;\;\; \text{and} \;\;\; \dot{\xi} \neq \dot{\xi}(\xi) \tag{13}$$

The conditions in (13) allow a simple reformulation of (12). We choose to express only the $\ddot{x}$ term in (12) by $\dot{\xi}^2 \dfrac{d^2}{d\xi^2}$ so that (12) may then be rewritten as

$$m_s \dot{\xi} \frac{d}{d\xi} \left[\dot{\xi} \frac{dx}{d\xi} \right] + \bar{k}x + b\dot{x} = F \tag{14}$$

Hamilton's extended principle (Sorek and Blech [9]) is now written as a governing functional of the system expressed by equation (12) (or (13) and (14)) in the form

$$\frac{1}{2} \int_{\tau_o}^{\tau_f} [m_s \dot{x}^2 - \overline{k}x^2 - 2(\overline{bx - F})x]\, dt \tag{15}$$

The first and second terms in the integrand represent the kinetic and potential energies respectively. The third term represents the dissipative work done by the non-conservative force terms (denoted as ($\overline{}$)), which are not varied because of being generalized dissipative force terms (Sorek and Blech [9]). By comparing (14) with (1), we note the following analogies

(1) for 1–D	h	x	T	$\dfrac{\partial h}{\partial x}$	β	S	$Q - \beta H_s$
(14)	x	ξ	m_s	$\dot{\xi}\dfrac{dx}{d\xi}$	$\overline{k}$	b	F

Accordingly, the expression analogous to (15) takes the form

$$\frac{1}{2} \int_{\tau_o}^{\tau_f} \left[T(\nabla h)^2 + \beta h^2 + 2\left(\overline{S\,\frac{\partial h}{\partial t} + Q - \beta H_s}\right) h \right] dt \tag{16}$$

Thus the energy integrant in (16) is replaced by a local element goal function $\hat{\phi}^e$ assembled at elemental control points and defined by

$$\hat{\phi}^e = \frac{1}{2} \sum_{\overline{m}=1}^{m^\bullet} \left[W_{\overline{m}} T^e (\nabla \eta_{\overline{m}})^2 + \beta^e \eta_{\overline{m}}^2 + 2\left(\overline{S^e\,\frac{\partial \eta_{\overline{m}}}{\partial t} + Q^e - \beta^e H_s}\right) \eta_{\overline{m}} \right] \tag{17}$$

where $W_{\overline{m}}$ is a weighting coefficient supplied by the user to reflect his confidence in elements with more accurate estimates (related to measurements).

The summation of $W_{\overline{m}}$ over the entire region is bounded, i.e.

$$\sum_{\overline{m}=1}^{\overline{M}} W_{\overline{m}} = 1 \tag{18}$$

Note that in (17) we replace h (as in (16)) with $\eta_{\overline{m}}$ (eq. (10)) to indicate the change in energy from the datum measured value.

We now define a Hamiltonian, $\hat{H}^e$ for each element

$$\hat{H}^e = \hat{\phi}^e + \lambda^e\, f^e \tag{19}$$

where λ^e denotes the Lagrangian multiplier function associated with the element, and f^e expresses the behavior of the hydraulic head within the element. In view of (1), f^e is expressed by

$$f^e \equiv \frac{\partial \hat{h}}{\partial t} = \frac{1}{S^e} \{T^e \nabla^2 \hat{h} - [Q^e - \beta^e (H_s - \hat{h})]\} \quad ; \quad S^e \neq 0 \tag{20}$$

where, T^e, S^e, Q^e, β^e are parameters related to the element.

We now define an elemental functional of the form

$$\hat{\chi}^e = \hat{\chi}(t, \hat{h}, T^e, \omega^o, Q^e, \beta^e, \alpha^e, h_\Gamma, {}^e Q_\Gamma^e, h_o^e)$$

$$= \int_{\tau_o}^{\tau_f} \int_{R^e} \left[\hat{H}^e - \lambda^e \frac{\partial \hat{h}}{\partial t} \right] dR^e \, dt + \int_{\tau_o}^{\tau_f} \int_{\Gamma^e} B_\Gamma d\Gamma^e \, dt + \int_{R^e} I_o \int_{\tau_o}^{\tau_f} dR^e \tag{21}$$

where, B_Γ denotes the effect of the boundary given by

$$B_\Gamma = \frac{i}{2} \alpha^e (\hat{h} - h_\Gamma^e)^2 - Q_\Gamma^e \hat{h} \tag{22}$$

I_o denotes a function assigned to conditions at τ_o or τ_f time levels, defined by

$$I_o = \frac{1}{2} (\hat{h} - h_o^e)^2 \tag{23}$$

The variation of $\hat{\chi}^e$ defines a search for extremum, with respect to the parameters $\hat{h}, T^e, S^e, Q^e, \beta^e, \alpha^e, H_\Gamma^e, Q_\Gamma^e, h_o^e$ Hence, for each element we obtain the following algebraic set of equations

$$\frac{\partial \hat{\chi}}{\partial r} = 0 \; ; \; (r \equiv h_j, S^e, T^e, Q^e, \beta^e, \alpha^e, h_\Gamma^e, Q_\Gamma^e, h_o^e) \tag{24}$$

This is not a closed set as λ^e is still unknown. In order to determine λ^e, the proposed method now consists of solving the set (24) for a sequence of λ^e's iteratively until the constraint stated in (10) is satisfied.

At this point we can make use of the results of the sensitivity analysis presented earlier. The sensitivity analysis may indicate to which of the above r-parameters the predicted water levels are more sensitive. We start with an initial guessed value for all r parameters and check their variations after a few runs of (24). If the prior indication of insensitivity of a specific parameter meets its minor variations resulted in using (24), we regard it as constant, thus reducing the amount of work involved.

REFERENCES

1. Ahmed, N.U. and Teo, K.L., Optimal control of distributed parameter systems, Elsevier North Holland, N.Y., 1981.

2. Carrera, R.J., Estimation of aquifer parameters under transient and steady-state conditions, Ph.D. dissertation, Dept. of Hydrology and Water Resources, University of Arizona, Tucson, Arizona, 1984.

3. Cobb, P.M., McElwee, D. and Butt, M.A., Analysis of leaky aquifer pumping test data: An automated numerical solution using sensitivity analysis, Ground Water, vol. 20, no. 3, pp. 325-333, June 1982.

4. McCuen, R.H., Component sensitivity: A tool for the analysis of complex water resource systems, Water Resourc. Res., 9(1), 243-246, 1973.

5. McElwee, C.D. and Yukler, M.A., Sensitivity of groundwater models with respect to variations in transmissivity and storage, Water Resourc. Res., 14(3), 451-459, June 1978.

6. McElwee, C.D., Sensitivity analysis of ground water models. Fundamentals of transport phenomena in porous media, NATO/ASI, July 14, 1985.

7. Oblow, E.M., Sensitivity theory for reactor thermal-hydraulic problems, Nucl. Sci. Eng. 68, 322-337, 1978.

8. Pontryagin, L.S., Boltyanskic, V.G., Gamkrelidge, R.V. and Mishchenko, E.F. The mathematical theory of optimal processes, John Wiley, N.Y., 1962.

9. Sorek, S. and Blech, J., Finite-element technique for solving problems formulated by Hamilton's principle, J. Comp. Struct. 15(5), 1982.

10. Vemuri, V., Dracup, J.A., Erdmann, R.C. and Vemuri, N., Sensitivity analysis method of system identification and its potential in hydrologic research, Water Resourc. Res., 5(2) 341-349, 1969.

11. Yeh, William, W-G., Review of Parameter Identification Procedures in Groundwater Hydrology: The inverse problem, Water Resourc. Res., 22(2), 95-108, 1986.

Probabilistic Approach to Hydraulic Fracturing in Heterogeneous Soil

T. Sato, T. Uno
Department of Civil Engineering, Gifu University, Yanagido, Gifu 501-11 Japan

INTRODUCTION

Many technical terms exist relating to hydraulic fracturing, such as piping, quick sand, boiling, channelling, etc. The interaction between pore water and soil grain is one of functions in this problem. Attention, however, has been paid to soil mass movement because of unknown values of the soil-water interaction.

The critical hydraulic gradient is a well-known index that can predict the possibility of hydraulic fracturing. This originates from the equilibrium of a soil column vertically responding to gravity and seepage forces (Terzaghi[1]). The concept is limited to the general failure in a homogeneous soil. The study attempts to expand it to piping, which is extremely affected by erosion in heterogeneous soil.

A new index called the "local safety factor" is introduced to evaluate the failure probability for the piping in a subregion of a heterogeneous soil mass. The value is given by the ratio between the averaging hydraulic gradient through the soil mass and the critical by Terzaghi[1]. The GO-game (Otsuka[2]) is available in simulating the progression of failure zone.

LOCAL SAFETY FACTOR FOR PIPING

Terzaghi's definition is effective for piping due to heave that is classified into a general failure, and the safety factor is given by

$$F_s = i_{cr}/i \tag{1}$$

in which i_{cr} is the critical hydraulic gradient and i the spatially averaging value on hydraulic gradient. The i_{cr} value is uniquely determined from the soil parameters those who are the specific gravity of soil particle and the void ratio in REV (Bear[3]).

A soil element breaks down on semi-microscopic scale at the first stage of the piping, and the rupture zone successively progresses to the whole area. The boiling or quick-sand appears near the final stage of piping. The numerical treatment needs for a soil heterogeniety observed in a semi-microscopic scale.

A soil element is divided into N_b subregions and the local safety factor for piping is estimated by the use of the maximum entropy theory. There is no information about the exact value except for the macroscopic safety factor F_s. If there are n_i subregions having the local safety factor of ξ_i, the two equations are given as follows;

$$\sum n_i = N_b \tag{2}$$

$$\sum \xi_i n_i = N_b F_s \quad (i=1,2---) \tag{3}$$

The number of combination ways in which ξ_i is assigned to n_i subregions is given by

$$w = N_b! / \prod n_i! \tag{4}$$

The way of ξ_i occurrence is unknown, but it is reasonable to assume that it takes place to make the Eq.(4) a maximum value. This approximation is based on the maximum entropy theory.

The use of the Lagrangian multipliers α and β yields a estimate of n_i as follows;

$$n_i = \exp[-(\alpha + \beta \xi_i)] \tag{5}$$

If ξ_i is an continuous quantity, the probability of which a subregion takes the ξ_i value is computed by

$$p_i = \beta \exp(-\beta \xi_i) \tag{6}$$

Equation (6) indicates the probability of failure occurrence in subregions. The β value is given by

$$\beta = 1/F_s \tag{7}$$

There are difficulties in the observation of the hydraulic gradient within the subregion. In additions, it is difficult to identify a exact value of soil parameters those who fluctuate every subregion. Equation (6) gives a prospect of predicting the hydraulic fracturing in subregions without those difficulties.

COMPUTER SIMULATION FOR RUPTURE PROGRESSION

The Go-game model, that **was** originally developed to simulate the occurrence of earthquake (Otsuka[2]), is used in simulating the progression of hydraulic fracturing.It is **the** basic concept that the failure progression depends on the random process according to the ξ_i value. The game starts by putting a black stone, that shows a subregion rupture, on the initial point (see the p-point in Fig.1). Although it is important to find out the initial rupture point, it is given from the prior information in this study. The 2nd stage has the four points to be checked which direction the rupture progresses. A white stone plays a role of a barrier in the game. The game goes on if at least one black stone appears and ends up if the white are put on the all points to be checked.

The probability of the black stone occurrence is given by the P_F value.

$$P_F=\int_0^1 \xi_i d\xi=1-e^{-\beta} \tag{8}$$

The P_F value describes the probability of hydraulic fracturing at a subregion under the regional safety factor of F_s. Random numbers from 0 to 1 are produced at each stage and assigned to the four points surrounding the rupture. According to Eq.(8), a decision is made which stone to be put.

If the 2nd stage becomes as shown in Fig.1, the initial rupture expands to p-j and p-k directions and stops at i and l-points. The computation is done at j and k-points next stage. The GO-game is continued until rupture stops, in other words, no black stones appear.

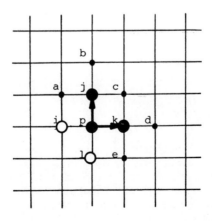

Figure 1. Computer simulation for rupture progression.

RESULTS AND DISCUSSIONS

The relationship between β and P_F is shown in Figure 2. The reciprocal β is a regional safety factor. According to this result, it is clear that the possibility of local failure occurrence is high rate even when the regional safety factor is evaluated as 1.0.

The GO-game was done by the use of the 1000 subregions (100 in horizontal and 10 in vertical). It is assumed that the initial rupture occurs at the middle of the bottom subregions. The game was repeated by 100 times, and the rupture progression was pursued untill 10 stages at each repetition.

Figure 3 shows the computation results at the final stage. The figure of (a) and (b) are for F_s=1.0, (c) and (d) for 2.0 and (e) and (f) for 5.0. Various types of the progression are observed from the results. The boiling is recognized as a vertical succession of the rupture zones. The probability of boiling occurrence is also estimated as 0.39 for P_F=1.0, 0.05 for 2.0 and 0 for 5.0, respectively.

This computation has still some problems to be overcome. It does not take account into the anisotropy of the soil and the transient value on ξ_i. The efforts will be continued to develop the computation method.

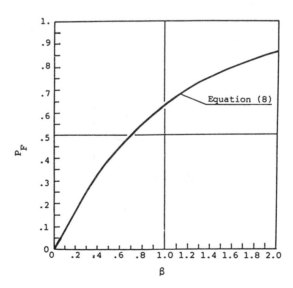

Figure 2. Probability of rupture occurrence, P_F.

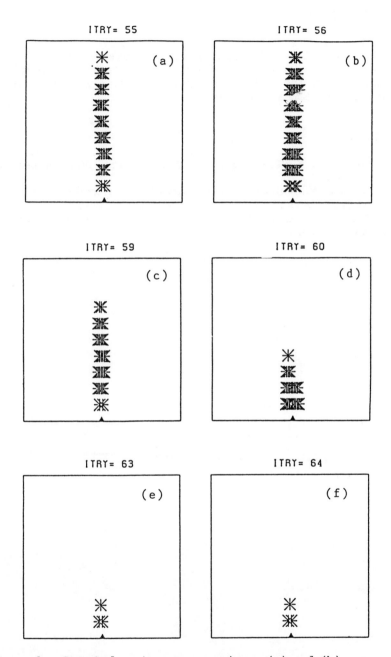

Figure 3. Computed rupture progression : (a) and (b) are for P_F=1.0, (c) and (d) for 2.0, and (e) and (f) for 5.0.

ACKNOWLEDGMENTS

The authors are gratefully acknowledged to Professor M. Otsuka (Kyushu University) for his sending the relating papers on the GO-game model. The study was partially supported by the Grant-in-Aid for Scientific Research (C) of the Japanese Education Ministry.

REFERENCES

1. Terzaghi, K.:Theoretical Soil Mechanics, John Wiley and Sons, 1943.
2. Otsuka, M.:A Simulation of Earthquake Occurrence, Physics of the Earth and Planetary, Interios 6, pp.311-315, 1972.
3. Bear, J.:Dynamics of Fluids in Porous Media, American Elsevier, 1972.

Domain Decomposition for Randomly Heterogeneous Porous Media

G. Gambolati, G. Pini, F. Sartoretto

Dipartmento di Metodi e Modelli Matematici per le Scienze Applicate, Università degli Studi di Padova, Italy

INTRODUCTION

In recent years several methods based on the domain decomposition idea introduced in 1869 by Schwartz, have been developed to solve partial differential equations (e.g. Laplace equation). Non-overlapping subdomains with irregular boundaries are easily accomodated and different parameter values normally used. The interest for this approach stems from the fact that several parallel processors may simultaneously compute the solutions in the various subregions.

The final global solution is achieved by iterations, especially of the conjugate gradient type (Quarteroni and Sacchi-Landriani, 1982). A current line of investigation is the study of efficient preconditioners to accelerate the convergence of the conjugate gradients (Glowinski et al., 1983 and 1988; Bjorstad and Widlund, 1986; Bramble et al., 1986; Chan, 1987; Canuto et al., 1988, Meurant, 1988).

In the present communication we develop a domain decomposition type method which is different from the one we have referred to above in that in each subdomain we do not solve the original problem but an adjoint one. Moreover a new objective is pursued by our method which is particularly suited for non homogeneous systems where the scale of heterogeneity is smaller than the subdomain scale. The actual heterogeneity microstructure is reflected by the adjoint solutions and ultimately transferred to the global solution by integration of the former over the boundary of the subregions, hereafter called macromesh or macrogrid, as opposed to the finer micromesh or microgrid of the heteronegeity field. Judicious assembling of these boundary integrals will eventually lead to an algebraic system with a size equal to the number of nodes of the macromesh, which incorporates the essential information of the heterogeneous structure. Two considerations make this approach particularly attractive in view of the available computer technology:

1) in randomly heterogeneous domains where the scale of heterogeneity is much smaller than the scale of variation of the solution, the overall problem may be realistically decomposed into a number of dimensionally smaller problems with great saving of CPU time and computer storage;

2) the solution to the local adjoint problems, which are all independent, may be achieved simultaneously in a multiprocessor environment with further significant reduction of CPU times. Thus a problem with a very high number of microheterogeneities can be realistically (and repeatedly, if needed, e.g. in the solution of stochastic differential equations) dealt with.

A simple example of a 2-D steady flow problem over a piecewise constant randomly generated permeability field is provided to illustrate the numerical features and demonstrate the feasibility of the new domain decomposition approach.

WEAK FORMULATION AND ADJOINT SOLUTIONS

Consider the 2-D steady problem of groundwater flow:

$$L(h) = \frac{\partial}{\partial x_i}(K_{ij} \frac{\partial h}{\partial x_j}) - f(x) = 0 \qquad (1)$$

with the usual meaning for the symbols, subject to Dirichlet and/or Neumann type boundary conditions.

Let $W(x)$ be an arbitrary weighting function which is C° continuous. Applying twice Green's first identity to (1) leads to the weak formulation:

$$\int_{\Omega} \frac{\partial}{\partial x_i}(K_{ij}\frac{\partial W}{\partial x_j})h\, d\Omega \; - \; \int_{\Omega} fW\, d\Omega \; +$$

$$\int_{\Gamma} K_{ij}\frac{\partial h}{\partial x_j}n_i W\, d\Gamma \; - \; \int_{\Gamma} K_{ij}\frac{\partial W}{\partial x_j}n_i h\, d\Gamma \; = 0 \qquad (2)$$

where Γ is the boundary of the flow domain Ω.

Let us choose $W(x)$ so as to satisfy the formal adjoint operator L^* over Ω :

$$L^*(W) = \frac{\partial}{\partial x_i}(K_{ij}\frac{\partial W}{\partial x_j}) = 0 \qquad (3)$$

Then (2) reduces to:

$$\int_{\Gamma} K_{ij}\frac{\partial h}{\partial x_j}n_i W\, d\Gamma = \int_{\Gamma} K_{ij}\frac{\partial W}{\partial x_j}n_i h\, d\Gamma + \int_{\Omega} fW\, d\Omega \qquad (4)$$

Eq. (4) may be easily interpreted by simply stating that the boundary integral of the head flux weighted with W is equal to the integral flux of the adjoint solution weighted with h plus the weighted integral of the distributed source/sink.

DOMAIN DECOMPOSITION AND BOUNDARY INTEGRALS

We assume that K_{ij} is piecewise constant over a microstructure which is much smaller than the scale of variation for h, f and the boundary geometry and proceed as follows. Discretize Ω into macro-elements consistent with a proper description for h, for instance triangular elements (Figure 1).

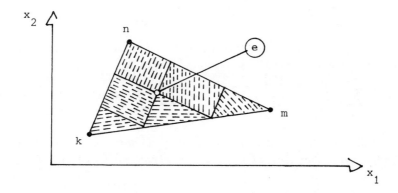

Figure 1. Triangular macro-elements with a microstructure of nonhomogeneous permeability

Each triangle e contains a permeability field which is piecewise constant at a scale smaller than the triangle sides. Let W_k^e, W_m^e and W_n^e be the solutions over e to the adjoint equation (3) satisfying the following boundary conditions:

a) $\quad W_r^e(x_p) = \delta_{rp}$ $\qquad\qquad\qquad r,p \in \{k,m,n\}$

b) $\quad W_r^e(x)$ varies linearly along the element sides.

The adjoint solutions are computed over each triangle by traditional finite elements. If M is the total number of macro-elements, 3M local independent f.e. problems are to be solved and this can be accomplished very efficiently using the concurrent processors of a parallel machine.

Integral (4) is written three times for each triangle:

$$\int_{\Gamma^e} K_{ij}^e \frac{\partial h^e}{\partial x_j} n_i^e W_r^e d\,\Gamma^e = \int_{\Gamma^e} K_{ij}^e \frac{\partial W_r^e}{\partial x_j} n_i^e h^e d\,\Gamma^e + \int_{\Omega^e} f^e W_r^e d\,\Omega \quad (5)$$
$$r \in \{k,m,n\}$$

Next, if p_k is the number of triangles which share the internal node k, we observe that:

$$\sum_e^{p_k} \int_{\Gamma^e} K_{ij}^e \frac{\partial h^e}{\partial x_j} n_i^e W_k^e d\,\Gamma^e = 0 \qquad\qquad (6)$$

due to the flux continuity along the element sides.

If node k belongs to the boundary of Ω where the flux q is given we have instead:

$$\sum_e^{p_k} \int_{\Gamma^e} K_{ij}^e \frac{\partial h^e}{\partial x_j} n_i^e W_k^e d\,\Gamma^e = - \int_{\Gamma_k^e} q W_k^e d\,\Gamma_k^e \qquad (7)$$

where Γ_k^e is the boundary segment contiguous to node k. Hence for each internal node eq. (5) yields:

$$\sum_e^{p_k} \int_{\Gamma^e} K_{ij}^e \frac{\partial W_k^e}{\partial x_j} n_i^e h^e d\,\Gamma^e + \sum_e^{p_k} \int_{\Omega^e} f^e W_k^e d\,\Omega = 0 \qquad (8)$$
$$k=1,2,\ldots$$

Additional equations hold for nodes lying on the Neumann type boundary of Ω with the right-hand side given by (7).

Now assume that h^e varies linearly on each triangle

side between the nodal values h_k, h_m and h_n and denote by:

$$q_k^e = K_{ij}^e \frac{\partial w_k^e}{\partial x_j} n_i^e \qquad (9)$$

the flux of W_k^e across Γ^e. Eq. (8) turns into:

$$\sum_e \int_{\Gamma^e} q_k^e (N_k h_k + N_m h_m + N_n h_n) d\,\Gamma^e + \sum_e \int_{\Omega^e} f^e w_k^e d\,\Omega = 0 \qquad (10)$$
$$k=1,2,\ldots$$

where N_k, N_m and N_n are the customary "roof" or "pyramidal" basis functions defined over element e.

The fluxes q_k^e are numerically computed from the (known) adjoint solution w_k^e. Hence the integrals in (10) are easily evaluated by some quadrature formula thus giving rise to an algebraic equation involving h on node k and the surrounding nodes. For equations corresponding to nodes lying on the Neumann boundary, contribution (7) is added. The domain decomposition procedure ends up with a linear system in the head values over the nodes of the macromesh. If A is the related coefficient matrix, the a_{km} term is provided by:

$$a_{km} = \sum_e \int_{\Gamma^e} q_k^e N_m d\,\Gamma^e \qquad (11)$$

and results from the assemblage of the local 3x3 matrix A^e related to triangle e:

$$a_{rp}^e = \int_{\Gamma^e} q_r^e N_p d\,\Gamma^e \qquad r,p \in \{k,m,n\}$$

It is easily seen that $a_{rp}^e \neq a_{pr}^e$ and hence A is unsymmetric. Moreover, $N_k + N_m + N_n = 1$ over Γ^e and since $W_k^e + W_m^e + W_n^e = 1$ satisfies eq. (3) over Ω^e, we have:

$$\sum_p a_{rp}^e = \sum_r a_{rp}^e = 0 \qquad r,p \in \{k,m,n\} \qquad (12)$$

Property (12) holds for the global matrix A as well.

If K_{ij} is uniform inside the triangle e, it may be proven that A is equal to the stiffness matrix of the f.e. method applied over the macromesh. Hence the numerical error affecting the present domain decomposition solution is essentially controlled by the size of the macromesh even though the adjoint problems are very accurately solved.

PRELIMINARY NUMERICAL RESULTS

The present numerical algorithm is applied to a boundary
value problem of the form (1) with f = 0 over a square
region Ω (Figure 2).

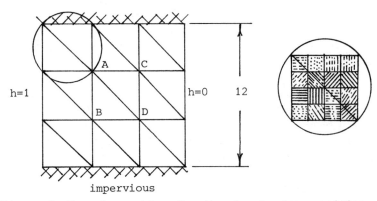

Figure 2. Example problem for the domain decomposition
method.   Each   macrosquare   contains   16
different permeability values.

Domain Ω is discretized into 9 square macroelements
(further   decomposed   into   2   macro-triangles),   each   one
presenting 16 different values of the piecewise constant
permeability (Figure 2). The size of Ω is 12 (arbitrary
units) while   the   size   of   the   elementary   permeability
microstructure is 1. Consequently Ω has 144 areas with
different K values. The isotropic permeability has been
generated by a homogeneous second order stationary random
process assuming for K a lognormal distribution with zero
mean and autocorrelation function C(d) given as (Tompson et
al., 1987):

$$C(d) = (1 - \frac{d}{\lambda}) \exp(-d/\lambda) \qquad (14)$$

where d is the distance and λ is the correlation length. In
our example $\lambda \leqslant 1$ provides a non-correlated or negatively
correlated random sample. Several experiments have been
performed   with   various   permeability   realizations
corresponding to random processes with standard deviations σ
equal to 0.5, 1 and 1.5 and λ equal to 0.1, 1, 6, 12 and 48.
Only a few results will be presented herein.

The adjoint solution W_k^e are determined over the macro-triangles (Figure 2) by standard triangular f.e. with element sizes $\leqslant 1$. Integrals (11) are carried out by Simpson rule using q_k^e values derived numerically from the discrete solution W_k^e. The domain decomposition solution on nodes A, B, C and D (Figure 2) is given in Table 1 for σ =0.5, 1, 1.5 and λ = 12, 6, 1 and is compared with the f.e. solution obtained with a nodal spacing $\Delta x = 1$. The results refer to progressively better estimates for W_k^e derived from local microgrids with a Δx ranging from 1 to 0.06125. Careful inspection of Table 1 shows that:

1) A good f.e. estimate of W_k^e, needed to provide reliable predictions for q_k^e, requires a local nodal spacing smaller than the scale of heterogeneity. This is particularly true for higher σ's and, subordinately, for smaller λ's.

2) If σ is large and λ is small, the permeability contrast over Ω may be several orders of magnitude large. In such conditions also the global standard f.e. method may require a refined mesh to provide accurate predictions of h.

3) The domain decomposition algorithm converges to a solution which is generally different from the global f.e. solution. This is consistent with the respective truncation errors, related to the macromesh size for the former and to the microstructure size for the latter.

4) Since the error for h is controlled by the macromesh, it may be not worth computing W_k^e with a very high accuracy. However, depending on the actual values of σ and λ , reliable estimates of q_k^e may require a f.e. mesh which may be finer than the heterogeneity scale.

CONCLUSION

A domain decomposition algorithm using triangular subdomains has been developed to solve steady flow problems in randomly heterogeneous porous media. In each triangle three adjoint solutions are sought by traditional finite elements. These solutions, which incorporate the actual non-homogeneities of

Table 1. Solution to a steady flow problem by the domain decomposition method and comparison with traditional f.e. results for lognormal permeability distributions characterized by 144 distinct values.

Δx	A F.E.	A D.DEC.	B F.E.	B D.DEC.	C F.E.	C D.DEC.	D F.E.	D D.DEC.
$\sigma = 0.5$	$\lambda = 12$				$K_{min} = 0.757$		$K_{max} = 3.886$	
1	0.415	0.433	0.429	0.416	0.169	0.177	0.157	0.161
0.5		0.397		0.407		0.171		0.154
0.25		0.401		0.405		0.171		0.154
0.125		0.402		0.404		0.171		0.155
$\sigma = 0.5$	$\lambda = 6$				$K_{min} = 0.672$		$K_{max} = 3.527$	
1	0.372	0.351	0.303	0.325	0.109	0.113	0.119	0.120
0.5		0.370		0.314		0.119		0.119
0.25		0.370		0.320		0.122		0.123
0.125		0.370		0.324		0.120		0.124
$\sigma = 0.5$	$\lambda = 1$				$K_{min} = 0.258$		$K_{max} = 2.826$	
1	0.432	0.421	0.424	0.411	0.141	0.124	0.137	0.130
0.5		0.406		0.439		0.151		0.157
0.25		0.411		0.426		0.147		0.153
0.125		0.414		0.419		0.145		0.151
0.06125		0.416		0.416		0.144		0.149
$\sigma = 1$	$\lambda = 1$				$K_{min} = 0.067$		$K_{max} = 7.984$	
1	0.476	0.438	0.454	0.393	0.159	0.106	0.149	0.112
0.5	0.477	0.427	0.456	0.521	0.156	0.180	0.148	0.188
0.25	0.477	0.433	0.457	0.482	0.155	0.166	0.148	0.177
0.125		0.436		0.462		0.160		0.170
0.06125		0.437		0.452		0.157		0.167
$\sigma = 1.5$	$\lambda = 1$				$K_{min} = 0.017$		$K_{max} = 22.559$	
1	0.509	0.431	0.470	0.309	0.171	0.072	0.153	0.075
0.5	0.509	0.441	0.475	0.617	0.167	0.207	0.153	0.208
0.25	0.509	0.441	0.477	0.536	0.164	0.178	0.152	0.186
0.125		0.441		0.497		0.165		0.174
0.06125		0.440		0.476		0.159		0.168

the permeability field and may be evaluated simultaneously on a parallel machine, provide adjoint fluxes to be subsequently integrated over the subdomain boundaries and properly assembled over the nodes of the macromesh. A final algebraic unsymmetric system is thus arrived at, whose solution gives the head on the nodes of the subregions or macromesh. The formulation fully includes the effects of the heterogeneity distribution at the microscale. The approach appears to be particularly suited to randomly heterogeneous media where the heterogeneity scale is much smaller than the basin scale, to a computer with a parallel architecture and to a problem which is to be repeatedly solved for many realizations of a random permeability field (e.g. solution of stochastic differential equations with the permeability given as a random process).

Acknowledgments. The authors are very grateful to Shlomo P. Neuman who introduced the basic idea of domain decomposition underlying the present development while on sabbatical leave at the Department of Mathematical Models in Applied Sciences of the University of Padova. This work has been in part supported by the Italian CNR, Gruppo Nazionale per la Difesa dalle Catastrofi Idrogeologiche, Linea di Ricerca no. 4, and Progetto Finalizzato "Sistemi Informatici e Calcolo Parallelo".

REFERENCES

1. Bjorstad P.E. and Widlund O.B., Iterative methods for the solution of elliptic problems on regions partitioned into substructures, SIAM J. Numer. Anal., 23, 6, 1097-1120, 1986.
2. Bramble J.H., Pasciak J.E. and Schatz A.H., An iterative method for elliptic problems on regions partitioned into substructures, Mat. Comp., 46, 174, 361-369, 1986.
3. Canuto C., Hussaini M.Y., Quarteroni A. and Zang T.A, Spectral methods in fluid dynamics, Springer-Verlag, New York, 1988.
4. Chan T.F., Analysis of preconditioners for domain decomposition, SIAM J. Numer. Anal., 24, 382-390, 1987.
5. Glowinski R., Dinh A.V. and Periaux J., Domain decomposition methods for non linear problems in fluid dynamics, Comput. Meth. Appl. Mech. Eng., 40, 27-109, 1983.
6. Glowinski R., Golub G.H., Meurant G. and Periaux J.,

eds., Proc. First Int. Symp. on Domain Decomposition Methods for Partial Differential Equations, SIAM, Philadelphia, 1988.

7.   Meurant G., Domain decomposition preconditioners for the conjugate gradient method, Calcolo, 25, 103–119, 1988.

8.   Quarteroni A. and Sacchi-Landriani G., Parallel algorithms for the capacitance matrix method in domain decompositions, Calcolo, 25, 76–102, 1982.

9.   Tompson A.F.B., Ababou R. and Gelhar L.W., Applications and use of the three-dimensional turning band random field generator in hydrology, Tech. Rept. No. 313, Parsons Laboratory, MIT, Cambridge, MA 02139, 1987.

10. Schwartz H.A., Uber einige Abbildungsu afgaben, Ges. Math. Abh.,11, 65–83, 1869.

Uncertainty Analysis Methods in Groundwater Modeling

D.C. Mckinney(1), D.P. Loucks

School of Civil and Environmental Engineering, Cornell University, Ithaca, New York 14853 USA

Groundwater simulation models are used extensively to investigate contaminated aquifers and design remediation plans for their cleanup. The potential for simulation models to help analysts understand the response of aquifers to alternative management plans and remediation designs is directly affected by the reliability of these models. One measure of reliability is the degree of uncertainty associated with the model's predictions resulting from uncertainty in the model parameters. This prediction uncertainty can be measured by the variance of the state variables (hydraulic head and concentration) computed by the model. A potential source of parameter uncertainty is a lack of knowledge about model inputs or coefficients resulting from the spatial variability of aquifer flow and transport properties. Due to the physical processes which form them, the flow and mass transport properties of aquifers tend to exhibit a high degree of spatial variability. It is this type of parameter uncertainty and its effect on the uncertainty of model predictions which is examined in this paper.

FIRST-ORDER UNCERTAINTY ANALYSIS

Simulation model prediction variance is computed by a first-order uncertainty analysis of the model finite element equations. First-order uncertainty analysis is a time domain perturbation method of approximating the solution of stochastic partial differential equations. The method uses a Taylor series expansion of the system state variables in terms of the system parameters to compute the first two moments (mean and covariance) of the state variables. First-order analysis requires the computation of state variable sensitivity matrices, whose elements measure the responsiveness of a state variable at one location to changes in a model parameter at another location.

Let $y(u)$ be a vector of p system state variables which depends on u a vector of q system parameters. System state variables can be defined implicitly as a vector functional relationship or set of system state equations, $f(y,u) = 0$. The system state variables, $y(u)$, can be approximated by a first-order Taylor series expanded about some nominal values of the parameter vector, $\hat{u}$

---

[1] Currently at Department of Civil Engineering, University of Texas at Austin, Austin, TX, USA

$$y(u) \approx y(\hat{u}) + \frac{\partial y}{\partial u} \cdot (u - \hat{u}) \tag{1}$$

where $\frac{\partial y}{\partial u}$ is a (p x q) sensitivity matrix of derivatives of the state variables with respect to the parameters. Sensitivity theory methods for computing the sensitivity matrices [1, 2] from the implicit form of the system state equations were used in this work. If the system parameters u are stochastic variables or random functions, first-order approximations of the mean and covariance of y can be found from Equation 1, where the nominal values of the parameters are the mean values of the stochastic parameters. The first-order approximation of the mean state variables is

$$\hat{y} = E\{y(u)\} \approx y(\hat{u}) \tag{2}$$

The first-order mean is the vector resulting from the use of the mean parameters. Using the first-order approximations for y and its mean, the first-order approximation of the state variable covariance matrix is

$$P_y = \text{Cov}(y,y) = E\{[y - \hat{y}][y - \hat{y}]'\} \approx \frac{\partial y}{\partial u} P_u \frac{\partial y'}{\partial u} \tag{3}$$

where $P_u = \text{Cov}(u,u)$ is the covariance matrix of the system parameters. The uncertainty in the state variables depends, to first-order, on the uncertainty in the parameters and on the sensitivity of the state variables to changes in the parameters.

GROUNDWATER FLOW AND MASS TRANSPORT MODEL

The system state variables for flow and mass transport within a groundwater system are hydraulic head, velocity, and concentration. The governing partial differential equations are conservation of fluid and contaminant mass, and conservation of fluid linear momentum. Vertically integrated partial differential equations governing the steady flow of an incompressible fluid and the transient transport of a sorbing-desorbing, dilute contaminant dissolved in that fluid were considered. A two-dimensional simulation model was used to illustrate the uncertainty analysis. The equations representing flow and mass transport in the system are [3]

$$\nabla \cdot bK \cdot \nabla h - Q_p + Q_r = 0 \tag{4}$$

$$v = -\frac{K}{n} \cdot \nabla h \tag{5}$$

$$bR\frac{\partial c}{\partial t} - \nabla \cdot nbD \cdot \nabla c + nbv \cdot \nabla c - Q_p c + Q_r c_r = 0 \tag{6}$$

where h is the hydraulic head [L], b is the aquifer thickness [L], K is the hydraulic conductivity [L/T], Q_p and Q_r are extraction and recharge rates [L/T], respectively, v is the Darcy velocity vector [L/T], n is the effective porosity, c and c_r are concentrations in the formation and recharge waters [M/L^3], respectively, D is the hydrodynamic dispersion tensor $[L^2/T]$, and R is the retardation coefficient. The Galerkin finite element method was used to discretize spatial derivatives in these equations, while a fully implicit finite difference scheme was employed for the temporal derivatives.

First-order uncertainty analysis requires the groundwater simulation model to be cast in an implicit function framework. The system state equations for the groundwater flow and mass transport model are the time discretized finite element equations, which can be written compactly in a general implicit form as

$$f(y, u) = A(y, u)y - b(y, u) = 0 \qquad (7)$$

where the matrix A and vector b are derived from the finite element equations [2] and include appropriate boundary and initial conditions. The vector of system state variables is $y = [h, \hat{v}, c^1, ..., c^T]'$, where h, $\hat{v}$, c^1, ..., c^T are vectors of predicted hydraulic head, velocity, and concentration, respectively, and T is the number of time steps in the simulation. The vector of system parameters is $u = [K_x, K_y, b, n]'$, where K_x, K_y, b, and n are vectors of x and y direction conductivity, thickness, and porosity, respectively. Any of the system parameters may be known with less than absolute certainty causing randomness to enter the system. This randomness or uncertainty is propagated through the system state equations and appears in the system state predictions.

UNCERTAINTY ANALYSIS EXAMPLE

Following is an example which uses first-order uncertainty analysis method to compute the mean and standard deviation of head and concentration. The results of the first-order analysis are compared to Monte Carlo simulation and spectral perturbation results. In the first-order analysis, mean state variables were computed by solving the groundwater simulation model using mean parameter values. The state variable covariance matrix was computed using the covariance matrix of the parameters and the state variable sensitivity matrix computed by sensitivity theory methods.

The finite element mesh for this hypothetical example, shown in Figure 1, consists of 231 nodes and 200 elements each 10 cm by 10 cm. Hydraulic conductivity, the only stochastic aquifer property in the example, was assumed to be log-normally distributed and follow an isotropic covariance model given by [4]

$$C(r) = \sigma_{lnK}^2 \frac{\pi r}{2\lambda} K_1(\frac{\pi r}{2\lambda}) \qquad (9)$$

where r is the separation distance between two points, K_1 is a modified Bessel function, σ_{lnK}^2 and λ are the variance and correlation length of the log-conductivity stochastic process, respectively.

Boundary conditions for this steady flow, transient transport example are constant head on the left ($h=100$ cm) and right ($h=0$ cm) sides, and no flow on the top and bottom of the aquifer. A line source of contaminant (1 mg/l) is distributed across three nodes and constant concentration boundaries (0 mg/l) are maintained on the left and right hand sides of the aquifer. The remaining aquifer simulation parameters are thickness = 1.0 cm, porosity = 1.0, μ_{lnK} = -

4.605 cm/s, σ_{lnK} = 0.99 cm/s, λ = 17.4 cm, longitudinal and transverse dispersivities = 10 and 4 cm, respectively, no molecular diffusion or adsorption, initial concentration = 0.0 mg/l. The system was simulated for 2.5 hours with 0.5 hour time steps. The mean and standard deviation of the head and concentration predictions were computed by both the first-order and Monte Carlo methods. The situation described here has been used in Monte Carlo simulations of steady state flow [5].

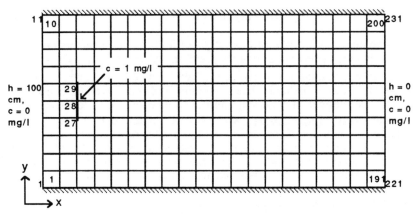

Figure 1. Example aquifer boundary conditions.

The mean head resulting from this example, illustrated in Figure 2 by plotting the first-order mean head along the aquifer centerline at y = 50 cm, is linear and ranges from 100 on the left to 0 cm on the right boundary. Figure 2 also illustrates the head standard deviation which varies smoothly from 0 cm on the left to a maximum of 11.4 cm in the center and back down to 0 cm on the fixed head right boundary. The first-order and Monte Carlo results are very similar except for small variations near the center of the aquifer resulting from the finite number of realizations used to compute the Monte Carlo solution. The first-order and Monte Carlo head standard deviation computed in this example were compared with published Monte Carlo results [5] and found to be in complete agreement. The head standard deviation was also compared with the spectral perturbation result for this essentially one-dimensional flow field [6]

$$\sigma_h = \sigma_{lnK}\lambda J \qquad (10)$$

where J = 0.5 is the average head gradient in the x direction. The head standard deviation resulting from this expression is σ_h = 8.61 cm, agreeing with σ_h = 8.64 cm and 8.40 cm for the average first-order and Monte Carlo results respectively.

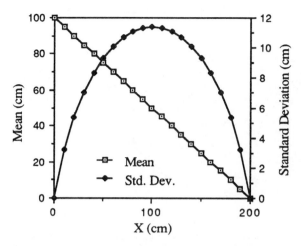

Figure 2. First-order head, mean and standard deviation.

A contour plot of the mean concentration from the first-order method after 2.5 hours is shown in Figure 3. First-order concentration standard deviation contours, shown in Figure 4, are symmetric about the plume centerline, with two maxima of 0.185 mg/l at x = 50 cm, y = 30 and 70 cm respectively. The maxima occur at locations of greatest concentration gradient as expected since spectral perturbation theory shows that concentration standard deviation is directly proportional to concentration gradient [7]. The first-order and Monte Carlo solutions were nearly identical except for a slight asymmetry about the plume centerline and smaller maxima in the Monte Carlo method.

Expressions for concentration standard deviation in heterogeneous porous media [6] can be used to give a qualitative assessment of the first-order and Monte Carlo standard deviation results computed here. For this example the expression for the concentration variance along the plume centerline, where the y-direction concentration gradient is zero, is [7]

$$\sigma_c^2(x, y=50 \text{ cm}) = \frac{2\sigma_{\ln K}^2 \lambda^3 G_x(x)^2}{3\gamma^2 \alpha_T} \tag{11}$$

where $G_x(x)$ is the x-direction concentration gradient and γ is a flow factor. At centerline distances, x, where the y-direction concentration gradient is almost zero, a comparison of the square root of this expression and the standard deviation computed by the first-order and Monte Carlo methods can be made. The results of this calculation are shown in Figure 5. The spectral perturbation result and the first-order and Monte Carlo values agree at locations where the concentration gradient in the y direction is almost zero near the centerline; in Figure 5, this is the region x > 80 cm. Near the source the spectral results differ widely from the others; since in this region both x- and y-direction concentration gradients are large and vary greatly.

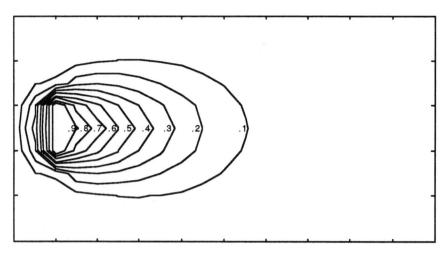

Figure 3. First-order mean concentration (mg/l).

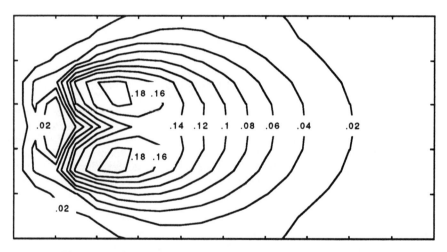

Figure 4. First-order concentration standard deviation (mg/l).

CONCLUSION

Uncertainty present in groundwater simulation model predictions due to the spatial variation of aquifer properties (hydraulic conductivity) was examined through the use of first-order uncertainty analysis. The variance of predicted state variables was used as a measure of uncertainty in model predictions. An example demonstrates the computation of the mean and variance of the simulation model predictions by this method. The results show that the first-order method is consistent with Monte Carlo and spectral perturbation results. These results indicate that the first-order uncertainty analysis method provides an accurate means of assessing groundwater simulation model prediction uncertainty.

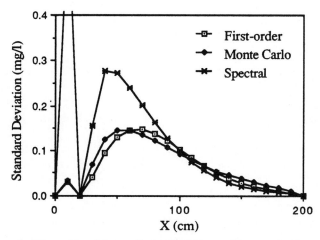

Figure 5. First-order, Monte Carlo, and spectral concentration standard
deviation (mg/l).

REFERENCES

1. Ahlfeld, D. P., J. M. Mulvey, G. F. Pinder, and E. F. Wood,
   Contaminated groundwater remediation design using simulation,
   optimization, and sensitivity theory, 1. Model development, Water Resour.
   Res., Vol. 24, No. 3, pp. 431-441, 1988.
2. McKinney, D. C., Predicting Groundwater Contamination: Uncertainty
   Analysis and Network Design, Ph.D. diss., Cornell Univ., Ithaca, N. Y.,
   1990.
3. Bear, J., Hydraulics of Groundwater, McGraw Hill, New York, 1979.
4. Vanmarcke, E., Random Fields, M. I. T. Press, Cambridge, 1983.
5. Smith, L., and R. A. Freeze, Stochastic analysis of steady-state
   groundwater flow in a bounded domain, 2: Two dimensional simulations,
   Water Resour. Res., Vol.15, No.6, pp. 1543-1559, 1979.
6. Bakr, A. A., L. W. Gelhar, A. L. Gutjahr, and J. R MacMillan, Stochastic
   analysis of spatial variability in subsurface flows 1. Comparison of one- and
   three-dimensional flows, Water Resour. Res., Vol.14, No.2, pp. 263-271,
   1978.
7. Vomvoris, E. G. and L. W. Gelhar, Stochastic prediction of dispersive
   contaminant transport, Rep. CR-811135-01-2, U.S. Environmental
   Protection Agency, Washington, D. C., 1986.

Three - Dimensional Simulation of Solute Transport in Inhomogeneous Fluvial Gravel Deposits using Stochastic Concepts

P. Jussel, F. Stauffer, Th. Dracos

Institute of Hydromechanics and Water Resources Management Fedral Institute of Technology, Zurich, CH - 8093 Zurich, Switzerland

ABSTRACT

A 3-dimensional flow and solute transport model has been developed to perform numerical tracer experiments in inhomogeneous aquifers. The finite-element flow model is optimized for calculating domains with a large number of elements with the aid of a supercomputer. A special interpolation technique based on tetrahedra is used to determine the velocity field for the transport model. Examples of numerical tracer experiments are compared with the results of the spectral methods for determining macrodispersivities.

INTRODUCTION

Hydromechanic dispersion in aquifers, a phenomenon which must be taken into account in many groundwater contaminant problems, is mainly caused by the variation of hydraulic conductivity. Various conceptual models exist to calculate the parameter describing this phenomenon, namely the dispersivity or the macrodispersivity, for given spatial structures of the hydraulic conductivity. Because these approaches make considerable simplifications of the real situation, the actual spatial structure of hydraulic conductivity has been evaluated by investigating outcrops of fluvial gravel deposits above the groundwater level. Several classes of mostly lens-shaped inhomogeneities with different geometric features and hydraulic properties were distinguished in order to numerically generate artificial aquifers with the same stochastic structure. A new 3D flow and solute transport model is presented which allows numerical tracer experiments in the artificial aquifers to be performed.

CONCEPTS FOR CALCULATING MACRODISPERSION

Stratified formations

As shown in Molz et al.[10], the macrodispersivity as a function of time can be calculated deterministically from the vertical profile of hydraulic conductivity of a perfect layered aquifer.

Random hydraulic conductivity with defined covariance function

Gelhar et al.[5] and Dagan [4] used the spectral representation to approximately calculate the macrodispersivity tensor for a anisotropic, but statistically homogeneous log hydraulic conductivity field which can be characterized by its covariance function.

Hydraulic conductivity field based on outcrop evaluations

Investigations of fluvial gravel deposits show neither a perfect stratification nor the hydraulic conductivity field can completely be described by an anisotropic covariance function. The investigated deposits consist rather of a kind of gravel matrix with mostly lens-shaped inclusions of certain extents and hydraulic properties. Macrodispersivity in such formations can only be calculated by a number of numerical tracer experiments.

DESCRIPTION OF FLUVIAL GRAVEL DEPOSITS

Unweathered and large outcrops in pits of fluvial gravel deposits near Zurich have been studied in order to determine the actual spatial structure of hydraulic conductivity in this kind of deposit. Based on geological investigations (Huggenberger et al.[7]), the deposits can be interpreted as being composed of six characteristic types of geological elements which can be distinguished by their grain size, colour and geometry. Fig.1 shows a part of a typical outcrop in the Huentwangen gravel pit (20 km north of Zurich).

20 undisturbed and about 70 disturbed samples were taken to determine the hydraulic conductivity and porosity of each geological element. The following list gives their main characteristics. Noticable is the extraordinary high hydraulic conductivity of open framework gravels. For detailed information see Jussel [8].

1. Grey gravel: Sandy, poorly-sorted gravel with little silt, grey colour. (K = 0.14 mm/sec)
2. Brown gravel: Sandy and silty, poorly-sorted gravel, containing pebbles. The silt causes a brown colour. (K = 0.03 mm/sec)
3. Sand lenses: Well-sorted sand with strong homogeneity within the lens. (K = 0.3 mm/sec)
4. Open framework gravel: Well-sorted gravel with practically no sand or silt; occurs usually in very thin horizontal or inclined lenses. (K = 300 mm/sec)
5. Open framework/bimodal couplets: More or less regular alternations of inclined open framework zones with bimodal gravels. (K of bimodal gravels = 0.23 mm/sec)
6. Silt lenses: Well-sorted silt, brown colour. (K < 0.001 mm/sec)

Grey and brown gravel are considered to be a kind of matrix, whereas the rest of the elements are lens-shaped inclusions in the matrix. The mean length of the lenses varies from 1.3 m (open framework-zones) to about 10 m (open framework/bimodal couplets). It is an interesting observation that the lengths of each type of lenses are lognormally distributed. The mean relation length/maximal thickness of the lenses is almost constant at about 17.

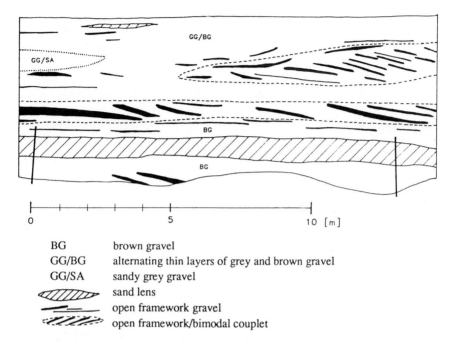

BG	brown gravel
GG/BG	alternating thin layers of grey and brown gravel
GG/SA	sandy grey gravel
	sand lens
	open framework gravel
	open framework/bimodal couplet

Figure 1: Sketch of the geological elements of a part of an outcrop

3D-FINITE-ELEMENT FLOW MODEL

The finite-element flow model is based on a cubic topology: The modeled domain consists of a rectangular block which is subdivided in hexahedral finite elements. Each element can be distorted in order to reproduce the typical lens-shape geometry of the classified inhomogeneities. Anisotropy within an element is taken into account by treating the hydraulic conductivity as a tensor.

The local matrices of the finite elements are calculated by subdividing the (distorted) hexahedra into five tetrahedra which can be done in two patterns (see fig.2). Tetrahedral volume coordinates are used for the shape function defining the hydraulic head within the tetrahedron (Pelka [11]). According to the Galerkin method of the weighted residuals the weighting function is made equal to the shape function. This procedure allows the weighting function over the tetrahedron volume to be integrated analytically. Local matrices of hexahedra are calculated by the superposition of the five local matrices of the tetraheda. Both possible tetrahedral patterns are used to avoid asymmetry; the two local hexehedral matrices can be averaged arithmetically (Huebner et al.[6]). This discretization method is supposed to produce less discretization errors than finite difference methods because each node is connected to 18 neighbouring nodes compared to 6 when using a 7-point finite difference scheme.

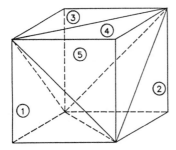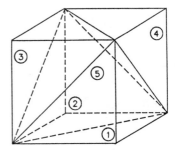

Figure 2: The two possibilities of subdividing hexahedra into five tetrahedra

The conjugate gradient method turned out to be the most efficient method for the solution of the system of linear equations. Two types of this iterative method have been used:

1) Conjugate gradient method with incomplete Cholesky factorization (Braess [3]). The incomplete Cholesky factorization preconditioning reduces the number of iterations and the storage requirements are about 50*N floating point numbers (N = number of nodes). However, most of the calculations cannot be vectorized.
2) Conjugate gradient EBE (=element-by-element) method (Bartelt [1]). Diagonal preconditioning needs more iterations and the storage requirement is about 60*N floating point numbers. Matrix-vector products are calculated at the hexahedron element level which allows to perform the multiplications for independent blocks of hexahedra in a parallel manner.

The first method is advantageous especially for poorly conditioned small problems which can be solved with a conventional computer. The EBE-method is well suited to big problems (more than about 20,000 elements) which are solved on a supercomputer because all CPU time-consuming calculations are vectorized. For big problems, the EBE-method is about twice as fast as the first method. Test calculations with the EBE-method with 125,000 elements needed 120 to 360 sec CPU-time, depending on the condition number of the problem, while a well conditioned test problem with 200x50x50 = 500,000 hexahedral elements needed less than 10 min CPU-time on a CRAY2.

PARTICLE TRACKING TRANSPORT MODEL

Propagation of particles
In agreement with the flow model, the particle tracking transport model is based on the tetrahedral discretization. Particles which represent a certain tracer mass are moved in discrete advective steps. Local dispersion will be introduced in a program version which is currently being developed. After each advective step the two tetrahedra of both tetrahedral patterns in which the particle lies have to be found. By calculating the tetrahedral volume coordinates for the particle position we check whether the particle is located within the considered tetrahedron or not. Because of the cubic topology only a few tetrahedra must be inspected to find the correct tetrahedron.

Interpolation of velocities

For each tetrahedron of both tetrahedral patterns a velocity is calculated from the hydraulic heads in its 4 edges. At every point of the flow domain 2 velocities from the 2 tetrahedral patterns are defined. A test run which can be compared with an analytical solution showed that it is best to average the absolute value of the velocities arithmetically and to average their directions separately.

The fluvial gravel deposits described here contain highly permeable thin lenses. Because of the extremely high velocity gradients in the vicinity of these lenses a special interpolation technique must be used for a realistic transport simulation. The proposed interpolation method is based on a local coordinate system, centered at the particle position (see fig.3). The interpolated velocity is calculated from the velocity at the origin, and from the velocities of six points located on the axes of the coordinate system at a distance equal to about one-half of a hexahedron element length. A smearing out of large gradients in the velocity field in the vicinity of strong hydraulic conductivity changes is avoided by weighing the velocities of the outer points with a function of the ratio of hydraulic conductivities between the particle position and the outer points. In a test run the interpolated velocities were compared with the analytically calculated velocity distribution for a highly permeable cylinder in a homogeneous formation. The differences between interpolated and analytically calculated velocities were small.

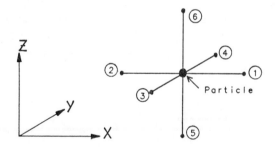

Figure 3: Local coordinate system for the velocity interpolation (1 to 6: outer points)

Adjustment of advective step length

The special interpolation technique provides information about the velocity gradients at the position of the particle. This information is used to adjust the advective step length. In case of small velocity gradients a large advective step is made and vice versa. Because of the strong changes of hydraulic conductivity (and consequently of velocity) overshooting of particles from a hexahedron of low into a hexahedron of high hydraulic conductivity and vice versa has to be avoided. For this reason the particle is stopped and a new advection step starts when it enters a hexahedron with a considerably different hydraulic conductivity.

EXAMPLES OF NUMERICAL TRACER EXPERIMENTS

Up to now two large domain numerical tracer experiments were performed with the model presented. Both domains consist of 50x50x50 = 125,000 hexahedral elements. In spite of the different spatial structure of hydraulic conductivity both have a geometric mean of hydraulic conductivity of 1.0 mm/sec and a standard deviation of its natural logarithm sigma(lnK) of 1.0. As an initial condition, 1600 particles were located in a yz-plane in a rectangular domain which is large compared to the square of the correlation length. Local dispersion was neglected in both examples.

Example 1: Statistically homogeneous log hydraulic conductivity field

Analogous to Tompson et al.[13] a numerical tracer experiment was performed in an isotropic statistically homogeous log hydraulic conductivity field with exponential covariance which was generated with the turning bands method (Tompson et al.[12]). The element length is 0.5 m and the correlation length is 1.0 m.

Example 2: Highly permeable thin lenses in a homogeneous formation

As a simple model of open framework gravel lenses embedded in grey gravel, highly permeable (K=300 mm/sec) lenses with approximately the same geometry as open framework gravel lenses are statistically distributed within a homogeneous (K=0.85 mm/sec) 3-dimensional domain. The correlation length is about 1.4 m in horizontal and 0.03 m in vertical direction. Fig.4 shows a vertical cross section of the 12.5x12.5x2.5 m domain. The black zones are highly permeable lenses and the hydraulic head contours are shown for a flow from the left to the right.

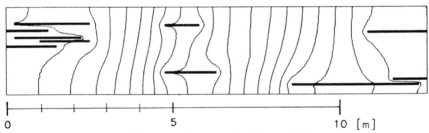

Figure 4: Vertical cross section through the domain of example 2.

The results of the two numerical tracer experiments are shown in fig.5 in terms of the longitudinal displacements variance of tracer mass as a function of the transport distance, both made dimensionless by the horizontal correlation length. The numerically calculated variances of example 1 are always lower than the values predicted by Dagan [4] for the isotropic case whereas the asymptotic value for the macrodispersivity (inclination of the curve) is practically identical with the asymptotic macrodispersivity predicted by Gelhar [5]. Because some particles left the modeled domain very early in example 2 only a rough estimate could be made for the variance for dimensionless transport distances larger then 3 by estimating the variance from the breakthrough curve. In example 2 the variance increases more strongly than predicted by Dagan [4] for the anisotropic case (anisotropy factor Lvert/Lhor=0.022). Gelhar [5] predicts the same macrodispersivity as in the isotropic case which is also smaller than the observed value.

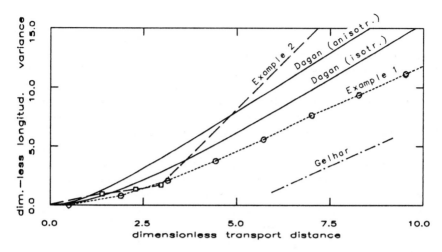

Figure 5: Longitudinal displacements variance of tracer mass as a function of transport distance; comparison of numerical tracer experiments with results of Gelhar [5] and Dagan [4].

DISCUSSION

The flow and transport model presented here turns out to be a useful tool for calculating macrodispersivities for various spatial structures of hydraulic conductivity. The numerical tracer experiments which have been performed up to now show that the asymptotic macrodispersivity is larger in an aquifer in which highly permeable thin lenses are embedded than in an aquifer which can completely be described by its covariance function of the hydraulic conductivity. Additional numerical tracer experiments in domains with a spatial structure of hydraulic conductivity which is close to the actual one are necessary to check the validity of this observation and to estimate the macrodispersivity as a function of the transport distance for the investigated kind of gravel deposit.

ACKNOWLEDGEMENT

This work was financially supported by the Swiss National Science Foundation, grant nr. 2.118-0.86.

REFERENCES

1. Bartelt, P. Finite Element Procedures on Vector/Tighly Coupled Parallel Computers, Diss. ETH, Zurich, 1989.
2. Bear, J. Hydraulics of Groundwater, McGraw-Hill, New York, 1979.
3. Braess, D. PRECG - ein Programm zur schnellen Loesung von grossen linearen Gleichungssystemen mit positiv definiten, duenn besetzten Matrizen, Bochum, 1989 (unpublished).

4. Dagan, G. Flow and Transport in Porous Formations, Springer Verlag, Berlin and Heidelberg, 1989.
5. Gelhar, L.W. and Axness, C.L. Threedimensional stochastic analysis of macrodispersion in aquifers, Water Res. Res., Vol.19, pp. 161-180, 1983.
6. Huebner, H.K. and Thornton A.E. The Finite Element Method for Engineers, John Wiley & sons, 1982.
7. Huggenberger, P., Siegenthaler, C. and Stauffer, F. Grundwasserstroemung in Schottern; Einfluss von Ablagerungsformen auf die Verteilung der Grundwasserfliessgeschwindigkeit, Wasserwirtschaft, Vol.78, pp. 202-212, 1988.
8. Jussel, P. Stochastic description of typical inhomogeneities of hydraulic conductivity of fluvial gravel deposits, Proceedings of the Int. Symposium on Contaminant Transport in Groundwater, Stuttgart, 1989. A.A. Balkema, Rotterdam and Brookfield, 1989.
9. Meyer, P.D., Valocchi, A.J., Ashby, S.F. and Saylor P.E. A Numerical Investigation of the Conjugate Gradient Method as Applied to Three-Dimensional Groundwater Flow Problems in Randomly Heterogeneous Porous Media, Water Res. Res., Vol.25, pp. 1440-1446, 1989.
10. Molz, F.J., Guven, O. and Melville, J.G. An examination of scale-dependent dispersion coefficients, Groundwater, Vol.21, pp. 715-725, 1983.
11. Pelka, B. Modelle zur Berechnung mehrschichtiger Grundwasserleiter auf der Basis von Finiten Elementen, Mitteilungen Institut fuer Wasserbau und Wasserwirtschaft Rheinisch-Westfaelische Technische Hochschule Aachen, Aachen, 1988.
12. Tompson, A.F.B., Ababou, R. and Gelhar, L.W. Application and Use of the Three-Dimensional Turning Bands Random Field Generator: Single Realization Problems, MIT Report Number 313, 1987.
13. Tompson, A.F.B., Vomvoris, E.G. and Gelhar, L.W. Numerical Simulation of Solute TransportInRandomlyHeterogeneousPorousMedia:Motivation,ModelDevelopment,and Application, Lawrence Livermore Laboratory, 1987.

Numerical Experiments on Dispersion in Heterogeneous Porous Media

P. Salandin(*), A. Rinaldo(**)

(*) Institute of Hydraulics "G. Poleni", University of Padova,v, Loredan 20, I-35131 Padova, Italy (**) Department of Civil and Environmental Engineering, University of Trento, Mesiano di Povo, I-38050 Trento, Italy

ABSTRACT

The paper discusses numerical simulations of dispersion processes of inert solutes in two-dimensional random log-permeability fields $Y(x)$. A suitable number of realizations is generated preserving a given spatial correlation structure. For each realization a finite element model solves the flow equation and a particle-tracking method solves the transport equation. Ensemble averaging yields then the statistics. The range of σ_Y^2 investigated is 0.2÷2.0. The results show that Dagan's linear theory yields acceptable results in an unexpectedly broad range of σ_Y^2. Interestingly, a linearization of the flow equation yields larger deviations from the linear theory than the corresponding fully nonlinear model, thereby suggesting that previous conclusions drawn on the limitations of the linear theory might be restrictive.

INTRODUCTION

During the last two decades field findings and theoretical results made clear that chemical transport in geologic media is strongly influenced by spatial variations in hydraulic conductivity. A first-order approach on Fickian and non-Fickian dispersion in heterogeneous aquifers[1,4,6] produces analytical solutions for the dispersion plumes. A fundamental validation showed that the linear theory yields a good agreement with the results of the large scale field study known as the Borden site experiment[5]. The spatial variance tensor $X(t)$ of the ensemble mean concentration resulting from the instantaneous injection of solute mass in a random permeability field is[4]:

$$\frac{d^2 X(t)}{dt^2} = \frac{2}{(2\pi)^m} \iint < \hat{u}\,(k',0)\cdot\hat{u}^{*T}\,(k'',t)\,exp\,(ik''\cdot X')>$$

$$x\,exp\,(ik''\cdot U - k''^{T}\cdot D_d\cdot k'')t\,dk'\,dk''. \tag{1}$$

Here m (=2,3) is the number of space dimensions, D_d is the molecular dispersion tensor, $\hat{u}^*(k,t)$ is a spatial Fourier transform of random fluctuation in the Eulerian velocity field about the mean value U (assumed constant), $\hat{u}^*$ is the complex conjugate of $\hat{u}$, X' is the fluctuation of a particle displacement about its mean trajectory Ut, <> represents expectation and T transpose. This result rests on the hypotesis that the Lagrangian velocity field is statistically homogeneous[4].

In order to solve analytically equation (1) two simplifications are required: *i*) the fluctuation X' of particles trajectories are assumed negligible, so that $X'\approx 0$ and *ii*) the Eulerian velocity field is inferred from the Poisson equation instead of Laplace's. These hypoteses are in principle correct only if the variance of the log-conductivity field σ_Y^2 is small ($\sigma_Y^2 \ll 1$). Although the good agreement with the Borden site experimental results[5] ($\sigma_Y^2 \approx 0.2$) demonstrates the validity of the linear theory, application to formations characterized by large σ_Y^2 is not warranted.

A few partially nonlinear solutions are analytically obtained. Dagan[2] demonstrates that the first-order approximation ii) for the head variance is robust even for $\sigma_Y^2 \approx 1$. Neuman[7] removes i) by adopting Corrsin's conjecture and quasilinearizing the result assuming $\mathbf{X}(t) \approx 2\mathbf{D}(\infty)t$ in a recursive scheme to solve equation (1). Rubin[9], using an analytical covariance of the Eulerian velocity field, obtains the statistics of the dispersion process by particle-tracking simulations based on Gaussian conditioning.

In all studies only one of the two linear hypoteses is removed.

In this paper we apply a Monte Carlo approach using a fully nonlinear numerical scheme for the solution of flow domain and a particle-tracking solution to the transport equation to obtain plume evolutions for relatively large σ_Y^2 in a 2-D domain. Comparisons with the linear theory substantiate the discussion.

THE NUMERICAL SIMULATION OF THE STOCHASTIC TRANSPORT

The Monte Carlo method is applied as follows. For each successive step a random 2-D log-normal transmissivity field with assumed mean and spatial covariance structure is generated in a rectangular domain by using a multivariate normal random generator. The velocity field is obtained from finite elements solution of boundary value problem[8]. The particles' continuous movement is discretized in time: only one particle is relased from a fixed initial position for each random transmissivity field, so that ergodic requirements are met. This procedure is repeated for as many particles as desired.

The Monte Carlo approach applied here is similar to the method employed by Smith and Schwartz[11]. The major differences are: 1) the random generator used allows an *a priori* definition of the extent of the integral scale; 2) major accuracy is posed in the domain discretization, in order to overcome the loss of information in the presence of large transmissivity contrasts between adjacent blocks; 3) the problem is posed in dimensionless form; 4) the boundary (deterministic) conditions effect on the calculated velocity variance is verified for each run.

The equations are written in dimensionless form with respect the lenght of correlation l_Y (exponential factor in the log-transmissivity covariance structure here assumed[1,10] as $C_Y = \sigma_Y^2 \exp(-|x|/l_Y)$) and the mean velocity $\mathbf{U} = <T>\mathbf{J}/n$, where $<T>$ is the expected value of the isotropic and weakly stationary transmissivity field $T(\mathbf{x})$. The mean head gradient $\mathbf{J} = (J,0)$ and the porosity n are assumed constant. The equations solved in order to obtain the random velocity field are

$$\nabla^2 h + \nabla h \cdot \nabla Y = \mathbf{J} \cdot \nabla Y \quad \text{and} \quad \mathbf{V} = -T\,(\nabla h - \mathbf{J})/n \tag{2}$$

where $h(\mathbf{x})$ is the head fluctuation about the mean value $-Jx_1$, $\mathbf{V}(\mathbf{x})$ is the Eulerian velocity and $Y = ln(T)$ is as usual the log-transmissivity field. If molecular dispersion is neglected (i.e. Peclet number $Pe = \infty$), the infinitesimal increment of total particle's displacement $\mathbf{X}_t(\mathbf{x};t)$ is given by

$$d\mathbf{X}_t = \mathbf{V}\,dt. \tag{3}$$

The domain is $36\,l_Y \times 18\,l_Y$ and the major length is in the x_1 direction of mean flow (Fig.1). Following Dagan[3], we define a *scale of measure* as l_Y. A discrete spatially correlated transmissivity field is generated over 648 square blocks (dimension $l_Y \times l_Y$) with given exponential covariance. In this manner we realized an accurate discretization of the exponential covariance structure, by using blocks of length much smaller then the integral scale. It should be noted that in a previous work[10] it has been suggested that the choice of analytical expression for the autocorrelation structure of random conductivity field is quite irrelevant, provided that the integral scale is properly defined. In order to obtain an accurate random velocity field, each block is subdivided in 18 triangular elements: this number is chosen upon strict numerical testing to warrant numerical stability and accuracy. The total number of elements is 9504 and the nodes are 4905. The deterministic boundary conditions imposed are: no flux for $x_2=0$ and $x_2=18$ l_Y and unit specific discharge in the x_1 direction. Only one first type boundary condition is required since we imposed $h=0$ for the node at $\mathbf{x}=(0,0)$. The particle-tracking is carried out only in a central part of the entire domain not affected by the deterministic boundary conditions: the time step choosed is $t^*=tU/l_Y=0.05$. The number of Monte Carlo steps is, for each plume simulation, fixed at 1500 realizations. This number provided a good approximation for the log-transmissivity covariance structure (Fig.2): in this manner the problem input for all values of σ_Y^2 used is consistent.

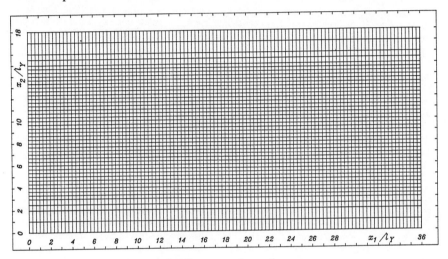

Fig.1 - *Two-dimensional mesh used in the proposed experiments.*

Eulerian velocity results are represented in the same scale of measure used by defining the field of transmissivity. One can see (Fig.3 and 4) that the variance of velocities $u_{11}(0)$ and $u_{22}(0)$ respectively in the x_1 and x_2 direction is influenced near the boundary by the deterministic conditions imposed: this influence grows with the value of σ_Y^2 (considerably beyond $3l_Y$) where the non-affected area reduces. At $\sigma_Y^2 \approx 2$ a meaningful portion of the velocity field is affected by boundary conditions: for this reason the corresponding results obtained are inconclusive.

The longitudinal and transversal components of the spatial variance tensor of expected concentration are illustrated in Fig.5 for σ_Y^2 equal to 0.2, 0.4, 0.8, 1.2 and 1.6, in comparison with Dagan's analytical solution[1]. The differences are limited also for the upper values of σ_Y^2.

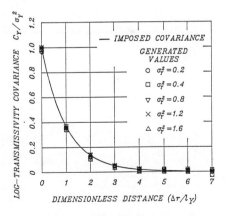

Fig.2 - *Comparison of the imposed log-transmissivity covariance structure with the computed values after 1500 Monte Carlo simulations for several values of σ_Y^2.*

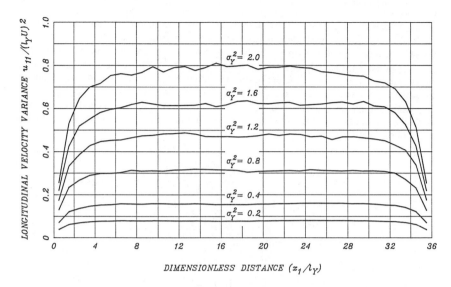

Fig.3 - *Computed longitudinal velocity variance $u_{11}(0)$ versus x_1 coordinate for several values of σ_Y^2.*

To explain this unexpected result the calculations have been repeated using (in order to solve the flow domain) the same linear approximation *ii*) of the analytical solution. The approach is unchanged unless for the solution of the equations (2) which are substituted by

$$\nabla^2 h = \mathbf{J} \cdot \nabla Y \quad \text{and} \quad \mathbf{V} = e^{<Y>} \mathbf{J}/n - e^{<Y>}(\nabla h - \mathbf{J}Y)/n. \qquad (4)$$

Needless to say the linearization in equations (4) leads to substantial reductions in the computational burden.

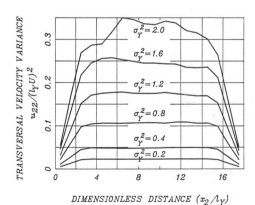

Fig.4 - *Computed transversal velocity variance $u_{22}(0)$ versus x_2 coordinate for several values of σ_Y^2.*

In Fig. 6 the previous numerical results are compared with the results obtained by solving in the same conditions equations (4) and (3): the corresponding Dagan's analytical results are also indicated. To emphasize the differences, a consistent comparison is made on the numerical estimate of the dispersion coefficients ($D_{jl}=0.5 \ dX_{jl}/dt$). The longitudinal dispersion obtained by relaxing the hypothesis *ii*) deviates significantly from the fully non linear scheme and Dagan's linear result for $\sigma_Y^2 > 0.2$. This fact suggests that the first order approximations *i*) and *ii*) in the analytical solution are somewhat counteracting. In all cases (Fig.7) the transversal dispersion coefficients show a good agreement with Dagan's results.

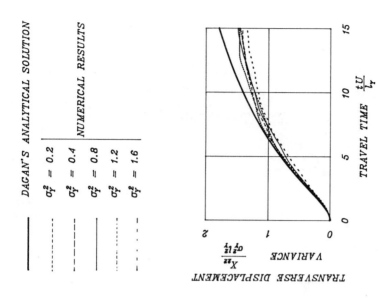

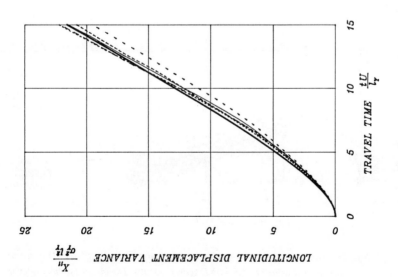

Fig. 5 - Dimensionless comparison of computed longitudinal and transversal displacement variance X_{jj} versus travel time for several value of σ_Y^2 (fully non-linear case) with Dagan's[1] results.

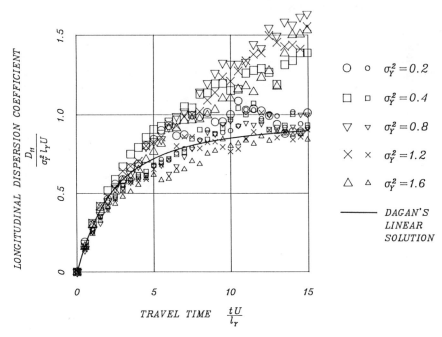

Fig. 6 - *Dimensionless comparison of computed longitudinal dispersion coefficient D_{11} versus travel time for several value of σ_Y^2 with Dagan's[1] results. The larger symbols refer to partially non-linear case and the smaller to fully non-linear case.*

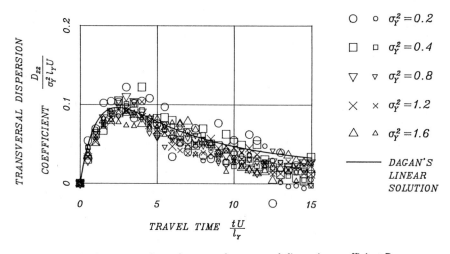

Fig. 7 - *Dimensionless comparison of computed transversal dispersion coefficient D_{22} versus travel time for several value of σ_Y^2 with Dagan's[1] results. The larger symbols refer to partially non-linear case and the smaller to fully non-linear case.*

DISCUSSION AND CONCLUSIONS

We have examined nonlinear effects in the calculation of macrodispersion coefficients in a random conductivity field by using a numerical scheme. In the numerical experiments the log-transmissivity variance σ_Y^2 ranges from 0.2 to 2.0. Several controls (on the domain discretization and on the effect of deterministic boundary conditions) are made to assure numerical accuracy. In the experiments stationarity of the Lagrangian velocity field is not required and only the ergodic hypoteses needs be obeyed. The results extend the validity of the first order Dagan's solution almost to formations in which σ_Y^2 grows up to 1.6. In effects it seems that the linear approximations in the analytical solution (on the particles' trajectory and the velocity random field) yield counteracting effects. The numerical method used is time-consuming and can be perfectioned to improve efficiency. Newerthless the aim of the research was not to create a efficient numerical scheme but rather to contribute to ongoing discussions on the role of nonlinear effects on dispersion in heterogeneous media. Further research is progressing to ascertain conclusively the ranges of numerical error and to limit running computer time.

Acknowledgments. This research is supported by Italian Ministry for Public Education (fondi 40%; project Fenomeni di Trasporto di Inquinanti nel Ciclo Idrologico) and GNDCI (Linea 4, "Vulnerabilità degli acquiferi").

REFERENCES

1. Dagan,G. Solute transport in heterogeneous porous formations, J.Fluid Mech., Vol.145, pp.151-177, 1984.
2. Dagan,G. A note on the higher-order corrections of the head covariances in steady aquifer flow, Water Resour.Res., Vol.21(4), pp.573-578, 1985.
3. Dagan,G. Statistical theory of groundwater flow and transport: pore to laboratory, laboratory to formation, and formation to regional scale, Water Resour.Res., Vol.22(9), pp. 120S-134S, 1986.
4. Dagan,G. Theory of solute transport by groundwater, Annual Rev.Fluid Mech., Vol.19, pp. 183-215, 1987.
5. Freyberg,D.L. A natural gradient experiment on solute transport in a sand aquifer, 2, Spatial moments and the advenction and dispersion of nonreactive tracers, Water Resour.Res., Vol.22(13), pp.2031-2046, 1986.
6. Neuman,S.P., Winter,C.L. and Newman,C.M. Stochastic theory of field-scale Fickian dispersion in anisotropic porous media, Water Resour.Res., Vol.23(3), pp.453-466, 1987.
7. Neuman,S.P. and Zhang,Y.K. A quasilinear theory of non-Fickian and Fickian field scale dispersion: 1. Statistically isotropic porous media, Water Resour.Res., Vol.00(00), 1990.
8. Pinder,G.F. and Gray,W.G. Finite element simulation in surface and subsurface hydrology, Academic Press, London, 1977.
9. Rubin,Y. Stochastic modeling of macrodispersion in heterogeneous porous media, Water Resour.Res., Vol.00(00), 1990.
10. Salandin,P. and Rinaldo,A. The influence of the form of the log-conductivity covariance on non-Fickian dispersion in random permeability fields, Int.J.Num.Meth.Eng., Vol.27, pp. 185-193, 1989.
11. Smith,L. and Schwartz,F.W. Mass transport, 1, A Stochastic analysis of macroscopic dispersion, Water Resour.Res., Vol.16(2), pp.303-313, 1980.

Numerical Experiments on Hydrodynamic Dispersion in Network Models of Natural Media

A. Rinaldo(*), R. Rigon(**), A. Marani(**)
(*) Department of Civil and Environmental Engineering, University of Trent, Mesiano di Povo, I- 38050,Trento, Italy (**) Department of Environmental Sciences, University of Venice,Calle Larga S. Marta,II- 30124 Venezia, Italy

ABSTRACT

Dispersion of tracers in random network is considered. Network models of natural media here include: i)channel networks, possibly of fractal nature, whose geometrical characters have recently attracted interest in hydrological literature; and ii) network models of porous media. For a random tube network exact rules are derived for tracer motion under the combined action of molecular diffusion and heterogeneous convection, which allow the calculation of first-passage time distribution of the tracer at fixed locations. The latter are connected to the dispersive character of the microdispersion processes wich are related in a rational manner not only to the heterogeneity of the convection field but also to the geometrical features of the network. It is shown that the formulation of the transport for relevant hydrological processes at different scales can be viewed under a unifyng approach of kinematic nature. In particular, we show a significant characterization of the hydrologic response for Hortonian channel networks.

INTRODUCTION

The interplay of diffusion and convection has a substantial effect in shaping the characters of dispersion processes in nature. A further dispersive effect, which is gaining increasing attention , is due to the connection of the system, i.e. the lengths, bifurcations and domain of attraction of the elements in network models possibly underlying the geometry of natural media. A distinction may be drawn among random networks (modeling, somewhat arbitrarily, porous media as a random resistor network[5]), deterministic or even fractal networks [1,7]. The latter characters have been recognized as peculiar to natural channel networks. Recent contributions [1,2,3,4] have provided new significant inroads into a unifying approach for transport processes based on travel (first-passage, arrival or residence) time distributions. It is granted in this study that: i) travel time distributions may be related in a rational manner to the "concentration" (or Eulerian) approaches based on the solution of the mass and momentum balance equation in an Eulerian framework and blend all sources of uncertainty into a unique curve [1]; ii) inert solutes and carrier flow behave dynamically in the very same way, such that first-passage time distribution may be referred both to carrier and solute without distinction[2]. iii) the travel time distribution at the outlet of a channel network after an instantaneous pulse is the geomorphologic unit hydrograph (GUH) which is the core of the hydrologic response[3,6]; iv)it is possible to extend travel time concepts to understand mass arrival of sorptive solute in heterogeneous media[4]. In this paper we derive exact rules for tracer (carrier) motion which allow the calculation of first -passage time distribution at the outlets of a network of assigned geometry. Numerical experiments are presented to emphasize the relative role and the mutual interactions of dispersive and convective effects at the scale of the organized network.

TRAVEL TIME DISTRIBUTIONS

Travel time distributions f(t) at the outlet of a system whose input mass is not localized but it is rather distributed over the domain (e. g. the case of watershed drained by a channel network) are given by[3,6]:

$$f(t) = \sum_{\gamma \in \Gamma} p(\gamma) f_{x_\omega} * f_{x_\omega+1} * \dots * f_{x_\Omega}(t) \tag{1}$$

where γ is the collection of states $< x_\omega, x_{\omega+1}, ..., x_\Omega >$ such that the transitions are $x_\omega - > x_{\omega+1} - > ... - > x_\Omega$; Γ is the collection of all possible paths from the source to the outlet; $*$ denotes convolution; $p(\gamma)$ is the path probability defined by [3]

$$p(\gamma) = (A_{x_\omega}/A) p_{x_\omega, x_\omega+1} \cdots p_{x_{\Omega-1}, x_\Omega} \tag{2}$$

where A_{x_ω} is the contributing area (or, domain of attraction) draining into the channel state x_ω of the given path γ; A is the total area of the domain of source input; $p_{x_\omega, x_{omega+1}}$ is the transition probability from the state x_ω to state $x_{\omega+1}$ (note that in geomorphological descriptions of river networks transitions from $x_\omega to x_{\omega+2}, x_{\omega+3}$ are admissible). In general $p_{x_j, x_{j+1}}$ is the number of stream x_j draining into state x_{j+1} divided by the total number of streams x_j. It is both expedient and convenient to work in the Laplace-transform domain. Let $\hat{f}(s)$ be the L-transform of $f(t)$:

$$\hat{f}(s) = \sum_{\gamma \in \Gamma} p(\gamma) \prod_{x_\omega \in \gamma} \hat{f}_{x_\omega}(s) \tag{3}$$

which generalizes the first passage probability density obtained for random resistor networks or network models of porous media[5] in that the sum over all the paths Γ from the inlet to the outlet is weighed by the path probability.

The question on whether eq. (3), which gives an exact rule for computation of first passage distributions, is amenable to analytic solution rests on the definition of $f_{x_\omega}(t)$ which is the travel time probability density that a tracer injected at the inlet of the state x_ω reaches the endpoint of the reach in a time t. In the study of hydrodynamic dispersion in network model of porous media[5], the microscopic rule for the tracer motion was specified by the combined influence of convection and molecular diffusion, the travel distance being fixed by the scales of the pores then arranged in a tube network. This cannot be the case in basin scale transport. As it has been argued [1,2,3,6] that $f(t) \propto \frac{d}{dt} < M_s(t) >$, where $< M_s(t) >$ is the expected mass within the transport volume at time t, we propose that the " microscopic" rule in channel reaches be described by convection and turbulent diffusion. Molecular diffusion prompts, in fact, a rapid sampling of the heterogeneity of the turbulent velocity profiles and the resulting longitudinal dispersion, after a Taylor period extending to travel distances of the order of fourty of hydrodynamic radii, is in the form:

$$D_T \propto u^* d \qquad (4)$$

where u^* is the shear velocity and d a meaningful hydraulic length (normally the hydraulic radius). The probability $g(x,t)dx$ for a particle of being in $(x, x + dx)$ at t is the described by the general model of longitudinal " developed" turbulence, by the Fokker-Planck equation [8]:

$$\partial g/\partial t + < u > \partial g/\partial x = D_T \partial^2 g/\partial x^2 \qquad (5)$$

where x is the longitudinal coordinate along the channel and $< u >$ is the mean convection velocity. This completes our assumptions as[1,2,8]

$$f(t) = -\frac{d}{dt} \int_{x_\omega} g(x,t)dx \qquad (6)$$

The crux of the matter is therefore solution of eq.(5) in the L-transform-domain. To this respect interesting questions arise on the role of boundary conditions on absorbing or reflecting barriers[8]. We argue (Rinaldo A., Rigon R. and Marani A., in preparation) that for reaches x_ω of length $L(\omega, \Omega, \Omega)$ being the order of the basin[7], a suitable L-transform for $Pe = (< u > L(\omega,\Omega)/D_T) > 10$ and $L(\omega, \Omega) > 10l$ (l is the integral scale of turbulent diffusion) is:

$$\hat{f}_{x_\omega}(s) = e^{-L(\omega,\Omega)\theta(s)} \qquad (7)$$

where

$$\theta(s) = (- < u > +\sqrt{< u >^2 + 4sD_T})/2D_T$$

Substitution yields:

$$\hat{f}(s) = \sum_{\gamma \in \Gamma} p(\gamma)e^{-\sum_{x_\omega \in \gamma} \theta(s)L(\omega,\Omega)} \qquad (8)$$

It is interesting to note that eq.(8) for u is a moment generating function for the hydrologic response and that it is suited to a nice analytic inversion when $< u >, D_T \neq f(\omega)$, i. e. convection and dispersion are basin constant. Although somewhat unphysical, this model is a useful tool to get inroads into the basic mechanism of generation of the hydrologic response. Mean travel time is, in general:

$$E < T >= \sum_{\gamma \in \Gamma} p(\gamma) \sum_{x_\omega \in \gamma} \frac{L(\omega, \Omega)}{< u >}$$

$$VAR < T >= \sum_{\gamma \in \Gamma} p(\gamma) \sum_{x_\omega \in \gamma} \frac{L(\omega, \Omega) D_T^2}{2 < u >^3} + V(\Gamma) \frac{L^2(\omega, \Omega)}{< u >^2}$$

Where $V(\Gamma)$ is a function of the topological structure of the network. The analytic expression of eq.(1) for constant $< u >, D_T$ is:

$$f(t) = \frac{1}{\sqrt{2 D_T} \sqrt{2 \Pi t^3}} \sum_{\gamma \in \Gamma} p(\gamma) \sum_{x_\omega \in \gamma} L(\omega, \Omega) e^{-((L(\omega, \Omega) - < u > t)^2 / 4 D_T t)} \qquad (9)$$

which is our result for the computation of travel time distributions in a structured network.

As equation (5) decays to a model of pure convection as $D_T \rightarrow 0$, i. e. $Pe \rightarrow \infty$, it is interesting to perform numerical experiments in fractal channel networks of different order[7]. The geometrical setup is illustrated in Figure 1. Subsequent figures illustrate the response of the same basin under different dynamical conditions. In spite of the extreme regularity of the network model and the consequent irregularity of the instantaneous unit hydrograph the shape of the finite time response is quite regular to resemble a natural catchment[7].

CONCLUSIONS

Longitudinal dispersion in network models is predicted through the definition of an exact L-transform of the first passage distribution at the outlet. Numerical experiments and theory show interesting features of the breakthrough curves. The results suggest that the dispersive character of the transport at the scale of the network is dominated by its geometry rather than by the characters of the dispersive processes dominant at the meso and micro-scales.

REFERENCES

1. Dagan,G and V. Nguyen, A comparison of travel time and concentration approaches to modeling transport by groundwater.,J. Contam. Hydrology, Vol. 4(1),79-92, 1989.

2. Rinaldo, A., A. Bellin and A. Marani, On mass response functions, Water Resour. Res., Vol. 25, 1603-1617, 1989.

3. Rodriguez-Iturbe and Valdes, The geomorphological structure of hydrologic response, Water Resour. Res., Vol. 15, 1409-1420, 1979.

4. Cvetkovic,V.D., and A. M. Shapiro, Mass arrival of sorptive solute in heterogeneous porous media, water Resour. Res, Vol. 26, 000-000, 1990.

5. De Arcangelis,L., J. Koeplik, S. Redner and D. Wilkinson, Hydrodynamic dispersion in network models of porous media, Physical Review Letters, Vol. 57(8), 996-999, 1986.

6. Gupta,V.J., E. Waymire and C. T. Wang, A representation of an instantaneous unit hydrograph from geomorphology, Water Resour. Res., Vol 16,855-862, 1980.

7. Marani A.., R. Rigon, A. Rinaldo, A note on fractal channel networks, in review,1990.

8. Cox,D.R., and H.D. Miller, The Theory of Stocastic Processes, Methuen & Co., London,1965.

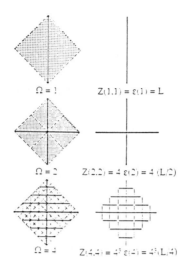

Figure 1. The basin chosen for the simulation is known as Peano basin. for every change of scale two new links are drawn for each preexisting link. their length is half that of the receiving link entered at its midpoint. This basin perform fractal features which have recently gained interest in hydrologic literature[7]: in particular it is a curve that covers a whole area, i. e. its fractal dimension is $D_A = 2$.

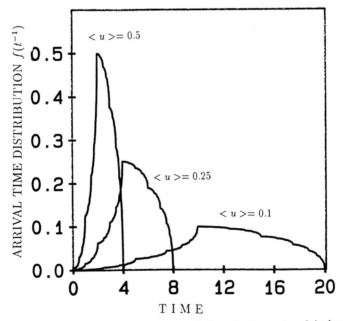

Figure 2. GUH for null dispersion and variable drift. The order of the basin is $\Omega = 8$. Lengths are normalized to the maximum distance from the outlet of the basin, $L(\Gamma)$. Time is an independent parameter.

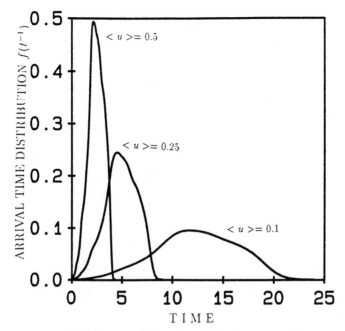

Figure 3. GUH for non null dispersion within the streams. The case of fixed shear velocity, $D_T = 7.8 \times 10^{-3}[L(\Gamma)]^2/[Time]$. and variable drift.

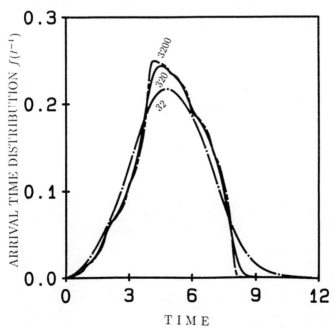

Figure 4. GUH. The case of fixed drift, $< u > = 0.1 \, [L(\Gamma)]/[Time]$, and variable shear velocity. Numbers are the corresponding Pe values.

SECTION 8 - PLANNING AND OPTIMIZATION FOR GROUNDWATER PROJECTS

Computational Aspects of the Inverse Problem

J. Carrera(*), F. Navarrina(**), L. Vives(*), J. Heredia(*), A. Medina(**)

(*) Departmento de Ingenierí del Terreno (**) Departamento de Matemática Aplicada III E.T.S.I. Caminos. Universidad Politécnica de Cataluna Jodi Girona Salgado, 31. 08034 Barcelona, Spain

ABSTRACT

The use of automatic parameter estimation techniques is becoming increasingly extended. As a result, inverse algorithms are expected to solve larger and larger problems, which leads to a need for computationally efficient and robust methods for solving the inverse problem. These may require first and second order derivatives of state variables and/or calibration criteria. Efficient methods for computing such derivatives are presented. More specifically, a methodology for computing the exact hessian matrix of objective functions arising from the inverse problem is described.

INTRODUCTION

Automatic parameter estimation methods are becoming standard tools in groundwater modelling because they allow fast evaluation of alternative conceptual models of hydrological stystems. As a result, inverse algorithms are expected to solve problems with increasingly large meshes. Thus, it is not surprising that computational cost is the most bounding factor in applications of inverse modelling to real cases.

The inverse problem is usually posed as the minimization of an objective function with respect to model parameters subject to the state equations. The most widely spread methods for minimizing the objective function can be classified under two general groups: gradient search methods, which require computation of the derivatives of the objective function with respect to model parameters, and Newton methods, which require computing the Hessian matrix of the objective function. Evaluation of gradient and/or Hessian matrix of the objective function are extremely time consuming tasks, accounting for most of CPU time. While second order methods tend to converge in much less iterations than gradient search methods, they require much more CPU time per iteration.

In summary, the most critical computational aspects of the inverse problem include the evaluation of first and/or second order derivatives and the use of efficient minimization methods. In order to address the first aspect, we present the general framework for computation of first and second order derivative, including a formulation for relatively unexpensive evaluation of the Hessian matrix. Regarding the second topic, we have performed extensive comparisons between gradient search and second order methods. They are not shown here because of space constraints. However, they confirm the validity of Cooley [1] conclusions regarding the superiority of second order methods, which converge more consistently and usually in less iterations than gradient search methods.

FORMULATION OF THE INVERSE PROBLEM

Only a brief description of the inverse problem is given here. Our only objective at this point is to set the frame for the presentations to be made in the remaining of this paper. Details on inverse methodologies can be found in Yeh [2] and Carrera [3].

Statement of the problem. Basic formulations

In subsequent derivations and discussions, we will concentrate on groundwater flow, which is governed by the difussion-type equation:

$$\nabla(\boldsymbol{T} \, \nabla h) + q = S\frac{\partial h}{\partial t} \quad on \ \Omega \tag{1}$$

where $\boldsymbol{T}$ is the transmissivity tensor; h is head; q is the areal recharge; ∇ is the "del" operator (divergence or gradient); S is the storage coefficient; t is time. Equation (1) is solved on the domain Ω subject to the boundary conditions:

$$(\boldsymbol{T} \, \nabla h).\boldsymbol{n} = \alpha(h - H) + Q \quad on \ \Gamma \tag{2}$$

where $\boldsymbol{n}$ is the unit vector normal to Γ and pointing outwards; H is a external head; Q is prescribed flow; α is a coefficient controlling the type of boundary condition ($\alpha = 0$ for prescribed flow; $\alpha = \infty$ for prescribed head; and $\alpha \neq 0, \infty$ leakage condition). Initial conditions have to be specified on Ω, though they are often chosen as the steady-state equivalent of (1).

Solution of the flow equation requires knowledge of the parameters, $\boldsymbol{T}$, S, q, H, Q and α, over the entire flow domain and its boundary. We will refer to these as "**aquifer or physical parameters**". Actually, equations (1)-(2) are solved by numerical methods, so that one only has to know representative discrete values of such parameters over subregions of the aquifer (elements, cells, etc). Even though some of the physical parameters are often known, the number of unknown discrete values may still be much too high to be estimated. Thus, we face the first issue of inverse modelling, **parametrization**, which consists in expressing the values of aquifer parameters as functions of a, hopefully small, number of unknown variables. We will refer to these as **model parameters**. Having expressed the aquifer parameters as functions of model parameters, the solution of the flow equation can be obtained by various numerical methods. Those using fully implicit schemes would lead to expressions of the form:

$$(\boldsymbol{A} + \frac{\boldsymbol{D}}{\Delta t}) \, \boldsymbol{h}^k = \boldsymbol{b}^k + \frac{\boldsymbol{D}}{\Delta t}\boldsymbol{h}^{k-1} \tag{3}$$

during the transient state, and to

$$\boldsymbol{A} \, \boldsymbol{h}^0 = \boldsymbol{b}^0 \tag{4}$$

during the steady-state. Here, $\boldsymbol{A}$ and $\boldsymbol{D}$ are symmetric n×n conductance and storage matrices, respectively, n is the number of nodes, $\boldsymbol{h}^k$ is the vector of nodal heads, $\boldsymbol{b}$ is the vector of nodal sinks or sources and Δt is the time increment. The superscript k is used to denote time step number and the null subscript denotes steady-state variables. The actual expressions for A, D and b depend on the method of solution, but all of them depend on the model parameters. In the case of using the Finite Element Method with linear triangles, A, D and b are given by:

$$A_{mn} = \sum_e A_{mn}^e \quad ; \quad D_{mn} = \sum_e D_{mn}^e \quad ; \quad b_m = \sum_{\Gamma_m} Q_m^\Gamma + \sum_e \frac{q^e A_e}{3} \quad (5)$$

where e indicates element number, Γ_m represents boundary segment adjacent to node m, Q_m^Γ is the inflow through such segment for node m, A_e is element area, q^e represents areal recharge. For triangular elements, matrices A_{mn}^e and D_{mn}^e are given by:

$$A_{mn}^e = \frac{1}{4A_e} \left[T_{xx}^e b_n b_m + T_{xy}^e (b_n c_m + b_m c_n) + T_{yy}^e c_n c_m \right] \quad (6)$$

$$D_{mn}^e = \frac{\delta_{nm} S^e A_e}{3} \quad (7)$$

where
$$b_m = y_n - y_p \quad ; \quad c_m = x_p - x_n \quad ; \quad A_e = (b_m c_n - b_n c_m)/2 \quad (8)$$

and m, n and p are a counterclockwise permutation of the node numbers of element e, δ_{nm} is Kronecker's delta, and T^e and S^e are element transmissivity and storage. Similar expressions can be written for other solution methods. The **inverse problem** can now be stated as follows: *Given point measurements of heads and prior information about the model parameters, obtain the values of the latter which are close to the former and lead to an optimum match between measured and computed heads.*

The objective function

Various statistical formulations of the inverse problem have been used by differents researchers, including least squares ([4], [5]), Bayesian Estimation [6] Maximum Likelihood Estimation [7]. However, they all lead to posing the problem as the minimization of an objective function of the form:

$$J = J_h + J_p \quad (9)$$

where

$$J_h = (h_m - h_m^*)^t V_h^{-1} (h_m - h_m^*)$$
$$J_i = (p - p^*)^t V_p^{-1} (p - p^*) \quad (10)$$

and h_m^* is the vector of head measurements and p^* is the vector of prior estimates of parameters, p; J_h is often called model fit criterion and J_p measures the plausibility of type parameters. The vector of computed heads in measurement points, h_m, is composed of several subvectors, h_m^k, which comprise the heads computed in all measurement points at time k. These can be expressed as:

$$h_m^k = B \, h^k \quad (11)$$

where B is an interpolating matrix and h^k is the vector of nodal heads at time k.

Minimization algorithms

There are several methods for minimizing the objective function (11). Newton methods are derived by finding the minimum of the second order approximation of the objective function. This leads to an equation of the form:

$$H(p_i)d_i = -g_i \quad (12)$$

where $\boldsymbol{H}$ is the Hessian matrix (its computation is the subject of this report) and $\boldsymbol{d}_i$ is the updating vector. In practice, computation of the Hessian matrix is so expensive that a first order approximation is used instead. Parameters are updated according to

$$\boldsymbol{p}_{i+1} = \boldsymbol{p}_i + \rho\boldsymbol{d}_i \tag{13}$$

where $0 \leq \rho \leq 1$. Usually ρ equals 1, but is reduced if $\boldsymbol{d}_i$ is exceedingly large [8]. The main drawback of this method is that the solution of (12) often leads to a too large step size. This results in unstable convergence. There are several alternatives to overcome such difficulty. The most widely used alternative is the one proposed by Marquardt [9], which consists of substituting (12) by

$$(\boldsymbol{H} + \mu\boldsymbol{I})\,\boldsymbol{d} = -\boldsymbol{g} \tag{14}$$

where μ is a positive parameter. The effect of including this parameter is to change the updating direction from the Gauss-Newton solution towards the direction of the gradient, while reducing its size.

In summary, Newton algorithms consist of the following steps:
1. Initialization. Define initial parameters $\boldsymbol{p}^1$ and set i=1.
2. Compute heads (eq. 3 and 4) and J (eq. 9).
3. Compute the Hessian matrix (several algorithms are discussed in next section).
4. Solve equation (14) for $\boldsymbol{d}^i$.
5. Update model parameters (eq. 13) and μ.
6. If convergence has been achieved, stop. Otherwise, return to step 2.

Criteria for updating Marquardt's parameter μ deserve attention. If μ is too large, convergence rate is slowed down. On the other hand, a too small μ may lead to instabilities. This issue is extensively discussed in [1] and [8]. For the sake of robustness, it is convenient to devise several criteria both for increasing and decreasing μ. In our implementation, μ is increased when a iteration fails to reduce the objective, when the newly computed objective function is close to the quadratic approximation of J in the previous iteration, or when $\boldsymbol{g}$ and $\boldsymbol{d}$ are nearly orthogonal. Otherwise, μ is increased, becoming zero is it falls below a threshold value (see [10] for details). Criteria for convergence (step 6) also deserve attention. If they are too tight, exceedingly many iterations may be devoted to approaching the exact minimum, which is a nearly impossible task because of roundoff errors. On the other hand, too loose convergence criteria may lead to erroneously stop the iterative process before convergence.

Gradient search algorithms are motivated by avoiding computation of the Hessian matrix. The updating direction $\boldsymbol{d}_i$ is given by:

$$\boldsymbol{d}_i = -\boldsymbol{H}_i\boldsymbol{g}_i \quad (Quasi-Newton\ methods) \tag{15}$$

$$\boldsymbol{d}_i = -\boldsymbol{g}_i + \beta_i\boldsymbol{d}_{i-1} \quad (Conjugate\ gradient\ methods) \tag{16}$$

where, now, $\boldsymbol{H}_i$ is an approximation of the Hessian matrix that is updated in every iteration, and β^i is a scalar that may depend on $\boldsymbol{g}_i$, $\boldsymbol{g}_{i-1}$ $\boldsymbol{d}_{i-1}, \boldsymbol{p}_{i-1}$ and $\boldsymbol{p}_i$. Various expressions for $\boldsymbol{H}_i$ and β_i are given in Fletcher [8]. As quasi-Newton methods, we have used those of Broyden (BFGS) and Davidon-Fletcher-Powell (DFP). As conjugate gradient methods, we have used that of Fletcher-Reeves (FR). In our experience, convergence is aided by using these methods sequentially. That is, one method (say FR) is used until it fails to significantly reduce the objective function; a different method (say BFGS) is then used, etc. This process can be repeated until convergence is reached.

The iterative algorithm is very similar to that of Newton methods and can be summarized in the following steps:
1. Initialization. Define initial parameters, p^1 and set $i = 1$.
2. Compute heads according to equations (3)-(4) and J (eq. 9).
3. Compute the gradient as discussed in next section.
4. Compute the updating direction using equations (15) or (16).
5. Minimize the objective function along d_i. Efficient approximate methods methods for this minimization are described in [11] and [12].
6. Update model parameters according to equation (13).
7. If convergence has been reached, stop. Otherwise return to step 2.

DERIVATIVES OF THE OBJECTIVE FUNCTION

General formulation

Assume that one is interested in obtaining the derivatives of a function J:

$$J = J(h, p) \tag{17}$$

where h is a n-dimensional vector of state variables (heads in the case of groundwater flow) and p is a m-dimensional vector of parameters. Assume further that h and p are functionally related through n state equations that can be written as:

$$\psi(h, p) = 0 \tag{18}$$

The lagrangian of J subject to constraints (18) is given by:

$$\mathcal{L} = J + \lambda^t \psi \tag{19}$$

where λ is a n-dimensional vector of Lagrange multipliers, also called adjoint-state, and the superindex t indicates that it is transposed. The derivatives of $\mathcal{L}$ with respect to p, which equal the total derivatives of J with respect to p since ψ is identically zero, are:

$$\frac{d\mathcal{L}}{dp} = \frac{dJ}{dp} = \frac{\partial J}{\partial p} + \frac{\partial J}{\partial h} \cdot \frac{\partial h}{\partial p} + \lambda^t \left(\frac{\partial \psi}{\partial p} + \frac{\partial \psi}{\partial h} \cdot \frac{\partial h}{\partial p} \right) \tag{20}$$

where the derivatives of λ have been omitted because they would be multiplied by ψ, which is zero. The adjoint-state is chosen as the solution of the following linear sistem of equations:

$$\frac{\partial J}{\partial h} + \lambda^t \frac{\partial \psi}{\partial h} = 0 \tag{21}$$

This simplifies (21), which becomes:

$$\frac{dJ}{dp} = \frac{\partial J}{\partial p} + \lambda^t \frac{\partial \psi}{\partial p} \tag{22}$$

This expression has been used often in order to obtain the gradient of the objective function or other variables in the groundwater literature ([11], [12], [13]). The main advantage of the method is that, having solved Equation (21) for λ, direct substitution in (22) yields the gradient. This requires a relatively low amount of computer time. The alternative method is direct derivation of J, which yields,

$$\frac{dJ}{dp} = \frac{\partial J}{\partial p} + \frac{\partial J}{\partial h} \cdot \frac{\partial h}{\partial p} \tag{23}$$

where the jacobian matrix $(\partial h/\partial p)$ can be computed from the m sistems of equations that result from taking derivatives of the state equations:

$$\frac{\partial \psi}{\partial p} + \frac{\partial \psi}{\partial h} \cdot \frac{\partial h}{\partial p} = 0 \tag{24}$$

If m (dimension of p) is large, the cost of (24) can become very large compared to (21).

Second derivatives can also be computed by direct derivation or by the adjoint state. Direct derivation of (22) yields:

$$\frac{d^2 J}{dpdq} = \frac{\partial^2 J}{\partial p\partial q} + \frac{\partial^2 J}{\partial h\partial p} \cdot \frac{\partial h}{\partial q} + \frac{\partial^2 J}{\partial h\partial q} \cdot \frac{\partial h}{\partial p} + \frac{\partial^2 J}{\partial h^2} \cdot \frac{\partial h}{\partial p} \cdot \frac{\partial h}{\partial q} + \frac{\partial J}{\partial h} \cdot \frac{\partial^2 h}{\partial p\partial q} \tag{25}$$

The evaluation of (25) can be very computer demanding because the last term requires computation of the second derivatives of heads with respect to model parameters. These can be obtained by solving the m^2 systems of equations that result taking derivatives of (24):

$$\frac{\partial \psi}{\partial h} \cdot \frac{\partial^2 h}{\partial p\partial q} = -\frac{\partial^2 \psi}{\partial h\partial p} \cdot \frac{\partial h}{\partial q} - \frac{\partial^2 \psi}{\partial h\partial q} \cdot \frac{\partial h}{\partial p} - \frac{\partial^2 \psi}{\partial p\partial q} - \frac{\partial^2 \psi}{\partial h^2} \cdot \frac{\partial h}{\partial p} \cdot \frac{\partial h}{\partial q} \tag{26}$$

Again, if m is large, solution of the m^2 linear systems in equation (26) may become a very expensive task.

The proposed approach circumvents this difficulty by taking further advantage of the adjoint state methodology. Taking derivatives of (20), while neglecting again the derivatives of λ because they are multiplied by a zero factor, leads to:

$$\frac{d^2 J}{dp\,dq} = \frac{\partial^2 J}{\partial p\partial q} + \frac{\partial^2 J}{\partial p\partial h} \cdot \frac{\partial h}{\partial q} + \frac{\partial^2 J}{\partial h\partial q} \cdot \frac{\partial h}{\partial p} + \frac{\partial^2 J}{\partial h^2} \cdot \frac{\partial h}{\partial p} \cdot \frac{\partial h}{\partial q} + $$

$$\lambda^t \left(\frac{\partial^2 \psi}{\partial p\partial q} + \frac{\partial^2 \psi}{\partial p\partial h} \cdot \frac{\partial h}{\partial q} + \frac{\partial^2 \psi}{\partial h\partial q} \cdot \frac{\partial h}{\partial p} + \frac{\partial^2 \psi}{\partial h^2} \cdot \frac{\partial h}{\partial p} \cdot \frac{\partial h}{\partial q} \right) + \left(\frac{\partial J}{\partial h} + \lambda^t \frac{\partial \psi}{\partial h} \right) \frac{\partial^2 h}{\partial p\partial q} \tag{27}$$

The last term in this equation is zero (Equation (21)). The result is an expression which is far less expensive (only requires solving $m + 1$ systems) than the previous one (requires solving $m^2 + m$ systems). Many of the terms in (25) and (27) can be dropped in applications to linear problems, which lowers significantly the cost of computing the Hessian matrix. Although this formulation is not a second order pure adjoint variable method, it can be shown to be more efficient [16].

Application to groundwater models

The application to groundwater flow models follows directly from this methodology. Specifically, function J in equations (17) to (27) is the objective function (9), and the state equations are the equivalent of (3)-(4), which should be understood as a column of $n \times N_t$ [or $n \times (N_t + 1)$ if steady-state solution is required] equations (n being the number of nodes, and N_t the number of time steps). Similarly, the vector of state variables, $\boldsymbol{h}$, should be understood as a vector comprising N_t (or $N_t + 1$) subvectors, $\boldsymbol{h}^k$. The derivatives of the objective function and state equations are given by:

$$\frac{\partial J}{\partial \boldsymbol{h}^k} = 2 \sum_{l=1}^{N_t} \boldsymbol{B}^t \boldsymbol{V}_{kl}^{-1} (\boldsymbol{h}_m^l - \boldsymbol{h}_m^{*l}) \tag{28}$$

where the sum over l is not required if the correlation among head errors at different observation times can be neglected and $\boldsymbol{V}_{kl}^{-1}$ is the block of $\boldsymbol{V}_h^{-1}$ corresponding to time steps k and l. Similarly,

$$\frac{\partial J}{\partial \boldsymbol{p}} = 2 \boldsymbol{V}_p^{-1} (\boldsymbol{p} - \boldsymbol{p}^*) \tag{29}$$

Analogously, the second derivatives are given by

$$\frac{\partial^2 J}{\partial \boldsymbol{h}^k \partial \boldsymbol{h}^l} = 2 \boldsymbol{B}^t \boldsymbol{V}_{kl}^{-1} \boldsymbol{B} \tag{30}$$

$$\frac{\partial^2 J}{\partial \boldsymbol{p} \partial \boldsymbol{q}} = 2 \boldsymbol{V}_p \tag{31}$$

Regarding the state equations the only non-zero head derivatives are:

$$\frac{\partial \boldsymbol{\psi}^l}{\partial \boldsymbol{h}^k} = - \frac{\boldsymbol{D}}{\Delta t^{k+1}} \qquad if \qquad l = k + 1$$

$$\frac{\partial \boldsymbol{\psi}^l}{\partial \boldsymbol{h}^k} = \left(\boldsymbol{A} + \frac{\boldsymbol{D}}{\Delta t^k} \right) \qquad if \qquad l = k \tag{32}$$

The derivatives with respect to model parameters are:

$$\frac{\partial \boldsymbol{\psi}^k}{\partial \boldsymbol{p}} = \left(\frac{\partial \boldsymbol{A}}{\partial \boldsymbol{p}} + \frac{1}{\Delta t^k} \frac{\partial \boldsymbol{D}}{\partial \boldsymbol{p}} \right) \boldsymbol{h}^k - \frac{\partial \boldsymbol{b}^k}{\partial \boldsymbol{p}} - \frac{1}{\Delta t^k} \frac{\partial \boldsymbol{D}}{\partial \boldsymbol{p}} \boldsymbol{h}^{k-1} \tag{33}$$

And similar expressions can be written for the steady-state equation. Derivatives of $\boldsymbol{A}$, $\boldsymbol{D}$ and $\boldsymbol{b}$ with respect to $\boldsymbol{p}$ can be obtained in a straightforward manner from (5) through (8). For example, the derivative of $\boldsymbol{A}$ with respect to log-transmissivity in zone i is given by:

$$\frac{\partial A_{mn}}{\partial log - (T_i)} = \sum_{e \in zone \ i} \frac{1}{4 A_e} (b_n b_m + c_n c_m) T^e . ln(10) = ln(10) \sum_{e \in zone \ i} A_{mn}^e \tag{34}$$

where the summation is only extended over the elements belonging to zone i, and it has been assumed that the transmissivity is isotropic.

Second derivatives of state equations are:

$$\frac{\partial^2 \boldsymbol{\psi}^l}{\partial \boldsymbol{h}^k \partial \boldsymbol{h}^m} = 0 \tag{35}$$

$$\frac{\partial^2 \boldsymbol{\psi}^l}{\partial \boldsymbol{p} \partial \boldsymbol{h}^k} = - \frac{1}{\Delta t^{k+1}} \frac{\partial \boldsymbol{D}}{\partial \boldsymbol{p}} \qquad if \qquad l = k + 1$$

$$\frac{\partial^2 \boldsymbol{\psi}^l}{\partial \boldsymbol{p} \partial \boldsymbol{h}^k} = \frac{\partial \boldsymbol{A}}{\partial \boldsymbol{p}} + \frac{1}{\Delta t^k} \frac{\partial \boldsymbol{D}}{\partial \boldsymbol{p}} \qquad if \qquad l = k \tag{36}$$

$$\frac{\partial^2 \psi^k}{\partial p \partial q} = \left(\frac{\partial^2 A}{\partial p \, \partial q} + \frac{1}{\Delta t^k} \frac{\partial^2 D}{\partial p \, \partial q} \right) h^k - \frac{\partial^2 b^k}{\partial p \, \partial q} - \frac{1}{\Delta t^k} \frac{\partial^2 D}{\partial p \, \partial q} h^{k-1} \qquad (37)$$

where the second derivatives of A, D and b with respect to model parameters can be obtained by direct derivation of (5) through (8). Inasmuch as they depend quasi-linearly on the parameters, most of terms arising from (37) are zero. The only non-zero terms are those corresponding to the term αH in vector b, associated to mixed boundary conditions, and those corresponding to log-transformed parameters. For example,

$$\frac{\partial^2 A_{mn}}{\partial \log - (T_i)^2} = \sum_{e \in zone \ i} \frac{1}{4 A_e} (b_n \, b_m + c_n \, c_m) \, T^e \, ln^2(10) = ln^2(10) \sum_{e \in zone \ i} A^e_{mn} \qquad (38)$$

The fact that (38) is similar to the state equations simplifies its computation. Computation of these and higher order derivatives for general FEM formulations including shape variations of elements are discussed in [14]. Equations (28) through (38) can now be used to formulate the adjoint-state, sensitivity and hessian equations.

Adjoint-state equations. Direct substitution of (28) and (32) into (21) leads to:

$$\lambda^{t^k} \left(\frac{A + D}{\Delta t^k} \right) = -2 \sum_l B^t \, V^{-1}_{kl} \, (h^l_m - h^{*l}_m) + \lambda^{t^{k+1}} \frac{D}{\Delta t^k + 1} \qquad (39)$$

It should be noticed that the form of (39) is very similar to that of the state equations (3). The only differences are that vector b has been substituted by the residuals term (28), that equation (39) must be solved backwards in time, and that it is transposed. The latter is a minor problem because, if A and D are symmetric they coincide with their transposes. Otherwise, the LU decomposition of the coefficient matrix has to simply be transposed. The steady-state equivalent of (39) is

$$\lambda^{t^o} A = -2 \sum_l B^t \, V_{ol} \, (h^l_m - h^{*l}_m) + \lambda^{t^1} \frac{D}{\Delta t^1} \qquad (40)$$

which keeps the same structure as (4) except for the presence of the last term in (40).

Sensitivity equations. Substitution of the steady-state equivalents of (32) and (33) into (24) leads to:

$$A \frac{\partial h^o}{\partial p} = \frac{\partial b^o}{\partial p} - \frac{\partial A}{\partial p} h^o \qquad (41)$$

for the steady, and to

$$\left(A + \frac{D}{\Delta t^k} \right) \frac{\partial h^k}{\partial p} = \frac{\partial b^k}{\partial p} + \frac{D}{\Delta t^k} \frac{\partial h^{k-1}}{\partial p} - \frac{\partial A}{\partial p} h^k - \frac{\partial D}{\partial p} \left(\frac{h^k - h^{k-1}}{\Delta t^k} \right) \qquad (42)$$

for the transient state. It should be noted again that the structure of (41) and (42) is identical with that of the state equations. Therefore the LU decomposition of the coefficient matrices, required for the solution of the direct problem, can be used without further modification for solving (41) and (42).

Second order sensitivity analysis. The equations for the second derivatives of heads with respect to model parameters can be obtained by direct substitution of (35) through (37) into (26). For steady state, this yields:

$$A \frac{\partial^2 h^o}{\partial p \, \partial q} = \frac{\partial^2 b^o}{\partial p \, \partial q} - \frac{\partial A}{\partial p} \cdot \frac{\partial h^o}{\partial q} - \frac{\partial A}{\partial q} \cdot \frac{\partial h^o}{\partial p} - \frac{\partial^2 A}{\partial p \, \partial q} h^o \qquad (43)$$

and similar equations can be written for the transient state.

Exact hessian. Classical approach. If the second derivatives of heads have been computed, equation (25) can be used for computing the Hessian matrix. For simplicity, we shall write it as:

$$H = \frac{d^2 J}{dp}\, dq = H_1 + H_r \tag{44}$$

where

$$H_1 = 2 \sum_k \sum_l J_m^{t^k} V_{kl}^{-1} J_m^l + 2\, C_p \tag{45}$$

$$H_r = 2 \sum_k \sum_l (h_m^k - h_m^{*k})^t\, V_{kl}^{-1}\, B\, \frac{\partial^2 h^l}{\partial p\, \partial q} \tag{46}$$

and

$$J_m^l = B\, \frac{\partial h^l}{\partial p} \tag{47}$$

In this expression it is interesting to notice that if the residuals can be neglected (i.e., assume that h_m^k equals h_m^{k*}), then the hessian matrix equals H_1, which is its first order approximation.

Hessian matrix. Adjoint-state approach. The derivation of the hessian matrix follows straight from (27). We shall write it as:

$$H = H_1 + \sum_k H_2^k + \sum_k H_3^k \tag{48}$$

where H_1 is the first-order approximation to the hessian matrix, given by Equation (45) and H_2^k and H_3^k are given by:

$$H_2^k = \lambda^{t^k} \left[\frac{\partial^2 A}{\partial p\, \partial q}\, h^k + \frac{\partial^2 D}{\partial p\, \partial q} \left(\frac{h^k - h^{k-1}}{\Delta t^k} \right) + \frac{\partial^2 b^k}{\partial p\, \partial q} \right] \tag{49}$$

$$H_3^k = \lambda^{t^k} \left[\frac{\partial A}{\partial p} \cdot \frac{\partial h^k}{\partial q} + \frac{\partial A}{\partial q} \cdot \frac{\partial h^k}{\partial p} + \right.$$
$$\left. \frac{1}{\Delta t^k} \frac{\partial D}{\partial p} \left(\frac{\partial h^k}{\partial q} - \frac{\partial h^{k-1}}{\partial q} \right) + \frac{1}{\Delta t^k} \frac{\partial D}{\partial q} \left(\frac{\partial h^k}{\partial p} - \frac{\partial h^k}{\partial p} \right) \right] \tag{50}$$

for the transient state, and by

$$H_2^o = \lambda^{t^o} \left[\frac{\partial^2 A}{\partial p\, \partial q}\, h^o + \frac{\partial^2 b^o}{\partial p\, \partial q} \right] \tag{51}$$

$$H_3^o = \lambda^{t^o} \left[\frac{\partial A}{\partial p} \cdot \frac{\partial h^o}{\partial q} + \frac{\partial A}{\partial q} \cdot \frac{\partial h^o}{\partial p} \right] \tag{52}$$

for the steady-state. Inasmuch as all the operations involved in the evaluation of these matrices are similar to those required in most inverse problem algoritms, programming of these expressions does not require use of any special tool. Some implementation details are discussed in [15].

SUMMARY AND CONCLUSIONS

Efficient methods for computation of first and second order derivatives of state variables and scalar functions of state variables have been revised. In particular, we have presented a method for computing the Hessian matrix of an objective function at a relatively low cost. This method should prove very useful in stochastic and inverse modelling programs.

REFERENCES

[1] Cooley R. L.: A comparison of several methods of solving nonlinear regression groundwater flow problems. Water Resources Research, 21(10), 1525-1538, 1985.

[2] Yeh W. W. G.: Review of parameter identification procedures in groundwater hydrology: The inverse problem. Water Resour. Res., 22(2), 95-108, 1986.

[3] Carrera J. and S. P. Neuman: Estimation of aquifer parameters under transient and steady-state conditions, 1. Maximum likelihood method incorporating prior information. Water Resour. Res., 22(2), 199-210, 1986a.

[4] Yoon, Y.S. and W.W.-G. Yeh: Parameter Identification in an Inhomogeneous Medium with the Finite Element Method. Soc. Pet. Eng. Jour., 217-226, 1976.

[5] Cooley, R.L.: Incorporation of prior Information on parameters Into Nonlinear Regression Groundwater Flow Models,1. Theory. Water Resour. Res., 18(4), 965-976, 1982.

[6] Neuman S. P. and S. Yakowitz: A statistical approach to the inverse problem of aquifer hydrology, 1. Theory. Water Resour. Res., 15(4), 845-860, 1979.

[7] Carrera, J. and S.P. Neuman: Estimation of aquifer parameters under transient and steady-state conditions,1. Maximum likelihood method incorporating prior information. Water Resour. Res., 22(2), 199-210, 1986(a).

[8] Fletcher R.: Practical Methods of Optimization. Unconstrained Optimization (Ed. John Wiley &Sons), Vol. 1, 1981.

[9] Marquardt D. W.: An algorithm for least squares estimation of non-linear parameters. SIAM Jour. Appl. Math., 11, 431-441, 1963.

[10] Carrera J. and J. Heredia: Invert-4 user's gide. E.T.S.E. Camins, Barcelona, 1987.

[11] Neuman S. P.: A statistical approach to the inverse problem of aquifer hydrology, 3. Improved solution method and added perspective. Water Resour. Res., 16(2), 331-346, 1980.

[12] Carrera J. and S. P. Neuman: Estimation of aquifer parameters under transient and steady-state conditions, 2. Uniqueness, stability and solution algorithms. Water Resour. Res., 22(2), 211-227, 1986b.

[13] Townley L. R. and J. L. Wilson: Computationally Efficient Algorithms for Parameter Estimation and Uncertainty Propagation in Numerical Models of groundwater Flow. Water Resour. Res., 21(12), 1851-1860, 1985.

[14] Navarrina, F., E. Bendito and M. Casteleiro: High Order Sensitivity analysis in Shape Optimization Problems. Computer Methods in Applied Mechanics and Engineering, 75, 267-281, 1989.

[15] Carrera J. and J. Heredia: Inverse modelling of Chalk River Block, NAGRA NTB 88-14, 117 pp., 1988.

[16] Hareg, E.J., K.K. Choi and V. Komkov: Design Sensitivity Analysis of Structural Systems.Acadamic Press, Orlando, 391 pp., 1986.

Hydrogeological and Optimization Models for an Agricultural Development Project in the Farafra Area (Egypt)

A. Bertoli(*), G. Ghezzi(**), G. Zanovello(***)

(*)INC- IL NUOVO CASTORO 9, Piazzale Flaminio Rome, Italy (**) Consultant Geologist, Pisa, Italy (***) Consultant Engineer, Padova, Italy

ABSTRACT

The company INC - Il Nuovo Castoro performed during the years 1983-1985 a technical-economic feasibility study for the agricultural development of about 20.000 hectares in the Farafra Area (Western Desert of Egypt) through the exploitation of the artesian water from the multilayered aquifer system of the Nubian Sandstones.

The drilling and discharge test in deep wells , the hydrogeological study , the soil surveys , the agronomical and marketing studies allowed to realize a hydrogeological model and an optimization model , which were used as an integrated decision support system in order to identify the most convenient development scheme.

A simple and easy to handle hydrogeological model was choosen in order to perform a large number of simulations. It has been applied to different exploitation hypotheses of the main productive horizons of the aquifer system and to different well profiles , and then integrated with the operation model of the well fields.

Once the drawdown mechanism of the piezometric level was known , a detailed optimization model allowed to identify the least cost and maximum afficienty scheme , assuming as variables the water requirement and its increase in course of time (both in the projet area and at regional level) , the profiles of the well , the horizons exploited and the water distribution and application procedures.

INTRODUCTION

A Feasibility Study for the reclamation of 50,000 acres (about 20,000 hectares) in the Farafra Oasis in the Egyptian New Valley has been carried out during 1983-1985 by INC - Il Nuovo Castoro in fulfilment of a Consultancy Agreement with the General Authority for Rehabilitation Projects and Agricultural Development (GARPAD).

The scope of the Farafra Project was defined within the national strategy of reclamation of desert areas by making use of water sources different from the Nile supply wherever natural resources are available since agricolture in the Delta Region and the Nile Valley cannot support the demographic pressure pushed by the annual rate of population grouth in Egypt.

The New Valley region comprises the four main oases in the Western Desert namely Kharga, Dakhla, Baharya and Farafra, which can draw irrigation water from a huge ground water multi layered reservoir in the Nubian Sandstone.

Since the early sixties important land reclamation works have been carried out in Kharga and Dakhla areas and works of minor importance in Baharia area.

The Farafra area is considered a future centre of development in the Western Desert because its soil and water potentials left, in practice, underdeveloped till now and the favourable location with respect to Cairo city.

Present rate of extrations from the Nubian Sand-stones aquifers induces ground water depletion. Thus any increase in extractions in any one oasis will involve decline of the piezometric level all over the reservoir trough infinite time.

However, water itself is not the limiting factor but rather the cost of pumping it to the surface. Finally the project has been restricted to a total area of 22,000 acres (about 9,000 hectares) subdivided into 5 irrigation districts selected on the base of soil and hydrogeological conditions. Three alternatives have been considered with farm type and size (family,medium and large farms), cropping pattern and irrigation method (surface , sprinkler and drip irrigation) strictly correlated.

HYDROGEOLOGICAL OUTLINE

The hydrogeological characteristics of the Farafra Oasis were defined in the framework of the regional hydrogeological system of the Nubian Sandstone Complex in the New Valley (Egypt), on the basis of the data collected during the field investigations, flow test run in 11 exsisting wells , and the drilling and testing of three new exploratory wells.

Stratigraphy and structure

The outcropping and buried lithostratigraphical succession of the Oasis (Fig. 1) mainly consists of a marine sequence (limestone, chalks, shales) unconformably overlying the continental Nubian Sandstone Complex (alternations of sandstones, siltstones, and shales).

The age of Nubian Sandstone Complex ranges from pre-Cabrian to Upper Cretaceous; the total thickness, inferred on the basis of seismic profiles, may reach as far as 2000 m. Within the study area the investigated thickness reached about 1000 m after the drilling of the three exploratory wells.

From the structural point of view the Farafra depression is located on the eastern flank of a NE-SW trending uplift bounded due NW by a NNE-SSW trending fault. To the west of this fault a downthrown block (Ain Dalla Basin) is present.

Hydrogeology

a) The aquifer of the Nubian Sandstone System is a multilayer aquifer bound by impermeable (to semipermeable?) beds of Maestrichtian age (Dakhla shale and Chaik formation) which form the cap rock of the system. Its lower boundary is conventionally taken at the top of the crystalline basement which, however, is not necessarily impermeable.

Within the project area the producing system is known down to a maximun depth of 1200m. , which means over a thickness of about 1000 m.

The sequence may be considered (Fig. 2), from a practical point of view, as a three water bearing horizons aquifer, separated by aquicludes (or aquitards). The total thickness of the third water bearing horizon is unknown.

The net pay values of the first and the second water bearing horizons, evaluated on the basis of all the available borehole lithologs, have been reported on Isopach Masps.

Both the aquifers show the maximum net pay values in the Arabyia Project Area: up to 124 m for the 1st water bearing horizon and to 352 m for the 2nd one; on the basis of the values supplied by the new borehole SMle, which is the deepest exploratory well of area, the ascertained net pay of the penetrated third aquifer is 240 meters.

The Net Pay Isopach Maps drawn from the available lithologs represent the calculation key for the Trasmissivity values to be adopted in the modelling phase, assuming a costant permeability value for each water bearing horizon.

b) The aquifer system is confined and the deep wells in the Farafra area, are free flowing with head well pressure ranging from 2.6 (first w.b.h.) to 9.2 Bars (third w.b.h.).

c) The flow tests performed in the selected wells confirm the high potential productivity of the confined Nubian Sandstone Aquifer System in the Farafra Oasis area.

The parameters retained as input for the hydrogeological model are summarized in TABLE 1. They are of quite stisfactory quality but they are scanty and localized in few areas.So , their use is not suitable as input for a detailed and sophisticated model.

TAB. 1
INPUT PARAMETERS FOR GROUNDWATER AND OPTIMIZATION MODEL

DATA	1st w.b.h.	2nd w.b.h.	3rd w.b.h.
Thickness (m)	Variable according to the Isopach Map	Variable according to the Isopach Map	240
Permeability (m/s)	2.2×10^{-5}	1.2×10^{-4}	1.2×10^{-4}
Transmissivity (m^2/s)	Permeability x thickness	Permeability x thickness	2.75×10^{-2}
Storage	1.2×10^{-4}	4×10^{-4}	2.8×10^{-4}
Well formula	$So = 11,000 + 25,000^2$	$So = 32Q + 371Q^2$	
		Wells tapping 2nd + 3rd w.b.h.	
		$So = 33.4Q + 990 Q^2$	

DECISION SUPPORT MODEL

For the project altenatives evaluation an articulated and easy to handle decision support model (DSM) was required. In fact it was necessary to examinate in detail the cost flows and the efficiency of many possible development alternatives considering many parameters and taking into account all direct and indirect effects. In particular time variation and discontinuity of parameters were to be simulated.

The DSM is divided in two modules.

a) HYDROGEOLOGICAL MODULE
The scope of the model is to give a detailed drawdown response in the short and medium time in a relatively small area with well determined extractions closely varying in the time.
Many regional models were prepared in the past for the simulation of the hydrogeological behavior of the whole Nubian Sandstone aquifer.   In these model the finite difference method was used based on Darcy law and conservation of mass.
The hydrogeological model used is based however on the application of Theis formula. This model , which considers the spatial effect of pumping from many wells from an undefined aquifer in transient state conditions, was recognized, in our case, as more suitable than a finite difference Darcy model, because of scarsity of geometrical data and hydraulic parameters. This model allows in fact single well simulation, multiple well fields interference calculations, self caused drawdown

and well losses evaluation, that is essential for the determination of the total drawdown and consequent pumping head.

The model is based on the following considerations : the future level in any well of the project area is function of various elements; the regional interference drawdown caused by exstraction fields producing from the same aquifer, the local interference drawdown caused by the wells of the same wellfield , the self caused drawdown and well losses determined by the characteristics and the tubular dimensions of the well itself.

The interference values Di may be calculated adding the single components determined by th Theis formula:

$$Di = \frac{Q}{4 \, Tm} W(u) \qquad W(u) = \int_o^\infty e^{-u}/u \, du \; ; \qquad u = \frac{d^2 S}{4 T t}$$

in which Q is the extraction, t is the time step , d is the distance of any well from the considered one, T is the trasmissivity value and S is the storage value.

The self caused drawdown Ds is calculated by the same formula putting d=radius of the well and summing to it the well losses (CQ^2).

Spatial change of trasmissivity coming from isopach maps of the porous layers and variations in the time of discharge are considered.

The previous considerations are translated into a computer program whose input data are:
-the geometrical boundaries of the zone having an homogeneous trasmissivity;
-the trasmissivity and the storage of the single zones;
-the coordinates of the single wells or of the area of extraction;
-the discharge of any well or wellfield varying year by year;
-the initial water head.

The output of program are water head and drawdown , point by point and year by year.

The following extractions for present and future production of the nearesta reas were considered in the model , with various progressions in the years:
-West Farafra   64.000 m3/d starting from 1988
-Arabia project   43.000 m3/d starting from 1986
-Baraka plain   320.000 m3/d starting from 1991
-Kerawein plain 470.000 m3/d starting from 1991
-Abu Munkar   48.000 m3/d starting from 1986
-South Marzouk 320.000 m3/d starting from 1991
-Sheik Marzok   330.000 m3/d starting from 1991

The Sheik Marzouk area , which is the project area in the vicinity of the Farafra oasis, is divided into 5 districts.

As the potential productivity of the first aquifer is rather poor many simulations were performed considering various alternatives of extraction from the only second aquifer and from both the second and the third aquifer with some hypotheses for future development in the external areas.

The five wellfields may be composed each of 6 to 14 wells , according to their characteristic discharge , which may be 110, 150 or 180 1/s. The well spacing goes so from 1.250 x 1.250 m. to 1.600 x 1.600 m. Two types of well design were designed: type A (384.1 mm. diameter) and type b (320.4 mm. diameter) and two different depths : 750 m to exploit only the second aquifer and 1.150 m. to add also the third aquifer.

With the characteristics of wells and wellfields described before many simulations on Theis model were carried out to detrmine the sensitivity of local interference drawdown to spacing and specific discharge of wells.

It was noted that the drawdown depends on the aquifer exploited and not much on the configuration of the wellfield ; in fact the greater discharges of few production wells are quite well compensated by their greater spacing. This drawdown component has a faible gradient in the first 5 years and after is quite costant; it reaches the value of about 20 m after 35 years for the second aquifer and about 10 m for the second + third aquifers.

The sum of self caused drawdown and well losses for any well was calculated by the formula:

$$Dw = \frac{0{,}183\, Q_{av}}{T} \log \frac{2.25\, T\, t}{S\, r_w^2} + (1 + delta*t)\, CQ^2 \;.$$

where: Q_{av} is the yearly average discharge
   T   is the transmissivity
   t   is the time
   S   is the storativity
   rw   is the radius of the liner
   C   is the well loss coefficient
   Q   is the maximum discharge of the pump
   delta is the deterioration coefficient $=0.006$ per year

The self caused + well losses values for the second aquifer vary from about 10 m of wells type A with Q=110 1/s to about 45 m of wells type B with Q=180 1/s; while for the second + third aquifer from about 8 m of wells type A with Q=110 1/s to about 40 m of wells type B with Q=180 1/s. In the peak month the increase of drawdown (local interference + self caused) is about 7 - 10 m for the second aquifer and 4 - 5 m for the second + third.

The total well drawdown is the sum of the regional aquifer drawdown, the local interference one, the self caused one and the well losses. The average calculated value after a transient period of 30 years, in the hypothesis of null extraction from the external areas, are (in meters):

	2nd aquifer	2nd+3rd aquifer
regional drawdown	70-80	40-50
local interference	20	10
self caused d.d. well losses	10-45	8-40
peak well lowering	7-10	4-5
total	107-155	62-105

The installation of pumps is necessary in any case after some years since , in the peak month, the drawdown limit for pumping is 32-71 m.

b) OPTIMIZATION MODULE
The economical optimization model had the object to evaluate the cost of S.Marzouk project for different alternatives and conditions of exploitation, in order to choose the best one from economical and from functional point of view.
Because of the non-linearity and discontinuity of the cost functions and the large number of possible technical solutions a dynamic programming procedure was choosen; a data base of technical and economical elements feeds the computing model.

The model first computes the basic technical features of any solution and compares it with the technical ties; then computes the cost of wells,of pumps with annexed thermal units for energy production, of pumping energy, of maintenance of pump and units, and sums them to the other costs of the project to evaluate the total cost flows at different discount rates.

All the calculations are made by a computer interactive program which reads and elaborates the regional drawdown for any district (from hydrogeological module), the local interference drawdown for any aquifer , the self caused drawdown, the water requirement for any district , the elementary costs of wells, civil works, pumps, thermal units, energy , the cost flows of the other components of the project, the rates of maintenance cost.

Particular attention was given to energy costs; in fact pumping is necessary after a few years, when the water level drops near the soil level. The estimated time for pump replacement is 5 years. Thermal units for energy production have been estimated to have 10 years life. The power of single pumps can reach 339 Kw under the most unfavourable condition.

The expected life of wells and civil works has been considered 50 years; as the project life is 35 years, the residual values are always computed.

c) MAIN RESULTS

36 technical solutions were selected among the possible ones after the first runs of the DSM model. The economic performance of them emphasized with high confidence the most convenient strategy of development: the exploitation of both second and third aquifers is economically more favourable, because the cost of deeper wells is more than balanced by the smaller cost of pumps and pumping energy; the type A wells are more profitable and can be conveniently exploited at 150 - 180 l/s with pumping heads of not more than few of meters after many years of exploitation.

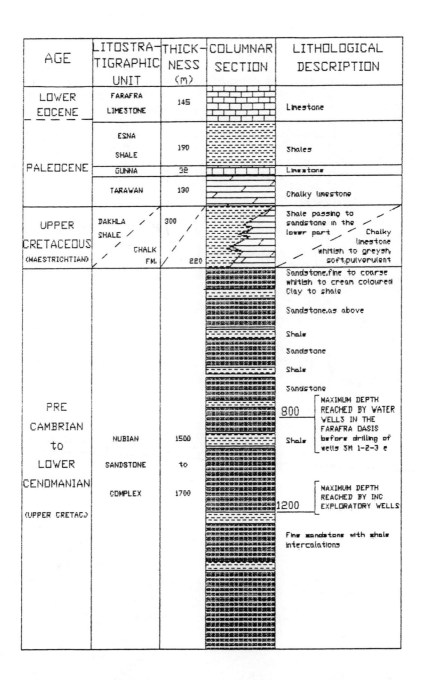

AGE	LITOSTRA-TIGRAPHIC UNIT	THICK-NESS (m)	COLUMNAR SECTION	LITHOLOGICAL DESCRIPTION
LOWER EOCENE	FARAFRA LIMESTONE	145		Limestone
PALEOCENE	ESNA SHALE	190		Shales
	GUNNA	38		Limestone
	TARAWAN	130		Chalky limestone
UPPER CRETACEOUS (MAESTRICHTIAN)	DAKHLA SHALE / CHALK FM.	300 / 220		Shale passing to sandstone in the lower part / Chalky limestone / whitish to greyish soft,pulverulent
PRE CAMBRIAN to LOWER CENOMANIAN (UPPER CRETAC.)	NUBIAN SANDSTONE COMPLEX	1500 to 1700		Sandstone,fine to coarse whitish to cream coloured Clay to shale Sandstone,as above Shale Sandstone Shale Sandstone 800 ⌐MAXIMUM DEPTH REACHED BY WATER WELLS IN THE FARAFRA OASIS Shale before drilling of ⌐wells SM 1-2-3 e 1200 ⌐MAXIMUM DEPTH REACHED BY INC EXPLORATORY WELLS Fine sandstone with shale intercalations

Fig. 1 · Farafra area · LITHOSTRATIGRAPHIC SEQUENCE

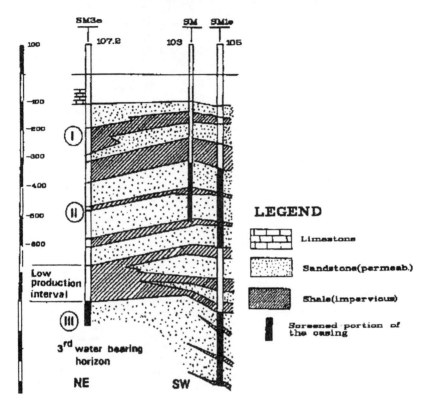

FIG 2 · Sheikh Marzouk Area
WELL CORRELATION SECTIONS

Models and Data Integration for Groundwater Decision Support

G. De Leo(*), L. Del Furia(*), C. Gandolfi(**), G. Guariso(**)

(*) Dipartimento di Elettronica, Politecnico di Milano, Italy (**) Istituto di Iraulica Agraria, Università degli Studi di Milano, Italy

ABSTRACT

A flexible software environment to support routine activities of groundwater managers is presented in the paper. Within such environment the user can interactively access a comprehensive database containing hydrogeological and water quality data, as well as several groundwater flow and pollution simulation models, stored in a modelbase. Integration of the data and model bases is provided by an intelligent module which is capable of matching data and models to the specifications of a given problem and of automatically selecting the candidate model(s) and the appropriate set(s) of input data. Traditional programming techniques are used along with those derived from the recent developments in Expert Systems and Artificial Intelligence. The overall software package is designed according to the paradigms of flexibility, modularity and adaptability. It allows a user with elementary knowledge on software and mathematical models to quantitatively analyse a wide set of groundwater flow and pollution problems, without the burden of data preparation and with the support of a powerful set of graphical routines for result presentation.

INTRODUCTION

Although the literature on environmental modelling has been steadily growing in the last twenty years, the use of these models in public agencies in charge of environmental planning and management is spreading at a much lower speed. This is mainly due to the difficulties faced by non experienced user in dealing both with the mathematical formulation of such models and with their software implementation. In recent years, a first step towards the solution of these problems has been accomplished by using tools developed in the context of artificial intelligence and software engineering. Their adoption can clearly contribute in creating an environment which is more friendly for the user and which allows him to easily access and use the appropriate models to solve his application problem. This is particularly true in the field of groundwater resources management, where problems complexity and lack of adequate data often hinder the effective use of the available mathematical models. Therefore a package for the support of groundwater managers' activities must be the result of a largely interdisciplinary effort, aimed at integrating techniques and expertises from different fields (computer sciences, hydrogeology, data base management, aquifer modelling, decision making theory).

This paper presents a software package, the Groundwater Manager's Toolkit (GMT), which provides the water manager with a flexible working environment, including a database system and a wide set of models, commonly used to analyse groundwater problems. More complex models, possibly developed specifically for the user's situation, can be easily integrated and effectively utilized within the environment.

To promote its diffusion and utilization, the package has been implemented on a Personal Computer, which, in our opinion, will be the type of hardware more easily available to environmental managers still for some years in the future.

GMT provides the user with a wide range of groundwater flow and pollution simulation models: from the simplest and widely accepted ones which give a rough estimate of the behaviour of the system (single layer, homogeneous, confined aquifer) to others, more sophisticated, which can provide accurate solutions to more complex problems (heterogeneous, unconfined aquifers, complicated boundary conditions). Moreover, given the specifications of a problem, the package automatically searches for the required data and manipulates some structural knowledge, representing the expert experience, in order to advice the user about the most appropriate model choices.

The package has the classical structure of decision support systems (see, for instance, [1]) and is made up of the following modules:

- a <u>Database</u>, which contains data about the wells (wells location, piezometric measures and quality parameters);

- a <u>Modelbase</u>, which contains the groundwater models currently available within the system;

- a <u>Knowledge base</u>, where information about the models is stored through frames;

- a <u>System Definition module</u>, which elaborates the structural informations on models and data, and through which the former three modules communicate with each other;

- a <u>I\O module</u>, which contains a number of programs for data acquisition and display.

All these modules will be discussed in more detail in the following sections.

THE DATABASE

The database contains the groundwater quality and quantity data, as well as the informations on the hydrogeologic characteristics of the aquifer, in a number of spatially distributed measurement points. Each of them is represented in the database by a data structure including both static and dynamic attributes. The former are:

- identification attributes, such as measurement point code, name and type

- geographical attributes, such as absolute geographical coordinates, which allow to locate the well inside the region under study;

- aquifer hydrogeologic attributes, such as hydraulic conductivity, storage coefficient and thickness in the case of a confined aquifer; hydraulic conductivity, effective porosity and bottom level in the case of unconfined aquifer.

The dynamic attributes are associated with time series data, such as time series of piezometric head and of pumping rate, and time series of physical, chemical and biological parameters of the aquifer (all these parameters are important to define the characteristics of the water and its possible uses). A fully interactive menu-driven user interface allows the user to perform all the classical database management operations, such as data retrieval and maintanance, and report preparation.

THE MODELBASE

The Modelbase plays an important role in the GMT package. The models represent the processors of information, which must extract all the details of the problem and make them explicit. In its present version the modelbase of the package contains four mass transport models and four groundwater flow models. These models solve the classical mass transport and flow equations under different initial and boundary conditions using analytical or numerical methods (e.g. finite difference or finite element). Only two dimensional models were considered for this first implementation.

Analytical models have some distinct advantages over numerical ones. They are faster and require only a few parameters, but imply hypotheses which are never met in real cases. They are thus useful to show which are the critical areas to be tested with greater accuracy and on which more detailed measurements are needed. Furthermore, the user can see what soil characteristics would be necessary for a certain situation to occur, or not occur and they can also be used in the process of calibration and comparison of numerical models. These different types of models thus correspond to different level of decision making and may all be useful at different stages of the decision process.

Flow simulation models

Only under very simple, idealized conditions, the flow equation can be solved analytically. One such situation is the single perfect well in a homogeneous, isotropic and confined aquifer of infinite horizontal extension. The solution [2] can be extended to several wells by means of the superimposition principle [3]. The analytical solution can be found even for other boundary conditions (rectangular domain with recharge and impervious boundary [4]).

When the geometry of the aquifer is no longer regular or when homogeneity and isotropy do not apply, numerical methods are required. In the GMT package a finite difference method is used for the solution of the flow equation in the general case of a confined or unconfined aquifer, both homogeneous and heterogeneous, both isotropic and anisotropic. For this purpose, the aquifer is divided into a regular grid of rectangular elements.

Altogether the following situations can be simulated:

- Flow in a confined, homogeneous aquifer with infinite lateral dimensions. The aquifer has a constant head recharge boundary condition.

- Flow in a confined, homogeneous aquifer with infinite lateral dimensions.

- Flow in a confined, homogeneous aquifer with finite lateral dimensions. The aquifer has a constant head recharge boundary and three impervious boundaries.

- Flow in a confined or unconfined, homogeneous or heterogeneous, isotropic or anisotropic aquifer under different boundary conditions.

Water quality simulation models

The two dimensional mass transport equation is considered under different boundary and source conditions. The equation can be solved analytically only if conditions such as homogeneity of the aquifer, parallel flow of constant velocity, constant retardation factor, reaction rate and dispersivities apply. When these conditions are not verified numerical methods must be used.

The following situations can be simulated:

- Solute transport in an infinite aquifer: the pollutant source is a strip with equal concentration along it.

- Solute transport in an infinite aquifer: the source concentration is modeled by a gaussian shaped curve.

- Mass transport in a finite aquifer with point pollutant source. The equation is solved by means of the finite difference method and the finite element method.

For more details on flow and transport equations, see for instance [3] and [4].

THE KNOWLEDGE BASE

For every available model, information concerning the application domain, the solution method, the input and output variables, the parameters required, the DOS name and mass memory unit   are stored in forms of frames such as the one shown in Figure 1.

These frames are accessed by the system definition module (see the next section) in order to evaluate which model best suits the problem defined by the user and which data are necessary to run it. In other words, this part of the system stores all the rules to be followed for the correct application of the various flow and water pollution models.

```
Problem: two dimesional groundwater flow
Aquifer: confined
Domain: infinite
Properties: homogeneous, isotropic
DOS name: C:\models\unconf1.exe
Parameters:  -- initial piezometric head
             -- wells location and pumping rate
             -- aquifer thickness
             -- hydraulic conductivity
             ..................................................
Output: piezometric head at the end of the simulation
Description: analytical solution of the two dimensional....
```

Figure 1 - Instantiated frame of a model.

In order to modify and enlarge the Knowledgebase a Menu Editor is available. The Menu Editor displays both empty and instantiated frames and the user can add or delete an entry, thus enabling or disabling the choice of a particular model.

Any executable file may be added to the system and thus any independently developed model in almost any computer language, provided they conform to fixed input/output standards.

SYSTEM DEFINITION MODULE

The "bridge" between the Modelbase, the Database and the Knowledge base is represented by the System Definition module. This component allows the user to define in a very friendly way the domain to be analyzed and to query, in an "intelligent" and fast way, the database. This module is based on Artificial Intelligence techniques as well as traditional search and interpolation algorithms.

Once the user has specified the kind of problem to be solved (quality or quantity), the hydraulic characteristics of the aquifer, the location of the domain under study and the date to start from, a set of operations is necessary. In particular, one should:

- query the database about the correct set of input data determined by the nature of the problem and the domain characteristics;

- extract these data from the database, if they exist within the region and at the date specified;

- determine the region, which can be a sub-area of the one chosen by the user, where all the required data are available;

- prepare the data in such a format that they can be used by the models.

The System Definition module executes the above mentioned operations automatically and communicates with the user in order to inform him or her about the decisions made, to ask him or her informations about the boundary conditions, to advice him or her about the proper range of values of some parameters. The advantages of such solution are: an improvement in system flexibility, a reduction of the elaboration time, an increase in reliability through effective error checking mechanisms, the possibility to imbed vast domain knowledge and to handle complex situations.

As mentioned above, the system provides the user with an intelligent guide for choosing the models which suit his or her needs. The model selection represents the main task of the System Definition module and is based on the domain characteristics, the data stored in the database concerning the region under study and the peculiar features of each model, contained in the knowledge base. Furthermore, the system informs the user whenever a lack of informations or an inconsistency of the data occur. The search operation can also identify more than one model or no models. In the first case a brief description of each model is displayed, and the user is invited to choose the model he or she wants to use. In case no model is suited to the examined problem, the system informs the user on how some of the problem specifications (aquifer structure or boundary conditions) should be modified in order to be treated with at least one of the available models.

THE USER INTERFACE

The section of the software dedicated to communication with the user takes a large part of the code. The input data interface is based on menus for alternative choices, on fill-in forms for numerical and textual data and on graphics for the definition of the domain and boundary conditions.

The output of model results is mainly based on graphics. Several different representations are available. The first is the classical representation of the aquifer piezometric heads or of the values of a selected water quality indicator, in terms of contour lines (see Figure 2). This procedure reads the model output values from a file and interpolate them using an algorithm based on fifth order splines [6]. The user has a number of options on the resulting display, such as labelling some curves, showing the wells, and so on. The same type of plot may also be used to display the time at which a given pollutant concentration is reached in each point of the region under study. In case of accidents, for instance, this information may be very useful to determine the most critical areas and the effects of increased pumping from the sorrounding wells.

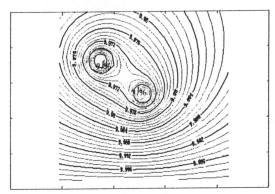

Fig 2 - Contour lines of the piezometric surface

A second routine provides a three dimensional representation of the piezometric surface or of the values of a water quality indicator, using a hidden line removal algorithm and allowing a rotation of the surface in order to improve the perception of its shape. The user may also display the flow velocity field on a regular plane grid, where the length of each vector is proportional, through a user selectable parameter, to the flow velocity.

Finally, a numerical output in a matrix form may also be displayed. A built-in editor allows modification of single values in case, for example, the user is unsatisfied with the results of the interpolation procedures, or some local information is available, which was not stored in the database.

All these representations may obviously be used both directly on measured data extracted from the database, or on the results obtained by executing the simulation models. At present, the graphic display allowed by the GMT package, may be sent directly to the computer screen (with EGA card) or to a plotter or a laser printer through a suitable software driver.

CONCLUDING REMARKS

The application of simulation models in practice   is a delicate matter since many groundwater decision makers are not very familiar with these techniques. In order to close the gap between theoretical and practical effectiveness of such models, a new and comprehensive approach, which can address and solve a wide range of groundwater management problems, has been designed and implemented in a PC software package. This may constitute an ideal support for routine activities of groundwater managers, since it includes programs for data input and output, several two dimensional models for aquifer dynamics and pollution and can incorporate in a simple and effective way additional models developed by the user.

Artificial Intelligence and symbolic programming techniques have lead to an easier access to data and models, to a better presentation of the results and to the automatization of the model choice and of data selection and preparation. This allows the user to concentrate his or her specific decision problems and to interactively and rapidly test and compare the effects of a number of management alternatives, using different solution procedures and for a wide range of aquifer configurations.

REFERENCES

1. Guariso, G. and Werthner, H. Environmental Decision Support Systems, Ellis Horwood, Chichester, 1989.

2. Theis, C. V. The relation between the lowering of the piezometric surface and the rate and duration of discharge of a well using groundwater storage, Trans. American Geophys. Union, Vol. 16, pp. 519-524, 1935.

3. Kinzelbach, W. Groundwater modelling. An Introduction with Sample Programs in Basic, Elsevier, Amsterdam, 1986.

4. Bear, J. and Verruijt, A. Modeling Groundwater Flow and Pollution, Riedel Publ. Co., Dordrecht, 1987.

5. Princeton Software Group, Groundwater Pollution and Hydrology Models, User's Manual, Version 1.9, Princeton, 1986.

6. Akima, H. A method of bivariate interpolation and smooth surface fitting for irregularly distributed data, ACM Trans. Mathematical Software, Vol. 4, pp. 148-159, 1978.

The Quasi-Linearity Assumption in Groundwater and Groundwater Quality Management Problems

T.Tucciarelli(*), G. Pinder(**)

(*) Facoltà di Ingegneria, Reggio Calabria, Italy
(**) College of Engineering and Mathematics,
University of Vermont, Burlington Vermont,
Canada

ABSTRACT

Groundwater and groundwater quality management problems are usually solved with standard optimization techniques. They require negligible installation costs, convexity of the feasible domain, and the ability to compute the derivatives of the constraint functions. A new hypothesis is proposed herein for the shape of the constraint functions. It allows the solution of the problem without some or all the aforementioned restrictions.

INTRODUCTION

Most of the groundwater and groundwater quality management problems require the following computation

$$\text{Minimize } f(\mathbf{x}) \quad \text{s.t. } \mathbf{g}(\mathbf{x}) \geq 0 \quad \mathbf{x} \geq 0 \quad \mathbf{x} \leq \mathbf{x}_{max} \tag{1}$$

where $\mathbf{x}$, the independent variables, are pumping rates of extraction or injection wells, $f(\mathbf{x})$ is the cost of the operation, $\mathbf{g}(\mathbf{x})$ is the set of constraint functions to be satisfied.

Suppose you want to drop the water heads in some observation wells by means of extracting water from the pumping wells. Call $\mathbf{h}$ the heads in the observation wells, $\mathbf{q}$ the pumping rates, $\mathbf{h}_{max}$ the maximum heads allowed in the observation wells during the operation period, **cost** the cost of the unit pumping rates for the operation period, and **costw** the installation cost of the wells. Assuming the transient time to be negligible and the objective function to be linear in the non-zero independent variables, one can write problem (1) in the form

$$\text{Minimize } \mathbf{cost} \; \mathbf{q} + \mathbf{costw} \; \mathbf{n}$$
$$\text{s.t. } \mathbf{h}_{max} - \mathbf{h}(\mathbf{q}) \geq 0 \tag{2}$$
$$\mathbf{q} \geq 0 \quad \mathbf{q} \leq \mathbf{q}_{max}$$

points will also be feasible. To identify a feasible point along the chosen **d** direction, we need only a few iterations to reach a sufficiently small interval by means of interval bisections.

Call B-region the domain of the points along any positive direction from a B-point. Observe in the example of fig. 1b) the location of the B-points after three line searches. All the points in the B_3 and B_4 regions have negative distance because the corresponding objective function is greater than the best one computed with the A_3 point. All the B_3, B_4 regions are not quasi-feasible. I can discard the B_3 and B_4 points and choose an other B-point to start a new line-search. After each iteration one reduces the quasi-feasible space and one is likely to improve the optimal point estimation. When the quasi-feasible space is negligible, one stops the algorithm. With the optimal point estimation one also knows the the upper boundary of the error, given by the maximum distance of the existing B-points.

THE B-POINTS GENERATION

The B-points are defined as the minimum point of the quasi-feasible space. The quasi-linearity (QL) algorithm computes a series of A-points on the feasibility boundary to make negligeable the maximum distance of the B-points with respect to the optimal objective function.

One can imagine the original variables space as defined by m A-points and one B-point, the origin. The jth A-point has the following coordinates

$$x^{Aj}_1 = x^{Aj}_2 = ... = x^{Aj}_{j-1} = x^{Aj}_{j+1} = ... = x^{Aj}_m = \infty \qquad x^{Aj}_j = 0$$

In this way all the non-positive points are interior to a A-region as well as all the infeasible points.

After the computation of a new A-point one must be able to update the number and the position of the B-points. First remember that the B-points are the local minima of the objective function f(**x**) in the quasi-feasible space, and the f(**x**) is monotonically increasing. The minimum condition requires that moving in any direction the objective function is improved only by loosing feasibility. It means, for each B_i point and j coordinate

$$x^{Bi}_j - dx_j < x^{Al}_j \qquad\qquad (3a)$$

$$x^{Bi}_n < x^{Al}_n \qquad\qquad n = 1,m \quad n \neq j \qquad\qquad (3b)$$

for at least one A_1 point.

The quasi-feasibility condition requires

$$x^{Bi}_j \geq x^{Al}_j \qquad\qquad (4)$$

for each A_1 point and at least one j coordinate. Combining (3) and (4) one gets

where $\mathbf{n}$ is a one-zero vector and $\mathbf{q}$ are the independent variables. The jth element of $\mathbf{n}$ is equal to zero if the jth pumping rate is zero, one otherwise. The relation $\mathbf{h(q)}$ is usually quite complex. It depends on the boundary conditions and on the physical parameters of the aquifer. If a finite-element model is used, the computation of the heads $\mathbf{h}$, given the pumping rates $\mathbf{q}$, requires the solution of large linear systems (see an example in Galeati and Gambolati [1]).This and other problems in the form (1) are quite peculiar in the operations research context (Luenberger [2]) for the following reasons:

a) The constraint functions are usually very complex and the derivatives of these functions with respect to the independent variables are even more complex. If a stochastic approach to the problem is used, the functions $\mathbf{g(x)}$ also contain the derivatives of the state variables (heads, concentrations) with respect to the physical parameters or the pumping rates (Gorelick [3]).

b) The feasible domain (the domain of points satisfying the constraints) can be non-convex and many local minima can exist.

c) The objective function $f(\mathbf{x})$ is usually a simple polynomial relation, but it is discontinuous at the zero values, due to the installation costs of the wells. The discontinuity of the objective function requires non-linear mixed-integer programs for the solution of the problem (Christofides et al.[4]). The resulting programs are very time-consuming on the computer and only find local minima due to (b).

d) The number of independent variables as defined earlier is very small (less then one hundred) and most of the optimal values are zero even if we neglect the installation costs (Ahlfeld et al.[5]).

e) A large numerical error for the optimal solution is tollerated. This is because the field implementation of the results will change the pumping rates anyway. An error of 10-20% is expected.

In spite of the above difficulties, the groundwater and groundwater quality management problems have been usually treated with standard optimization techniques. The main contribution of the hydrologists to the solution of optimization problems has been the adjoint state sensitivity analysis (Ahlfeld et al. [6]). Recently the problem has been solved by means of the simulated anealling technique (Kirkpatrick et al. [7]). The simulated annealing is able to locate the global minimum without any hypothesis on the shape of the constraint functions. The convergence of the algorithm relies on the choice of some critical parameters and the number of the required iterations can easily become extremely large.

The following paper introduces a methodology dedicated to finding the minimum of problem (1) by means of what we call quasi-linearity assumption. Quasi-linearity is a more general hypothesis than convexity and has an immediate physical meaning.

THE QUASI-LINEARITY ASSUMPTION

Consider an infeasible point in problem (2). This is a set of pumping rates that are not sufficient to drop the water heads below the maximum values. The quasi-linearity assumption states that to reach feasibility one must increase at least one of the pumping rates.

The constraint functions $g(x)$ can be thougth as a measure of the quality of the state variables. Quasi-linearity states that any increment of the independent variables produces an improvement of the state variables. This is true, for example, for any location of the wells in problem (2).

We say that problem (1) is quasi-linear if, given an infeasible point x_0, any point $x_0 - d$ k is also infeasible if d is strictly positive and k is a positive scalar. If the feasible domain is unbounded in any positive direction we can show that every convex problem is quasi-linear, but a quasi-linear problem can be non-convex.

We first show that non quasi-linearity contradicts the original assumption of convexity. Given an infeasible point x_0, it must be $g(x_0) \not\geq 0$. Due to the non quasy-linearity, it must exist at least a feasible point along a strictly negative direction from x_0. It implies, for some positive vector d and some positive scalar k_1, $g(x_0 - dk_1) \geq 0$. If the feasible domain is unbounded in any positive direction, there also exists a k_2 scalar, such that $g(x_0 + dk_2) \geq 0$. The conditions $g(x_0 - dk_1) \geq 0$, $g(x_0 + dk_2) \geq 0$ and the convexity hypothesis contradict the original assumption $g(x_0) \not\geq 0$.

On the other hand, fig. 1a) presents one case of a quasi-linear, non-convex feasible domain. We recognize the quasi-linearity because from any A-point on the feasibility boundary we can reach in any strictly negative direction only the infeasible points in the dashed area. We also observe that each A-point generates m (2 in the 2D case) B-points that are local minima for any monotonically increasing objective function of x. This minimum condition holds also if the function is discontinous at the zero values, as usually happens in problem (1).

In the example of fig. 1a) suppose the objective function to be linear. The points with a constant value of the objective function will lie on a straight line. If the line direction is the one proposed, we recognize two local minima M1,M2. Due to the non-convexity, the local minimum conditions are not sufficient to identify the global minimum.

IMPROVING THE OPTIMAL POINT ESTIMATION VIA COMPUTING A-POINTS AND REJECTING B-POINTS.

Call $Cost_{min}$ the minimum objective function already computed. The corresponding point is stored as the optimal point. Call quasi-feasible the space of the points with the objective function smaller than $Cost_{min}$ that are not infeasible due to any A-point previously computed. In fig. 1a) the quasi-feasible space after the first computed A-point is the space below the straight line and out of the dashed area, if we assume $Cost_{min}$ as the objective function of the point A . The points B_1 and B_2 represent the local minima of the objective function in the quasi-feasible space. Call distance h_j of the jth B-point the difference between $Cost_{min}$ and the objective function of the jth B-point. We choose among the computed B-points the one with maximum distance and we perform along a suitable positive direction a line search from the B point up to the feasibility boundary.

Due to the quasi-linearity assumption the line search is quite inexpensive. Moving in any positive direction from the chosen B-point we know that after the first feasible point, all the other

$$x^{Bi}_{j} = x^{Al}_{j} \qquad\qquad\qquad\qquad\qquad\qquad (5a)$$

$$x^{Bi}_{n} < x^{Al}_{n} \qquad n = 1, m \quad n \neq j \qquad\qquad\qquad (5b)$$

$$x^{Bi}_{p} \geq x^{Aq}_{p} \qquad\qquad\qquad\qquad\qquad\qquad (5c)$$

for at least one A_1 point and p coordinate, as well as all the A_q points. The relation (5a) must be satisfied by m A_1 points for the m coordinates of each B-point. Call these points support-points. Because the other m-1 coordinates of each A_1 point must satisfy condition (5b), only one coordinate for each A_1 point can satisfy codition (5a) and the m support points must be different.

After each line search, first identify all the B-regions where the new A-point is interior. The B-points corresponding to the other B-regions will not change, because the condition 5) continues to be satisfied. For each B_i region where the new A-point is interior, there are m support points. Changing each of the old m points with the new A-point, one obtains m possible sets of support points that define m possible (candidate) new B-points. For each possible set of B-support points the minimum value of each coordinate and the index of the corresponding point are computed. If all the indeces are different the candidate is really a B-point. If two or more indeces are equal the candidate fails, because it would require two equal support-points. If one or more coordinates of the different support points are equal, conditions 5) cannot be satisfied and the point is not a minimum; we call these points degenerate.

In fig. 2a) we can see the position of three B-points after the first line search is performed in the 3D space. Because the variables domain is defined by m A-points, the index of the first computed A-point is m+1. In fig. 2b) one can see the changes in the local minima of the quasi-feasible space due to a new point A_5.

The coordinates of the A_4 point are $x^{A4} = (1,2,1)$. The coordinates of the A_5 point are x^{A5} = (2,3,0.5). A_5 is not interior to B_3 in fig. 2a) and B_3 remains, with the index 5, in fig. 2b). The point B_1 in fig. 2a) disappears and generates three candidates. The support points of B_1 in fig. 2a) are A_4, A_2, A_3. In table I we can see the coordinates of the A-points and the indeces of the support points of the candidates generated by the new A_5 point. The coordinates of the candidates defined by the sets of support points are the same of the support point with the printed index. For example, the first coordinate of the first candidate is equal to the first coordinate of the A_5 point.

candidate
support points

	x_1	x_2	x_3		x_1	x_2	x_3
A_4	1.0	2.0	1.0	2-3-5	5	2	3
A_2	∞	0	∞	4-3-5	4	4	3
A_3	∞	∞	0	4-2-3	4	2	5
A_5	2.0	3.0	0.5				

Table I

The second candidate has two equal indeces for the x_1, x_2 coordinates and it is not a B-point. The first and third candidates in table I are the points B_1, B_4 in fig. 2b). With the same procedure, we can recognize that the point B_2 in fig. 2a) disappears and generates the points B_2, B_3 in fig. 2b).

If there are not degenerate points, the original quasi-feasible space is completely decomposed in the infeasible region below the new A-point and in the quasi-feasible regions above the new B-points. Observe that the new B-points generated from the new A-point are all interior to the old B-region. This is because the coordinates of the new A-point will be all greater than the corresponding coordinates of the old B-point . For the aforesaid reason the distances of the new B-points are smaller than the distance of the old B-point.

Going on with the A-points computation, one drops the distance of the B-points. When the distance of a B-point is small enough, I can discard the B-point, because the quasi-feasible space in the B-region is negligeable. I can also discard all the A-points that are not support-point of any B-point.

THE LINE SEARCH DIRECTION AND THE DISCARDING CRITERIUM.

To reduce to zero the quasi-feasible space one must create infeasible space via the A-points and drop the objective function. To reach the first goal one must move from the inital B-point along all the positive direction components. If the objective function is linear along the non-zero coordinates, the same linear coefficients are used as direction components.

Because most of the optimal variables will be zero, the previous direction is not very likely to drop the objective function. To reach this goal it is more convenient to preliminarily perform a first line search where the direction components corresponding to zero coordinates of the original B-point are kept equal to zero. The result of this line search is used only to update the objective function without computing any new A-point.

Each time one drops the objective function, one also drops the distances of all the B-points. When the distance of a B-point becomes too small, one discards the point because the improvement he can reach in the objective function is negligeable for all the points inside the B-region.

Different criteria can be used to discard a B-point. One follows. Compute the ratio between the distance and the minimum value of the objective function corresponding to all the quasi-feasible points. This value is given by the minimum objective function of the existing B-points. When the ratio is smaller than a given parameter, discard the B-point. Using this criterium, one can guarantee that the percent error in the optimal point evaluation is smaller than the given parameter. I assign a parameter that is small with respect to the field error.

Observe that the proposed metodology doesn't give any information about the sensitivity of the optimal variables with respect to the objective function. If the sensitivity is large, the optimal solution can be very different from the computed one. The difference between the real operation cost and the computed one depends on the objective function structure and on the pumping rate error. If the objective function is linear, the difference is simply proportional to the pumping rate error. Even if the real optimal solution is very different from the computed one, the field application error will be greater than the difference between the computed and the real optimal objective function. If a field error of 20% is forecasted, a 5% approximation in the optimal objective function is still negligeable.

CONCLUSIONS

A new methodology for the solution of groundwater management and groundwater quality management problems has been presented, based on what we called quasi-linearity assumption. Quasi-linearity states that an infeasible point remains infeasible if one does not improve at least one variable. Using this hypothesis it is possible to solve the optimization problem without computing the constraint function derivatives and without neglecting the installation cost of the wells.

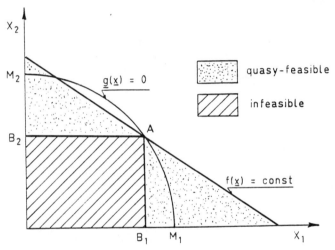

Fig. 1a) Infeasible and quasi-feasible space after one line search.

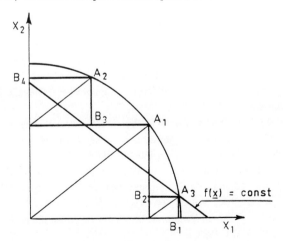

Fig. 1b) B-point locations after three line searches.

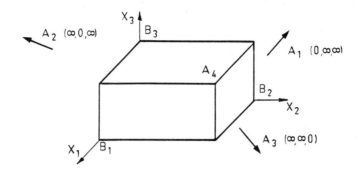

Fig. 2a) Locations of a new A-point and three B-points in 3D.

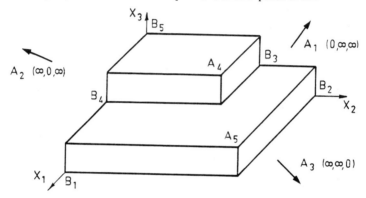

Fig. 2b) Locations of two new A-points and five B-points in 3D.

REFERENCES

1. Galeati, G. and Gambolati, G., Optimal Dewatering Schemes in the Foundation Design of an Electronuclear Plant, Water Resour. Res., Vol. 24, pp 541-552, 1988.
2. Luenberger, D.G., Linear and Nonlinear programming, Holdison Wesley, New York, 1984.
3. Gorelick, S.M. ,Optimal Groundwater Quality Management Under Parameter Uncertainty, Water Resour. Res., Vol. 23, pp 1162-1174, 1987.
4. Christofides, N., Mingozzi, A., Toth, P. and Sardi, C. Combinatorial Optimization, Wiley Intrrscience, London and New York, 1979.
5. Ahlfeld, D. P., Mulvey, J.M., Pinder, G.F. and Wood, E.F. , Contaminated Groundwater Remediation Design Using Simulation, Optimization and Sensitivity Theory 1: Model Development, Water Resour. Res., Vol. 24, pp 430-442, 1988.
6. Ahlfeld, D. P., Mulvey, J.M., Pinder, G.F. and Wood, E.F. , Contaminated Groundwater Remediation Design Using Simulation, Optimization and Sensitivity Theory 2: Analysis of a Field Site, Water Resour. Res., Vol. 24, pp 443-452, 1988.
7. Kirkpatrick, C.D., Gelatt, C.D. and Vecchi, M.D., Optimization by simulated annealing, Science, Vol. 220, pp 671-674, 1983.

Optimization of Groundwater Pumping Strategies of the Wuwei Basin by Integrated Use of Systems Analysis and Groundwater Management Model

Y. Zhou

Institute of Environmental Geology, Beijing, China. TNO Institute of Applied Geoscience, Delft, The Netherlands

ABSTRACT

An optimal groundwater pumping strategy was designed for Wuwei basin by integrated use of systems analysis and a groundwater management model. The systems analysis was applied to analyze the various aspects related to groundwater pumping and a management model was formulated and used to optimize the designed pumping alternatives. The optimal strategy can be served as a guide for the groundwater exploitation in the basin.

INTRODUCTION

Wuwei basin is one of the most important agricultural bases in the northwest of China. Groundwater is an important water source for the agricultural irrigation. In the past 20 years, groundwater levels in the basin were continuously decreasing with the increasing groundwater abstraction. As a consequence, some ecological and environmental problems, such as salinization and desertification, have occurred. Therefore, an optimal groundwater pumping strategy needs to be designed for the basin to get maximum groundwater for irrigation while the environment must be kept in good condition.

In the ideal case, the optimization of the groundwater pumping strategies includes the following aspects:
- establishment of the objectives;
- identification of the constraints;
- design and optimization of the alternatives;
- evaluation and choice of the optimum.

All those aspects should be included into a groundwater management model. However, this is always too complicated to be executed in practice. Instead, a sub-optimal procedure could be carried out by integrated use of systems analysis and groundwater management modelling. In the procedure, systems analysis is employed to analyze the various aspects related to groundwater pumping, and a groundwater management model is used to optimize the designed alternatives.

An optimal groundwater pumping strategy was designed for Wuwei basin using the procedure. A groundwater management model based on the response matrix approach was formulated. The model maximizes the total amount of annual groundwater pumpage subjected to the physical, environmental, and social constraints. Two alternative groundwater pumping strategies were optimized using the management model. An optimal strategy was chosen after evaluation of the alternatives.

GROUNDWATER RESOURCES SYSTEM IN THE BASIN

Wuwei basin, located in the northwest of China, is a faulted sedimentary basin in the Qilian mountain fronts and consists of several river alluvial fans and a river drainage plain (Fig. 1). The basin has an arid climate with very little rainfall (about 200 mm/year) but very large evaporation (about 2300 mm/year), and is surrounded by deserts in the east. Shiyang river stems from Qilian mountains, flows through Wuwei basin and disappears in Minqin basin due to seepage to groundwater and evaporation losses.

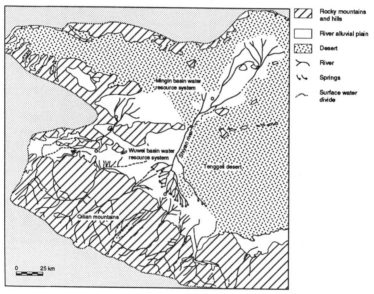

Fig.1 Water resources systems and location of Wuwei basin

The porous aquifer system in Wuwei basin is an important underground reservoir for the agricultural irrigation. The aquifer is composed of gravels and sands and it is characterized with three distinct sub-areas: river alluvial fans, spring discharge zone and groundwater evaporation zone.

The river alluvial fans are the groundwater recharge zone of the aquifer. The seepage from rivers is the only groundwater recharge since the rainfall can hardly form the groundwater recharge. The springs occur in the area where the river alluvial fans interfingered with silt and fine sand lenses. The baseflow of Shiyang river in the lower river reach is formed mainly by spring discharges. The groundwater evaporation zone is located in the northeastern plain. Since the aquifer is bounded with impermeable rocky hills in the north, naturally groundwater flows towards the surface, the depth of the water table is less than 3 m, and the evaporation forms natural discharge of groundwater.

DESIGN OF THE GROUNDWATER PUMPING STRATEGIES

Pumping well fields
111 well fields are designed based on the irrigation system (Fig. 2). Each designed well field represents the total pumpage from the real wells in that cell.

Time period
The potential of the groundwater pumpage in the basin was investigated for three typical years: normal year (1979), dry year (1972) and wet year (1983). Each year is a cycle of the irrigation and thus is divided into three unequal-length time periods: non irrigation period (3 months), major irrigation period (5 months) and minor irrigation period (4 months). Only the pumpages at major and minor irrigation periods are used as the decision variables.

Pumping strategies
In the basin the possibility to increase the groundwater pumpage is to reduce the evaporation losses of groundwater in the evaporation zone. So, two pumping strategies were investigated. One is to pump groundwater keeping the current groundwater level regime (strategy I). Another is to maximize groundwater pumpage lowering groundwater level in the evaporation zone (strategy II). Each of them will be optimized under three typical years.

FORMULATION OF THE GROUNDWATER MANAGEMENT MODEL

Establishment of the objective
The goal of the groundwater management in the basin is to get maximum groundwater pumpage for irrigation. Therefore, the objective was chosen to maximize the annual total amount of the groundwater pumpage, i.e.

$$\text{Maximize } Q = \sum_{j=1}^{M} \sum_{n=1}^{N} c(j,n)q(j,n) \tag{1}$$

Where: Q is sum of the pumpages of M wells for N time periods;
q(j,n) is pumpage of well j during period n, and
c(j,n) is weighting coefficient of the decision variable q(j,n).

Identification of the constraints
The constraints subjected to groundwater pumping in Wuwei basin can be distinguished into three groups: ecological environmental constraint, social environmental constraint, and constraint of groundwater state equation. All those constraints can be incorporated in the following equations:

$$s_0(k,n) + \sum_{j=1}^{M} \sum_{i=1}^{n} q(j,i)\beta(k,j,n-i+1) \leq s_{max}(k,n) \quad \text{for all k and n} \tag{2}$$

$$-e(i) \sum_{j=1}^{M} q(j,1)a(i,j) + d(i) \sum_{j=1}^{M} q(j,2)a(i,j) \leq 0 \quad \text{for all i} \tag{3}$$

Where:
$s_0(k,n)$ is additional drawdown of cell k at the end of time period r caused by the uncontrolled stresses;

$\beta(k,j,i)$ is system response coefficient, consisting of response matrix;

$s_{max}(k,n)$ is the maximum allowance of groundwater drawdown in cell k at the end of time period n, depending on the ecological and social constraints;

$d(i), e(i)$ are percentage of groundwater pumpage during major and minor irrigation periods to total annual pumpage, respectively, in irrigation district i;

$a(i,j)$ is the percentage of pumpage in cell j allocated to irrigation district i.

The objective function (Equation 1) and constraints 2 and 3 form a linear programming management model for Wuwei basin.

OPTIMIZATION OF THE GROUNDWATER PUMPING STRATEGIES

Generation of the response coefficients $\beta(k,j,i)$
A two-dimensional transient finite difference model was calibrated for generation of the response matrix. The modelled area is discretized into 217 polygon cells (Fig. 2). The spring discharge zone is discretized with smallest cells while the largest cells are used near the boundaries. The month is used as time step in the modelling.

For generation of the response coefficients, the unconfined aquifer system was linearized using Jacob's procedure [4]. To avoid the influence of initial groundwater flow field, first, using a pumpage of $1{\times}10^6$ m^3/month as a unit impulse, a pseudo response matrix was generated under the homogeneity of the initial and boundary conditions without any natural recharge and discharge; then, the pseudo response matrix subtracted by the drawdowns induced by the initial flow field results in the real response matrix.

Simulation of the additional drawdown $s_0(k,n)$
As discussed by many authors, e.g. Bear [1], Heidari [3], the function of the initial groundwater flow field, changes of the boundary conditions and natural recharges and discharges must be modelled by a "capture term", i.e. the additional drawdown. In case of Wuwei basin, the additional drawdown for three typical years were simulated using the numerical model.

Data for constraints
Maximum allowance of drawdown $s_{max}(k,n)$ To ensure the priority of using Shiyang river discharge by Minqin basin, in the spring discharge zone, $s_{max}(k,n) = 0$ for $n = 1,2$.

To protect from salinization and desertification, in the groundwater evaporation zone, the experiments and observation show that the optimal depth of groundwater level should be 4 m. The difference between the optimal depth and the initial depth is used as $s_{max}(k,2)$. Only 5 per cent of the saturated aquifer thickness could be depleted during the major irrigation period, which is used as $s_{max}(k,1)$.

Allocation of pumpage among two irrigation periods   Based on the water demand for irrigation in two irrigation periods, 60 and 40 per cent of the total pumpage are needed during the major and minor irrigation periods, respectively, i.e. $d(i) = 0.6$ and $e(i) = 0.4$.

Weighting values of the objective function $c(j,n)$ Since there are 5 months in major irrigation period and 4 months in minor irrigation period, the weighting values of the objective function are $c(j,1) = 5$ and $c(j,2) = 4$.

Optimal groundwater pumping strategies
By operation of the management model, six groundwater pumping alternatives were optimized for two pumping strategies in three typical years.

Table 1   Evaluation results of the optimal pumping strategies

Typical year	Strategy	Maximum pumpage	Groundwater Supply	Demand	Deficit	Shiyan river inflow to Minqin basin
Normal	N-I	392	402	442	-40	275
year	N-II	495	444	442	+2	335
Dry	D-I	334	354	442	-88	240
year	D-II	413	382	442	-60	290
Wet	W-I	470	485	442	+43	310
year	W-II	552	497	442	+55	375

Unit: $1 \times 10^6 \ m^3/year$.

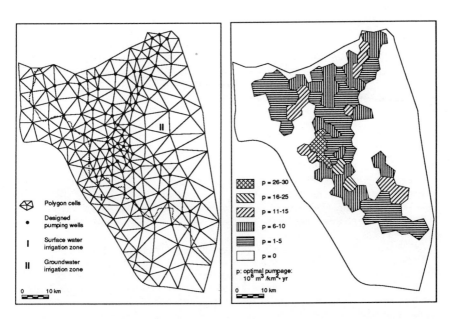

Fig. 2 Discretization of the modelled   Fig. 3 Distribution of the optimal
       area and designed well fields   pumpage of the strategy N-II

For evaluating the optimal pumping strategies, two criteria were used. One is the deficit of water supply. Another is the amount of Shiyang river inflow to the Minqin basin. The evaluation results are listed in Table 1. It can be clearly seen that the strategy II is much better than the strategy I, and thus the pumping strategies N-II, D-II and W-II can be chosen as optimal pumping schemes for normal years, dry years and wet years, respectively.

The spacial distribution of optimal pumpage can be visualized by representing the annual optimal pumpage per square kilometer. Figure 3 shows the distribution of the annual optimal pumpages of the strategy N-II. The groundwater pumping conditions in the basin can be divided into six subareas. The largest intensity of groundwater pumpage is located along the river courses while the smallest intensity is in the spring discharge zone and the area nearby the desert. There is no groundwater pumping in the surface water irrigation zone and desert.

CONCLUSIONS

Groundwater resources systems optimization is a complicated system engineering project. It could be performed by integrated use of system analysis and groundwater management modelling. The optimization procedure includes the establishment of the objectives, identification of the constraints, design and optimization of the alternatives, and evaluation and choice of the optimum.

The procedure was applied to the Wuwei basin for optimization of the groundwater pumping strategies for the agricultural irrigation. Two pumping strategies were optimized for three typical years. The pumping strategies under lowering the groundwater level in the evaporation zone was chosen as the optimal pumping strategies after the evaluation. The results show that reducing groundwater evaporation in the evaporation zone can increase the groundwater pumpage. Under operation of the optimal pumping strategies, the groundwater level decline in Wuwei basin could be controlled and the priority of using Shiyang river discharge by the Minqin basin could be guaranteed. The optimal strategy provides a guide for the groundwater exploitation in the basin.

REFERENCES

1. Bear, J., Hydraulics of Groundwater. McGraw-Hill, New York, 1979.
2. Gorelick, S.M., A Review of Distributed Parameter Groundwater Management Modelling Methods. Water Resour. Res., 19(2), 1983.
3. Heidari, M., Application of Linear System's Theory and Linear Programming to Groundwater Management in Kansas, Water Resour. Bull., 18(6), 1982.
4. Jacob, C.V., Notes on Determining Permeability by Pumping Tests Under Water-Table Conditions. U.S.G.S. Mimeograph Report, 1944.
5. Maddock, T. III, Algebraic Technological Function from a Simulation Model. Water Resour. Res., 8(1), 1972.

Studies of the Relationship Between Soil Moisture and Topography in a Small Catchment

B. Erichsen, S. Myrabø

University of Oslo, 0316 Oslo 3, Norway

ABSTRACT

This paper discusses results from a field study in a small catchment in Norway. The main object here is to describe the relation between topography and soil moisture. To describe the pathways and the processes generating rainfall to runoff, descriptive methods are evolved in hillslope hydrology. The response-area model has shown to give a good explanation to the observations made in natural catchment in Norway. To evolve the model and test it, a lot of different methods are used. These methods often require much work and are difficult to transfer to other catchment. We therefore present a method which uses digital terrain models (DTM) to deduce the response-area and their possible variation in time and space. The response-area is estimated as a function of drainage area, gradient and curvature. The method is verified with field mapping and measurements of the groundwater level in 105 observation sites. The results show that DTM can be used to estimate response-area in a catchment with a certain accuracy.

INTRODUCTION

Hillslope hydrology is a part of hydrology which has become of increasingly great importance for the development of runoff models and for the understanding of the hydro-chemical processes, among other things in connection with a growing pressure on nature due to pollution such as acid rain, agricultural manure and refuse disposal(Myrabø [10]). In order to explain the pathways of the water, from the stage where it as precipitation reaches the ground and until it is actually in the stream, there have within hillslope hydrology been developed descriptive models of the processes that occur; socalled conceptual models. Hillslope hydrology studies have taken place in different parts of the world and under different climatic conditions. This has resulted in a development of different conceptual models, valid in the areas in which they are developed. These are, however, difficult to transfer directly to new areas. The models are more closely described by Myrabø [8]. Previous work has shown that the response-area model gives a good explanation to the observations we make in natural fields in Norway. To develop the model and to verify it, a set of different methods are used. The methods most employed in connection with the response-area model is to map saturated areas in the field (Myrabø [9]), direct measurements of groundwater and vadose water variations (e.g. Anderson and Burt [2]; Myrabø [12]), plus mapping of the vegetation where vegetation is used as an indicator to moisture conditions (Gurnell and Gregory [5]). These methods are often both demanding and difficult to transfer to other areas. Thus we will here introduce a method to employ digital terrain models in order to deduce the response-area and their potential variation area in a catchment. DTM has been and is now to an increasing extent used within a wide range of studies. The areas of application reach from pure quantitative studies of topography (e.g. Evans [4]; Zevensbergen and Thorne [14]) to the modelling of hillslope evolution. DTM has also been used in hydrology to deduce parametres for flood generating areas for use in flood models (Heerdegen and Beran [6]). The grid size employed in previous work is 20 metres or more, which in our opinion is far too large for detailed studies.

EXPERIMENTAL AREA

The catchment used in this study is a small part of Sæternbekken experimental area (Erichsen and Nordseth [3]), 10 km vest of Oslo, Norway. The catchment has an area of 0.001 km^2 and has a median elevation of 250 m.a.sl. It is covered by a thin layer of till unless small area of bare rock. A topographic map is shown in figure 1. The vegetation consists generally of billberry and pine. For closer description of the catchment and its instrumentation, see Myrabø [11].

METHODS

In a field the motion of the water will be directed by the force of gravity and will therefore seek lower-lying areas. In natural terrain the moist areas will be hollows in the terrain and lower-lying areas with a large drained area. In addition soil characteristics will be decisive for the distribution of moisture in the field. In this paper we shall limit ourselves to see what effect topography has on the distribution of moisture. The variables chosen were drained area, gradient and curvature.

Primary parametres

The digital data of height were collected by levelling of the catchment. These data were then transferred to a regular grid by means of the interpolation procedure GINTP in the plot package UNIRAS. The procedure interpolates irregular distributed data to a regular grid using weighted average interpolation, slope calculation, quadratic interpolation and smoothing. The grid size was chosen to 2X2 metres in order to include small scaled variations.

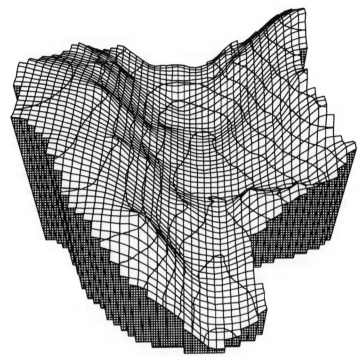

Figure 1: A 3-D topograpic map of the reseach area

Secondary parametres

A computer programme was then used to estimate slope, aspect and curvature
(both lengthwise and and crosswise of the slope) from 8 neighbouring grids.
Values of curvature were estimated with different resolution from 2 to 18 metres.
For both types of curvature the values stabilized at a 10 metre resolution. From
this, curvature values were estimated both with a fine resolution (2 metres)
and a more coarse resolution (10 metres) in order to cover the small and the
large scaled variation. The algorithms used are described by Zevenbergen and
Thorne [14]. Drained area was calculated by means of a personally developed
algorithm. This was done in two steps:

1) A drained area in each grid was estimated by starting at the limit of the
catchment and letting each grid drain to a lower-lying grid. Which one of the
lower-lying grids a grid drain to is estimated from the calculated aspect.

2) Finally a search through the result matrix was carried out in order to detect
local "tops" not discovered in the first step.

To group the different terrain variables in physically explicable categories a
cluster analysis of all variables was carried out. The variables were standardized.
The cluster analysis chosen was a non-hierarchical cluster analysis based on
nearest centroid sorting (Anderberg [1]).

Groundwater maps were used to verify the results from the cluster analysis. The groundwater level in the catchment is measured with a hourly resolution in 4 wells. In addition the groundwater level is measured simultaneously in 105 PVC tubes at different discharges. In this work the groundwater level in the 105 tubes, at 5 different moisture conditions, is used. The point values for groundwater level were transferred to a regular grid by means of the above mentioned GINTP in UNIRAS.

Uncertainty

Interpolation By interpolating grid values from measured point values or vector data one makes a filtration of the data. Some information will be lost. How useful the result will be, depends among other things on the grid size. As a basis we have used 2x2 metres in order to obtain small scaled variations. Interpolation is a very subjective act and the result is not right or wrong, rather more or less useful (Monmonier [7]).

Meeting drainage In some cases two neighbouring grids will drain to each other. This may often occur in flat areas where small variations of height will decide the aspect. In this catchment this only happened in one case. The error was therefore corrected interactively.

Drainage between individual grid The method has a weakness in that water from a grid cell only can drain down to another one. One now tries to develop an algorithm that makes it possible for water from a grid cell to drain to various neighbouring cells.

RESULT

By treatment of the data general statistics were estimated (tab. 1). The highly varying order of magnitude between the variables made it necessary to standardize the variables before the cluster analysis. The correlation between the different variables is reproduced in table 2. The correlation shows that the intercorrelation is small. Because of this we chose to use all variables further on in the cluster analysis to cover different physical aspects.

Table 1. General statistics

Variable	Mean	Std Dev	Minimum	Maximum	N
SLOPE	-.17	.08	-.4156	-.0169	1233
ASPECT	209.30	73.61	3	349	1233
PL1	.15	2.43	-20.33	12.70	1233
PL2	.17	1.27	-6.94	5.48	1233
PR1	-.43	3.03	-14.94	20.49	1233
PR2	-.35	1.67	-5.61	6.59	1233
DRAINAGE AREA	31.17	132.27	0	1107	1233

PL1 = Plan curvature, 2 m resolution PL2 = Plan curvature, 10 m resolution
PR1 = Profile curvature, 2 m resolution PR2 = Profile curvature, 10 m resolution

Table 2. Correlation between the variables

	SLOPE	ASP	PL1	PL2	PR1	PR2
SLOPE	1.000	.009	.140	.254	-.025	-.059
ASP	.009	1.000	-.027	-.059	.143	.193
PL1	.140	-.027	1.000	.748	-.500	-.352
PL2	.254	-.059	.748	1.000	-.432	-.462
PR1	-.025	.143	-.500	-.432	1.000	.746
PR2	-.059	.193	-.352	-.462	.746	1.000
ANT	.244	-.164	.012	.119	-.069	-.102

N of cases: 1233

In the cluster analysis we chose as a starting point to divide the data in 10
categories. The cluster analysis gave 4 main and 6 smaller categories (tab.3).
If we plot the different clusters on a map, we will see the spatial distribution
of the clusters. From figure 2 and from the results of the cluster analysis the
following groups may be connected to the catchment:

1) The drainage net itself
2) Flat areas next to streams with a large drained area
3) Slopes with good drainage
4) Flat remote areas with small drained area

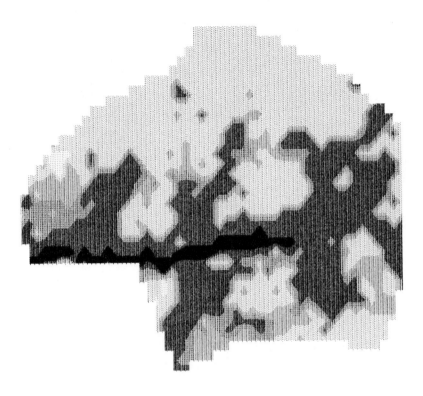

Figure 2: The respons-area map resulting from the envolved DTM model. The grey tone shows the probability for the area to get saturated. Black areas will be saturated first. Due to the interpolation procedure the figure 2 show a smaller part of the catchment than the figure 3

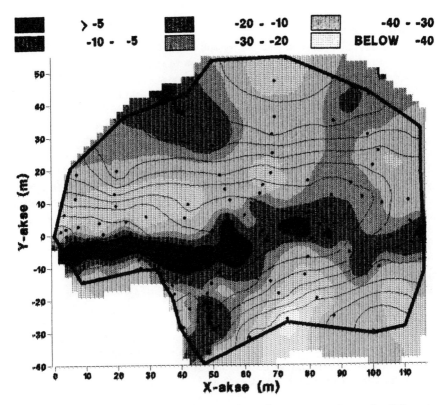

Figure 3: A 2-D view of the reseach area, where the gray tone shows the different groundwater levels(cm) when Q is 1.1 l/s. The groundwater tubes are marked with black pints.

Table 3. Final Cluster Centers.

No	ZSLOPE	ZPL1	ZPL2	ZPR1	ZPR2	ZANT	Cases
1	-.5207	2.5700	1.4946	-3.5609	-2.2609	-.1822	16
2	.8185	.1050	.3034	.1794	.1399	-.1339	452
3	-.4553	-.4046	-.8476	.9653	1.2724	-.1923	189
4	-.6496	-3.4216	-3.0380	2.1620	1.9377	-.1988	16
5	1.4308	.2205	.8154	-.1288	-.2680	5.7852	26
6	.9913	-.6277	-.3818	-.9331	-1.3547	4.6566	10
7	-.7535	-.7368	-.8219	-.4348	-.4500	-.1958	173
8	1.0140	2.7022	2.2907	-1.4275	-.7877	-.0721	16
9	-.4185	-7.1889	-4.8150	5.8864	3.7916	-.2331	3
10	-.5931	.4488	.4529	-.4459	-.5993	-.1756	332

The result of one of the interpolated groundwater maps is shown in figure 3.

DISCUSSION AND CONCLUSION

Comparison of the cluster map (fig.2) and the groundwater map (fig.3) shows that we get roughly the same pattern for the response-area. The biggest differences are due to the fact that the interpolation of the groundwater data does not consider the terrain. Because of this we get the same groundwater depth at the tops of a ridge as on both sides of it if we only have measurements of hollows on both sides of the ridge. Likewise, we will on the groundwater map have a large distribution of a small moist area in a hollow, when there are no other points of measure nearby. On the other hand some of the driest areas on the groundwater map, which are not found on the cluster map, are due to coarse till or bare rock. One may therefore improve the already developed DTM model by taking soil depth and permeability into consideration. As a conclusion one may say that the developed model can estimate the response-area with a certain accuracy, and that simple improvements can give a very satisfactory result.

The areas of application and the utility value of this model are diverse. From the discussion above one may notice large improvements of the groundwater maps by using the model in interpolation (Myrabø & Erichsen [13]). Usage of the model in levelled catchment will give a good basis for estimation of the response area by making groundwater observations in a small number of points and by only few observations. This will give a unique basis for the use of the response-area model in a catchment. Furthermore one may use a nation-wide map, digitally, with good resolution, to estimate response-areas in large water systems. By either continuous measurements of the groundwater level or the stream flow one may thus at any time know the distribution of the response-area/ moisture in a catchment (Myrabø [12]).

In order to improve the methodology for the description of the response-area/ moisture map from DTM, other parametres must be included. First of all it is therefore relevant to improve the method by including soil depth and permeabilty/soil type. Furthermore, work will be made to transfer the method to other grid sizes, so that the analysis can be made on standard digital map series. This may increase the areas of application considerably. Consequently this model opens up for new possibilities regarding the estimation of parametres that are necessary for conceptual models, particularly the response-area model within fields such as hydrology, soil erosion and hydro-chemistry. Both economically and as far as work is considered this makes it possible operatively to use more physically correct, distributed hydrological models.

REFERENCES

[1] Anderberg, M. R. Cluster analysis for application. Academic Press, 1973.

[2] Anderson, M. G. and Burt, T.P. Toward more detailed monitoring of variable source areas. Water res. res. Vol.14, pp. 1123- 1131, 1978.

[3] Erichsen, B. and Nordseth, K. Sæternbekken Øvings- og forsøksfelt. NHK rap. 18, 1985.

[4] Evans, I.S. An integrated system of terrain analysis and slope mapping. Zeitschrift fur Geomorphologie., Supplementband, 36. pp 274-295, 1980.

[5] Gurnell, A. M. and Gregory, K. J. Vegitation charactristics and the prediction of runoff : analysis of an experiment in New Forest, Hampshire. Hydrological Processes, 1. 125-142, 1987.

[6] Heerendegen, R. G. and Beran, M. Quantifying source areas through land surface curvature and shape. Jour. of hydrology, Vol. 57, pp. 359-373, 1982.

[7] Monmonier, M. S. Computer assisted cartograph, principles and prospects. Prentice-Hall, 1982.

[8] Myrabø, S. Skråningshydrologi - avløpsstudier i et lite nedbørfelt. Rapportserien Hydrologi, rap. 8. 1986.

[9] Myrabø, S. Runoff studies in a small catchment. Nordic Hydrology, Vol. 17, pp. 335-346, 1986.

[10] Myrabø, S. Avrenningsprosessenes betydning for langtransportert forurensning av våre vassdrag og deres økosystem. Vann i Norden, Vol. 1, 1987.

[11] Myrabø, S. Automation in hillslope hydrology. NHP-rap. no 22, 1988.

[12] Myrabø, S. Temporal and spacial scale of respons-area and groundwater variation, in prep., 1990.

[13] Myrabø, S. and Erichsen, B. Use of a DTM model in hillslope hydrology, in prep., 1990.

[14] Zevenbergen, L. W. and Thorne, C. R. Quantitative analysis of landsurface topography. Earth surf. proc. and landforms, Vol. 12, pp. 47-56, 1987.

Systematic Pumping Test Design for Estimating Groundwater Hydraulic Parameters using Monte Carlo Simulation

J.M. McCarthy(*), W.W-G. Yeh(**), B. Herrling(***)
(*)Insitute für Hydromechanik, Universität Karlsruhe, West Germany (**) Department of Civil Engineering, UCLA,L.A., CA, USA (***)Institute für Hydromechanik, Universitaät Karlsruhe, West Germany

ABSTRACT

A Monte Carlo method is developed for the evaluation of pumping test designs for the identification of hydraulic parameters. The results indicate that the proposed method is an effective tool which can be used to systematically estimate the best well locations for a pumping test. Each set of pumping test observations is judged by a combination of how accurately the parameters identified from the observations can predict a prescribed management objective, and the expected uncertainty in the parameter estimates. The management objective used is the prediction of the steady state drawdown at a proposed production well.

INTRODUCTION

In order to systematically design a pumping test we presume the following procedure will be used to carry out the test and utilize its results.

1. Design the pumping test.
2. Perform the pumping test to collect observations.
3. Calibrate the computer simulation model by solving the inverse problem of parameter identification.
4. Predict the aquifer's response to proposed future aquifer management by a management model using the calibrated parameters.

Parameter identification via least square fitting and its variations, often identify parameters that very accurately reproduced the pumping test observations. However, this does not guarantee that the parameters will accurately simulate the desired management objective.

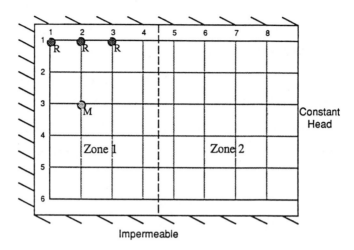

Figure 1: Synthetic aquifer configuration.

Table 1:   Map of the management prediction error (in meters) assuming the pumping test pump well can be at any of ...te.the 48 nodes in the domain.

y-nodes	x-nodes							
	1	2	3	4	5	6	7	8
1	3.38	2.79	2.09	1.80	0.07*	0.36	0.65	1.14
2	1.11	0.96	0.97	1.09	0.11	0.42	0.73	1.23
3	0.98	0.92^M	0.96	1.08	0.13	0.49	0.82	1.32
4	1.15	1.12	1.25	1.63	0.14	0.46	0.77	1.28
5	1.58	1.71	2.34	2.96	0.11	0.35	0.63	1.12
6	1.98	2.33	3.34	3.50	0.07*	0.28	0.53	0.97

Figure 1 and Table 1 are given to illustrate the above mentioned problem. Figure 1 shows the synthetic aquifer used for this research. It is a two-dimensional, horizontal, confined, isotropic aquifer with 48 finite difference nodes. The aquifer is divided into two transmissivity zones, has three recharge wells simulating a constant flux boundary condition in the northwest and a constant head boundary condition in the east. The remaining boundaries are impervious. Three parameters were identified, the zone 1 and zone 2 transmissivities (T_1 and T_2), plus the constant flux boundary condition. The numerical experiment involved simulating 48 different pumping tests (one for each node in the finite difference grid). Each test had one pumping well, one observation well (at the same location), a duration of three days, and a total of 30 observations. The test was simulated once for each well location. In each case the "true" parameter vector ($\mathbf{P^o}$) was used to generate the "true" observations which were then corrupted with white noise. A Gauss Newton parameter identification procedure was used to identify the best parameters for each test. These parameter estimates were then used to generate the steady state drawdown at a proposed production well (management objective) at (2,3). The error in these predictions is given in Table 1. A superscript M indicates the production well location, while the * indicates the best well locations for the pumping test.

The best locations for the pumping test well are at (5,1) and (5,6). These locations provide substantially better predictions then performing a pumping test near the site of the proposed pumping well. This result was surprising at first glance, however, after analyzing the parameter sensitivity of the pumping tests and the management objective, the explanation became quite clear. The management objective is very sensitive to the transmissivity value in zone 2. Since the management objective is a steady state solution, all the water taken from the production well, minus the boundary recharge, must come from the river, through zone 2. In contrast, a three day pumping test at (2,3) is not long enough to be strongly effected by the zone 2 transmissivity. The flux boundary and zone 1 transmissivity must be accurately estimated in order to best fit the pumping test observations at (2,3). Accuracy in estimating the zone 2 transmissivity is sacrificed in order to better estimate the other parameters, therefore, larger errors are observed in these predictions. On the other hand, a pumping test with the well at (5,1) or (5,6) is strongly sensitivity to T_2 so that T_2 is accurately estimated and the management objective, accurately predicted.

LITERATURE REVIEW

The literature dealing with pumping test design has been relatively sparse and can be divided into two main groups. One is designed to maximize the reliability of the parameter estimates (Hsu and Yeh [1], Nishikawa and Yeh [2]) and the other is designed to maximize the prediction reliability (Yeh and Sun [3], McCarthy and Yeh [4]). Chavent [5] provides a review of the relationship between identifiability and the well posedness of the inverse problems governed by partial differential equations.

In contrast, this research uses Monte Carlo simulation to generate parameter estimates as a function of the assumed errors in the hydraulic head measurements (McCarthy, [6]). From these realizations, the covariance matrix can be estimated to evaluate parameter reliability and the prediction accuracy is easily evaluated. Each are important components in the pumping test design problem and the modeler should have the opportunity to weight the various reliability measures. In addition, it quite simple to combine boundary conditions and hydraulic parameters into one parameter set for the system.

MATHEMATICAL MODELS

Figure 1 shows the assumed geometry and boundary conditions of the sample problem. The parameters to be identified (**P**) are the transmissivities in zones 1 and 2, as well as the constant flux boundary condition shown. The management objective is to predict the steady state drawdown at a proposed production well.

The pumping test design approach will use two simulation models, a parameter identification model and a Monte Carlo model. The simulation models are for calibration against the pumping test observations and prediction of the management objective. The implicit finite difference method is used to solve the simulation models and the modified Gauss Newton method is used to solve the parameter identification problem.

Calibration Model

Two-dimensional, heterogeneous, isotropic confined flow in porous media is governed by the following equation:

$$\frac{\partial}{\partial x}\left(T\frac{\partial h}{\partial x}\right)+\frac{\partial}{\partial y}\left(T\frac{\partial h}{\partial y}\right) = Q_h + S\frac{\partial h}{\partial t} \qquad (1)$$

subject to the following general initial and boundary conditions:

$$h(x,y,0) = h^\circ(x,y) \qquad (x,y) \in \Omega$$

$$h(x,y,t) = f_1(x,y,t) \qquad (x,y) \in \partial\Omega_1 \qquad (2)$$

$$T\frac{\partial h}{\partial n} = f_2(x,y,t) \qquad (x,y) \in \partial\Omega_2$$

where:

$h(x,y,t)$	=	hydraulic head [L];
$T(x,y)$	=	transmissivity [L^2/T];
S	=	storage coefficient [unitless];
$Q_h(x,y)$	=	pumping test pump rate [L^3/T/L^2];
x,y	=	space variables [L];
t	=	time variables [T];
Ω	=	flow region;
$\partial\Omega$	=	boundary of aquifer ($\partial\Omega_1 \cup \partial\Omega_2 = \partial\Omega$);
$\frac{\partial}{\partial n}$	=	normal derivative to the boundary;
h°, f_1, f_2	=	specified functions.

Management Model

The management (or prediction) model is a steady state version of Equation 1.

MONTE CARLO PUMPING TEST DESIGN

The Monte Carlo algorithm will proceed as follows for one realization of a particular pumping test design alternative.

1. Choose a "true" parameter set $\mathbf{P}^\circ$.
2. Generate the "true" observations $\mathbf{h}(\mathbf{P}^\circ)$.
3. Corrupt $\mathbf{h}(\mathbf{P}^\circ)$ with noise to simulate field observations, **obsh**.
4. Use the Gauss Newton algorithm to identify the set of parameters, $\mathbf{\check{P}}$ that best fits the corrupted observations, **obsh**.

5   Use $\tilde{\mathbf{P}}$ to predict the management objective and calculate the management prediction error, $E_p = h_p (\tilde{\mathbf{P}}) - h_p(\mathbf{P}°)$.

This process produces one realization of the parameter and management prediction error distributions, for a particular pumping test design.

For one alternative pumping test design, the process is repeated until the modeler feels there are sufficient realizations to statistically characterize the parameters and management prediction errors. The process is repeated for each potential pumping test design and the distributions are compared to determine the best design.

Because the design is approximate, finding a single best design is not as important as identifying the regions of best well locations and the trends in choosing these well sites. The parameter uncertainty criteria (minimize the determinant or trace of the covariance matrix by choice of the pumping test design) and the management prediction error criterion, should be considered both individually and together. This will provide the modeler with a less compact, but more informative set of data by which to choose a best design.

The achilles heel of the methodology is the observation error generation. In a real case there will be a combination of modeling, measurement and numerical errors. For this research a normal observation error structure is assumed, however, it seems unreasonable to assume the errors have the same distribution through time. Therefore, the error is assumed to be correlated in time, by prescribing the standard deviation for any observation at time t, to be a percentage of the average drawdown over the entire aquifer at time t. This is not a perfect representation of the observation errors but it does appear to be more realistic than an error distribution that is uniform over both time and space.

NUMERICAL EXPERIMENTS

Two numerical experiments are presented in order to illustrate the Monte Carlo approach for pumping test design. The examples are designed to show the following.

1.   Assume there are no wells in the aquifer and we must choose the best sites for a pumping well and four observation wells.
2.   Evaluate the sensitivity of the design to different presumed "true" parameter values.

The synthetic aquifer is shown in Figure 1 and the following numerical values are assumed for these experiments.

production well pump rate (Q_p)	9000 m³/day
production well location	(2,3)
pumping test pump rate (Q_h)	4000 m³/day
pumping test duration	3 days
observations per well per day	10
parameter ranges	

zone 1 transmissivity	$1000 \leq P^1 \leq 10,000 \text{ m}^2/\text{day}$
zone 2 transmissivity	$500 \leq P^2 \leq 5,000 \text{ m}^2/\text{day}$
boundary recharge per well	$-500 \leq P^3 \leq +500 \text{ m}^3/\text{day}$
assumed "true" parameter vector ($\mathbf{P^o}$)	$\begin{pmatrix} 2,500 \\ 1,500 \\ 100 \end{pmatrix}$
aquifer storativity (S)	0.0002

Observation errors at time t are assumed to have a standard deviation of 5% of the average drawdown over the aquifer at time t. Pumping and observation well locations are the decision variables considered in these numerical experiments. The algorithm adds one well at a time, considering each node as a potential well location and therefore, a different pumping test design. For each possible well location, observations are generated and randomly corrupted. A Gauss Newton algorithm is then used to identify the parameters that best fit the corrupted observations. Using these parameters, the management objective is predicted, and prediction error calculated. This process is repeated 20 times (Monte Carlo) in order to obtain sample distributions for each parameter, as well as, the management prediction error for a particular design. Once the sample distributions have been generated for each possible well location, the sites can be sorted and ranked. The three pumping test design criteria considered are;

C1. Minimize the management prediction error;

C2. Minimize the determinant of the covariance matrix;

C3. Minimize the trace of the covariance matrix.

Example

This example identifies the best locations for one pumping well and four observation wells. It assumes that the best location for a pumping well with one observation well (at the same location) is the best location for a pumping well with any number of observation wells.

Table 2 shows the sites and their ranking for the best five pumping well locations for each of the three pumping test design criteria. It is interesting that the criteria do not identify the same areas of the aquifer, although, each identifies areas that seem justifiable. The pattern developed by sorting the sites and studying their relationship, are very useful to the modeler. The site that stands out is (5,1). It is the only location that ranks high for all three criteria. The area preferred by measures of the covariance matrix (C2 and C3) are very well define, however, they each provide very poor predictions (C1). (5,1) seems like a reasonable location since it is near an impermeable boundary and a parameter boundary as well as in the low transmissivity zone. More importantly, the results show quite clearly the aquifer regions that are good sites for the pumping well to be placed and the locations that are poor.

Table 2:   Best locations for one pumping well and up to four observation wells. The pumping well location is chosen along with the first observation well.

Observation Well Number	Criterian	Best Five Locations				
Existing Wells		1	2	3	4	5
1	C1	(5,6)	(5,1)*	(5,2)	(5,5)	(5,3)
None	C2	(1,3)	(2,3)	(5,1)*	(3,3)	(1,4)
	C3	(2,3)	(3,3)	(1,3)	(2,2)	(3,2)
2	C1	(4,6)	(5,6)	(3,6)	(1,6)	(2,6)
(5,1)	C2	(4,1)*	(1,1)	(5,2)	(3,1)	(4,2)
	C3	(5,2)	(4,1)*	(6,2)	(7,1)	(6,1)
3	C1	(1,1)*	(2,1)	(1,2)	(2,2)	(1,3)
(5,1)	C2	(1,1)*	(2,1)	(1,2)	(2,2)	(3,1)
(4,1)	C3	(1,6)	(1,5)	(2,6)	(2,5)	(1,4)
4	C1	(6,1)	(5,2)*	(2,1)	(6,6)	(3,2)
(5,1)	C2	(5,2)*	(1,6)	(1,5)	(2,6)	(1,4)
(4,1)	C3	(1,6)	(2,6)	(1,5)	(3,6)	(2,5)
(1,1)						

The remainder of Table 2 shows the design patterns when adding the next three wells. In each case the patterns seem reasonable and quite interesting. From these results it appears that a very good design would be to place the pumping well at (5,1) and place the observation wells at (5,1), (4,1), (1,1) and (5,2). However, the well sites (1,6) and (7,1) appear to be very good observation locations as well. A pumping well at (2,3) may also be a good design.

When a new well is added, the existing well estimates are no longer (necessarily) the best well sites for the system. Therefore, the well locations were systematically varied (incremental search) in their neighborhood to check the local optimality. In no case was there a significant reason to change well locations, therefor, no results are presented for the incremental search.

CONCLUSIONS

Monte Carlo simulation is an effective tool that is quite simple to program and use. The numerical experiments performed in this study indicate that many simulations are required for the convergence of the parameter statistics and the management prediction errors. However, when designing a pumping test it is only important for the model to correctly sort the different pumping test designs. This seems to be accomplished within 10 to 20 realizations. Therefore, the Monte Carlo approach appears to be a promising and practical approach for pumping test design.

From the preliminary results of the study, the following conclusions are drawn:

1. Monte Carlo simulation allows the modeler to easily combine the boundary conditions and hydraulic parameters into one parameter set for the pumping test design to consider.

2. When evaluating the reliability of a pumping test design, both the parameter reliability and prediction reliability should be considered.

3. When evaluating the reliability of the parameters, both the determinant and/or the trace of the covariance matrix should be considered.

4. By using a number of different criteria, the pumping test design appears to become more robust. Thus the best observation well locations are not a strong function of the estimated "true" parameter vector.

ACKNOWLEDGEMENTS

This work was supported in part by the Deutsche Forschungsgemeinshaft, Schwerpunktprogramm "Anwendungsbezogene Optimierung und Steuerung".

REFERENCES

[1] Hsu, N.S. and Yeh, W.W-G. Optimum Experimental Design for Parameter Identification in Groundwater Hydrology, *Water Resour. Res.*, 25(5), 1025-1040, 1989.

[2] Nishikawa, T. and Yeh, W.W-G. Optimal Pumping Test Design for the Parameter Identification of Groundwater Systems, *Water Resour. Res.*, 25(7), 1737-1747, 1989.

[3] Yeh, W.W-G. and Sun, N.Z. An Extended Identifiability in Aquifer Parameter Identification and Optimal Pumping Test Design, *Water Resour. Res.*, 20(12), 1837-1847, 1984.

[4] McCarthy, J.M. and Yeh, W.W-G. Optimal Pumping Test Design for Parameter Estimation and Prediction in Groundwater Hydrology, to be published in *Water Resour. Res.*, 1990.

[5] Chavent, G. Identifiability of parameters in the output least squares formulation, Chapter 6, in *Identifiability of Parametric Models*, Edited by E. Walter, Pergamon Press, 1987.

[6] McCarthy, J.M. Systematic Pumping Test Design for Estimating Groundwater Hydraulic Parameters, using Monte Carlo Simulation, Institute für Hydromechanik, Universität Karlsruhe, West Germany, 1989.

Computing Animation Techniques in Groundwater Modelling

R. Horst, W. Pelka

Lahmeyer International GmbH, Lyoner Strasse 22, D-6000 Frankfurt/Main, West Germany

Abstract

Results of transient (non-steady-state) groundwater flow and transport models are usually evaluated and presented in the form of isoline maps at characteristic time steps and hydrographs at characteristic locations in the model area. While the isoline maps contain spatial but static information, the hydrographs present dynamic but non-spatial information.

To obtain even a very imperfect impression of the temporal and spatial behaviour of the flow and transport process, many maps and hydrographs have to be compared concerning spatial differences and temporal changes. The usual approach to interpretation becomes even more complicated and unsatisfactory if acceleration or deceleration of the development in flow and transport occur in several parts of the model area and at varying points in time.

Advanced graphic animation techniques in connection with CAD offer completely new methods in the evaluation, interpretation and presentation of non steady-state model results. The temporal changes in flow or transport can be displayed as a continuous process, for example in the form of isolines, cross-sections and three-dimensional views.

Introduction

The output of numerical models, especially of models for time-dependent processes, consists of a large amount of data. For each node and each time step of a transient groundwater flow and transport model, piezometric head or pressure, velocity and concentration or temperature will be computed and stored on a mass storage device. The total amount of the result can easily exceed 1 Million numbers or ever more.

In the early days of numerical modeling the user received a frustrating pile of printed lists of model results. Obviously the evaluation of this type of output was quite time consuming, if possible at all. Very soon the first step towards a more user-friendly output was made with post-processing software converting the numerical results into graphical information in the form of line printer plots and (later) pen plotter drawings.

Today the results of non steady-state models usually are evaluated, interpreted and presented in the form of plots with

- isolines at characteristic time steps
- hydrographs at characteristic locations in the model area.

In order to obtain an idea of the temporal and spatial behaviour of the flow and transport process, or the interdependence of influences of different boundary conditions, many isoline maps have to be plotted and compared. Acceleration and delays in the development of flow or transport can hardly be determined this way.

Usually this kind of information is displayed in the form of hydrographs showing the temporal changes of groundwater level or concentration at certain (hopefully characteristic) points in the model area. Unfortunately the spatial information is lost.

Each plot can only present two dimensions of the result, the two spatial coordinates or additionally the time coordinate and one spatial coordinate. Thus, in evaluating the results the choice has to be made between the spatial distribution at a certain point in time or the dynamic behaviour at a certain point in space.

The analyst will in any event receive only a very incomplete impression of the real spatial and temporal behaviour of the groundwater flow and the concentration distribution. This can be a severe disadvantage, if certain changes of boundary conditions at different times are superimposed upon complex flow and transport situations within the model area.

Advanced computer animation techniques, which represent state-of-the-art in weather forecast simulations or architectual planning, offer completely new possibilities also in the evaluation, interpretation and presentation of non steady-state groundwater modelling results. The spatial and temporal changes of the flow and transport field can be displayed in a contiuous process, adding the otherwise lost dimension. Complete information of the spatial and temporal changes in the model area is provided simultaneously.

Processing of the numerical data for the animation

Every 'film' or 'animation' is a series of pictures projected one after the other with a high frequency (at least about 10 pictures per second) creating for the human eye the illusion of smooth motion.

Fig. 1 presents a post-processing procedure for the animation of models results based entirely on standard PC-hardware and software, converting the numerical model results to a continuous 'film' or 'animation'.

The numerical model results are usually stored in mass storage files. Each file contains e.g. piezometric heads, pressures, velocities, concentration or temperatures for each time-step. These numerical results are passed through a post-processing program converting the numerical information of each time-step into graphical information which can be read by the CAD software (in the present system a .dxf-file for AUTOCAD). Additional non-graphic information, e.g. the time and date of the processed time-step is included for later reference.

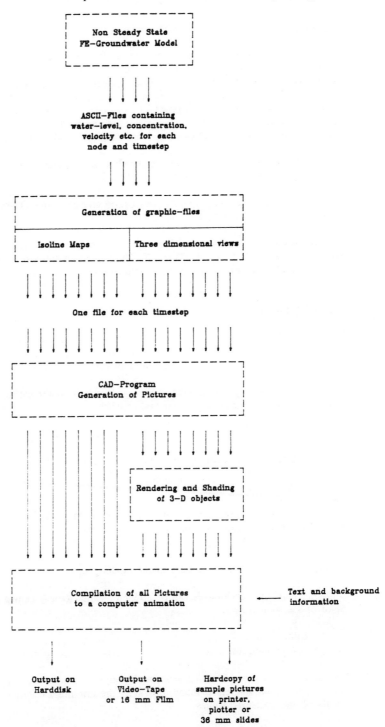

Fig. 1: Process from numerical results to animation

The flow and transport results may be prepared for

- isoline maps (piezometric head, concentration, temperature ...)
- isoarea maps (piezometric head, concentration, temperature ...)
- three-dimensional views (piezometric head, concentration, temperature ...)
- cross-sections (piezometric head, concentration, temperature ...)
- vector maps (velocities or mass flow)
- differences (piezometric head, concentration, temperature ...)
- combinations of isolines, three-dimensional views or vector fields (the possibility of

making transport and its dependence on the flow field visible is one of the most exciting applications of this method)

The CAD system reads the graphical information and generates the respective drawings (one for each computed time-step).

Each drawing is converted into a 'slide' which is transferred to the animation program, in the present case AUTOFLIX. The animation software is used for the assembly of all slides to form a 'film' or 'animation'.

Controlling the animation

The 'film' or 'animation' is displayed by the animation software on a high-resolution colour-monitor. The sequence of slides is displayed at high frequency on the monitor, leading to the illusion of smooth and continuous motion. The analyst may

- control the frequency, e.g. to pass over periods of minor interest at high speed, or to watch critical periods very slowly and carefully

- reverse the direction of motion to repeat a certain sequence

- stop the animation at a certain or several time-steps

- go through the animation or parts of it in single steps

providing full control over the running animation. With screen-projectors the whole team may watch the animation and discuss the results.

Application of animation techniques in the model project

Computer animation techniques can be applied advantageously at two important stages of a model project

- model calibration
- prognostic simulations.

For calibration of the model, measured and computed results are compared until an optimum set of parameters is found, providing the minimum differences between calculated and measured values.

With measured reference data of reasonable spatial and temporal resolution, measured and computed results can be superimposed in an animation, providing continuous information about measured and computed results as well as the spatial and temporal deviations. The interaction of parameter changes and boundary conditions can be observed, and reasons for spatial and temporal deviations between measured and computed values can be investigated and the development traced.

By using animation techniques the calibration can be completed more efficiently and cost effectively. The quality of the calibration is improved and thus the prognostic accuracy of the model.

In the evaluation of prognostic simulation runs, a certain scenario usually has to be compared with a reference situation or with other cases. Superimposing the actual scenario on the reference situation or on different alternatives in an animation provides full spatial and temporal information about the differences and deviations. The influence of changing boundary conditions can be observed as they propagate into the model area.

Most complicated flow and transport situations become transparent, and the history of a certain flow or transport pattern becomes obvious. The evaluation of non steady-state model results becomes faster, safer and more efficient.

Thus animation techniques will develop into a new advanced tool for model calibration and prognostic simulation, which complements the traditional 'hardcopy' approach.

Application of animation techniques in real world projects

Fig. 2 shows a well field for potable water supply near the River Rhine and groundwater contamination of industrial origin. From time to time rising concentrations of chloride hydrogen carbonates were found in some of the wells. Using computer animation techniques it was possible to prove that

- the origin of the pollution was the industrial site
- the rise of concentration in the wells uccurred a certain nearly constant period after

a characteristic sequence of water levels of the River Rhine.

Fig. 3 shows the River Lech and the head race of a river power station. During the re-filling of the head race after rehabilitation, the groundwater level was rising in a nearby village. At the same time, rainfall and storm water flows in the River Lech were experienced in the project area. Applying animation techniques, it was possible to distinguish and quantify three influences on the rise of the groundwater level in the village, making use of the phase differences and travel times of the different boundary conditions.

Summary and future outlook

The application of animation techniques enhances the efficiency and reliability of model calibration and reduces the cost of evaluating the prognostic simulation results of non steady-state groundwater flow and transport models. Animation techniques will therefore develop into a new advanced tool for model calibration and prognostic simulation, which complements the traditional 'hardcopy' approach.

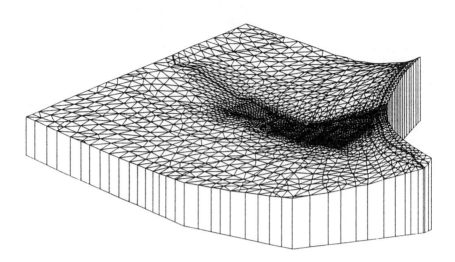

Fig. 2: Well field near the River Rhein

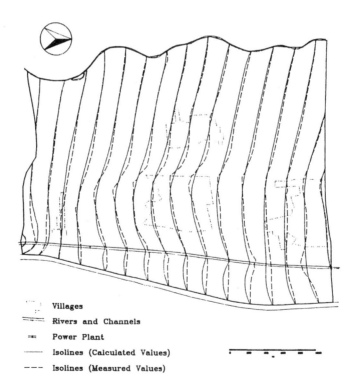

⋯   Villages

═══   Rivers and Channels

⚌   Power Plant

───   Isolines (Calculated Values)

‐ ‐ ‐   Isolines (Measured Values)

Fig. 3: Village near the River Lech

Only a few years ago animation techniques would have required the application of super-computers. Because of the enormous advances in hardware and software development during the past decade, the advantages of computer animation techniques are coming within the reach of every scientist and engineer.

Today animation techniques are implemented by post-processing software, using and converting the results of the numerical model which has to be run before on the same or a different computer. With the expected advance of hardware within the coming years, it can be expected that animation processing will take place in real-time, parallel to the processing of the numerical computation.